Biological Interactions on Materials Surfaces

Biological Interactions on Materials Surfaces

Understanding and Controlling Protein, Cell, and Tissue Responses

David A. Puleo and Rena Bizios
Editors

Springer

Editors
David A. Puleo
Center for Biomedical Engineering
University of Kentucky
Lexington, KY 40506-0070
USA
puleo@uky.edu

Rena Bizios
Department of Biomedical Engineering
University of Texas
San Antonio, TX 78249-0669
USA
rena.bizios@utsa.edu

ISBN 978-0-387-98160-4 e-ISBN 978-0-387-98161-1
DOI 10.1007/978-0-387-98161-1
Springer Dordrecht Heidelberg London New York

Library of Congress Control Number: 2009926044

Cover Illustration: (**Left image**): Nanogold beads used to identify proteins in a dual-protein film. The proteins were patterned on PDMS surfaces that were coupled to 1.4 nm gold particles. The film was imaged by tapping mode AFM. (**Right image**): Histology example of a section obtained from an implant in a rabbit femur trabecular bone 4 weeks postimplantation. The implant material is collagen-coated nanoporous Ti. Stain: acid fuchsin; original magnification, × 10. The histology was performed by Drs. Milena Fini, Gianluca Giavaresi, and Roberto Giardino, Istituti Ortopedici Rizzoli, Bologna, Italy.

Printed on acid-free paper

Springer is part of Springer Science+Business Media (www.springer.com)

Preface

Biomaterials are used in numerous in vivo applications, ranging from joint and dental implants to vascular grafts and heart valves. Recent advances in fields such as tissue engineering have expanded the scope of uses for biomaterials as an integral part of implant devices. Depending on particular aspects/specifications for each such application, chemical, mechanical, and electrical properties undoubtedly contribute to the performance of biomaterial/prosthetic devices. In all cases, however, success or failure of implants in vivo critically depends on the biological interactions (molecular, cellular, and tissue) at the implant/tissue interface. Advances in molecular biology and biochemistry, cell biology, developmental biology, wound-healing physiology, materials science and engineering, and novel laboratory techniques, as well as in advanced laboratory and clinical instrumentation, have provided incentives and capabilities for renewed and ever-increasing interest in the tissue/biomaterial interfacial milieu. These new approaches have the potential for elucidating the mechanisms of important physiological processes pertinent to new tissue formation and for enhancing integration of implants in the surrounding tissues. As scientists and engineers meet these challenges, they are opening new scientific frontiers and are making seminal contributions that steer the biomaterials/implant device field in new directions of great potential and promise.

Biological Interactions on Materials Surfaces was motivated by the need to bring to the attention of the scientific community the latest developments in current understanding of protein, cell, and tissue interactions with biomaterials (possessing both conventional as well as nanoscale features). In addition to discerning and judicious reviews, established and renowned experts provided insightful evaluation of the current state of the art in, and projected their opinions on the future of, their respective research areas. Multidisciplinary perspectives in the field are represented in contributions from authors from North America, Europe, and Asia.

The present book is intended to provide valuable insight for scientists, engineers, and medical researchers seeking not only to understand but also to control tissue–biomaterial interactions for the in vivo success of implant biomaterials/devices and tissue-engineering constructs. It is our hope that the scientific information presented in this book will also stimulate further research, resulting in important contributions toward development of the next generations of devices used to replace, augment, or perform the functions of diseased and damaged tissues and organs. Clinical applications of such knowledge will, undoubtedly, have a major impact on the health care and welfare of many patients.

David A. Puleo — Lexington, KY
Rena Bizios — San Antonio, TX

Acknowledgment

It has been a pleasure to work with our colleagues who contributed chapters in this book. We take this opportunity to thank all the authors for sharing aspects of their research endeavors as well as for providing discerning reviews of their respective fields of expertise. Furthermore, we are thankful for their insightful projections into the future pertinent to understanding a critically important milieu and of controlling cellular-, molecular-, and genetic-level events at the tissue/implant interface.

Contents

5 Biomolecule–Nanomaterial Interactions: Effect on Biomolecule Structure, Function, and Stability

Ravindra C. Pangule, Shyam Sundhar Bale, Dhiral A. Shah, Amit Joshi, Prashanth Asuri, Jonathan S. Dordick, and Ravi S. Kane

6 Phage Display as a Strategy for Designing Organic/Inorganic Biomaterials

Sharon Segvich and David H. Kohn

7 Extracellular Matrix-Derived Ligands for Selective Integrin Binding to Control Cell Function

Timothy A. Petrie and Andrés J. García

10 Cell and Tissue Interactions with Materials: The Role of Growth Factors

Christopher C. Gibson, David A. Puleo, and Rena Bizios

11 In Vitro and In Vivo Monocyte, Macrophage, Foreign Body Giant Cell, and Lymphocyte Interactions with Biomaterials

James M. Anderson

12 Development and Differentiation of Neural Stem and Progenitor Cells on Synthetic and Biologically Based Surfaces

Erin N. Boote Jones, Donald S. Sakaguchi, and Surya K. Mallapragada

13 Toward Osteogenic Differentiation of Marrow Stromal Cells and In Vitro Production of Mineralized Extracellular Matrix onto Natural Scaffolds

Ana M. Martins, Catarina M. Alves, Rui L. Reis, Antonios G. Mikos, and F. Kurtis Kasper

14 Biomimetic Nanophase Materials to Promote New Tissue Formation for Tissue-Engineering Applications

Xiaohua Liu, Ian O. Smith, and Peter X. Ma

Short Biographical Sketches

David A. Puleo received B.S. and Ph.D. degrees in Biomedical Engineering from Rensselaer Polytechnic Institute. In 1991, he joined the University of Kentucky as a junior faculty member to establish a track in biomaterials in the Center for Biomedical Engineering. In addition to being promoted through the ranks to Professor, he served as Director of Graduate Studies and currently is Director of the Center, overseeing all administrative and educational matters. He also holds an appointment in the Center for Oral Health Research in the College of Dentistry. Dr. Puleo has developed and teaches courses ranging from introductory biomaterials to advanced offerings on events at the tissue–implant interface. He has taught undergraduates, graduates, orthopedic surgery residents, and postgraduate dental students.

Dr. Puleo's research interests focus on applying understanding of cell–biomaterial interactions to develop materials and surface modification strategies for the purpose of controlling cellular responses at the tissue–implant interface. He is active in a variety of professional organizations, including the Society for Biomaterials, the Biomedical Engineering Society, and the International and American Associations for Dental Research. He is an Associate Editor for the *Journal of Biomedical Materials Research Part B (Applied Biomaterials)* and is on the editorial board of the *Journal of Biomedical Materials Research Part A*. He has served on and chaired numerous review panels centered on biomaterials and tissue engineering at the National Institutes of Health (NIH) and the National Science Foundation (NSF) and has reviewed for several other domestic and foreign organizations. Dr. Puleo received a Research Initiation Award from the NSF, and he was awarded the *Bourses de stage de recherche scientifique of the Programme québécois de bourses d'excellence* administered through the Ministère de l'Éducation of the Gouvernement du Québec. He is coauthor of *An Introduction to Tissue–Biomaterial Interactions*, a textbook written specifically for undergraduates and first semester graduate students transitioning into the field of biomaterials. Dr. Puleo is a Fellow of the American Institute for Medical and Biological Engineering (AIMBE).

Rena Bizios is a Peter T. Flawn Professor in the Department of Biomedical Engineering at the University of Texas at San Antonio (UTSA), San Antonio, TX. She received her B.S. (cum laude) degree in Chemical Engineering from the University of Massachusetts, M.S. degree in Chemical Engineering from the California Institute of Technology, and Ph.D. degree in Biomedical Engineering from the Massachusetts Institute of Technology.

During her career in academia, Dr. Bizios has taught various undergraduate and graduate fundamental engineering and biomedical engineering courses as well as developed new courses for biomedical engineering curricula. She has coauthored a textbook (entitled *An Introduction to Tissue–Biomaterial Interactions*) and authored/coauthored numerous scientific publications and book chapters. Her research interests include cellular engineering, bone tissue engineering, tissue regeneration, biomaterials (including nanostructured ones), and biocompatibility. She has given numerous presentations at scientific conferences and invited seminars/lectures in academic institutions and industry. She has also organized and/or co-chaired many symposia and sessions at national/international conferences. Dr. Bizios is a member of many professional societies; she has been an active participant (including elected officer positions) in the Society for Biomaterials, the Biomedical Engineering Society, and the American Institute of Chemical Engineers. She is a member of the editorial board of the *Journal of Biomedical Materials Research Part A*. She has participated in various NSF Review Panels and similar national-level review committees, and she has participated in (and even chaired some) NIH Study Sections.

Dr. Bizios received the *Outstanding Alumna in Engineering Award* of the Society of Women Engineers, College of Engineering, University of Massachusetts, Amherst, MA (1985), the *Rensselaer Alumni Association Teaching Award*, Rensselaer Polytechnic Institute (1997), and the *Clemson Award for Outstanding Contributions to Literature* (the Scientific Literature of Biomaterials) from the Society for Biomaterials (1998). She was Jubileums Professor at Chalmers University of Technology, Göteborg, Sweden (fall of 2002), Chercheur Associé (spring of 2003), and Directeur de Recherche Associé, Centre National de la Recherche Scientifique, Faculté de Médicine Saint-Louis Lariboisiére, Université Paris VII, Paris, France (fall of 2005). She is a Fellow of the American Institute for Medical and Biological Engineering (AIMBE), an International Fellow of Biomaterials Science and Engineering of the International Union of Societies for Biomaterials Sciences and Engineering, and a Fellow of the Biomedical Engineering Society (BMES).

1

Protein Adsorption to Biomaterials

David Richard Schmidt, Heather Waldeck, and Weiyuan John Kao,

Within milliseconds after biomaterials come in contact with a biological fluid such as blood, proteins begin to adhere to the surface through a process known as protein adsorption. Protein adsorption is initially strongly influenced by protein diffusion, but protein affinity for the surface becomes critically important and, over time, higher-affinity proteins can be replaced by lower-affinity proteins in a dynamic process. By the time cells arrive, the material surface has already been coated in a monolayer of proteins; hence, the host cells do not "see" the material but "see" instead a dynamic layer of proteins. Multiple parameters influence protein adsorption to a substrate surface including the chemical and physical properties of both the protein and the material surface, as well as the presence of other proteins on the surface.

Many methods have been developed in the last several decades to study protein adsorption to biomaterial surfaces. These new techniques provide information about the type and conformation of adsorbed proteins from multicomponent solutions such as blood serum. Nanomaterials as well as functional group immobilization and novel, stimuli-sensitive polymer surfaces have provided new alternatives for the study and modulation of protein adsorption, with insight into the mechanisms underlying protein adsorption and subsequent cell adhesion. However, a molecular-level understanding of all aspects of protein adsorption is still incomplete. The future of this field, however, is bright as new technologies offer great promise for further elucidation of protein adsorption.

Abbreviations

AFM	Atomic force microscopy
ATR-FTIR	Attenuated total reflectance-Fourier transform infrared spectroscopy
ELISA	Enzyme-linked immunosorbent assay
FTIR	Fourier transform infrared spectroscopy
HA	Hydroxyapatite
IR	Infrared

D.R. Schmidt, H. Waldeck, and W.J. Kao • School of Pharmacy, University of Wisconsin-Madison, Madison, WI, 53705, USA; Department of Biomedical Engineering, College of Engineering, University of Wisconsin-Madison, Madison, WI, 53705, USA

D.A. Puleo and R. Bizios (eds.), *Biological Interactions on Materials Surfaces*,
DOI 10.1007/978-0-387-98161-1_1, © Springer Science+Business Media, LLC 2009

MALDI-ToF/MS	Matrix-assisted laser desorption/ionization time-of-flight mass spectrometry
PEG	Polyethylene glycol
PEO	Polyethylene oxide
pI	Isoelectric point
PLGA	Poly(lactic-*co*-glycolic acid)
PLLA	Poly(L-lactic acid)
PNIPAAm	Poly(*N*-isopropylacrylamide)
RGD	Arginine–glycine–aspartic acid
SAM	Self-assembled monolayer
SEIRA	Surface-enhanced infrared absorption
SEM	Scanning electron microscopy
SPR	Surface plasmon resonance
STM	Scanning tunneling microscopy
ToF-SIMS	Time-of-flight secondary ion mass spectrometry
XPS	X-ray photoelectron spectroscopy
2D	Two dimensional
3D	Three dimensional

1.1. Introduction

The host response to a biomaterial is critical in the determination of the success of biomedical implants. This response is in turn dictated by the function of cells that respond to the implant; however, one of the most important aspects of such events is that the cells of the body do not interact directly with the implant material itself but with proteins bound to the surface of the implant material. Almost immediately upon coming into contact with blood, a biomedical implant is coated with serum proteins in a process known as protein adsorption. "Adsorption" means adherence of a molecule to the surface of a solid and should not be confused with the term "absorption," which indicates that the molecule is brought into the solid. Consequently, once cells finally get to the surface they no longer "see" the biomaterial surface itself but instead "see" a dynamic coating of proteins [1–3].

The nature and activity (that is, the ability to interact with cells and other biological molecules) of these proteins once they are adsorbed dictates the initial cellular, and subsequent, host responses. Cells interact with proteins through direct binding to receptors on the cell membrane but do not have receptors for a material surface alone without a coating of proteins. Consequently, the presence of proteins allows cells to adhere to surfaces; the presence of increased amounts of protein may lead to increased cell adhesion, but this is not always the case [1–3].

In subsequent sections of this chapter, the underlying concepts of protein adsorption will be discussed in greater detail, followed by a presentation of more recent discoveries in the field of protein adsorption and future directions for this critical area for biomaterials.

1.2. Fundamentals of Protein Adsorption

1.2.1. General Concepts

Before addressing the underlying properties that influence protein adsorption at solid surface interfaces and specific protein examples, it is important to discuss the concepts relating to protein adsorption in some detail. To begin with, protein adsorption to a biomaterial is influenced by the bulk concentration of the protein in solution. The practical application of this concept is that higher concentration of a single protein in a liquid solution surrounding a material generally leads to more protein on the material surface. However, this phenomenon gets considerably more complicated when more than one type of protein is present in solution because, in this case, more than one protein becomes involved. This phenomenon will be discussed later in the section "The Monolayer Model and Protein–Protein Interactions". Another important aspect in protein/material interactions is the movement of the protein in solution, which is principally governed by the rate of diffusion. In turn, the rate of diffusion is dictated primarily by the size of the protein. Faster (that is, smaller) proteins tend to get to the surface first and adsorb onto that surface before slower moving (that is, bulkier/bigger) proteins arrive.

Yet another important aspect, and perhaps the most important one for long-term considerations, is the affinity of a protein for the substrate surface. Affinity of a protein for a surface describes how likely that protein will adsorb and how strongly it will adhere to the substrate surface. Compared with proteins with a lower affinity, proteins with a higher affinity for a material are more likely to adsorb and stay on the substrate surface. This interaction can vary tremendously depending on the protein and the material surface. Proteins that adsorb to a material form a number of bonds on that surface; proteins with high affinity have more and/or stronger bonds while proteins with low affinity tend to have weaker and/or a smaller number of such bonds [1, 2, 4].

Taking these three aspects together, namely, protein concentration, rate of diffusion, and affinity, one can come to the conclusion that, in a multicomponent solution, proteins are in competition for a surface. Arriving at the surface first, however, does not necessarily mean that a protein will stay there permanently. Proteins adsorbed on a material surface replace each other over time in a process called the Vroman effect, named for the well-known scientist (also poet and painter) who initially discovered and studied this phenomenon. The bonds between a protein and a material surface established during protein adsorption are not completely static; they can be broken and reformed randomly over and over. However, a single protein is highly unlikely to desorb (the process opposite of adsorption) off a material surface because that event requires that all bonds between the protein and surface are broken simultaneously. In contrast, another protein may form bonds with the sites that have become available as bonds are broken between an already adsorbed protein and the material surface. If the new protein has higher affinity with surface sites, this protein will take over the specific site from the lower-affinity protein (Figure 1.1). Over time, the higher-affinity protein can replace the previously preadsorbed, lower-affinity protein; thus, a dynamic protein layer that changes over time is formed on the material surface [1, 2, 4].

1.2.2. Protein Properties

Overview of Relevant Protein Properties

Once a protein arrives at a material surface, it interacts with that surface through intramolecular bonds, such as hydrophobic interactions (nonpolar domains of the protein molecule avoid polar regions of the material surface or vice versa in a process referred to as the hydrophobic effect); ionic bonds (bonds between positive and negative charges); and charge transfer (a stabilizing charge is transferred between two molecules). The prevalence of these types of bonds between proteins and material surfaces is strongly influenced by the properties of individual proteins, which are summarized in Table 1.1. Protein size influences protein adsorption because it affects the rate of protein diffusion; the size of a protein molecule also partially determines the affinity of the protein. Larger proteins may have more binding sites for interaction with regions on the material surface and thus they may adsorb readily.

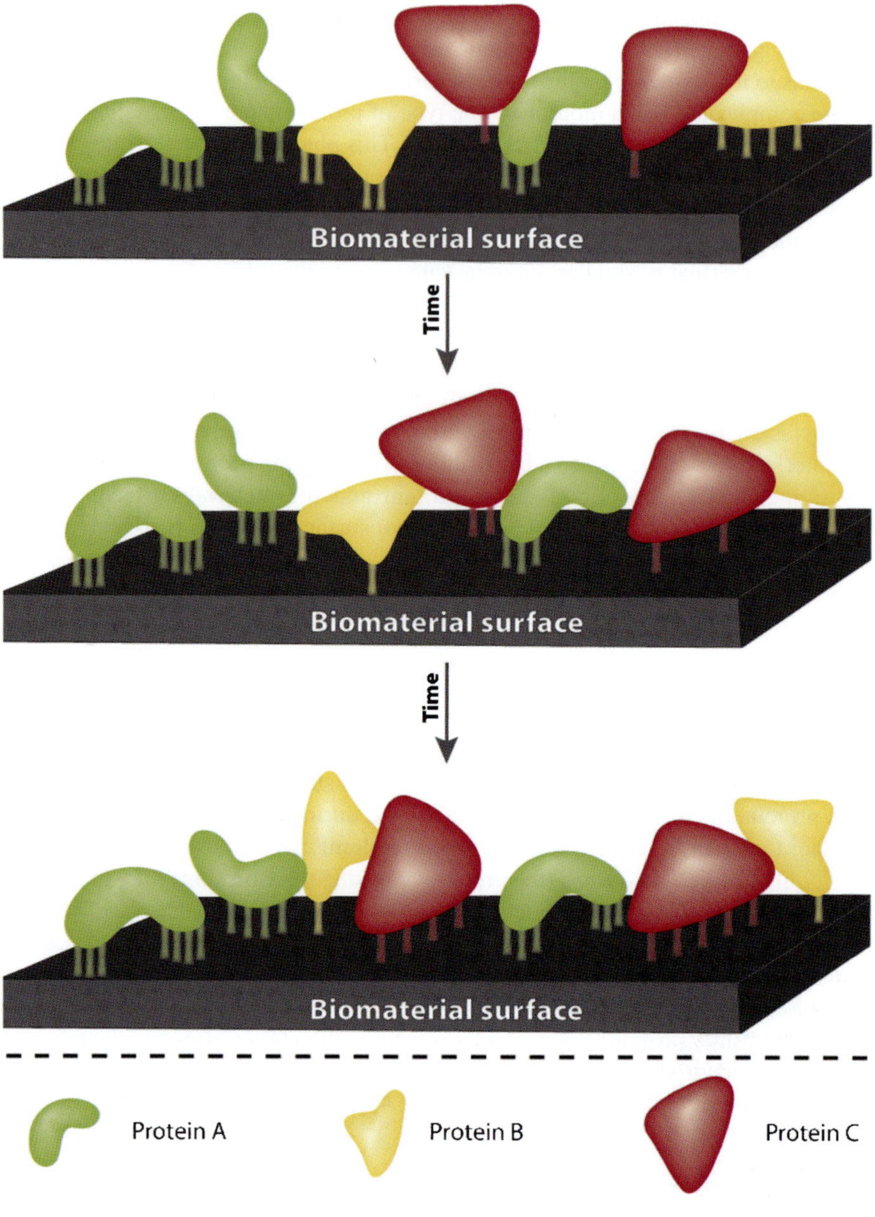

Figure 1.1. Schematic of the sequential adsorption of proteins as described by the Vroman effect. Initially, many protein molecules in various conformations are adsorbed onto the biomaterial surface. On the left portion of all three frames are two proteins A (green) in different conformations, which change over time. In the center of the figure, different proteins B (yellow) with multiple bonds are replaced over time by a larger, higher-affinity protein C (red) that arrived later on the substrate surface.

A number of other factors come into play when considering the underlying properties of the amino acid constituents of proteins. Since proteins consist of sequences of amino acids, their properties affect the adsorption properties of protein molecules. Amino acids may be charged (depending on the pH of their environment), and thus be more polar;

Table 1.1. Protein properties that affect protein adsorption on material substrates [1, 2] (adapted from [1]).

Protein property	Description
Hydrophilicity/hydrophobicity	Since more hydrophilic side chains are present on the outside of a protein molecule in aqueous media, they may interact with substrate surfaces. Polar (hydrophilic) domains tend to adsorb to polar material surfaces, while hydrophobic domains tend to adsorb to hydrophobic material surfaces
Size	In a multicomponent solution, smaller proteins diffuse more quickly and arrive at the substrate surface faster. Smaller proteins, however, form fewer contact points with the material surface than larger proteins
Charge	Charged proteins preferentially adsorb on substrate surfaces of the opposite charge. On most surfaces, however, proteins adsorb the most at their p*I* where they are at neutral charge
Structural stability/rigidity	Proteins that are less structurally stable exhibit greater unfolding upon adsorption onto a material surface and form more contact points/bonds. Proteins that unfold quickly form bonds with the surface before other proteins from a multicomponent solution arrive at that surface

such amino acids are more hydrophilic because they are attracted to the polar water molecules in their environment. These charged hydrophilic amino acids then have a tendency to be on the outside of a protein and thus can interact preferentially with the polar regions of the material surface. Positively charged domains preferentially adsorb to negatively charged surface regions and vice versa; however, this scheme of interactions is not the dominant process for adsorption because many other aspects influence the protein adsorption process. Generally, proteins tend to adsorb more readily near their isoelectric point (p*I*; that is, when the environmental pH is such that the protein has no net charge). This adsorption event was theorized to be due to reduced electrostatic repulsion with other proteins on the material surface. Hydrophilic domains tend to preferentially interact with hydrophilic surface regions and vice versa. Hydrophobic domains tend to be buried within the three-dimensional (3D) structure when proteins are in solution; however, these domains may become exposed during unfolding while the protein interacts with a material surface. Depending on the extent of conformation of the protein, pertinent amino acid sequences may not be completely hidden within the protein structure [1, 2, 4].

In addition to specific amino acid sequences, proteins have shapes and structures that may also influence their adsorption on material surfaces. "Soft" proteins, that is, proteins with lower than average thermodynamic stability, less internal bonding, and/or crosslinks (such as disulfide bonds), tend to adsorb more easily than "hard" proteins, which have a higher level of thermodynamic stability compared with "soft" proteins. "Relatively" softer proteins tend to adsorb strongly to surfaces due to the easier unfolding of protein molecules that have lower levels of thermodynamic stability [1, 2]. As a protein unfolds, more domains may be revealed and be available for subsequent interactions with the material surface. The structure of a protein may be crucial to its adsorption because specific binding sites may only be exposed when the protein molecules are in a specific conformation. A protein with specific activity in bulk solution may lose its activity once it undergoes a conformational change

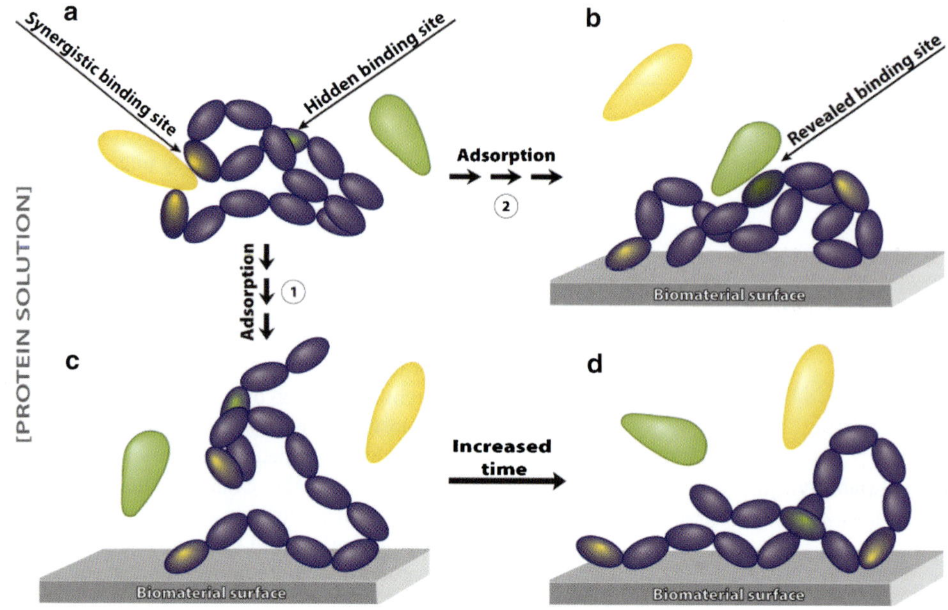

Figure 1.2. Schematic depiction of protein conformational changes upon adsorption of proteins on material surfaces. The protein in this illustration has a binding site that requires a specific structure of two regions of the molecule (A). Once the protein is adsorbed, these conformational epitopes are no longer functional because the two regions are far apart (C). Alternatively, the protein may have a hidden biding site that is revealed but becomes available for binding to another molecule once the protein has unfolded upon adsorption on the material surface (B). Over time, the adsorbed protein may continue to unfold, thereby exposing additional binding sites (D).

upon adsorption to a material surface; pertinent protein unfolding and changes in protein activity may occur over some period of time (Figure 1.2) [1, 2, 4].

The Monolayer Model and Protein–Protein Interactions

Adsorption of a specific protein on a substrate surface is also affected by the presence of other proteins in the solution. Excepting coagulation, most blood serum proteins do not adhere to one another on a surface. Exceptions to this rule occur because the conformation of the adsorbed protein may expose new binding sites for subsequent protein–protein interactions; this process is hypothesized to occur when extracellular matrix proteins adsorb on material surfaces. Since most proteins do not adsorb in a conformationally favorable fashion to mediate subsequent protein–protein binding, protein–protein bonds reflect a small percentage of the total protein layer compared with protein–surface bonds. Consequently, monolayers and near-monolayers are the rule in protein adsorption on material surfaces [1, 2, 4, 5].

The presence of other proteins on a material surface may induce lateral interactions between adsorbed protein molecules. Many proteins (such as enzymes) have specific binding sites that need to be accessible for effective function; moreover, the presence of other proteins on the material surface can sterically hinder target proteins from interacting with their respective binding sites and thus affect bioactive function. Additionally, proteins adsorbing on a material surface may unfold and thus expose previously hidden bioactive epitopes; however, the presence of other proteins on a material surface can sterically contain this unfolding process, rendering respective epitopes unavailable for binding. Lateral protein interactions become more apparent in situations involving increases from a lower to greater concentration of protein in the bulk solution; in this case, an increased number of proteins on the material surface may interact with one another [1, 2, 4, 5].

1.2.3. Surface Properties

Overview of Material Surface Properties Relevant to Protein Adsorption

The properties of individual proteins are important in protein adsorption, but are only half the story. The material surface itself is another determinant of the quantity, type, conformation, and function of proteins that preferentially adsorb there. Material properties of surfaces that affect protein adsorption are listed in Table 1.2. It should be noted that, even before proteins adsorb onto a material surface, water molecules from the aqueous biological environment interact with the material. In the case of hydrophobic materials, a "shell" of water molecules forms in which these molecules interact with each other more than with the hydrophobic surface. One hypothesis postulates that these surface-surrounding water molecules represent a fairly ordered scenario with a decreased level of entropy; disruption of this layer with proteins is energetically favorable due to a concomitant increase in entropy. This increase in entropy is the primary motivating force behind protein adsorption to material surfaces that are hydrophobic. Consequently, enhanced protein adsorption and conformational change are observed on hydrophobic surfaces. Greater conformational change may also lead to a greater loss of protein activity in cases that require the protein to be in its native state. In contrast, since more water molecules can form hydrogen bonds directly on hydrophilic surfaces, competition between water molecules and proteins results in decreased protein adsorption and conformational change. Moreover, material surfaces may have a distribution of charge. Since like charges tend to repel one another, opposite charges promote adsorption and ionic bonding of proteins on such substrates. For example, since the majority of blood serum proteins are negatively charged, a net negative charge on a material surface may reduce adsorption of these proteins. This outcome, however, is complicated because proteins contain amino acids that are themselves either positively or negative charged; for this reason, even if a protein is net negatively charged, there may still be positively charged domains on that molecule.

Table 1.2. Surface properties that affect protein adsorption on material substrates [1, 2] (adapted from [1]).

Surface property	Description
Hydrophilicity/hydrophobicity	Hydrophobic surfaces tend to adsorb more proteins, while hydrophilic surfaces tend to resist protein adsorption
Charge	Opposite charges between the surface and protein promote increased protein adsorption, while like charges tend to reduce protein adsorption
Topography	Increased surface roughness and topological features provide increased material surface area for protein adsorption
Chemistry	The chemical composition of a substrate surface dictates the types of bonds between protein and material surface

Topographical features on a biomaterial surface may also strongly influence protein adsorption. Increased surface roughness provides more surface area for protein adsorption onto the biomaterial surface; this event may lead to a net increase in protein adsorption. Such protein adsorption outcomes may be achieved either intentionally (by incorporating material surface physical modifications such as grooves, etc., during manufacturing) or unintentionally (through creation of surface defects and flaws during the manufacturing processes). Surface defects of micrometer size can accommodate thousands to millions of blood serum proteins [1, 2, 6, 7].

Effect of Material Surface Functional Groups on Protein Adsorption

The chemical properties of a material surface can strongly influence adsorption of proteins and, through these proteins, subsequent adhesion of various cell types. Material surface chemistry and topography affect cell functions via the conformation and bioactivity of the adsorbed proteins. Examples of material surface functional groups and their observed effects on proteins and cells are listed in Table 1.3. It should be noted though that these are generalized observations and may vary depending on experimental conditions and the presence of a multicomponent protein solution. Since "protein surface concentration" is not the same as "protein surface activity," higher protein concentration at the material surface does not necessarily mean that these surfaces are more bioactive [8].

The influence of a number of surface functional groups on protein adsorption has been studied. Methyl ($–CH_3$), nonpolar, hydrophobic groups tightly bind fibrinogen (a protein involved in blood clotting), and immunoglobin IgG, a family of antibody proteins involved in the immune response. *In vivo*, the $–CH_3$ group induces high recruitment of inflammatory cells to the material surface, likely due to the high adsorption of IgG on material surfaces containing $–CH_3$. Hydroxyl groups –OH increase the hydrophilicity of a material surface and, thus, reduce the affinity of plasma proteins there. The –OH group induces changes in the conformation of fibronectin, thus exposing adhesive domains for cell "focal" (tight mechanical links) adhesion. Amine groups ($–NH_2$) are polar and, thus, hydrophilic; these groups are also

Table 1.3. The effect of material surface functional groups on proteins and cells (adapted from [8]).

Functional group	Properties	Effect on proteins and cells
$–CH_3$	Neutral; hydrophobic	Has high affinity/binding with fibrinogen; binds strongly with IgG; promotes increased leukocyte adhesion and phagocyte migration
–OH	Neutral; hydrophilic	Has decreased affinity for plasma proteins; induces exposure of cell adhesive domains on fibronectin; increases differentiation of osteoblasts
$–NH_2$	Positive; hydrophilic	Has high affinity for fibronectin; promotes increased myoblast proliferation; promotes differentiation of osteoblasts; promotes increased endothelial cell proliferation
–COOH	Negative; hydrophilic	Has increased affinity for fibronectin and albumin

Note: These are generalized observations and may vary depending on experimental conditions and the presence of other proteins in solution.

positively charged in blood serum and most other solutions in which the solvent is water. The $-NH_2$ group binds strongly to fibronectin and induces several responses from various cells (Table 1.3). Moreover, $-NH_2$ also triggers acute inflammatory reactions *in vivo*. Carboxyl ($-COOH$) groups are negatively charged in blood serum and other aqueous protein solutions, are hydrophilic, and interact preferentially with fibronectin and albumin [8, 9].

Multiple functional groups, such as the ones mentioned earlier in this section, can be found on the same material surface. In the case of mixed $-OH$ and $-CH_3$ groups, hydrophobic and hydrophilic influences are effectively in competition with each other. In this case, as the surface concentration of $-OH$ groups is increased, protein adsorption decreases (likely due to the increased hydrophilicity associated with increased $-OH$ groups) [9]. In the case of mixed $-NH_2$ and $-COOH$ functional groups, platelet adhesion was the lowest when these two functional groups were mixed in equal molar fractions making the net charge of the surface approximately neutral [8, 9]. It should be noted that, over time, the presence of water and other molecules in the surrounding environment may modify these material surface groups [8, 9].

Nonfouling Surfaces

While hydrophobic surfaces promote increased protein adsorption, highly hydrophilic surfaces are characterized by decreased protein adsorption. The phenomenon of decreased protein adsorption is known as nonfouling. Since serum protein adsorption on the surfaces of cardiovascular biomaterials is associated with blood clotting and inflammatory/immunological responses, a strategy in biomaterials research and applications is to use materials that inhibit serum protein adsorption. Although polysaccharides can also be used for such applications, by far the most widely used material is polyethylene glycol (PEG; also known as polyethylene oxide [PEO] depending on its molecular weight), as well as its derivatives and copolymers. PEG, a very hydrophilic material surface, was chosen and is used in numerous drug delivery, biomaterials, and tissue engineering applications primarily because of its low level of protein adsorption. The unique ability of PEG to greatly limit protein adsorption is due to the highly hydrophilic nature of its polymer coils. Once PEG comes into contact with water, these hydrophilic coils become surrounded by water molecules and, thus, get highly hydrated. When a protein arrives at the PEG surface, the adsorption process requires compression of these hydrated coils. Since this process requires energy (because water molecules must be removed), it is thermodynamically unfavorable. However, some small amount of protein still adsorbs and, given enough exposure time, protein adsorption may occur, but to a smaller degree [10, 11].

A recent example of protein adsorption research on polyethylene glycol involved the use of a plasma deposited PEG-like polymer called tetraethylene glycol dimethyl ether (tetraglyme [$CH_3O(CH_2CH_2O)_4CH_3$]). Tetraglyme-coated surfaces resist protein adsorption *in vitro*; these surfaces, however, still induced a foreign body reaction *in vivo*, most probably due to monocyte (a critical inflammatory/immune leukocyte) activation on the tetraglyme surface. Although resistant to other serum proteins, the blood complement factor C3 adsorbs to tetraglyme in significant amounts (approximately 10 ng/cm^2) [12]. Monocytes may interact with, and bind to, complement factors in the blood. Monocyte adhesion to tetraglyme was considerably reduced when C3-depleted serum (instead of normal serum) was used, suggesting that monocyte binding to tetraglyme is primarily mediated by monocytes binding to adsorbed C3. This result provides an explanation for the reported foreign body reaction to PEG-coated surfaces *in vivo* because significant C3 adsorption in such cases may be inducing adhesion and subsequent activation of monocytes [12]. In certain cases, monocyte adhesion to PEG hydrogels is comparable to that observed *in vitro* on tissue culture polystyrene culture plates, potentially due to the same C3-related mechanism [13].

1.2.4. Environmental Effects

The environment in which protein adsorption occurs can also affect the extent of, and conformation during, such interactions. Temperature can have a significant impact on protein adsorption; temperatures significantly above room temperature (for example, 80°C) can increase the amount of protein that is adsorbed. However, very high temperatures may lead to denaturation of proteins and loss of important bioactivity. The temperature at which temperature-driven denaturation of proteins (and thus changes in bioactivity and adsorption) occurs can vary considerably for different types of proteins due to differences in protein stability. Temperature-driven denaturation of proteins, particularly those with cysteine residues, can lead to protein aggregation on material surfaces. Since many published studies on protein adsorption were conducted at room temperature (22°C), for this reason, the difference between room and *in vivo* temperature (37°C) should be considered when analyzing the results of studies on protein adsorption.

Environmental pH can also affect protein adsorption because changes in the charge of both the material surface and the protein molecule may lead to variations in electrostatic interactions. Adsorption of a single protein is the highest at the p*I* but is reduced at other pH conditions. Decreases in protein adsorption due to either increases or decreases in the pH away from the p*I* lead to increased repulsion between like-charged protein molecules.

Lastly, the presence of detergents/surfactants (that is, molecules that are both hydrophilic and hydrophobic) can alter protein adsorption to a material surface by competing with proteins for adsorption sites or by binding directly to proteins. Adsorption of proteins in the presence of surfactants has not been well studied yet [4, 14].

1.2.5. Adsorption of Serum Proteins

There are more than 150 proteins in human blood serum; many of them have been used in studies of protein adsorption. These studies provided information regarding the sequence of protein adsorption on many material surfaces including glass, metal oxide

Table 1.4. Select human blood serum proteins and their main biological functions.

Blood serum proteins	Concentration in normal human blood (mg/ml)	Major function
Albumin	40	Maintains osmotic pressure; forms complexes with, and thus "carries," other serum molecules
Complement C3	1.6	Participates in complement system activation and function
Hemoglobin	0.01	Transports oxygen
High molecular weight kininogen	0.08	Participates in blood clotting
Immunoglobin (IgG)	15	Participates in the immune response
Fibrinogen	3	Participates in blood clotting
Fibronectin	0.3–0.4	Mediates cell adhesion
Factor XII	0.03	Participates in blood clotting
Vitronectin	0.3	Mediates cell adhesion

surfaces, and polymers such as polyethylene. The most widely studied serum proteins in protein adsorption studies are listed in Table 1.4. Pertinent studies reported in the literature provided evidence that serum protein adsorption on glass follows the sequence: albumin first, followed by IgG, fibrinogen, fibronectin, Factor XII, and high molecular weight kininogen. Since albumin is a small protein, but is present in high concentration in serum, it tends to adsorb first on material substrate surfaces. Albumin, however, has a relatively low affinity and so, over time, it is partially replaced by larger, higher-affinity proteins such as fibrinogen [1, 7].

1.3. Techniques for the Study of Protein Adsorption

Numerous techniques exist for the study of protein adsorption and for the characterization of biomaterial properties pertinent to such processes. This section will focus exclusively on techniques that directly monitor either adsorbing proteins or the adsorbed protein layer on material surfaces. Techniques that indirectly lead to information about protein adsorption (for example measuring and comparing the hydrophobicity of one surface versus another) may allow predictions of relative protein adsorption on those material surfaces but will not be covered in this section. Several of the more important techniques for the analysis of protein adsorption will be highlighted for discussion, and powerful emerging techniques will also be discussed in this section.

1.3.1. Established Techniques for the Study of Protein Adsorption

Various aspects of protein adsorption, including kinetics, thickness of the protein layer, conformation of adsorbed proteins, identity of adsorbed proteins, structure of the adsorbed protein layer, and/or the types of forces operating between the protein and the surface, have been the focus of research activities that either developed new or adapted existing methodologies for such studies. In the study of the chemical composition of adsorbed protein layers, spectroscopy/spectrometry methods have been very useful. Auger electron spectroscopy did not provide useful information on adsorbed proteins because this technique is very destructive to organic matter. On the other hand, X-ray photoelectron spectroscopy (XPS) has been used extensively in protein adsorption studies. In XPS, when a material surface is irradiated by a beam of monochromatic X-rays, photoelectrons are released and are then captured by a detector and analyzed. XPS provides detailed elemental (but not molecular) composition analysis, is relatively nondestructive, and is very useful for single-protein studies; however, this method is very limited for analysis of multicomponent solutions due to its inability to resolve individual proteins in an overly complex spectrum. Secondary ion mass spectrometry (SIMS) and time-of-flight (ToF)-SIMS have become widely used in studies of protein adsorption. ToF-SIMS offers significant advantages in protein resolution over SIMS. In both SIMS and ToF-SIMS, bombardment of the adsorbed protein layer with a focused beam of either ions or atoms results in the emission of secondary particles, which can then be analyzed to provide elemental and molecular information about the material surface. ToF-SIMS is similar to SIMS but includes an additional step. In this case, the secondary ion particles are accelerated through a field-free drift region; heavier ions travel more slowly than lighter ions, allowing for greater separation and increased sensitivity of detection of the ToF-SIMS over the SIMS technique [15–18].

Infrared (IR) spectroscopy has also been used to examine protein adsorption. An IR beam is reflected from a sample and a spectrum can be developed from the IR absorbance to different components of the adsorbed protein. Different chemical bonds have differing vibrational frequencies and thus absorb the IR beam at different frequencies; this property allows for the generation of an IR spectrum "fingerprint," which provides information about the chemical bonds, which, in turn, provide information about the protein conformation and load. IR has limited use in analyzing multicomponent protein solutions because the spectra in these cases can become very complicated and difficult to resolve.

Fourier transform infrared spectroscopy (FTIR) allows for the measurement of all wavelengths at once (rather than one at a time). A more advanced method called attenuated total reflectance (ATR)-FTIR uses total internal reflection of the IR beam back into the detector along the region of contact between the sample and the device. By combining IR with ATR, it is possible to greatly reduce the area of the material surface needed for analysis. Both techniques provide conformational and submolecular information about adsorbed proteins. IR spectroscopy can be used to analyze activity shifts in proteins by measuring the IR spectrum of a protein in one state and then subtracting this IR spectrum from another one obtained in a different (for example, active) state of the same protein, or, in the case of an enzyme, when it acts as a reaction intermediate. This spectrum difference between the two states contains the respective vibrational bands. An additional advantage of this technique is that since all redundant vibrational bands cancel out, the information thus obtained relates to the change in protein functionality only [15, 16, 18–20].

One of the oldest, and most popular, techniques for studying protein adsorption is the use of solute-depletion strategies. In such studies, changes in the bulk concentration of a protein are measured over time as the protein adsorbs onto a material surface. One assumption that must be made in this case is that all loss of protein from the bulk solution is due to its adsorption onto the material surface. The kinetics of protein adsorption onto a specific material surface at constant temperature can then be determined by plotting the change in bulk protein concentration versus time. Another similar technique that has been used extensively is to label proteins in solution with either fluorescent or radioactive probes. In this case, the concentration of adsorbed protein is determined by measuring either fluorescence or radioactivity, respectively, on the material surface. Protein adsorption can also be measured indirectly using immunoassays such as an enzyme-linked immunosorbent assay (ELISA). In an ELISA, an antibody directed against the protein is used to detect the presence of the adsorbed protein. Binding of the antibody to specific portions of the target protein indicates that the adsorbed protein is in a recognizable conformation, at least for that particular antibody. This approach provides information regarding the bioactivity of a protein rather than just confirming and quantifying the presence of proteins on material surfaces [19].

Several techniques have been developed to study the thickness of the adsorbed protein layer on material surfaces. Ellipsometry, a method that utilizes polarized light to nondestructively examine surfaces, is one of the most widely used such techniques. In a standard ellipsometry procedure, polarized light is directed at a surface at an angle, and is then reflected back to a detector. Changes in this reflected polarized light beam provide information about the interface, primarily the thickness of the adsorbed protein layer, which in turn can be directly correlated to the amount of adsorbed protein. Ellipsometry measures protein layer thicknesses in the nanometer range and is highly accurate. Surface plasmon resonance (SPR) is comparable to, and provides similar information as, ellipsometry; SPR, however, depends on excitation of surface plasmons (instead of changes in polarized light). Another method that has some similarities to ellipsometry is "neutron reflectivity," in which a beam of neutrons directed at the adsorbed protein layer is reflected back (in the form of reflected radiation) and is measured in a detector. The resulting reflected radiation profile provides information not only about the thickness of the adsorbed protein layer but also about the conformation of the adsorbed molecules. Neutron reflectivity is not widely used because of several drawbacks, which include very limited equipment availability, the need for a large specimen surface area, and the risk of making radioactive the material surface tested [15, 21].

1.3.2. Emerging Techniques for the Study of Protein Adsorption

Several emerging techniques for the study of protein adsorption offer improvements over the limitations of established techniques. Modern mass spectrometry-based techniques identify proteins from multicomponent solutions and exudates such as blood, serum, tears, and saliva, adsorbed either on biomaterials *in vitro* or on excised devices. These techniques can resolve many and different proteins of a large range of molecular masses. Tandem scanning probe microscopy techniques offer the unmatched ability to image a single protein adsorbed on a material surface. Lastly, new developments in surface IR techniques may offer the opportunity to study changes in protein conformation at the level of a single atomic bond.

The advent of more advanced tandem mass spectrometry techniques improved the sensitivities of ToF-SIMS/SIMS technologies and these techniques have tremendous potential for future studies of protein adsorption on biomaterial surfaces. To improve the resolution of mass spectrometry, researchers have used surface treatments (involving the use of matrix molecules such as nicotinic acid, dihydroxybenzoic acid, and sinapinic acid) to protect the protein sample and improve resolution of individual proteins. Matrix-assisted laser/desorption/ionization (MALDI) time-of-flight mass spectrometry (MALDI-ToF/MS) provides, in theory, unlimited mass resolution and extremely low detection limits. In this method, matrix molecules (such as nicotinic acid, dihydroxybenzoic acid, and sinapinic acid) are added to the adsorbed protein layer prior to irradiation with a pulse laser; the released ions are then carried to a detector for analysis (Figure 1.3). A mass-to-charge ratio for the peptide fragments can then be generated, compared with a database of known protein sequences, and statistically analyzed to determine the likelihood that a particular peptide fragment belongs to a specific protein. This technique can examine a wide mass range but, in practice (and depending on the specific mass spectrometer and equipment setup), the ability to resolve larger proteins can diminish because the large, slower-moving ions do not reach the detector quickly. The cut-off point for decreased sensitivity of larger proteins is highly dependent on the individual mass spectrometer used in the study [15, 22].

Another option is to digest adsorbed proteins with enzymes, blot the resulting solution directly onto a MALDI matrix plate, and analyze the digested proteins using a mass spectrometer. By first digesting the adsorbed protein *in situ*, mass spectrometers with a lower mass range are capable of detecting larger peptide fragments with greater sensitivity. For example, in our lab, adsorption of human and bovine serum on standard tissue culture polystyrene took place for 2, 24, and 96 h and was then analyzed via this digestion method. After washing and removing the nonadsorbed proteins, the adsorbed protein layer was digested with trypsin in 10% NH_4HCO_3 and 2% methanol in a trypsin resuspension buffer. The resultant protein fragments were further separated via reverse phase chromatography followed by MALDI-ToF/MS. The results were compared against the National Center for Biotechnology Information and Expert Protein Analysis System databases and provided evidence that albumin, vitronectin, and complement protein C3 were present on the polystyrene surface [unpublished data].

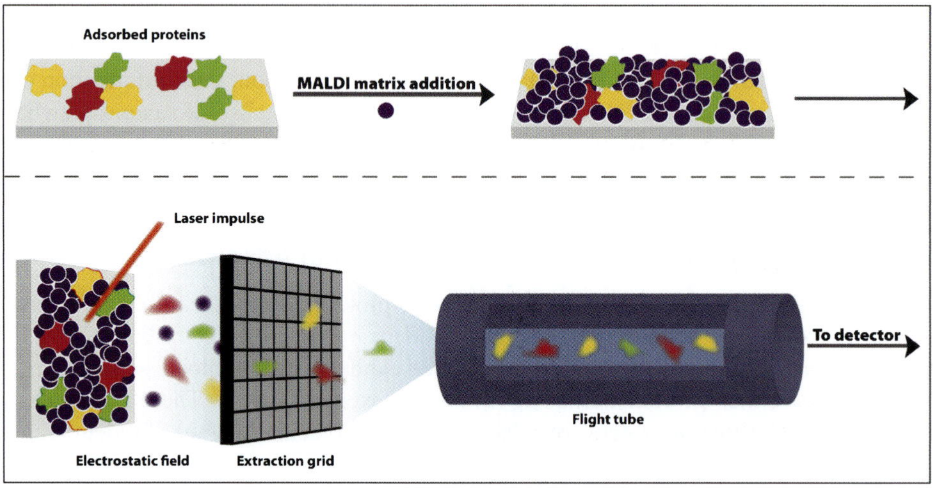

Figure 1.3. Schematic illustration of the major components of the MALDI-ToF/MS technique used for the analysis of proteins adsorbed on a material surface. The MALDI matrix is added to the adsorbed protein sample, which is then bombarded by a pulsed laser. Ion fragments (i.e., ionized peptide fragments from the adsorbed proteins that are generally smaller than the adsorbed protein molecules) are released from the MALDI plate and directed via an electrostatic field into the extraction grid. From there the ions pass though a flight tube to the detector, where proteins and peptide fragments can be analyzed and identified.

MALDI-ToF/MS can also be coupled with protein separation techniques such as two-dimensional (2D) gel electrophoresis to isolate specific bands of proteins from the gel; individual proteins can then be resolved using a mass spectrometer. Potential limitations of MALDI-ToF/MS include the inability to recognize proteins present at very low concentrations and the inability to identify specific proteins when using a MALDI matrix that does not effectively protect the protein sample. The selection of an appropriate MALDI matrix is often a process of trial and error [15, 22]. Future developments involving mass spectrometry will address some of these issues by utilizing tandem equipment strategies and improved protein-processing techniques.

Scanning electron microscopy (SEM), solid-state crystallography, and transmission electron microscopy have also been used to examine the structure of adsorbed proteins; all of these techniques, however, require extensive sample preparation, which may alter either the material surface or the adsorbed protein layer, particularly when the substrate surface interacts with a multicomponent, aqueous protein solution. These methods often require dehydration of the organic sample (which may alter the conformation of proteins), and coating with a conductive substrate (which can sometimes lead to additional artifacts due to changes in the structure of surfaces when dehydrated). Atomic force microscopy (AFM) improves upon other microscopic techniques by allowing imaging of the adsorbed proteins in their native state in nearly physiological conditions and in three dimensions. AFM functions by "feeling" the surface with a very sharp microscale cantilevered tip. More recently, strategies have been developed that combine multiple scanning-probe microscopy techniques to analyze single adsorbed proteins. AFM has been combined with scanning tunneling microscopy (STM) to accurately characterize single adsorbed proteins on conductive substrates. STM functions through measurement of the quantum mechanical tunneling current between a conductive surface and a very sharp metallic tip. Since the rate of tunneling electrons depends on the distance between the tip and surface, an image can be generated that captures extremely fine changes in height across the scanned surface area. AFM alone provides accurate information about the height of a protein, but STM can provide improved lateral information. Tandem STM/AFM has been used to characterize the morphology of metalloproteins such as azurin [15, 23, 24].

Surface-enhanced infrared absorption spectroscopy (SEIRA), which is an addition to traditional IR spectroscopy, has been applied to study interactions involving surface-immobilized proteins. SEIRA improves on the resolution of traditional IR by amplifying the signal of an adsorbed molecule by approximately two orders of magnitude. In addition to studies on protein recognition, SEIRA has been used to study the adsorption of nucleic acids and DNA. The high level of sensitivity of SEIRA potentially lends the technique to improved analysis of changes in protein functionality over the results obtained by the IR difference method, although it has yet to be used extensively to study adsorbed proteins [20, 25].

1.4. Recent Advances in Protein Adsorption

1.4.1. Protein Adsorption to Nanomaterials

Nanotechnology has exploded as a research direction in many fields of science; the study of protein adsorption on nanophase materials is no exception. Nanoscale materials, or nanomaterials, are defined as materials that have some component or feature (i.e., grains, surface texture, fibers, particles, etc.) with at least one dimension between 1 and 100 nm. Compared with conventional materials (such as a microphase metal or polymer) nanomaterials exhibit improved properties due to their nanoscale characteristics. Additionally, nanomaterials have vastly increased surface area-to-volume ratios compared with conventional materials [26]. Nanomaterials also better mimic physiological structures as most tissues *in vivo* have nanoscale features because they are assembled from nanoscale units of amino acids, proteins, and lipids, all of which are in the low end of the nanoscale range in terms of size (below 15 nm). In addition, the size of most blood proteins is in the range of 3–15 nm. The characteristic dimension of nanoscale materials is, therefore, nearly the size of, or smaller than, a serum protein molecule [26]. Thus, nanoscale biomaterials may promote protein adsorption, and subsequent cell attachment and function, in ways analogous to those promoted by tissues of the body and, potentially, much more effectively than the results obtained in conjunction with conventional materials. Many material surface properties relevant to protein adsorption can be modulated at the nanoscale level, since properties such as hydrophilicity/hydrophobicity, charge, topography, and chemistry can all greatly affect protein adsorption and conformation. Since such properties of nanoscale substrates vary from those of conventional material substrates, nanomaterials may induce controlled conformations of various proteins and, thus, present multiple and/or different cues and signals to cells.

Several literature reports have characterized protein adsorption to nanomaterials [26–30]. For example, adsorption of lysozyme and cytochrome c (an enzymatic heme molecule) to mesoporous (defined as having pore sizes between 2 and 50 nm) silica increased with pore diameter; specifically, using adsorption isotherms, the highest amount of lysozyme adsorbed increased from 13.4 μmol g^{-1} for the 3.54-nm diameter pores to 35.3 μmol g^{-1} for the 10.98-nm diameter pores with similar concentrations of cytochrome c. Additionally, cytochrome c adsorbed to the mesoporous silicate had greater bioactivity than the enzyme in solution. Sustained enzyme activity was also observed on a mesoporous carbon support substrate [27]. Additionally, a hydroxyapatite (HA) poly(L-lactic acid) (PLLA) composite made from nanocrystal hydroxyapatite mixed with PLLA promoted increased protein adsorption from fetal bovine serum as a function of increased nano-HA weight percentage content in the range 10–70% [26, 28]. Studies of the adhesion of osteoblasts on nanophase hydroxyapatite, titania, and alumina ceramics reported increased surface concentration of vitronectin, an adhesion protein known to mediate adhesion of osteoblasts [29, 30]. Compared with the respective conventional material, the nanophase ceramics contained many small pores (24–97 nm) where vitronectin may have preferentially adsorbed. Additionally, vitronectin displayed increased conformational unfolding on the nanomaterial (measured via Raman scattering), likely leading to exposure of more cell adhesive domains within this molecule that promoted the observed increased osteoblast adhesion [29, 30].

Researchers have sought to characterize the effect of nanoscale surface features on such phenomena as gross protein adsorption and changes in protein conformation. Recently, fibronectin (a large, cell adhesion molecule) interactions with a poly(lactic-*co*-glycolic acid) (PLGA) surface with nanoscale surface features was imaged for the first time using AFM [26, 31]. Fibronectin at concentrations of 0.5–5 $\mu g\ ml^{-1}$ was adsorbed to PLGA surfaces with 3D spherical "bumps" with sizes of 100, 200, and 500 nm. AFM images revealed small globules of fibronectin present on the 500-nm surface. The fibronectin molecules were more spread and elongated on the 200- and 100-nm surfaces; the greatest initial spreading was observed on the material surface with the 200-nm bumps [31]. This result indicates that there are specific surface feature sizes that promote enhanced spreading of specific protein molecules. It should be noted that since both the 100-nm and 200-nm spherical "bumps" are much larger than the "typical" (5–15 nm) size of proteins, these surface features can support adsorption of multiple protein molecules [26, 31].

1.4.2. Manipulating Protein Adsorption

Besides nanomaterial-directed strategies, researchers have continued to investigate the effect of other material surface properties on protein adsorption. Self-assembled monolayers (SAMs) present specific molecules on a material surface and are used in the study of protein adsorption. For the purpose of studying osteopontin (an extracellular matrix protein involved in inflammation) adsorption, various SAM terminal groups such as hydrophobic ($–CH_3$), hydrophilic but neutral (–OH), positively charged ($–NH_2$), and negatively charged (–COOH) were assembled on a gold material surface. SPR indicated that osteopontin adsorption to the $–NH_2$ surface was in the monolayer range, and comparable to that observed on the -COOH surface. Subsequent interactions of bovine aortic endothelial cells exhibited increased cell spreading on the surfaces tested in the following order: –COOH, $–CH_3$, gold, and $–NH_2$. Although the amount of total protein absorbed was similar on both the –COOH and $–NH_2$ surfaces, cell adhesion and spreading was much higher on the $–NH_2$ surface. This result provides evidence that the conformation and bioactivity of osteopontin on the positively charged surface was more favorable for cell binding. Liu *et al.* [32] speculated that the orientation of osteopontin on the –NH2 surface may have exposed the cell adhesive domain arginine–glycine–aspartic acid (RGD) in an orientation favorable to cell adhesion, while on the –COOH surface, RGD was oriented toward the material surface and not toward the adhering cells. Osteopontin is a very acidic, negatively charged protein that may have bound strongly and in a different orientation to the positively charged $–NH_2$ surface than to the negatively charged –COOH surface [32, 33].

Stimuli-responsive polymers (also known as "smart" polymers, defined as materials that respond to external stimuli by altering a specific property) have been developed to modulate protein adsorption in a dynamic and as-needed fashion. Since protein adsorption on material surfaces may vary tremendously due to hydrophilicity/hydrophobicity properties, switching these properties on a material surface may affect and/or change the adsorbed protein monolayer and, thus, subsequent cell interactions. Poly(*N*-isopropylacrylamide) (PNIPAAm), a thermoresponsive coating, has been used on traditional tissue culture polystyrene and in a microfluidic hotplate. Up to 37°C, the PNIPAAm surface is hydrophobic and promotes protein adsorption; when cooled to room temperature, this polymer undergoes a phase transition and its surface becomes hydrophilic, inducing decreased protein adsorption. PNIPAAm was used in a microfluidic hotplate to induce fast (in less than 1s) protein adsorption/desorption. A different type of stimuli-responsive polymer utilized monolayers with photocleavable 2-nitrobenzyl groups. Bovine serum albumin was adsorbed to this layer, limiting human embryonic kidney cell adhesion due to the inertness of albumin. When this material surface was exposed to ultraviolet light, the 2-nitrobenzyl groups were cleaved and albumin was released; then increased adsorption of fibronectin and human embryonic kidney cell adhesion was observed [34, 35].

1.5. Current Limitations and Potential Future Opportunities

Protein adsorption has been studied on many materials and under various conditions. In many of these cases, researchers reported changes in protein conformation and bioactivity upon adsorption; however, the underlying mechanisms correlating material surface chemistry, protein conformation, and protein bioactivity are still not fully understood.

Nanomaterials offer a new, and relatively untapped, opportunity to study these phenomena at the nanoscale. Coupling a wide range of proteins, nanomaterials, and cells with imaging techniques (such as AFM) will provide valuable information regarding specific aspects of these phenomena, potentially leading to improved understanding of protein-surface interactions. In addition to improving current understanding of protein adsorption, nanomaterials may also allow for improved control of protein adsorption (and, thus, subsequent cell adhesion and function pertinent to new tissue formation), leading to the design and formulation of new and improved biomaterials for use in clinical applications.

References

1. Dee KC, Puleo DA, Bizios R. Protein–surface interactions. In: Dee KC, Puleo DA, Bizios R, editors. An introduction to tissue-biomaterial interactions. Hoboken, NJ: John Wiley and Sons, 2002
2. Horbett TA. The role of adsorbed proteins in tissue response to biomaterials. In: Ratner BD, Hoffman AS, Schoen FJ, Lemons JE, editors. Biomaterials science. An introduction to materials in medicine. San Diego: Elsevier Academic Press, 2004, pp. 237–246
3. Anderson JM, Rodriguez A, Chang DT. Foreign body reaction to biomaterials. Semin Immunol 2008;20: 86–100
4. Nakanishi K, Sakiyama T, Imamura K. On the adsorption of proteins on solid surfaces, a common but very complicated phenomenon. J Biosci Bioeng 2001;91(3):233–244
5. Ramsden JJ. Puzzles and paradoxes in protein adsorption. Chem Soc Rev 1995;73–78
6. Kim MS, Khang G, Lee HB. Gradient polymer surfaces for biomedical applications. Prog Polym Sci 2008;33: 138–164

7. Sun S, Yue Y, Hunag X, Meng D. Protein adsorption on blood–contact membranes. J Membr Sci 2003; 222:3–18
8. Thevenot P, Wenjing H, Tang L. Surface chemistry influences implant biocompatibility. Curr Top Med Chem 2008;8:270–280
9. Roach P, Eglin D, Rhode K, Perry CC. Modern biomaterials: a review – bulk properties and implications of surface modifications. J Mater Sci: Mater Med 2007;18:1263–1277
10. Reintjes T, Tessmar J, Gopferich A. Biomimetic polymers to control cell adhesion. J Drug Del Sci Tech 2008;18(1):15–24
11. Ratner BD, Bryant SJ. Biomaterials: where we have been and where we are going. Annu Rev Biomed Eng 2004;6:41–75
12. Mayorga L, Ratner BD, Horbett TA. The role of complement adsorption and activation in monocyte adhesion to ultralow protein adsorption surfaces made by RF plasma deposition of PEO–like tetraethylene glycol dimethyl ether (tetraglyme). World Biomater Congr 2008:1162.
13. Schmidt DR, Kao WJ. Monocyte activation in response to polyethylene glycol hydrogels grafted with RGD and PHSRN separated by interpositional spacers of various lengths. J Biomed Mater Res 2007;83A(3):617–625
14. Miller R, Fainerman VB, Leser ME, Michel M. Kinetics of adsorption of proteins and surfactants. Curr Opin Colloid Interface Sci 2004;9:350–356
15. Merret K, Cornelius RM, McClung WG, Unsworth LD, Sheardown H. Surface analysis methods for characterizing polymeric biomaterials. J Biomater Sci Polym Ed 2002;6:593–621
16. Bhaduri A, Das KP. Proteins at solid water interface – a review. J Dispers Sci Technol 1999;20(4):1097–1123
17. McArthur SL. Applications of XPS in bioengineering. Surf Interface Anal 2006;38:1380–1385
18. Wahlgren M, Arnebrant T. Protein adsorption to solid surfaces. Tibtech 1991;9:201–208
19. Hlady V, Buijs J, Jennissen P. Methods for studying protein adsorption. Methods Enzymol 1999;309(26): 402–429
20. Ataka K, Heberle J. Biochemical applications of surface–enhanced infrared absorption spectroscopy. Anal Bioanal Chem 2007;388:47–54
21. Elwing H. Protein absorption and ellipsometry in biomaterial research. Biomaterials 1998;19:397–406
22. Gallagher WM, Lynch I, Allen LT, Miller I, Penney SC, O'Connor DP, Pennington S, Keenan AK, Dawson KA. Molecular basis of cell–biomaterial interaction: Insights gained from transcriptomic and proteomic studies. Biomaterials 2006;27:5871–5882
23. Silva LP. Imaging proteins with atomic force microscopy: an overview. Curr Protein Pept Sci 2005;6:387–395
24. Bonanni B, Andolfi L, Bizzarri R, Cannistraro S. Functional metalloproteins integrated with conductive substrates: detecting single molecules and sensing individual recognition events. J Phys Chem B 2007;111: 5062–5075
25. Garczarek F, Gerwert K. Integration of layered redox proteins and conductive supports for bioelectronic applications. Agnew Chem Int Ed 2000;39:1180–1218
26. Liu H, Webster TJ. Nanomedicine for implants: a review of studies and necessary experimental tools. Biomaterials 2007;28:354–369
27. Vinu A, Miyahara M, Ariga K. Assemblies of biomaterials in mesoporous media. J Nanosci Nanotechnol 2006;6(6):1510–1532
28. Wie G, Ma PX. Structure and properties of nano–hydroxyapatite/polymer composite scaffolds for bone tissue engineering. Biomaterials 2004;25:4749–4757
29. Webster TJ, Ergun C, Doremus RH, Siegel RW, Bizios R. Specific proteins mediate enhanced osteoblast adhesion on nanophase ceramics. J Biomed Mater Res 2000;51(3):475–483
30. Webster TJ, Schadler LS, Siegel RW, Bizios R. Mechanisms of enhanced osteoblasts adhesion on nanophase alumina involve vitronectin. Tissue Eng 2001;7(3):291.301
31. Miller DC, Haberstroh KM, Webster TJ. PLGA nanometer surface features manipulate fibronectin interactions for improved vascular cell adhesion. J Biomed Mater Res 2007;81A(3):678–684
32. Liu L, Chen S, Giachelli CM, Ratner BD, Jian S. Controlling osteopontin orientation on surfaces to modulate endothelial cell adhesion. J Biomed Mater Res 2005;74:23–31
33. Jandt KD. Evolutions, revolutions and trends in biomaterials science – a perspective. Adv Eng Mater 2007;9(12):1035–1050
34. Mano JF. Stimuli–responsive polymeric systems for biomedical applications. Adv Eng Mater 2008;10(6): 515–527
35. Nakanishi J, Kikuchi Y, Takarada T, Nakayama H, Yamaguchi K, Maeda M. Spatiotemporal control of cell adhesion on a self-assembled monolayer having a photocleavable protecting group. Anal Chim Acta 2006;578:100

2

Investigating Protein Adsorption via Spectroscopic Ellipsometry

Maria F. Mora, Jennifer L. Wehmeyer, Ron Synowicki, and Carlos D. Garcia

In this chapter, the basic concepts behind ellipsometry and spectroscopic ellipsometry are discussed along with some instrument details. Ellipsometry is an optical technique that measures changes in the reflectance and phase difference between the parallel (R_{P}) and perpendicular (R_{S}) components of a polarized light beam upon reflection from a surface. Aside from providing a simple, sensitive, and nondestructive way to analyze thin films, ellipsometry allows dynamic studies of film growth (thickness and optical constants) with a time resolution that is relevant to biomedical research. The present chapter intends to introduce ellipsometry as an emerging but highly promising technique, that is useful to elucidate the interactions of proteins with solid surfaces. In this regard, particular emphasis is placed on experimental details related to the development of biomedically relevant conjugated surfaces. Results from our group related to adsorption of proteins to nanostructured materials, as well as results published by other research groups, are discussed to illustrate the advantages and limitations of the technique.

Abbreviations and Symbols

Γ	Adsorbed amount
$\mathrm{d}\Gamma/\mathrm{dt}$	Adsorption rate
Δ	Phase difference
λ	Wavelength
ψ	Amplitude
AFM	Atomic force microscopy
BSA	Bovine serum albumin
CNT	Carbon nanotubes

M.F. Mora and C.D. Garcia • Department of Chemistry, The University of Texas at San Antonio, San Antonio, TX 78249 USA
J.L. Wehmeyer • Department of Biomedical Engineering, The University of Texas at San Antonio, San Antonio, TX 78249, USA
R. Synowicki • J. A. Woollam Co., Inc, 645 M Street, Suite 102, Lincoln, NE 68508, USA

D.A. Puleo and R. Bizios (eds.), *Biological Interactions on Materials Surfaces*,
DOI 10.1007/978-0-387-98161-1_2, © Springer Science+Business Media, LLC 2009

d	Thickness
DAAO	D-amino acid oxidase
DC	Direct current
DNA	Deoxyribonucleic acid
EMA	Effective medium approximation
Fib	Fibrinogen
HSA	Human serum albumin
IEP	Isoelectric point
k	Extinction coefficient
n	Refractive index
R_P	Parallel component of polarized light beam
R_S	Perpendicular component of polarized light beam
SDS	Sodium dodecyl sulfate
SE	Spectroscopic ellipsometry
t	Time

2.1. Introduction

Interaction of proteins with material surfaces is a common but rather complicated phenomenon [1]. One of the most remarkable consequences of this interaction is that materials coated with biomolecules display the properties of the adsorbed protein layer, rather than the material itself [2]. Consequently, understanding the protein adsorption phenomena is critical for the rational design of biologically active composites with sensing, biological, and electronic functions.

Among other substrates, nanomaterials are part of an industrial revolution that provides materials with unique properties (thermal, mechanical, electrical, biological, etc.) not found in conventional/microphased materials [3–7]. The combination of remarkable recognition capabilities of biomolecules with the unique properties of nanomaterials resulted in systems with significantly improved performance [8]. Apart from mechanical strength and light weight, most of the extraordinary biological properties of nanomaterials are linked to unique surface properties (surface area, surface roughness, energetics, and altered electron distributions) [9], which enable improved interactions between material surfaces and biological entities. Most importantly, the type and conformation of proteins adsorbed to nanomaterials proved to be a key factor in subsequent cellular responses [4]. Consequently, various experimental parameters must be optimized in order to control the biological activity of the bio/nano composite. Proteins, which tend to spontaneously accumulate at interfaces with materials, may undergo structural changes upon adsorption. The extent of such conformational changes induced by the sorbent surface depends on the material surface properties, the protein, the pH, and the degree of protein coverage of the surface. Understanding these conformational changes is probably one of the most important points for biomedical applications because they can *generally* be controlled. In this respect, the structural stability of the protein, defined by differential scanning calorimetry as "soft" or "hard" proteins, can provide the first indications about the driving forces behind, and consequences of, protein–biomaterial interactions [10, 11]. Further details regarding the adsorption of proteins to biomaterials can be found in other chapters of this book.

From the kinetics standpoint, the rate of protein adsorption at the solid/liquid interface comprises two steps: (1) transport of the solute molecules toward the interface and

(2) interaction with the sorbent material surface. The basic mechanisms of transport of molecules to the material surface are diffusion and convection by either laminar or turbulent flow. In the second step, the adsorbate molecules may attach at, or detach from, the sorbent material surface, giving rise to two fluxes, one forward and one backward. The relative contributions of each one of these fluxes to the overall adsorption process depend on both the attraction exerted by the material surface to the adsorbate and the solvent–sorbent interactions. Many energetic and entropic effects contribute to the free energy of the protein adsorption process. However, when proteins are adsorbed at the solid/liquid interface, the main driving forces of the interaction are electrostatic and hydrophobic ones. In the simplest case, the protein adsorption rate ($\mathrm{d}\Gamma/\mathrm{d}t$) may be considered as a first-order process, as described by Eq. (2.1):

$$\frac{\mathrm{d}\Gamma}{\mathrm{d}t} = k_{\mathrm{ADS}} C_{\mathrm{S}}, \tag{2.1}$$

where k_{ADS} is the adsorption rate constant and C_{S} is the concentration of solute (proteins, in the case under consideration). As the material surface coverage increases with time, k_{ADS} decreases, and an equilibrium is reached (Γ_{SAT}). When changes in the structure of the protein upon adsorption occur, the process could be described by considering an equilibrium reaction, in which the protein in solution can interact with the material surface, adsorb, and subsequently, undergo a structural/conformational change on that material surface. Such equilibria have been described for diverse proteins including lysozyme [12], albumin, and fibrinogen [13–18].

Kinetic and thermodynamic studies have indicated that significant conformational changes may occur as a protein adsorbs to a surface [19]. For this reason, many different techniques have been used to study not only the adsorption/desorption phenomena but also structural changes of such interactions [20]. Among others, Brewster angle microscopy, neutron and X-ray reflection, fluorescence and time-resolved fluorescence, circular dichroism, infrared spectroscopy, and electron microscopy were discussed in a book edited by Baszkin and Norde [21]. Studies of protein adsorption using mass spectrometry [22], confocal laser scanning microscopy [23], neutron reflection [24], atomic force microscopy (AFM; see other pertinent chapters in this book), scanning force microscopy [25], optical waveguide lightmode spectroscopy [19], quartz crystal microbalance [26, 27], surface plasmon resonance, total internal reflection fluorescence [28, 29], and capillary electrophoresis [30–33] have been also reported in the scientific literature.

2.2. Ellipsometry

Another technique that can provide information regarding the protein adsorption processes as well as the structure of the adsorbed protein layer is ellipsometry. Ellipsometry is an optical technique that measures changes in the reflectance and phase difference between the parallel (R_{P}) and perpendicular (R_{S}) components of a polarized light beam upon reflection from a surface. Using Eq. (2.2),

$$\tan(\psi)\mathrm{e}^{\mathrm{i}\Delta} = \frac{R_{\mathrm{P}}}{R_{\mathrm{S}}}, \tag{2.2}$$

the intensity ratio of R_p and R_s can be related to the amplitude (ψ) and the phase difference (Δ) between the two components [34]. Because ellipsometry measures the ratio of two values originated by the same signal, the data collected are highly accurate, and reproducible. Most importantly, no reference specimen is necessary. The changes in polarization measured by ellipsometry are extremely sensitive to film thickness (down to the monolayer level), optical constants, and film microstructure (such as surface roughness, index grading, and intermixing). This monolayer sensitivity is useful for real-time studies of layer-by-layer film deposition, including biological monolayers on a variety of substrates.

When substrates are flat, isotropic, and uniform, the interpretation of the ellipsometry results is relatively simple. Needless to say, biological substrates rarely meet these specifications, and often present multiple layers with different thicknesses, optical constants, and topographies. Because of the complexity of biological substrates, data must be obtained at multiple wavelengths and angles. Since the time of early applications, which were mainly focused on thickness quantification of oxide layers, ellipsometry has evolved to spectroscopic ellipsometry that is able to resolve details in the kinetics of layer formation for a variety of molecules [35] and even to investigate two-dimensional film thickness profiles with high spatial resolution and sensitivity [36–40]. Aside from providing a simple and nondestructive way to analyze thin, organic layers of biological interest, ellipsometry allows dynamic studies of film growth (thickness and optical constants) with a relevant time resolution.

2.3. Optical Models Used to Interpret Ellipsometric Results

Interpretation of ellipsometric measurements from raw data (ψ and Δ) is rather difficult and requires an optical model that describes the substrate microstructure in terms of refractive index (n), extinction coefficient (k), and thickness (d). This requirement is probably one of the biggest limitations of the technique, because the reliability of the calculated properties is only as good as the model used [41]. A second limitation (pertaining mostly to very thin films substrates), is that ellipsometry is not very sensitive to the value of n, because the n and d values of very thin films are highly correlated [42]. On the other hand, advances in instrumentation, matrix multiplication procedures, and modern computer applications enable modeling ellipsometric data with multilayer structures with better accuracy, reasonable time, and different optical models [43]. Most modern instruments provide comprehensive software packages with a built-in mean square error calculation that can be used to quantify the difference between the experimental and model-generated data.

The main objective of ellipsometry data analysis is to achieve an accurate description of the substrate using the simplest possible model. The procedure employed generally includes several iterations through four main steps: (1) modeling the dielectric function, (2) constructing an optical model that describes the overall behavior of the system, (3) fitting the collected spectra to the optical model, and (4) calculating the fitting error. If done correctly, this procedure also minimizes the uncertainty associated with the measurements. In other words, the conclusions obtained by ellipsometry are only as good as the optical model.

Depending on the electronic properties of the materials tested, different optical models can be applied to the data collected by ellipsometry [34, 44]. In the simplest case, when the layers can be considered transparent and homogeneous, the refractive index (n) as a function of the wavelength (λ) can be described using a classical Cauchy model (Eq. 2.3):

$$n_{(\lambda)} = A + \frac{B}{\lambda^2} + \frac{C}{\lambda^4}, \tag{2.3}$$

where A, B, and C are constants fitted by the model. Cauchy models have been used to study the structure of layers of biotin–avidin [45], albumin [46], ferritin [47], and polymers [48, 49], as well as binding of T-2 molecules to antibodies [50], and other small molecules [51, 52]. The interpretation of data related to biological molecules that absorb light (e.g., resonant electrons at 280 nm) requires the use of other models such as the Tauc–Lorentz model, which accounts for the unique band-gap of such amorphous materials.Tauc–Lorentz models have been used to describe the hemocompatibility of carbon thin films and their interaction mechanism with blood plasma proteins such as human serum albumin (HSA) and fibrinogen [53–57]. Similar models have also been used to evaluate the thrombogenicity of polysaccharide-coated surfaces [58].

These mathematical models used to describe each layer must be combined in a reasonable optical model that describes the overall behavior of the system. In this regard, most biomedically relevant substrates can be modeled by a small number of uniaxial layers with optical axes parallel to the normal axes of the substrate. Although other examples of these models abound in the literature, Figures 2.1 and 2.2 show two optical models used to describe biomedically relevant processes.

In the first case (*see* Figure 2.1), the model consists of a two-layer substrate (bulk silicon coated with a layer of amorphous hydrogenated carbon) and a protein layer (fibrinogen) as an overlayer [53]. An additional layer (four layers in total) was used to model the adsorption of β-casein at air/water and oil/water interfaces [60]. In situations where the

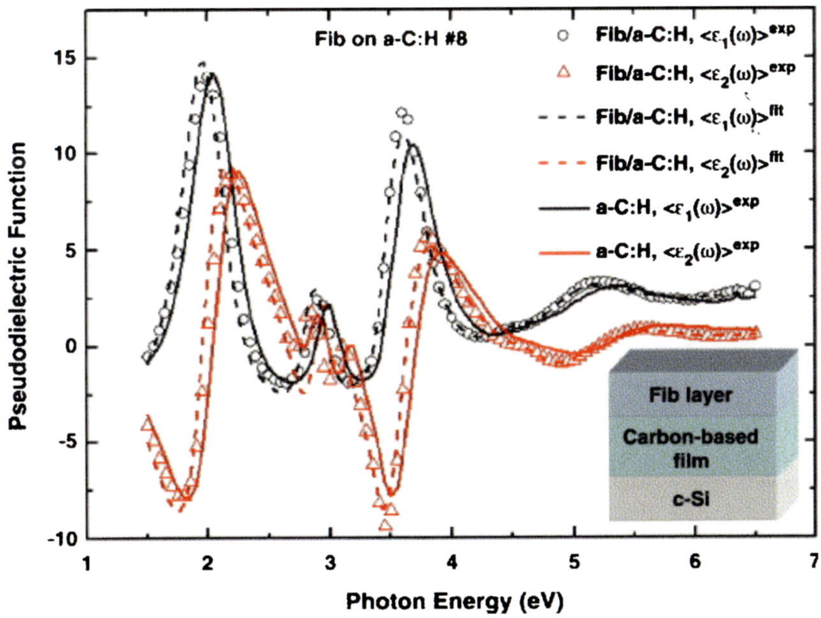

Figure 2.1. Three-layer model used to investigate the adsorption of fibrinogen (Fib) on hydrogenated carbon films. Reprinted from Ref. [53].

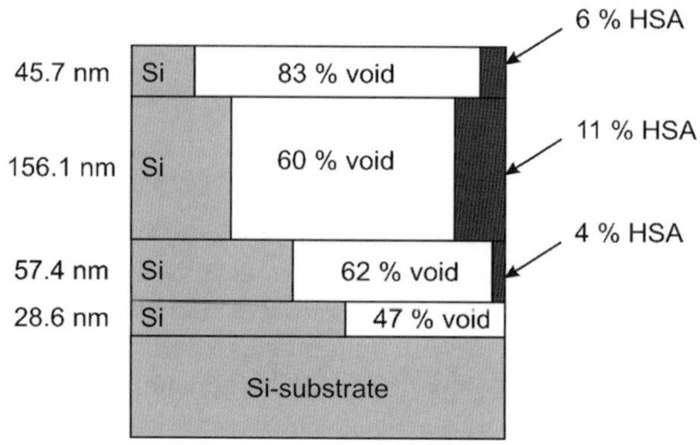

Figure 2.2. Five-layer model used to investigate the adsorption/penetration of human serum albumin (HSA) to porous silicon substrates. Reprinted from Ref. [59].

material surface and interface microstructures have to be accounted for, an effective medium approximation (EMA) model can be applied. EMA layers mix the optical constants of two or more materials and are extremely useful in modeling surface roughness [61], interface layers, and volume fractions in composite materials [62]. This is the case of the model illustrated in Figure 2.2, where a five-layer model, each layer with different composition (of silicon, void space, and protein), was used to investigate the adsorption of HSA to a porous silicon substrate [59]. Similar models were later used by Karlsson [46, 63] and Tsargorodskaya [64] to rationalize information related to the adsorbed protein amount as well as the concentration profile into the porous substrate. EMA layers also allow mixing materials, independently described by different optical models such as the Cauchy, Lorentz, or Tauc–Lorentz. EMA layers have been also applied to describe interactions between ferritin and gold [47].

Once the construction of an accurate optical model, describing the adsorbed protein layer (and the substrate), has been achieved, the thickness of a protein layer obtained by ellipsometry can be used to calculate the adsorbed protein amount (Γ, expressed in milligrams per meter squared) using Eq. 2.4:

$$\Gamma = \frac{\mathrm{d}(n - n_0)}{(\mathrm{d}n / \mathrm{d}c)}, \tag{2.4}$$

where n and n_0 are the refractive indices of the protein and the ambient environment, respectively [65]. In accordance with previous reports, the refractive index increment for the molecules in the layer ($\mathrm{d}n/\mathrm{d}c$) is generally assumed to be around 0.187 mL/g [66–69].

2.4. Instrument Considerations

In general, ellipsometers are relatively simple instruments. Figure 2.3 shows schematically (and not to scale) the main components of a rotating analyzer spectroscopic ellipsometer. Light from a Xe–arc lamp is directed through a monochromator (1), collimated,

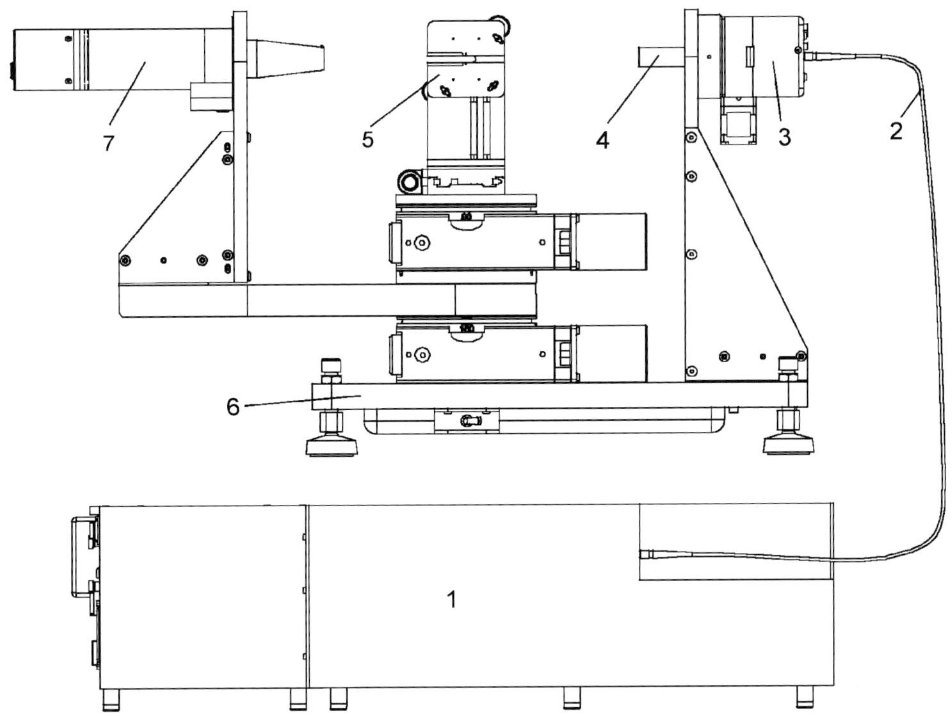

Figure 2.3. Schematic (not to scale) of the main components of a variable angle spectroscopic ellipsometer. (1) Monochromator, (2) fiber optic cable, (3) input unit, (4) alignment detector, (5) substrate stage, (6) goniometer base, and (7) detector unit. Note: The control box and the associated computer (for data acquisition and analysis) have been omitted from the diagram. (courtesy of J. A. Woollam Co., Inc.).

and passed through a polarizer (3). Then, the polarized beam interacts with the substrate (mounted on 5) at an oblique angle. Finally, the reflected beam passes through a second polarizer and enters a detector (7). Although many different instrument configurations are available, the data acquisition frequency is generally limited by the rotating element, light source intensity, selected precision of the measurement, wavelengths selected, and angles required for each experiment.

Many ellipsometric adsorption studies have been performed by measuring surfaces *ex situ* in air before and after protein adsorption [56, 70]. These experiments are very simple, allow many substrates (material surface/adsorbed protein) to be measured over a short period of time, and can be performed by minimally trained personnel. However, as pointed out by Arwin [59], this *ex situ* experimental approach involves rinsing and, sometimes, drying steps that introduce the uncertainty of possible desorption and denaturation of proteins before the pertinent measurements.

For this reason, several cells[1] have been designed enabling in situ ellipsometric measurements of protein adsorption processes. In this respect, a widely used design is an open cell assembled of fused silica slides [35] that could be mounted on the vertical substrate-stage of

[1] Unless otherwise noted, the word "cell" in this chapter has been reserved to denominate the chamber in which the adsorption experiment is performed.

the ellipsometer. Arwin's group also designed a cell for total internal reflection ellipsometry under flow injection conditions that can be also used without flow injection as well as with or without stirring [47]. Logothetidis [55] and other research groups [71] have recently reported very interesting results obtained with other cells suited for aqueous liquids (like salty solutions and buffers). Although these designs have proven to be extremely useful, they do not provide information to resolve the question of whether the mass transfer from the bulk or an interfacial process is the determinant step in the protein adsorption rate. To solve this problem, the adsorption experiment can be performed under stagnation conditions. In this case, the axis of the impinging jet intersects perpendicularly to the material surface that is being measured. Stagnation flow cells have been extensively applied to study adsorption kinetics of proteins [72–76], surfactants [77], and polymers [78] using reflectometry, a similar technique.

2.5. Material Surface Preparation

Unlike other techniques such as surface plasmon resonance, ellipsometry allows a wide variety of materials to be used as substrates. Gold [79], carbon [80], silicon [49], alumina [81], stainless steel [82], and titania [83] are only a few examples with biomedical relevance. Particularly important are micro/nanophased materials, which promote enhanced interactions with biological molecules. A major problem encountered when performing ellipsometric studies using nanomaterials is, however, that nanophase substrates are typically not suitable for ellipsometry. This is the case for ceramics [4], and material substrates prepared by either dip-coating [84] or electrophoresis [85], which result in rather opaque surfaces, with roughnesses that are several orders of magnitude larger than the nanofeatures themselves. Other techniques such as chemical vapor deposition [86] and direct-current reactive magnetron sputtering [87] offer versatility of fabrication conditions, crystal structure, composition, optical properties, bactericidal abilities, and effective ways to improve the reactivity of the obtained films. However, the cost and complexity of these techniques prevent them from being adopted for general use.

Often, the optical anisotropy observed in polymer thin films [88, 89], self-assembled layers [90], and Langmuir–Blodgett films [91] can be accounted for during the ellipsometric measurement and subsequent modeling. According to our experience, techniques such as sputtering, vaporization (for metallic surfaces), and sol-gel deposition [92–94] have the potential of producing nanometer-thick films with minimum instrument requirements and porosity values that are appropriate for ellipsometry. Other materials such as carbon nanotube (CNT) films can be deposited on a variety of substrates by spin-coating [95], spraying [96], chemical vapor deposition [97], and vacuum filtration [98]. Although they will not be discussed in this chapter, successful examples of ellipsometrically characterized nanostructured films abound in the literature [99–103].

2.6. Typical Protein Adsorption Experiment Followed by Ellipsometry

As mentioned in Sects. 2.2 and 2.3 of this chapter, the raw data generated using the ellipsometer are generally expressed in terms of the amplitude (ψ) and the phase difference (Δ) as functions of either time (t) or wavelength (λ). These data can be then interpreted using an optical model, which considers the optical constants (n and k) and thicknesses (d) of each layer. Figure 2.4 shows the raw data (ψ and Δ as functions of time) obtained during a typical dynamic adsorption experiment collected at two wavelengths (450 and 650 nm) using spectroscopic ellipsometry.

After fitting the data with an appropriate model, in which the thickness of the protein layer is variable, Figure 2.4 can be expressed in terms of either thickness of the protein layer or surface mass (Γ), and expanded to include the data collected during a subsequent desorption experiment. Figure 2.5 shows the final plot of the protein adsorption/desorption experiment.

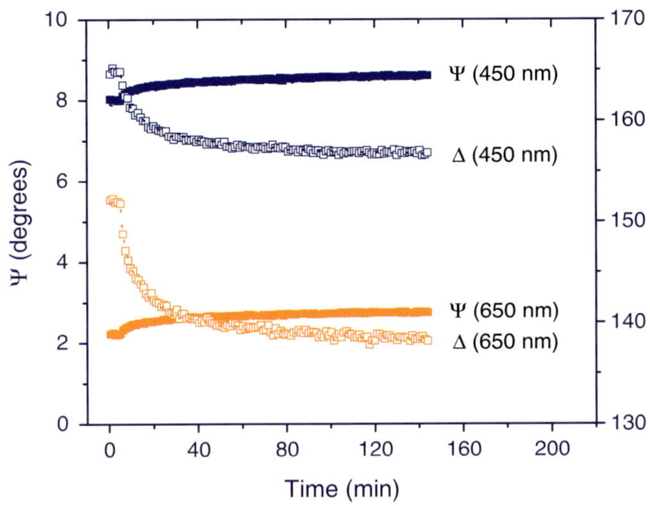

Figure 2.4. Typical amplitude (ψ) and phase difference (Δ) as functions of time (t) collected during an adsorption experiment of 0.1 mg/mL of DAAO to CNT by spectroscopic ellipsometry. Other pertinent conditions are described in Ref. [104].

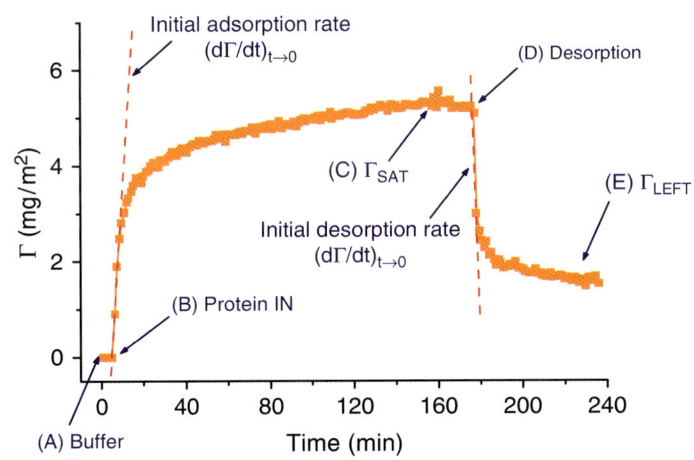

Figure 2.5. Typical protein adsorption/desorption experiment monitored by spectroscopic ellipsometry.

Figure 2.5 can be subdivided in five regions. In the first part (A), while only the background electrolyte is pumped through the cell, the initial thickness of the substrate is measured and the baseline stability is verified. In (B), the solution containing protein is introduced and the protein adsorption starts at an initial fast rate process, allowing the calculation of the maximum adsorption rate ($d\Gamma/dt$). As the protein molecules fill the substrate, the adsorption process slows down, reaching a plateau value (C). At this point, the adsorbed amount at saturation (Γ_{SAT}), for the designated experimental conditions, can be obtained. In the next region (D), a desorbing agent is introduced into the cell, the adsorbed amount decreases, and the initial desorption rate can be calculated. Finally, in region (E) the amount of protein that remains on the substrate (Γ_{LEFT}) can be determined. Additionally, a more accurate value for the optical constants of the substrate and the protein layer before and after desorption can be obtained upon performing spectroscopic scans during stages (A), (C), and (E), respectively.

2.7. Ellipsometric Determination of the Adsorption of Proteins to Nanomaterials

So far, this chapter has focused on the importance of studying protein adsorption to nanomaterials surfaces, the most remarkable features of ellipsometry and its instrumental and substrate requirements, as well as the type of data that can be collected in a typical ellipsometric experiment. In the following sections, the most recent results from our laboratory will be discussed. Emphasis will be placed on the adsorption of two proteins (bovine serum albumin [BSA] and D-amino acid oxidase [DAAO]) to two nanostructured materials (TiO_2 and CNT) deposited on Si/SiO_2 strips.

2.7.1. Adsorption of BSA to Nanostructured TiO_2

During the last decades, plain carbon and vanadium–steel orthopedic/dental implants have been gradually replaced by those made of stainless steel, cobalt–chromium alloys, titanium–platinum alloys, and polymeric materials, such as poly(tetrafluoroethylene), poly(methylmethacrylate), polyethylene, and silicones [105]. Because of their biocompatibility and mechanical properties, titanium and titanium-based alloys are one of the most popular materials for medical applications including bone and joint replacements, dental implants, and cardiovascular devices. Since titanium spontaneously generates a surface layer of TiO_2 when exposed to oxygen-containing environments (such as air or aqueous media), its biocompatibility is dominated by the interaction of cells, tissues, biological fluids, and the oxide layer instead of the metal itself [106].

This phenomenon has been widely recognized and it has motivated recent studies of the interaction of various proteins and TiO_2 [107, 108]. Among other proteins, BSA has been extensively studied. BSA is a globular protein with an isoelectric point (IEP) of 4.5–5.0, approximate molecular dimensions of $4 \times 4 \times 14$ nm, and a molecular weight of 66.5 kDa [109]. Because BSA generally undergoes significant structural changes upon adsorption to solid surfaces, it has been considered to be a "soft" protein [10, 11]. Besides its abundance in physiological fluids (such as blood plasma) and physiological functions attributed to albumin (control of osmotic pressure, buffer, and transport), BSA has been also considered a model protein for various biomedically related studies [110–112]. For these reasons, the mechanisms that regulate the adsorption of albumin to nanostructured TiO_2 surfaces were selected for further analysis in this chapter. In order to attain a better understanding of how nanostructured surfaces modulate protein–surface interactions, the "real-time" adsorption of BSA to nanostructured TiO_2 was investigated using spectroscopic ellipsometry [113]. For these experiments, the TiO_2-coated substrates were prepared using dip-coating techniques [92–94] and characterized by ellipsometry and atomic force microscopy (AFM). The thickness of the deposited TiO_2 films was between 1.5 and 3.5 nm. The ellipsometric data were modeled by three uniaxial layers (Si, bulk; SiO_2 d = 2.5 ± 0.5 nm; and TiO_2) with the optical axis parallel to the silicon wafer substrate. As shown in Figures 2.6 and 2.7, the agreement between the data generated by the optical model, the experimental data, and the topography of the nanostructured TiO_2 thin films (determined by AFM) was good.

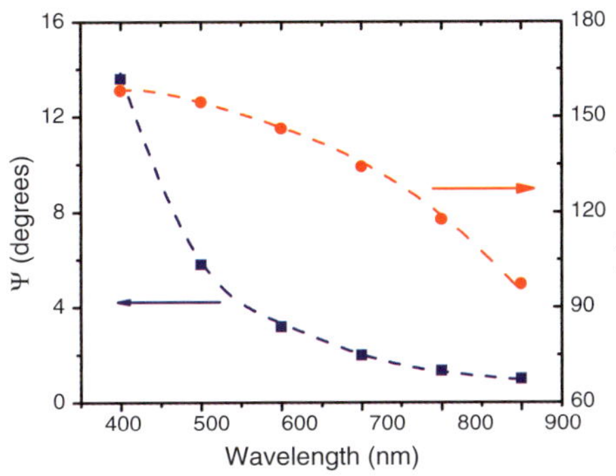

Figure 2.6. An example of data (Ψ, *blue*; and Δ, *red*) collected from a spectroscopic scan (*dots*) as well as data generated by the optical model (*lines*). Other pertinent conditions for this experiment are described in Ref. [113].

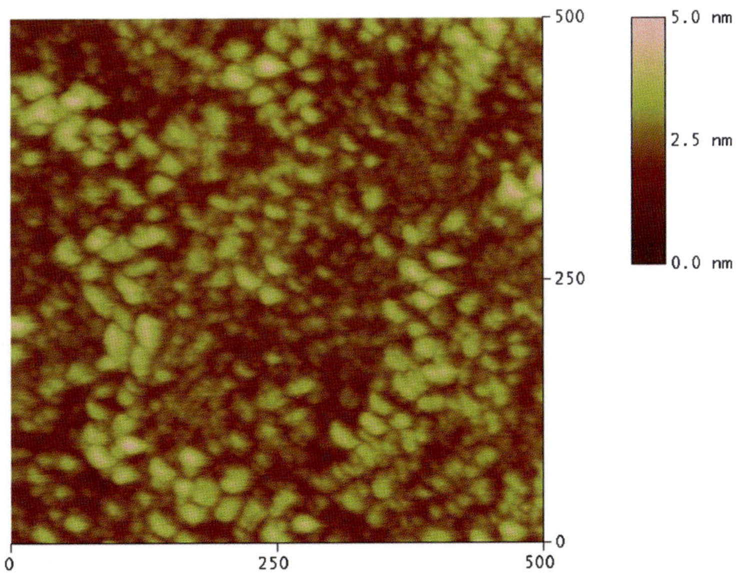

Figure 2.7. AFM image of a representative thin-film of TiO_2 deposited on a silica wafer. Other pertinent conditions for this experiment are described in Ref. [113].

As shown in Figures 2.8 and 2.9, both the adsorbed amount and the adsorption rate of BSA to nanostructured TiO_2 increased as a function of the protein concentration in the 0.001–0.1 mg/mL range.

Although the interaction of BSA on TiO_2 was not affected by the ionic strength of the protein solution, greater amounts of BSA were adsorbed at the IEP. This behavior can be explained by considering that, at the IEP, the adsorbed molecules minimize electrostatic repulsions, attain closer packing [114], and retain their native structure. The results collected by ellipsometry indicate that BSA adsorbed to nanostructured TiO_2 formed a monolayer with a (more or less) compact arrangement. In agreement with a rather general phenomenon in protein adsorption [115], formation of the BSA adsorbed layer was driven by a combination

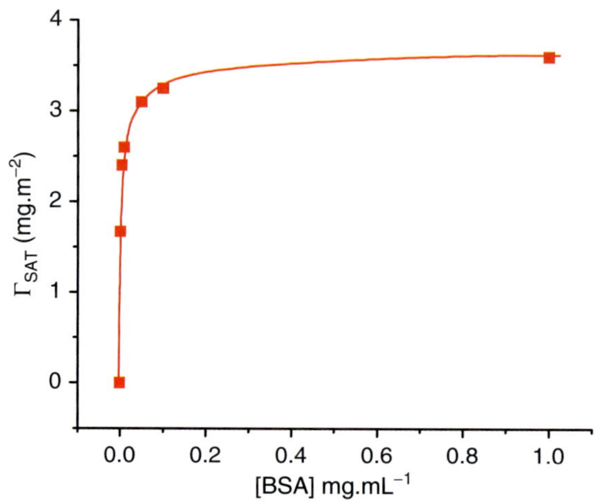

Figure 2.8. Effect of the protein concentration on the amount of BSA adsorbed on nanostructured TiO_2. Other pertinent conditions of this experiment are described in Ref. [113].

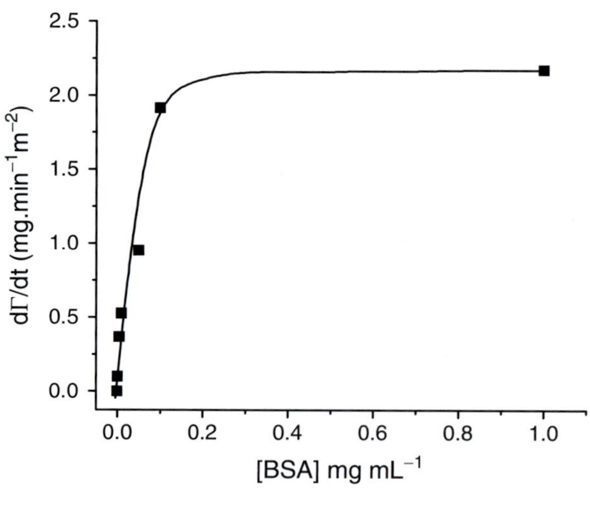

Figure 2.9. Effect of the protein concentration on the initial adsorption rate of BSA on nanostructured TiO_2. Other pertinent conditions of this experiment are described in Ref. [113].

of hydrophobic (mainly) and hydrophilic interactions. Data collected by ellipsometry indicated that protein layers in the 2–5-nm thickness range were obtained; this result suggests that, upon adsorption to nanostructured films of TiO_2, BSA undergoes (at least some) structural changes.

The aforementioned results highlight the utility of spectroscopic ellipsometry to investigate not only the optical properties of nanostructured materials, but also the adsorption of proteins to such layers. More information regarding adsorption of BSA to nanostructured TiO_2 can be found in Ref. [113].

2.7.2. Adsorption of Proteins to Carbon Nanotubes: Biosensing Applications

Sensitive, selective, and cost-effective analysis of biomolecules is important in clinical diagnostics and treatment. Among others, electrochemical biosensors based on enzyme-modified electrodes are very attractive because they integrate the selectivity of enzymatic reactions with highly sensitive electrochemical signal transduction [116–118]. Biosensors are currently applied in the clinical [116, 118], environmental [119, 120], agricultural [117], and pharmaceutical fields. Although different nanomaterials can be used as substrates [121], carbon nanotubes (CNT) have an enormous potential because they can act simultaneously as immobilization matrices and as electrochemical transducers [122–128]. In addition, CNT are stable over a large range of potentials, are catalytically active toward many electrochemical reactions [127–130], and provide a significant increase in electrode area [129]. Although considerable progress has been made by encapsulating or cross-linking enzymes [131–133], the analytical performance of CNT biosensors still suffers from some fundamental deficiencies such as slow response (≥10 s) and limited sensitivity (approximately micromolar).

In order to better understand the driving forces and consequences of the interaction of proteins with CNT, preliminary studies were performed by reflectometry using BSA as a model protein [74]. According to those results, BSA molecules arriving to the CNT surface adopted a preferred orientation with the positive and nonpolar patches of the protein facing the hydrophobic sorbent surface; this arrangement resulted in an attachment-controlled adsorption process. Even under electrostatically unfavorable conditions, dehydration of both the CNT surface and the nonpolar regions of BSA promoted adsorption on CNT. At steady state conditions, a layer of BSA adsorbed to CNT (at the IEP of the protein) resembled a close-packed monolayer of protein molecules. At pH values away from the IEP, repulsive protein–protein interactions prevailed over attractive surface–protein interactions, limiting the amount of BSA adsorbed to the CNT layer.

More recently, our group described the interaction of CNT with D-amino acid oxidase (EC 1.4.3.3, DAAO) [104]. DAAO is a dimeric protein of approximately 80.6 kDa that exhibits an elongated ellipsoidal framework with approximate dimensions of 11 nm (length) × 4 nm (width) [134] and an IEP in the 6.3 [135] to 7.0 [136] range. DAAO is of particular interest because it recognizes functional groups, instead of a specific analyte [137, 138]. Therefore, combining DAAO with a separation technique such as capillary electrophoresis should increase the versatility of the sensor, allowing detection of several analytes with similar structure. Although the biological role of DAAO in animals is not clear yet, recent progress in the detection of D-amino acids has linked DAAO to aging [139, 140] and pathological conditions such as schizophrenia [141, 142], epilepsy, Alzheimer's disease, and renal diseases [143]. Additionally, understanding the adsorption mechanisms of DAAO to solid surfaces would enable developing more efficient catalysts for biomedical applications [144–146]. Consequently, understanding the driving forces for the adsorption of DAAO to

CNT would enable the rational design of biosensors and biocatalysts, avoiding harsh immobilization conditions and trapping membranes [147, 148].

Dynamic adsorption/desorption experiments of DAAO to CNT were performed as a function of the protein concentration, pH, and ionic strength. In general, all adsorption experiments exhibited a similar general behavior: (1) adsorption of DAAO to the CNT surface was a single step process; (2) adsorbed DAAO was not significantly affected by rinsing the substrate with buffer; and (3) part of the adsorbed DAAO was removed by rinsing with 4 mM sodium dodecyl sulfate (SDS). An example of the data collected during such adsorption/desorption experiments is shown in Figure 2.5. Neither the desorption kinetics nor the amount of protein remaining attached to the CNT surface was affected by the amount of DAAO adsorbed (Γ_{SAT}). Based on these experimental observations, the overall adsorption process could be interpreted as the sum of two different populations of DAAO on the CNT surface: one that is removable by SDS (Γ_1), and another (Γ_2) that remains attached to the surface even after washing with surfactant.

Thickness values of the adsorbed DAAO obtained with spectroscopic ellipsometry analysis indicate that DAAO can adopt multiple orientations, either horizontal or tilted at different angles with respect to the CNT surface (Figure 2.10). It was also observed that higher amounts of DAAO were adsorbed at the IEP of the DAAO; moreover, the initial adsorption rate and the population of DAAO loosely attached (Γ_1) to the CNT surface increase as the bulk protein concentration increases. More importantly, measurements of the enzymatic activity of the adsorbed protein (Figure 2.10) provided evidence that the enzymatic activity correlated with the adsorbed amount of protein when the adsorption reaction was performed under attractive electrostatic conditions. However, CNT surfaces modified with DAAO at, or above, the IEP of the protein displayed lower enzymatic activity.

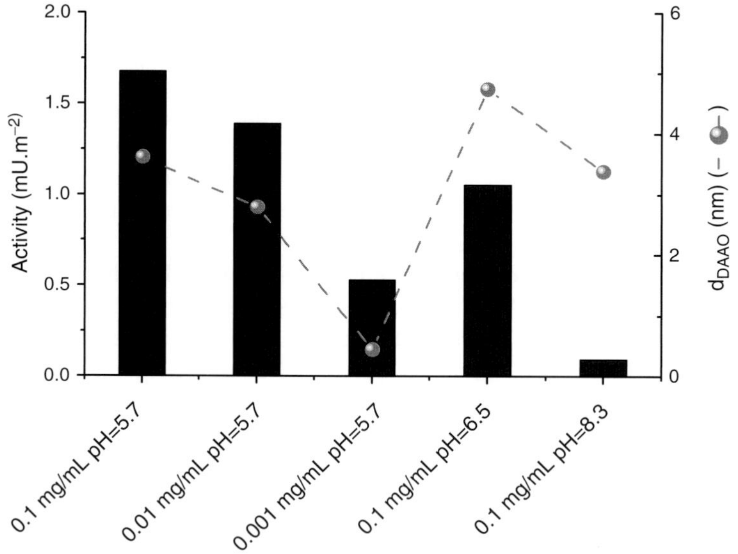

Figure 2.10. Comparison between the enzymatic activity and amount of DAAO adsorbed to CNT under different experimental conditions. The *dashed line* was included with the sole purpose of connecting the DAAO thickness values (d_{DAAO}). Other pertinent conditions for this experiment are described in Ref. [104].

If one considers that proteins generally display higher structural stability (and, therefore, a smaller tendency to spread at the IEP) [114], these results may suggest that, despite the amount adsorbed, DAAO could undergo small changes in orientation rather than changes in structural conformation on the CNT surface. These changes in orientation are responsible for the observed differences in the biological activity of the adsorbed protein; for this reason, not only the surface and the protein [149], but also the adsorption conditions dictate the biological activity of a protein adsorbed on a CNT surface. More information regarding adsorption of DAAO to CNT surfaces can be found in Ref. [104].

2.8. Innovative Applications

Recent developments in instrumentation and data analysis software have enabled ellipsometry to transcend the boundaries of traditional physics, engineering, and chemistry settings. In this respect, several biomedical applications have been recently reported. For example, Schulz et al. [150] demonstrated that spectroscopic ellipsometry is a suitable optical tool to investigate biological specimens such as liver tissue, human nails, and human skin. These researchers performed a hydration study that revealed changes of the optical constants upon hydration and dehydration of nails and liver. The very different dehydration time scales for nails and liver provided evidence of the importance of the keratin matrix as a water barrier [150]. Danny et al. later extended this study to investigate the optical properties of various layers of human skin. In this case, the evolution of ψ and Δ were described using a morphological model containing an effective medium approximation accounting for the water content of the skin, surface roughness of corneocytes, and the alternating lipid layers in the skin [151].

Cardenas et al. recently applied ellipsometry to investigate the amount, thickness, and structure of films formed by human whole saliva on alumina surfaces [81]. Their analyses were complemented by means of neutron reflectivity and AFM, and showed that saliva adsorbed rapidly on alumina. Such a film could be modeled by two layers: (1) an inner, dense, and thin region that formed a uniform layer and (2) a second layer, more diffuse and thicker, which protruded toward the bulk of the solution. The thickness of both layers of a salivary film formed on sapphire was found to be on the order of a few hundreds of Ångstroms [81]. Researchers from the same laboratory also investigated the adsorption of two salivary mucins (specifically, structure and topography) under conditions similar to those found in the oral cavity in terms of ionic strength, pH, and protein concentration [152]. In this case, the salivary protein film was described as a two-sublayer structure in which an inner, dense layer was decorated by large aggregates of proteins. The shape and height of these large aggregates largely depended on the type of substrata tested. Additionally, Santos et al. showed that adsorption of a human salivary mucin (MUC5B) was controlled by the type of substrata; in this case, film topography was similar to that of the larger aggregates present in the salivary films. According to these results, MUC5B molecules adsorbed on hydrophobic substrates were especially resistant to both elution with buffer solution and SDS. Therefore, these large mucins can be responsible for the increased resistance of the saliva films on hydrophobic substrates. Mucins could then protect the intraoral surfaces against surface-active components present in oral health care products. These results are also in line with the adsorption of other salivary proteins to biological materials [152].

Ellipsometry has also been applied in the development of various biosensors. Several of these sensors were reviewed by Arwin [153]. Attractive features for ellipsometric sensors are the high resolution of the thickness of the adsorbed protein layer and the possibility of performing in situ measurements using nonlabeled molecules. According to the type of sensing mechanism used, ellipsometric sensors were classified in terms of affinity layer (analyte interacts with a sensing layer deposited on a substrate, resulting in a change in thickness of the protein layer), matrix layer (analyte diffuses inside a thin layer deposited on a substrate, resulting in a change in the refractive index), integrating layer (analyte interacts with a surface resulting in an accumulated thickness change over time), and a homogeneous layer (optical properties of the layer change upon interaction with the analyte) [153]. Among other ellipsometric sensors, Demirel et al. [154] investigated the effects of several variables on the formation of self-assembled monolayers of 3-mercaptopropyltrimethoxysilane on Si surfaces. Such surfaces were then modified with oligodeoxynucleotides and used to detect hybridization by ellipsometry. Other studies involving DNA adsorption and subsequent interactions have been also reported [155–157]. Using a micropatterned panel of seven lectins, Carlsson et al. [158] discriminated different meat juices from cattle, chicken, pig, cod, turkey, and lamb. In this case, biorecognition was evaluated with null ellipsometry and the data obtained were related to lactoferrin, an internal standard. Furthermore, the patterns of lectins binding to the meat proteins were visualized by scanning ellipsometry [158].

One of the most exciting areas in ellipsometry research is the development of imaging ellipsometry, which enabled quantification and visualization of the lateral thickness distribution of thin protein layers formed on solid substrates [38]. Biosensors based on imaging ellipsometry combine the specificity of biomolecular interactions with protein-patterned surfaces and have the advantages of high spatial resolution, fast data acquisition, and simplicity of use [159]. In this respect, van Noort et al. [160] reported the fabrication of an affinity biochip with a matrix of 900 targets for detection of binding events of carbohydrates with lectins using imaging ellipsometry. More recently, an immunosensor based on imaging ellipsometry was developed for the detection of *Legionella pneumophila* [161]. The sensor was fabricated by sequential deposition of 11-mercaptoundecanoic acid, protein G, and a monoclonal antibody. Imaging ellipsometry was then used to detect binding of *L. pneumophila* to the antibody layer; the limit of detection in this case was approximately 10^3 colony-forming units/ml [161]. Similar results were obtained when *Salmonella typhimurium* [162] and *Yersinia enterocolitica* [163] were examined. Comparable approaches have been also used to detect *Arthrobacter oxydans* [164] and dengue virus particles [165], to investigate the orientation of human immunoglobulin G [166], and to visualize two neutralizing human monoclonal antibodies from patients infected with severe acute respiratory syndrome coronavirus [167].

Yu and Jin [79] recently combined ellipsometry with electrochemical methods for studying electrostatic interactions of proteins and solid surfaces. These researchers provided evidence that the rate of fibrinogen adsorption on a potentiostatic surface was faster than that observed on the non-potentiostatic surface and concluded that hydrophobic interactions were the major driving force for the observed adsorption of fibrinogen to gold. Descriptions of other biomedical applications of imaging ellipsometry abound in the literature [168–171].

2.9. Conclusions

In this chapter, the basic concepts behind ellipsometry and spectroscopic ellipsometry were discussed along with some pertinent instrument details. Particular emphasis was placed on experimental details related to the development of medically relevant bioconjugated surfaces. In this regard, ellipsometry has enabled collecting real-time data related to a wide variety of biological processes. When complemented with other techniques such as electron and fluorescence microscopy, circular dichroism, infrared spectroscopy, AFM, and quartz crystal microbalance, ellipsometry enables a rational interpretation of the microstructure of layers of protein adsorbed on material surfaces. Original papers investigating live tissues as well as imaging ellipsometry should open new possibilities for applications in the biomedical field. Clearly, opportunities abound for fundamental discovery as well as for breakthroughs in applications that involve proteins and biomaterials, biotechnology, and nanotechnology.

Acknowledgments

The authors would like to thank the Southwest Research Institute for proving access to their atomic force microscope. Financial support for this project was provided in part by The University of Texas at San Antonio, the National Institute of General Medical Sciences (NIGMS)/National Institutes of Health (1SC3GM081085), and the Morrison Trust.

References

1. Nakanishi K, Sakiyama T, Imamura K. On the adsorption of proteins on solid surfaces, a common but very complicated phenomenon. J. Biosci. Bioeng. 2001;91:233–244.
2. Lynch I, Dawson KA. Protein-nanoparticle interactions. Nano Today 2008;3:40–47.
3. Cheng M-D. Effects of nanophase materials (<20 nm) on biological responses. J. Environ. Sci. Health 2005;39:2691–2705.
4. Webster TJ, Ergun C, Doremus RH, Siegel RW, Bizios R. Specific proteins mediate enhanced osteoblast adhesion on nanophase ceramics. J. Biomed. Mater. Res. 2000;51:475–483.
5. Cheng F-Y, Wang SP-H, Su C-H, Tsai T-L, Wu P-C, Shieh D-B, et al. Stabilizer-free poly(lactide-co-glycolide) nanoparticles for multimodal biomedical probes. Biomaterials 2008;29:2104–2112.
6. Chung Y-C, Chen IH, Chen C-J. The surface modification of silver nanoparticles by phosphoryl disulfides for improved biocompatibility and intracellular uptake. Biomaterials 2008;29:1807–1816.
7. Deng C, Chen J, Chen X, Xiao C, Nie L, Yao S. Direct electrochemistry of glucose oxidase and biosensing for glucose based on boron-doped carbon nanotubes modified electrode. Biosens. Bioelectron. 2008;23:1272–1277.
8. Engel E, Michiardi A, Navarro M, Lacroix D, Planell JA. Nanotechnology in regenerative medicine: the materials side. Trends Biotechnol. 2008;26:39–47.
9. Liu H, Webster TJ. Nanomedicine for implants: a review of studies and necessary experimental tools. Biomaterials 2007;28:354–369.
10. Norde W. Driving forces for protein adsorption at solid surfaces. In: Malmsten M, (ed.) *Biopolymers at Interfaces*. New York: Marcel Dekker; 2003.
11. Norde W. My voyage of discovery to proteins in flatland and beyond. Colloids Surf. B Biointerfaces 2008;61:1–9.
12. Larsericsdotter H, Oscarsson S, Buijs J. Thermodynamic analysis of lysozyme adsorbed to silica. J. Colloid Interface Sci. 2004;276:261–268.
13. Giacomelli CE, Norde W. The adsorption-desorption cycle. Reversibility of the BSA-Silica system. J. Colloid Interface Sci. 2001;233:234–240.
14. Norde W, Zoungrana T. Surface-induced changes in the structure and activity of enzyme physically immobilized at solid/liquid interfaces. Biotechnol. Appl. Biochem. 1998;28:133–143.
15. Giacomelli CE, Norde W. Structural changes in proteins resulting from homomolecular exchange at solid surfaces. In: Hubbard AT, (ed.) *Encyclopedia of Surface and Colloid Science*. New York: Marcel Dekker; 2003.
16. van der Veen M, Stuart MC, Norde W. Spreading of proteins and its effect on adsorption and desorption kinetics. Colloids Surf. B Biointerfaces 2007;54:136–142.
17. Bernabeu P, Tamisier L, De Cesare A, Caprani A. Study of the adsorption of albumin on a platinum rotating disk electrode using impedance measurements. Electrochim. Acta 1988;33:1129–1136.

18. Zhang Y, Fung Y, Sun H, Zhu D, Yao S. Study of protein adsorption on polymer coatings surface by combining quartz crystal microbalance with electrochemical impedance methods. Sens. Actuators B Chem 2005;108:933–942
19. Gray JJ. The interaction of proteins with solid surfaces. Curr. Opin. Struct. Biol. 2004;14:110–115.
20. Sapsford KE, Ligler FS. Real-time analysis of protein adsorption to a variety of thin films. Biosens. Bioelectron. 2004;19:1045–1055.
21. Baszkim A, Norde W, (eds.). *Physical Chemistry of Biological Interfaces*. New York, NY: Marcel Dekker; 2000.
22. Griesser HJ, Kingshott P, McArthur SL, McLean KM, Kinsel GR, Timmons RBRB. Surface-MALDI mass spectrometry in biomaterials research. Biomaterials 2004;25:4861–4875.
23. Yang K, Sun Y. Optics-intrinsic double-circle phenomenon in protein adsorption visualized by confocal laser scanning microscopy. Biochem. Eng. J. 2008;39:258–266.
24. Lu JR, Zhao X, Yaseen M. Protein adsorption studied by neutron reflection. Curr. Opin. Colloid Interface Sci. 2007;12:9–16.
25. Reich Z, Kapon R, Nevo R, Pilpel Y, Zmora S, Scolnik Y. Scanning force microscopy in the applied biological sciences. Biotechnol. Adv. 2001;19:451–485.
26. Teichroeb JH, Forrest JA, Jones LW, Chan J, Dalton K. Quartz crystal microbalance study of protein adsorption kinetics on poly(2-hydroxyethyl methacrylate). J. Colloid Interface Sci. 2008;325:157–164.
27. Dolatshahi-Pirouz A, Rechendorff K, Hovgaard MB, Foss M, Chevallier J, Besenbacher F. Bovine serum albumin adsorption on nano-rough platinum surfaces studied by QCM-D. Colloids Surf. B Biointerfaces 2008;66:53–59.
28. Wertz CF, Santore MM. Adsorption and relaxation kinetics of albumin and fibrinogen on hydrophobic surfaces: single-species and competitive behavior. Langmuir 1999;15:8884–8894.
29. Wertz CF, Santore MM. Effect of surface hydrophobicity on adsorption and relaxation kinetics of albumin and fibrinogen: single-species and competitive behavior. Langmuir 2001;17:3006–3016.
30. Righetti PG, Gelfi C, Verzola B, Castelletti L. The state of the art of dynamic coatings. Electrophoresis 2001;22:603–611.
31. Verzola B, Gelfi C, Righetti PG. Quantitative studies on the adsorption of proteins to the bare silica wall in capillary electrophoresis: II. Effects of adsorbed, neutral polymers on quenching the interaction. J. Chromatogr. A 2000;874:293.
32. Castelletti L, Verzola B, Gelfi C, Stoyanov A, Righetti PG. Quantitative studies on the adsorption of proteins to the bare silica wall in capillary electrophoresis: III: effects of adsorbed surfactants on quenching the interaction. J. Chromatogr. A 2000;894:281–289.
33. Olivier JC, Taverna M, Vauthier C, Couvreur P, Baylocq-Ferrier D. Capillary electrophoresis monitoring of the competitive adsorption of albumin onto the orosomucoid-coated polyisobutylcyanoacrylate nanoparticles. Electrophoresis 1994;15:234–239.
34. Fujiwara H. *Spectroscopic Ellipsometry. Principles and Applications*. West Sussex, England: Wiley; 2007.
35. Arwin H. Spectroscopic ellipsometry and biology: recent developments and challenges. Thin Solid Films 1998;313–314:764–774.
36. Höök F, Vörös J, Rodahl M, Kurrat R, Böni P, Ramsden JJ, et al. A comparative study of protein adsorption on titanium oxide surfaces using in situ ellipsometry, optical waveguide lightmode spectroscopy, and quartz crystal microbalance/dissipation. Colloids Surf. B 2002;24:155–170.
37. Pak HK, LawBM . 2D imaging ellipsometric microscope. Rev. Sci. Instrum. 1995;66:4972–4976.
38. Jin G, Jansson R, Arwin H. Imaging ellipsometry revisited: developments for visualization of thin transparent layers on silicon substrates. Rev. Sci. Instrum. 1996;67:2930.
39. Linke F, Merkel R. Ellipsometric microscopy: developments towards biophysics. IEE Proc Nanobiotechnol 2004;151:95.
40. Linke F, Merkel R. Quantitative ellipsometric microscopy at the silicon–air interface. Rev. Sci. Instrum. 2005;76:063701.
41. Tompkins HG. *A User's Guide to Ellipsometry*. San Diego, CA: Academic Press; 1993.
42. Greef R. Ellipsometry in electrochemistry: a spectrum of applications. Thin Solid Films 1993;233:32–39.
43. Aspnes DE. Expanding horizons: new developments in ellipsometry and polarimetry. Thin Solid Films 2004;455–456:3–13.
44. Tompkins HG, Irene EA, (eds.). *Handbook of Ellipsometry*. Norwich, NY: W' Andrew; 2005.
45. Spaeth K, Brecht A, Gauglitz G. Studies on the biotin-avidin multilayer adsorption by spectroscopic ellipsometry. J. Colloid Interface Sci. 1997;196:128–135.
46. Karlsson LM, Schubert M, Ashkenov N, Arwin H. Protein adsorption in porous silicon gradients monitored by spatially-resolved spectroscopic ellipsometry. Thin Solid Films 2004;455–456:726–730.

47. Poksinski M, Arwin H. Protein monolayers monitored by internal reflection ellipsometry. Thin Solid Films 2004;455–456:716–721.
48. Feller L, Bearinger JP, Wu L, Hubbell JA, Textor M, Tosatti S. Micropatterning of gold substrates based on poly(propylene sulfide-bl-ethylene glycol), (PPS-PEG) background passivation and the molecular-assembly patterning by lift-off (MAPL) technique. Surf. Sci. 2008;602:2305–2310.
49. Goyal DK, Pribil GK, Woollam JA, Subramanian A. Detection of ultrathin biological films using vacuum ultraviolet spectroscopic ellipsometry. Mater. Sci. Eng. B 2008;149:26–33.
50. Nabok AV, Tsargorodskaya A, Holloway A, Starodub NF, Gojster O. Registration of T-2 mycotoxin with total internal reflection ellipsometry and QCM impedance methods. Biosens. Bioelectron. 2007;22:885–890.
51. Aroulmoji V, Aguié-Béghin V, Mathlouthi M, Douillard R. Effect of sucrose on the properties of caffeine adsorption layers at the air/solution interface. J. Colloid Interface Sci. 2004;276:269–276.
52. Nabok AV, Tsargorodskaya A, Hassan AK, Starodub NF. Total internal reflection ellipsometry and SPR detection of low molecular weight environmental toxins. Appl. Surf. Sci. 2005;246:381–386.
53. Logothetidis S, Gioti M, Lousinian S, Fotiadou S. Haemocompatibility studies on carbon-based thin films by ellipsometry. Thin Solid Films 2005;482:126–132.
54. Liu Y, Li Z, He Z, Chen D, Pan S. Structure and blood compatibility of tetrahedral amorphous hydrogenated carbon formed by a magnetic-field-filter plasma stream. Surf. Coat. Technol. 2007;201:6851–6856.
55. Lousinian S, Logothetidis S. Optical properties of proteins and protein adsorption study. Microelectron. Eng. 2007;84:479–485.
56. Lousinian S, Logothetidis S, Laskarakis A, Gioti M. Haemocompatibility of amorphous hydrogenated carbon thin films, optical properties and adsorption mechanisms of blood plasma proteins. Biomol. Eng. 2007;24:107–112.
57. Lousinian S, Kassavetis S, Logothetidis S. Surface and temperature effect on fibrinogen adsorption to amorphous hydrogenated carbon thin films. Diamond Relat. Mater. 2007;16:1868–1874.
58. Keuren JFW, Wielders SJH, Willems GM, Morra M, Cahalan L, Cahalan P, et al. Thrombogenicity of polysaccharide-coated surfaces. Biomaterials 2003;24:1917–1924.
59. Arwin H. Ellipsometry on thin organic layers of biological interest: characterization and applications. Thin Solid Films 2000;377–378:48–56.
60. Russev SC, Arguirov TV, Gurkov TD. β-Casein adsorption kinetics on air-water and oil-water interfaces studied by ellipsometry. Colloids Surf. B 2000;19:89–100.
61. Bae YM, Oh B-K, Lee W, Lee WH, Choi J-W. Study on orientation of immunoglobulin G on protein G layer. Biosens. Bioelectron. 2005;21:103–110.
62. Vinnichenko M, Gago R, Huang N, Leng YX, Sun H, Kreissig U, et al. Spectroscopic ellipsometry investigation of amorphous carbon films with different sp3 content: relation with protein adsorption. Thin Solid Films 2004;455–456:530–534.
63. Karlsson LM, Tengvall P, Lundström I, Arwin H. Penetration and loading of human serum albumin in porous silicon layers with different pore sizes and thicknesses. J. Colloid Interface Sci. 2003;266:40–47.
64. Tsargorodskaya A, Nabok AV, Ray AK. Ellipsometric study of the adsorption of bovine serum albumin into porous silicon. Nanotechnology 2004;15:703–709.
65. De Feijter JA, Benjamins J, Veer FA. Ellipsometry as a tool to study the adsorption behavior of synthetic and biopolymers at the air-water interface. Biopolymers 1978;17:1759–1772.
66. Lassen B, Malmsten M. Competitive protein adsorption studied with TIRF and ellipsometry. J. Colloid Interface Sci. 1996;179:470–477.
67. Kurrat R, Prenosil JE, Ramsden JJ. Kinetics of human and bovine serum albumin adsorption at silica-titania surfaces. J. Colloid Interface Sci. 1997;185:1–8.
68. Giacomelli CE, Esplandiu MJ, Ortiz PI, Avena MJ, De Pauli CP. Ellipsometric study of bovine serum albumin adsorbed onto Ti/TiO2 electrodes. J. Colloid Interface Sci. 1999;218:404–411.
69. Vinnichenko M, Gago R, Huang N, Leng YX, Sun H, Kreissig U, et al. Spectroscopic ellipsometry investigation of amorphous carbon films with different sp3 content: relation with protein adsorption. Thin Solid Films 2004;455–456:530–534.
70. Foose LL, Blanch HW, Radke CJ. Immobilized protein films for assessing surface proteolysis kinetics. J. Biotechnol. 2007;132:32–37.
71. Brétagnol F, Kylián O, Hasiwa M, Ceriotti L, Rauscher H, Ceccone G, et al. Micro-patterned surfaces based on plasma modification of PEO-like coating for biological applications. Sens. Actuators B 2007;123:283–292.
72. Riquelme BD, Valverde JR, Rasia RJ. Kinetic study of antibody adhesion on a silicon wafer by laser reflectometry. Optic. Laser Eng. 2003;39:589–598.
73. Elgersma AV, Zsom RLJ, Lyklema J, Norde W. Kinetics of single and competitive protein adsorption studied by reflectometry and streaming potential measurements. Colloids Surf. A 1992;65:17–28.

74. Valenti LE, Fiorito PA, Garcia CD, Giacomelli CE. The adsorption-desorption process of bovine serum albumin on carbon nanotubes. J. Colloid Interface Sci. 2007;307:349–356.
75. de Vos WM, Biesheuvel PM, de Keizer A, Kleijn JM, Cohen Stuart MA. Adsorption of the protein bovine serum albumin in a planar poly(acrylic acid) brush layer as measured by optical reflectometry. Langmuir 2008;24:6575–6584.
76. Hofs B, Brzozowska A, de Keizer A, Norde W, Cohen Stuart MA. Reduction of protein adsorption to a solid surface by a coating composed of polymeric micelles with a glass-like core. J. Colloid Interface Sci. 2008;325:309–315.
77. Atkin R, Craig VS, Wanless EJ, Biggs S. The influence of chain length and electrolyte on the adsorption kinetics of cationic surfactants at the silica-aqueous solution interface. J. Colloid Interface Sci. 2003;266:236–244.
78. Dijt JC, Stuart MAC, Fleer GJ. Reflectometry as a tool for adsorption studies. Adv. Colloid Interface Sci. 1994;50:79–101.
79. Yu Y, Jin G. Influence of electrostatic interaction on fibrinogen adsorption on gold studied by imaging ellipsometry combined with electrochemical methods. J. Colloid Interface Sci. 2005;283:477–481.
80. Logothetidis S. Haemocompatibility of carbon based thin films. Diamond Relat. Mater. 2007;16:1847–1857.
81. Cardenas M, Arnebrant T, Rennie A, Fragneto G, Thomas RK, Lindh L. Human saliva forms a complex film structure on alumina surfaces. Biomacromolecules 2007;8:65–69.
82. Vinnichenko M, Chevolleau T, Pham MT, Poperenko L, Maitz MF. Spectroellipsometric, AFM and XPS probing of stainless steel surfaces subjected to biological influences. Appl. Surf. Sci. 2002;201:41–50.
83. Advincula M, Fan X, Lemons J, Advincula R. Surface modification of surface sol-gel derived titanium oxide films by self-assembled monolayers (SAMs) and non-specific protein adsorption studies. Colloids Surf. B 2005;42:29–43.
84. Jihua Y, David SW, Keith CG, McQuillan AJ. Electronic states and photoexcitation processes of titanium dioxide nanoparticle films dip coated from aqueous Degussa P25 photocatalyst suspension. J. Appl. Phys. 2007;101:023714.
85. Besra L, Liu M. A review on fundamentals and applications of electrophoretic deposition (EPD). Prog. Mater Sci. 2007;52:1–61.
86. Vahlas C, Caussat B, Serp P, Angelopoulos GN. Principles and applications of CVD powder technology. Mater. Sci. Eng. R 2006;53:1–72.
87. Tanemura S, Miao L, Wunderlich W, Tanemura M, Mori Y, Toh S, et al. Fabrication and characterization of anatase/rutile-TiO2 thin films by magnetron sputtering: a review. Sci. Technol. Adv. Mater. 2005;6:11–17.
88. Losurdo M, Bruno G, Irene EA. Anisotropy of optical properties of conjugated polymer thin films by spectroscopic ellipsometry. J. Appl. Phys. 2003;94:4923–4929.
89. Winfield JM, Donley CL, Ji-Seon K. Anisotropic optical constants of electroluminescent conjugated polymer thin films determined by variable-angle spectroscopic ellipsometry. J. Appl. Phys. 2007;102:063505.
90. Styrkas DA, Keddie JL, Lu JR, Su TJ, Zhdan PA. Structure of self-assembled layers on silicon: combined use of spectroscopic variable angle ellipsometry, neutron reflection, and atomic force microscopy. J. Appl. Phys. 1999;85:868.
91. Lecourt B, Blaudez D, Turlet JM. Anisotropy in Langmuir–Blodgett films studied by generalized spectroscopic ellipsometry. Thin Solid Films 1998;313–314:790–794.
92. Fan Q, McQuillin B, Ray AK, Turner ML, Seddon AB. High density, non-porous anatase titania thin films for device applications. J. Phys. D Appl. Phys. 2000;33:2683–2686.
93. Biju KP, Jain MK. Sol-gel derived TiO2:ZrO2 multilayer thin films for humidity sensing application. Sens. Actuators B 2008;128:407–413.
94. Radha G, Ashok K. Bioactive materials for biomedical applications using sol-gel technology. Biomed. Mater. 2008:034005.
95. Jung HY, Jung SM, Suh JS. Horizontally aligned single-walled carbon nanotube field emitters fabricated on vertically aligned multi-walled carbon nanotube electrode arrays. Carbon 2008;46:1345–1349.
96. Barnes TM, van de Lagemaat J, Levi D, Rumbles G, Coutts TJ, Weeks CL, et al. Optical characterization of highly conductive single-wall carbon-nanotube transparent electrodes. Phys. Rev. B 2007;75: 23541001–2354110.
97. Elim HI, Ji W, Ma GH, Lim KY, Sow CH, Huan CHA. Ultrafast absorptive and refractive nonlinearities in multi-walled carbon nanotube film. Appl. Phys. Lett. 2004;85:1799–1801.
98. Fanchini G, Miller S, Parekh BB, Chhowalla M. Optical anisotropy in single-walled carbon nanotube thin films: implications for transparent and conducting electrodes in organic photovoltaics. Nano Lett. 2008;8:2176–2179.

99. Wakita K, Abe K, Shim Y, Mamedov N. Spectroscopic ellipsometry of powdered CuInS2 with nanowires. Thin Solid Films 2006;499:285–288.
100. Gilliot M, En Naciri A, Johann L, d'Orleans C, Muller D, Stoquert JP, et al. Application of spectroscopic ellipsometry to the investigation of the optical properties of cobalt-nanostructured silica thin layers. Appl. Surf. Sci. 2006;253:389–394.
101. Bhat RR, Genzer J. Using spectroscopic ellipsometry for quick prediction of number density of nanoparticles bound to non-transparent solid surfaces. Surf. Sci. 2005;596:187–196.
102. Losurdo M. Relationships among surface processing at the nanometer scale, nanostructure and optical properties of thin oxide films. Thin Solid Films 2004;455–456:301–312.
103. Takeda Y, Plaksin OA, Wang H, Kishimoto N. Optical nonlinearity of Au nanoparticles fabricated by negative ion implantation. Nucl. Instrum. Methods Phys. Res. B 2007;257:47–50.
104. Mora MF, Giacomelli CE, Garcia CD. Interaction of D-amino acid oxidase to carbon nanotubes: implications in the design of biosensors. Anal. Chem. 2009;81:1016–1022.
105. Williams DF. On the mechanisms of biocompatibility. Biomaterials 2008;29:2941–2953.
106. Giacomelli CE, Avena MJ, De Pauli CP. Adsorption of bovine serum albumin onto TiO2 particles. J. Colloid Interface Sci. 1997;188:387–395.
107. Sousa SR, Bras MM, Moradas-Ferreira P, Barbosa MA. Dynamics of fibronectin adsorption on TiO2 surfaces. Langmuir 2007;23:7046–7054.
108. Aubin-Tam ME, Hamad-Schifferli K. Structure and function of nanoparticle–protein conjugates. Biomed. Mater. 2008:034001.
109. McClellan SJ, Franses EI. Exclusion of bovine serum albumin from the air/water interface by sodium myristate. Colloids Surf. B 2003;30:1–11.
110. Kang F, Singh J. Conformational stability of a model protein (bovine serum albumin) during primary emulsification process of PLGA microspheres synthesis. Int. J. Pharm. 2003;260:149–156.
111. Aleksic M, Pease CK, Basketter DA, Panico M, Morris HR, Dell A. Investigating protein haptenation mechanisms of skin sensitisers using human serum albumin as a model protein. Toxicol. In Vitro 2007;21:723–733.
112. Boonsongrit Y, Abe H, Sato K, Naito M, Yoshimura M, Ichikawa H, et al. Controlled release of bovine serum albumin from hydroxyapatite microspheres for protein delivery system. Mater. Sci. Eng. B 2008;148:162–165.
113. Wehmeyer J, Bizios R, Garcia CD. Adsorption of BSA to nanostructured TiO2. 2008:submitted.
114. van der Veen M, Norde W, Stuart MC. Electrostatic interactions in protein adsorption probed by comparing lysozyme and succinylated lysozyme. Colloids Surf. B Biointerfaces 2004;35:33–40.
115. Haynes CA, Norde W. Globular proteins at solid/liquid interfaces. Colloids Surf. B Biointerfaces 1994;2:517–566.
116. D'Orazio P. Biosensors in clinical chemistry. Clin. Chim. Acta 2003;334:41–69.
117. Alaejos MS, Garcia Montelongo FJ. Application of amperometric biosensors to the determination of vitamins and alpha-amino acids. Chem. Rev. 2004;104:3239–3266.
118. Wang J. Electrochemical biosensors: towards point-of-care cancer diagnostics. Biosens. Bioelectron. 2006;21:1887–1892.
119. Rogers KR. Recent advances in biosensor techniques for environmental monitoring. Anal. Chim. Acta 2006;568:222–231.
120. Gooding JJ. Biosensor technology for detecting biological warfare agents: recent progress and future trends. Anal. Chim. Acta 2006;559:137–151.
121. Gómez-Hens A, Fernández-Romero JM, Aguilar-Caballos MP. Nanostructures as analytical tools in bioassays. Trends Anal. Chem. 2008;27:394–406.
122. Cai C, Chen J. Direct electron transfer of glucose oxidase promoted by carbon nanotubes. Anal. Biochem. 2004;332:75–83.
123. Zhang M, Smith A, Gorski W. Carbon nanotube-chitosan system for electrochemical sensing based on dehydrogenase enzymes. Anal. Chem. 2004;76:5045–50.
124. Lenihan JS, Gavalas VG, Wang J, Andrews R, Bachas LG. Protein immobilization on carbon nanotubes through a molecular adapter. J. Nanosci. Nanotechnol. 2004;4:600–604.
125. Lin Y, Taylor S, Li H, Fernando KAS, Qu L, Wang W, et al. Advances toward bioapplications of carbon nanotubes. J. Mater. Chem. 2004;14:527–541.
126. Liu Y, Qu X, Guo H, Chen H, Liu B, Dong S. Facile preparation of amperometric laccase biosensor with multifunction based on the matrix of carbon nanotubes-chitosan composite. Biosens. Bioelectron. 2006;21:2195–2201.

127. Qi H, Zhang C, Li X. Amperometric third-generation hydrogen peroxide biosensor incorporating multiwall carbon nanotubes and hemoglobin. Sens. Actuators B 2006;114:364–370.
128. Weber J, Kumar A, Kumar A, Bhansali S. Novel lactate and pH biosensor for skin and sweat analysis based on single walled carbon nanotubes. Sens. Actuators B 2006;117:308–313.
129. Zhang M, Gorski W. Electrochemical sensing based on redox mediation at carbon nanotubes. Anal. Chem. 2005;77:3960–3965.
130. Sánchez S, Roldán M, Pérez S, Fàbregas E. Toward a fast, easy, and versatile immobilization of biomolecules into carbon nanotube/polysulfone-based biosensors for the detection of hCG hormone. Anal. Chem. 2008;80:6508–6514.
131. Eggins BR. Sensing elements. In: Eggins BR, (ed.) *Chemical Sensors and Biosensors*. West Sussex, England: Wiley; 2002. pp. 98–106.
132. Krajewska B. Application of chitin- and chitosan-based materials for enzyme immobilizations: a review. Enzyme Microb. Technol. 2004;35:126–139.
133. López MS-P, López-Cabarcos E, López-Ruiz B. Organic phase enzyme electrodes. Biomol. Eng. 2006;23:135–147.
134. Mizutani H, Miyahara I, Hirotsu K, Nishina Y, Shiga K, Setoyama C, et al. Three-dimensional structure of porcine kidney D-amino acid oxidase at 3.0 A resolution. J. Biochem. 1996;120:14–17.
135. Yagi K, Ohishi N. Structure and function of D-amino acid oxidase – IV. Electrophoretic and ultracentrifugal approach to the monomer equilibrium. J. Biochem. 1972;71:993–998.
136. Tishkov VI, Khoronenkova SV. D-amino acid oxidase: structure, catalytic mechanism, and practical application. Biochemistry 2005;70:40–54.
137. Tessema M, Larsson T, Buttler T, Csoregi E, Ruzgas T, Nordling M, et al. Simultaneous amperometric determination of some mono-, di-, and oligosaccharides in flow injection and liquid chromatography using two working enzyme electrodes with different selectivity. Anal. Chim. Acta 1997;349:179–188.
138. Wang J, Chen G. Microchip capillary electrophoresis with electrochemical detector for fast measurements of aromatic amino acids. Talanta 2003;60:1239–1244.
139. Nagata Y, Akino T, Ohno K, Kataoka Y, Ueda T, Sakurai T, et al. Free D-amino acids in human plasma in relation to senescence and renal diseases. Clin. Sci. 1987;73:105–8.
140. D'Aniello A, D'Onofrio G, Pischetola M, D'Aniello G, Vetere A, Petrucelli L, et al. Biological role of D-amino acid oxidase and D-aspartate oxidase. Effects of D-amino acids. J. Biol. Chem. 1993;268:26941–26949.
141. Hall D, Gogos JA, Karayiorgou M. The contribution of three strong candidate schizophrenia susceptibility genes in demographically distinct populations. Genes Brain Behav. 2004;3:240–248.
142. Quan Z, Song Y, Feng Y, LeBlanc MH, Liu Y-M. Detection of D-serine in neural samples by saccharide enhanced chiral capillary electrophoresis. Anal. Chim. Acta 2005;528:101–106.
143. Hamase K, Morikawa A, Zaitsu K. Amino acids in mammals and their diagnostic value. J. Chromatogr. B 2002;781:73–91.
144. Pilone MS, Pollegioni L. D-amino acid oxidase as an industrial biocatalyst. Biocatal. Biotransform. 2002;20:145–159.
145. Betancor L, Hidalgo A, Fernandez-Lorente G, Mateo C, Rodriguez V, Fuentes M, et al. Use of physicochemical tools to determine the choice of optimal enzyme: stabilization of D-amino acid oxidase. Biotechnol. Prog. 2003;19:784–748.
146. Fernandez-Lafuente R, Rodriguez V, Mateo C, Fernandez-Lorente G, Arminsen P, Sabuquillo P, et al. Stabilization of enzymes (-amino acid oxidase) against hydrogen peroxide via immobilization and post-immobilization techniques. J. Mol. Catal., B Enzym. 1999;7:173–179.
147. Duran N, Rosa MA, Annibale A, Gianfreda L. Applications of laccases and tyrosinases (phenoloxidases) immobilized on different supports: a review. Enzyme Microb. Technol. 2002;31:907–931.
148. Rivas GA, Rubianes MD, Rodríguez MC, Ferreyra NF, Luque GL, Pedano ML, et al. Carbon nanotubes for electrochemical biosensing. Talanta 2007;74:291–307.
149. Karajanagi SS, Vertegel AA, Kane RS, Dordick JS. Structure and function of enzymes adsorbed onto single-walled carbon nanotubes. Langmuir 2004;20:11594–11599.
150. Schulz B, Chan D, Bäckström J, Rübhausen M. Spectroscopic ellipsometry on biological materials – investigation of hydration dynamics and structural properties. Thin Solid Films 2004;455–456:731–734.
151. Danny C, Benjamin S, Kathrin G, Heike Hedwig M, Michael R. In vivo spectroscopic ellipsometry measurements on human skin. J. Biomed. Opt. 2007;12:014023.
152. Santos O, Kosoric J, Hector MP, Anderson P, Lindh L. Adsorption behavior of statherin and a statherin peptide onto hydroxyapatite and silica surfaces by in situ ellipsometry. J. Colloid Interface Sci. 2008;318:175–182.
153. Arwin H. Is ellipsometry suitable for sensor applications? Sens. Actuators A 2001;92:43–51.

154. Demirel G, Çaglayan MO, Garipcan B, Piskin E. A novel DNA biosensor based on ellipsometry. Surf. Sci. 2008;602:952–959.
155. Nabok A, Tsargorodskaya A, Davis F, Higson SPJ. The study of genomic DNA adsorption and subsequent interactions using total internal reflection ellipsometry. Biosens. Bioelectron. 2007;23:377–383.
156. Wenmackers S, Pop SD, Roodenko K, Vermeeren V, Williams OA, Daenen M, et al. Structural and optical properties of DNA layers covalently attached to diamond surfaces. Langmuir 2008;24:7269–7277.
157. Cristofolini L, Berzina T, Erokhin V, Tenti M, Fontana MP, Erokhina S, et al. The structure of DNA-containing complexes suggests the idea for a new adaptive sensor. Colloids Surf. A 2008;321:158–162.
158. Carlsson J, Winquist F, Danielsson B, Lundström I. Biosensor discrimination of meat juice from various animals using a lectin panel and ellipsometry. Anal. Chim. Acta 2005;547:229–236.
159. Jin G, Tengvall P, Lundström I, Arwin H. A biosensor concept based on imaging ellipsometry for visualization of biomolecular interactions. Anal. Biochem. 1995;232:69–72.
160. van Noort D, Rumberg J, Jager EWH, Mandenius CF. Silicon based affinity biochips viewed with imaging ellipsometry. Meas. Sci. Technol. 2000;11:801–808.
161. Bae YM, Oh B-K, Lee W, Lee WH, Choi J-W. Immunosensor for detection of *Legionella pneumophila* based on imaging ellipsometry. Mater. Sci. Eng. C 2004;24:61–64.
162. Bae YM, Park K-W, Oh B-K, Lee WH, Choi J-W. Immunosensor for detection of *Salmonella typhimurium* based on imaging ellipsometry. Colloids Surf. A 2005;257–258:19–23.
163. Bae YM, Oh BK, Lee W, Lee WH, Choi JW. Immunosensor for detection of *Yersinia enterocolitica* based on imaging ellipsometry. Anal. Chem. 2004;76:1799–1803.
164. Marinkova D, Bivolarska M, Ahtapodov L, Yotova L, Mateva R, Velinov T. Plasmon microscopy and imaging ellipsometry of *Artrobacter oxydans* attached on polymer films. Colloids Surf. A 2008;65:276–280.
165. Pereira EMA, Sierakowski MR, Jó TA, Moreira RA, Monteiro-Moreira ACO, França RFO, et al. Lectins and/or xyloglucans/alginate layers as supports for immobilization of dengue virus particles. Colloids Surf. B 2008;66:45–52.
166. Wang Z, Jin G. Feasibility of protein A for the oriented immobilization of immunoglobulin on silicon surface for a biosensor with imaging ellipsometry. J. Biochem. Biophys. Methods 2003;57:203–211.
167. Qi C, Duan J-Z, Wang Z-H, Chen Y-Y, Zhang P-H, Zhan L, et al. Investigation of interaction between two neutralizing monoclonal antibodies and SARS virus using biosensor based on imaging ellipsometry. Biomed. Microdevices 2006;8:247–253.
168. Wang Z-H, Jin G. Silicon surface modification with a mixed silanes layer to immobilize proteins for biosensor with imaging ellipsometry. Colloids Surf. B 2004;34:173–177.
169. Schuy S, Faiss S, Yoder NC, Kalsani V, Kumar K, Janshoff A, et al. Structure and Thermotropic phase Behavior of Fluorinated Phospholipid Bilayers: A combined Attenuated Total Reflection FTIR Spectroscopy and Imaging Ellipsometry Study. J. Phys. Chem. B Biointerfaces 2008;112:8250–8256.
170. Srivatsa V, Neil B, Yanming Z, Russell C. Evanescent-imaging-ellipsometry-based microarray reader. J. Biomed. Opt. 2006;11:014028.
171. Beyerlein D, Kratzmüller T, Eichhorn KJ. Study of novel polymer architectures on solid surfaces by variable angle spectroscopic and imaging ellipsometry. Vib. Spectrosc 2002;29:223–227.

3

Atomic Force Microscopy Methods for Characterizing Protein Interactions with Microphase-Separated Polyurethane Biomaterials

Li-Chong Xu, Pranav Soman, Aashiish Agnihotri, and Christopher A. Siedlecki

Understanding the molecular scale interactions between proteins and microphase-structured polyurethane (PU) biomaterials can provide mechanistic insights into the blood/biomaterial problem. This knowledge is critically important to provide design parameters for developing new blood-contacting materials with improved hemocompatibility. This chapter introduces a series of novel atomic force microscopy (AFM) techniques for characterization of PU biomaterials under ambient and physiologically relevant conditions. Techniques for measuring the interaction forces between proteins and surfaces or proteins and proteins, for measuring the bioactivity of proteins, and for correlating biological activity with material surface properties are discussed. The measurements reveal a complex interfacial environment where the material surface is undergoing dynamic changes at the same time as proteins are undergoing time-dependent conformational and bioactivity changes. Combining the traditional in vitro techniques for biocompatibility with these new AFM techniques offers new insights into the fundamental relationships between PU microphase structures and the biological response to these materials.

Abbreviations and Symbols

Ab	Antibody
AFM	Atomic force microscopy
APS	3-aminopropyltrichlorosilane
BSA	Bovine serum albumin
BTS	n-butyltrichlorosilane

L.-C. Xu, P. Soman, A. Agnihotri, and C.A. Siedlecki • Departments of Surgery and Bioengineering, Biomedical Engineering Institute, College of Medicine, The Pennsylvania State University, Hershey, PA, 17033, USA

D.A. Puleo and R. Bizios (eds.), *Biological Interactions on Materials Surfaces*,
DOI 10.1007/978-0-387-98161-1_3,

Fib	Fibrinogen
HOPG	Highly ordered pyrolytic graphite
LDPE	Low-density polyethylene
mAb	Monoclonal antibody
OTS	octadecyltrichlorosilane
PBS	phosphate-buffered saline
PDMS	Polydimethylsiloxane
PTMO	Polytetramethyleneoxide
PC	Polycarbonate
PU	Polyurethane
PUU	Poly(urethane urea)
SAM	Self-assembled monolayer
A	the set point amplitude of the oscillation of the cantilever
A_0	the free amplitude of the oscillation of the cantilever
$A(T)$	the prefactor
d	the deflection of the cantilever (nm)
d_0	the initial deflection of the cantilever (nm)
E	the modulus (GPa)
E_a	the activation energy for protein unfolding (kT)
F	the adhesion force (nN)
F_e	the adhesion force at equilibrium (nN)
F_0	an empirical coefficient related to the initial interaction force (nN)
H	the domain height of fibrinogen (nm)
H_e	the equilibrium height of the domains at very long adsorption times (nm)
H_0	the initial height of molecule at the time of contact (nm)
h	the Planck constant
k	the Boltzmann constant
k_c	the spring constant of the cantilever (N/m)
k_s	the rate constant (s^{-1})
k_{off}	the thermal off-rate at zero force (s^{-1})
r_{sp}	the ratio of set point amplitude and free amplitude of oscillation
h_{indent}	the indentation depth (nm)
R	the radius of the tip (nm)
v	the Poisson ratio
T	the absolute temperature (K)
t	the contact time (s)
θ	the water contact angle (degree)
τ	the water adhesion tension (dyn/cm),
r_f	the loading rate (nN/s)
χ_β	the effective distance between the bound and transition states (Å)
Z_{pos}	the movement of the piezo during compression of the surface (nm)
Z_{defl}	the cantilever deflection (nm)

3.1. Introduction

When a biomaterial is placed into contact with blood or other tissues, proteins will immediately compete to adsorb onto the material surface; this event is then followed by cellular/platelet interactions with these adsorbed proteins [1–3]. These initially adsorbed proteins can mediate a series of subsequent interactions of cells and tissues with surfaces that may be either beneficial or detrimental to the performance of biomaterials [4]. For example, adsorption of albumin and globulins can affect the attachment of tissue cells [5] and decrease platelet adhesion and activation [6]. On the other hand, adsorbed fibrinogen may induce platelet adhesion and activation, and accelerate thrombus formation, which remains one of the major problems associated with the long-term use of blood-contacting medical devices [7, 8]. Thus, understanding the factors influencing protein interactions with surfaces can provide mechanistic insights into the blood/biomaterial problem and provide design parameters for new blood-contacting devices with improved hemocompatibility.

Protein adsorption to a surface is a complex process governed by the properties of the protein itself, the composition of the biological fluid, and, perhaps most importantly, the properties of the material substrate [9–12]. A variety of different interfacial forces exist between proteins and surfaces that can potentially influence protein adsorption, including van der Waals' interactions, electrostatic interactions, hydrogen bonding, covalent bonding, and particularly hydrophobic interactions [13–17]. Protein adsorption is also affected by solvent–protein interactions that provide an energetic basis to drive proteins from solutions to the material surface [18–20]. However, despite decades of intense research into this most fundamental area of biomaterials science, the relationships between surface properties and protein–surface interactions at the molecular level are still not well established.

Blood contains a multitude of proteins that have different biological functions and activity, and are present in vastly different concentrations. Albumin, immunoglobulins, and fibrinogen are the most abundant proteins in plasma and represent more than 50% of all plasma proteins. Albumin is generally considered to be inert toward platelet adhesion and activation [21], while fibrinogen is a central protein in the process of biomaterial-induced thrombosis [22, 23]. Fibrinogen is a multifunctional protein that is capable of promoting thrombus formation through binding of the platelet integrin receptor $\alpha_{IIb}\beta_3$ (GPIIb/IIIa), leading to platelet adhesion, activation, and aggregation [24], as well as being the precursor to fibrin, the important structural component in blood clotting/coagulation. Fibrinogen is also involved in bacterial adhesion to surfaces [25–27].

Fibrinogen is a symmetric molecule with two sets of three intertwined polypeptide chains termed Aα, Bβ, and γ chains (Figure 3.1) [28, 29]. Each fibrinogen molecule possesses three pairs of potential platelet-binding peptide sequences, two arginine–glycine–aspartic acid (RGD) sequences in each of the Aα chains (RGDF and RGDS) and a dodecapeptide sequence (HHLGGAKQAGDV) in each of the γ chains; the γ chain dodecapeptide sequence is the primary ligand for platelet adhesion to adsorbed fibrinogen [30, 31]. Although the amount of fibrinogen adsorbed on a material surface is one of the

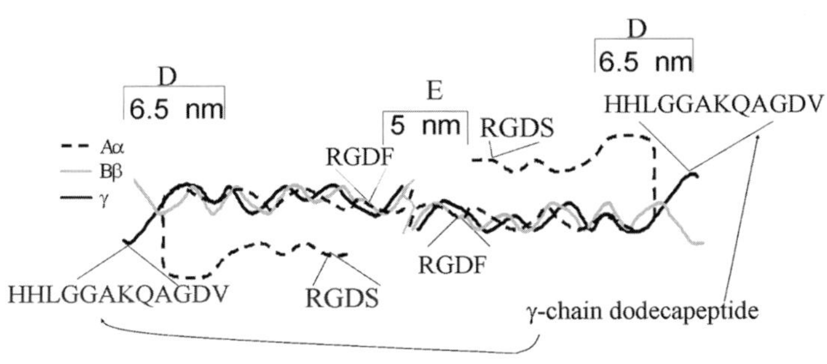

Figure 3.1. Schematic diagram of fibrinogen structure.

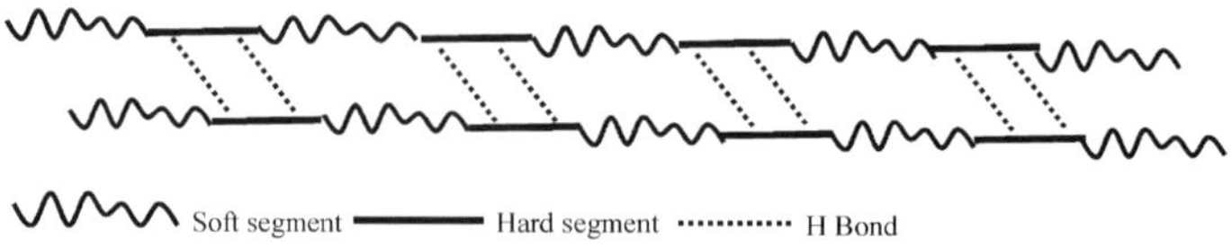

Figure 3.2. Schematic diagram of the PU structure.

important factors responsible for platelet adhesion, there is increasing evidence that the protein conformation and availability of platelet-binding sites in fibrinogen are more important in mediating platelet adhesion than just the amount of adsorbed protein [32, 33]. Studies have demonstrated that fibrinogen structure is dependent upon the substrate to which it is adsorbed, especially with regards to the surface chemistry, and that the material surface can potentially control the orientation/activity of adsorbing proteins thereby providing mechanisms for cellular responses to the adsorbed protein layer [34].

Segmented polymeric materials such as polyurethanes (PUs) are frequently used in a variety of blood-contacting medical devices due to their desirable mechanical properties, fatigue resistance, and acceptable hemocompatibility for many applications, including circulatory support devices, vascular grafts, and heart valves. PU materials are block copolymers constructed from a mixture of "soft" and "hard" segments that form a microphase-separated structure as a result of thermodynamic immiscibility of the polar hard segments and the relatively nonpolar soft segments (Figure 3.2) [35]. This microphase-separated structure is believed to contribute to the improved success of the polymer in blood-contacting applications, as it appears to influence protein adsorption, platelet adhesion, vascular cell attachment and proliferation, and bacterial adhesion. Yet, despite these observations linking microphase separation and biological responses, there is little insight into the molecular mechanisms underlying these fundamental protein–surface interactions.

A variety of techniques has been utilized to study protein adsorption on material surfaces in both physiologic and nonphysiologic environments. This extensive list of techniques includes ellipsometry [36, 37], surface plasmon resonance [38, 39], quartz crystal microbalance [40], radiolabeling [41], electrophoretic depletion measurements [20], circular dichroism [42], infrared spectroscopy [43], X-ray photoemission electron microscopy [44, 45], electron spectroscopy for chemical analysis, and time-of-flight secondary ion mass spectroscopy [46, 47]. Most of these techniques used in detecting proteins are indirect and measure the average properties of the biological events. The advent of atomic force microscopy (AFM) and related scanning probe microscopies offers new opportunities to directly examine protein adsorption on surfaces [48–50]. Utilizing a microprobe mounted on a flexible cantilever scanning across the surface, AFM can obtain topography images with molecular level resolution, and can also measure the interaction forces between proteins and surfaces or proteins and proteins at the piconewton scale. Moreover, AFM can perform these measurements of proteins under physiological conditions, opening the door to direct measurements of time-dependent, dynamic, biological processes at the molecular scale.

We have developed and utilized a series of novel AFM techniques to study the surface microphase-separated structures of polymers and to address the questions of protein adsorption and activity on these microphases. The aim of this endeavor is to eventually determine how the spatially dispersed chemical functionalities present in these important biomaterials contribute to the success of the material in biomedical device applications.

3.2. Surface Microphase Separation Structures of PU Materials

Biomedical-segmented PUs are generally synthesized from soft segment and hard segments. Materials used for soft segments are often polyethers, but, increasingly, PU materials are being made from polycarbonates (PCs) or mixed polytetramethyleneoxide (PTMO)/ polydimethylsiloxane (PDMS) precursors; for the hard segments, a diisocyanate with either a diol (resulting in a PU) or a diamine (resulting in a poly(urethane urea) [PUU]) are often used. The microphase structures of PUs depend on the chemistry and relative amounts of the two segments [51, 52] as well as the processing conditions [53]. Phase separation has been studied using a number of techniques, with small-angle X-ray scattering proving particularly useful [54–56]. However, most of these studies focus on the bulk properties of polymers under ambient environments. Obviously, these sorts of studies provide limited information that is relevant to blood–material interactions and subsequent biological responses. The material surface structure is believed to differ from the bulk and is influenced by the local environment [57]. It has long been suggested that exposure to aqueous environments results in significant rearrangements of the domains in PU biomaterials, leading to changes in phase separation and surface morphology that would not be detected by ambient environment techniques [58, 59].

3.2.1. Microphase Structure of PU Under Ambient Environments

High-resolution AFM imaging can illustrate the three-dimensional (3D) microphase-separated structure of polymers and also probe microphase dimensions and connectivity [60, 61]. Furthermore, and of particular importance to the biomedical problem, these studies can be extended to observe the dynamic rearrangement of these surface regions under physiologically relevant buffer conditions. Figure 3.3 illustrates the AFM height and phase images of a PUU film in an ambient environment. This particular PUU was synthesized using PTMO (2000 Da) end-capped with 4,4′-methylene di(p-phenyl isocyanate) and chain extended with ethylene diamine [62]. The PUU film surface was relatively smooth across the 500 × 500 nm^2

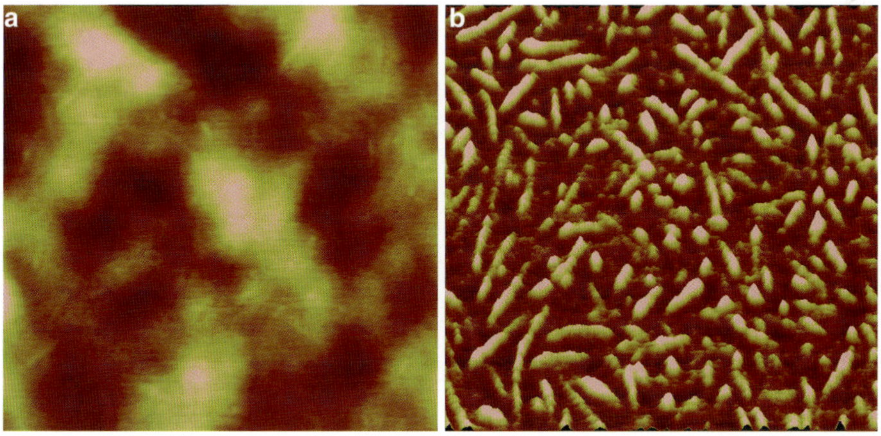

Figure 3.3. AFM (**a**) height and (**b**) phase images of PUU under ambient environment. Scan size: 500 × 500 nm^2. Scale: height 50 nm. Phase 25°.

scan area, with only a few occasional topographic features observed. The largest vertical features on the PUU surface were approximately 13.5 nm tall and the average roughness (R_q) was 2.1 nm. The phase image illustrates the distribution of mechanical properties in the film and shows the distribution of the microphase structures. The phase angle shifts of probe scanning across the polymer surface in tapping mode arise from differences in the mechanical properties of the film (principally viscoelasticity), with high phase angle shifts (lighter colors) representing the "harder" regions. In this PUU film, the hard microphases show spherical or cylindrical features that are dispersed in a soft segment matrix. The lateral dimensions of the hard domains are 15–50 nm (Figure 3.3b), similar in dimension to many plasma proteins, including fibrinogen.

Detection of the hard domains depends on the force applied by probe to the substrate, which is represented by the ratio of set point amplitude and free amplitude of oscillation ($r_{sp} = A/A_0$). At very light tapping force (Figures 3.4a and 3.4b; $r_{sp} = 0.95$ and 0.90, respectively), few or no phases are observed, suggesting the presence of a soft segment overlayer on the surface. As the imaging force increases (Figures 3.4c and 3.4d; $r_{sp} = 0.85$ and 0.75, respectively), randomly oriented cylindrical and spherical structures are observed, indicating that hard domains are being detected.

3.2.2. Microphase Structure of PU Under Aqueous Buffer Conditions

PU materials undergo significant reorganization and reorientation when placed in aqueous environments; AFM imaging is capable of demonstrating these changes directly. Figure 3.5 illustrates a series of sequential images of a PUU film following hydration in phosphate-buffered

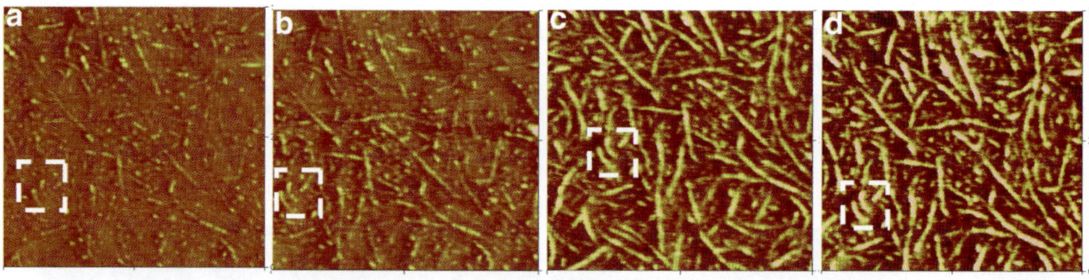

Figure 3.4. AFM phase images of PUU under different tapping forces. Scan size: 500 × 500 nm². Scale: Phase 25°. The *box* indicates a reference feature common in all images. (**a**) $r_{sp} = 0.95$; (**b**) $r_{sp} = 0.90$; (**c**) $r_{sp} = 0.85$; (**d**) $r_{sp} = 0.75$.

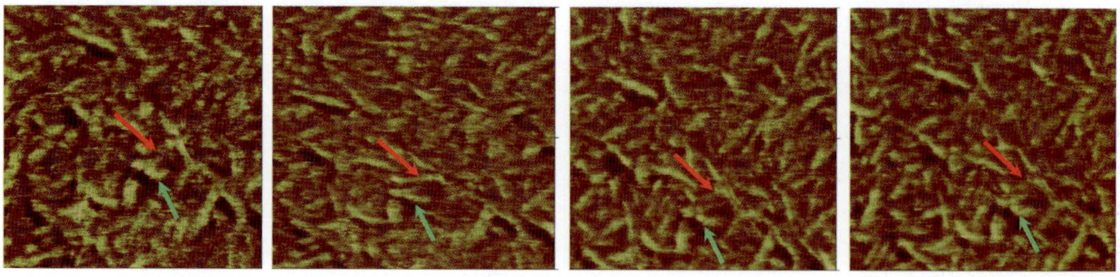

Figure 3.5. Sequential phase images of PUU under PBS. Scan size: 500 × 500 nm². Scale: Phase 20° buffer hydrated for 21 h, illustrating the hard domain reorientation and reorganization due to hydration.

saline (PBS) for 21 h. These images were captured at approximately 9-min intervals. Reorientation and reorganization of some of these hard domains even at this extended time point are indicated by the arrows in Figure 3.5. The reorientation results in hard domain enrichment on the surface as well as overall rougher surfaces [59, 63]. It is believed that reorientation of domains is a thermodynamic response to the aqueous environment; such reorientation minimizes surface energetics by relocating the more polar hard segment to the water/solid interface and moving the relatively less polar soft segments deeper into the material.

3.2.3. Micromechanical Analysis of PU Materials by AFM Indentation

AFM phase images illustrate the distribution of soft and hard domains in polymers through the indirect measurement of the "phase shift" in tapping mode. More quantitative information on the properties of phases can be observed from compression/adhesion measurements obtained by force curves. Figure 3.6 represents a typical force–distance curve as the probe approaches and compresses a PDMS-based PU surface at single location (Figure 3.6a). The indentation depth, $h_{\text{indent}} = Z_{\text{pos}} - Z_{\text{defl}}$ (Z_{pos} is the movement of the piezo during compression of the surface and Z_{defl} is the cantilever deflection), changes with the applied force in a nonlinear fashion, because the PU consists of both hard and soft domains distributed in the z-direction (Figure 3.6b). The large difference in the indentation depth to applied force relationships (as the probe indents deeper into the material) suggests that there are substantial changes in the mechanical properties of this film at different depths (Figure 3.6b). These properties can be quantified by calculating the modulus (E) from any arbitrary two points, (z_1, d_1) and (z_2, d_2), on the force curve using the Hertz model [64], which can be modified to the practical form that follows [65]:

$$\frac{1}{E} = \left[\frac{\left[(z_1 - z_2) - (d_1 - d_2)\right]}{\left(\frac{3k_c\left(1 - v^2\right)}{4\sqrt{R}} \right)^{2/3} \left[(d_1 - d_0)^{2/3} - (d_2 - d_0)^{2/3} \right]} \right]^{3/2}, \tag{3.1}$$

In this equation, k_c is the spring constant of the cantilever (0.32 N/m in this case), R is the radius of the tip, v is the Poisson ratio, and d_0 is the initial deflection of the cantilever. The indentation depth and modulus curves confirmed a soft segment layer on the material surface, as was suggested by the phase images at different tapping forces (Figure 3.4), with the thickness of the soft overlayer being about 5–10 nm in this case.

Generally, the AFM software can acquire an array of individual force–distance curves during scanning. We developed Matlab (Mathworks, USA) programs to analyze all curves and extracted the compression region data used for calculating the modulus from each curve. These data were used to produce modulus maps at different indentation depths. Figure 3.7 illustrates a series of these modulus maps at different depth levels for a PDMS-based PU material in an ambient environment. These maps show an increase in the amount of hard segment content at deeper levels in the material. At a depth of 3 nm, only about 11% of the material had a modulus consistent with hard domains, suggesting that the outer layers were primarily soft segment material. As the measurement depth increased, increasing amounts of

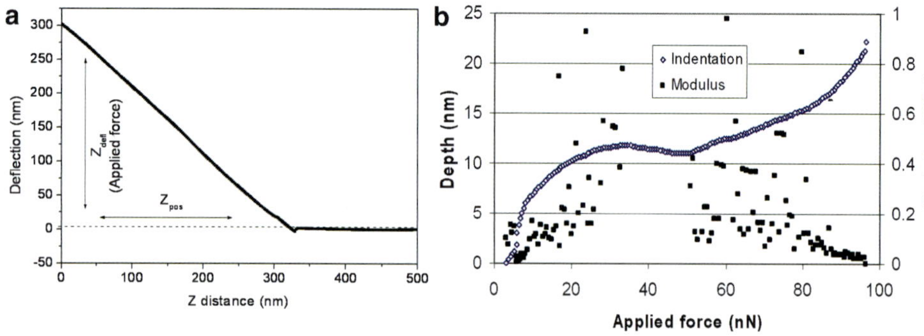

Figure 3.6. (**a**) Representative force–distance curve and (**b**) indentation and modulus curves as a function of the applied forces for a PDMS-based PU material.

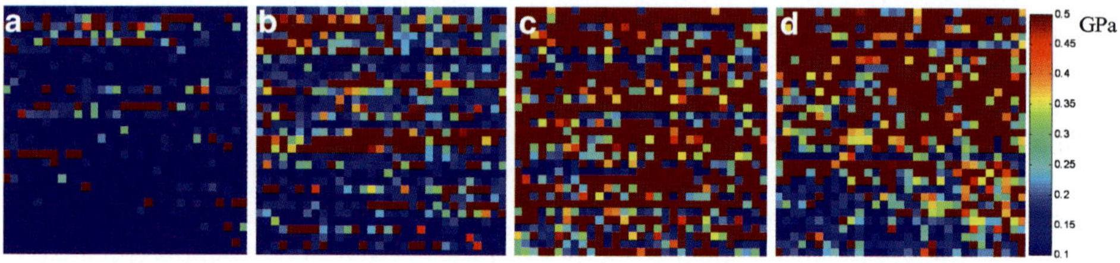

Figure 3.7. Modulus maps of PDMS-based PU showing the distribution of hard and soft segments at depths of (**a**) 3 nm, (**b**) 7 nm, (**c**) 11 nm, and (**d**) 15 nm. Cool colors represent soft segments while warm colors indicate hard segments.

hard domains were detected. For this sample, 11.2%, 40.0%, 76.8%, and 73.9% of the area tested possessed at modulus values greater than 0.2 GPa at depths of 3, 7, 11, and 15 nm, respectively. Similar measurements and analysis can be performed under aqueous conditions. Thus, this AFM indentation technique provided a powerful tool to produce a quantitative 3D network of micromechanical properties of polymer in air and under biologically relevant aqueous environments.

3.3. Protein Interactions with Hydrophobic and Hydrophilic Surfaces

Surface wettability (generally referred to as hydrophobicity/hydrophilicity) is one of the most important parameters affecting protein adsorption, platelet adhesion/activation, and blood coagulation, as well as vascular cell and bacterial adhesion [66–70]. Generally hydrophobic surfaces are considered to promote protein adsorption more than hydrophilic surfaces because of the strong hydrophobic interactions occurring at these interfaces, particularly when contrasted to the repulsive solvation forces arising from strongly bound water at the hydrophilic surface [14]. Wettability also affects the time-dependent conformational structures of proteins on material surfaces [71] as well as protein activity [42]. In this section, we introduce methodologies for observing and analyzing the conformational changes in fibrinogen when contacting either a hydrophobic or hydrophilic surface in an aqueous environments either through direct visualization of the protein structure [72] or by measuring the adhesive forces between proteins and material surfaces that reveal protein denaturation processes in terms of energy interactions [73].

3.3.1. Time-Dependent Conformational Changes in Fibrinogen Measured by AFM

The fibrinogen molecule can be modeled as two outer spherical domains (termed the D domains) having a diameter of 6.5 nm, connected by thin linear helical regions to a central E domain (Figure 3.1). The proposed structure and dimensions in this trinodular model have been confirmed through a combination of different techniques and have also been refined to a more detailed model [74]. The 3D molecular structure of fibrinogen has been visualized by AFM under aqueous conditions [28]; conformation of the protein is affected by the material surface properties. It was observed that structural deformation (spreading) of fibrinogen increased with material surface hydrophobicity; quantitative analysis of the different domains showed that the overall molecular length and widths of the individual D and E domains increased while the heights decreased when the material surface was more hydrophobic [75].

Various studies provided evidence that fibrinogen undergoes conformational changes following adsorption to synthetic material substrates [76–82] and that the transient exposure of functional epitopes (which result in platelet binding) appear to be both time and surface dependent. Using AFM, we measured the time-dependent changes of heights of the D and E domains of fibrinogen and calculated the free energy of protein spreading [72]. Two materials with vastly different surface wettability were used as the substrates: hydrophobic, highly ordered pyrolytic graphite (HOPG) and hydrophilic, muscovite mica. A fibrinogen solution was introduced to the sample and, following exchange for a fresh buffer solution, images of individual proteins were taken continuously for periods up to ~2 h. Spreading curves generated by plotting the height of the D and E domains of the individual fibrinogen molecules as a function of time demonstrated that the heights of both D and E domains decreased with time on the hydrophobic HOPG surface, indicating the expected protein spreading. On the hydrophilic mica surface, the height of the D and E domains actually increased with time (up to 60 min), eventually reaching a plateau height of 2.1 nm (Figure 3.8), greater than what was observed on the hydrophobic surfaces but still much lower compared with the dimension of 6.5 nm observed for fibrinogen in solution.

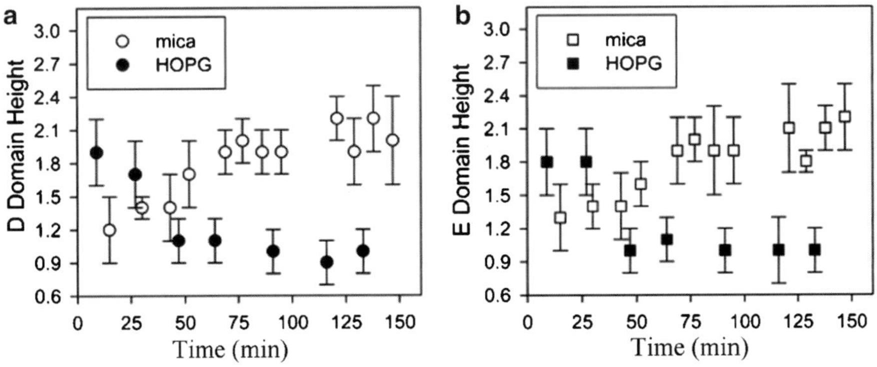

Figure 3.8. Spreading curves for (**a**) the D domain and (**b**) the E domain of fibrinogen on HOPG and mica surfaces. (Reprinted from [72]).

The spreading curves were fit to a simple exponential decay of height model, and were expressed in terms of domain height (H) at time (t) as:

$$H = H_e + (H_0 - H_e)\exp(-k_s t), \tag{3.2}$$

where H_e is the equilibrium height of the domains at very long adsorption times, H_0 is the initial height of the molecule at the time of contact, and k_s is the rate constant. Since spreading of proteins on a material surface is indicative of a protein denaturation process, the spreading rate can be used to calculate the Gibbs free energy of protein unfolding by the following equation:

$$k_s = \frac{kT}{h}\exp\left(\frac{-\Delta G}{kT}\right), \tag{3.3}$$

where k is the Boltzmann constant, h is the Planck constant, and T is the absolute temperature. Calculation using Eq. (3.3) yields a free energy of unfolding of ~37 kT for fibrinogen adsorbed on an HOPG surface; this result is similar to the value of 36 kT obtained by Wertz and Santore [80] for fibrinogen on a hydrophobic surface using macroscopic optical techniques.

Due to the long period of time required for AFM images acquired by scanning the material surface, the changes of protein structure at initial contact time (out to approximately 10 min) are not available through traditional imaging. Acquisition of AFM data at these early time points requires a different approach. In order to generate data for early time points of protein–material surface contact, proteins are attached to an AFM probe and interaction forces between this protein-modified probe and various material surfaces can be measured as a function of contact time; the approach provides insight into the early unfolding processes that occur when a protein contacts a synthetic material surface.

3.3.2. Effects of Surface Wettability and Contact Time on Protein Adhesion to Materials

We utilized a low-density polyethylene (LDPE) as a base material for preparation of modified surfaces spanning a range of water wettability. LDPE surfaces were treated with a glow discharge plasma cleaner for different periods of time (up to 150 min) to yield various levels of wettability as determined by advancing water contact angle measurements. Three test proteins (specifically bovine serum albumin [BSA], human fibrinogen, and Factor XII) were covalently coupled to AFM probes [83, 84]. Adhesion forces between protein and material surface were measured under PBS buffer using the AFM force mode.

The adhesion forces of proteins on the LDPE substrate showed a step-dependence on the wettability of surfaces. For surfaces with contact angle (θ) > ~60–65°, stronger adhesion forces were observed for all three proteins tested than for the surfaces with θ < 60°. One representative trend for fibrinogen is shown in Figure 3.9(a). Analysis of variance was performed for each of the protein–substrate combinations in order to test the statistical reliability of this observation. Poorly wettable (hydrophobic) surfaces always had statistically larger adhesive forces than highly wettable (hydrophilic) surfaces, confirming the observation of a step change in adhesive forces at/or around a water contact angle of ~60° for each protein tested. This observation is similar to the results of Sethuraman et al. [70], who reported a step in adhesion forces between protein-modified AFM probes and self-assembled

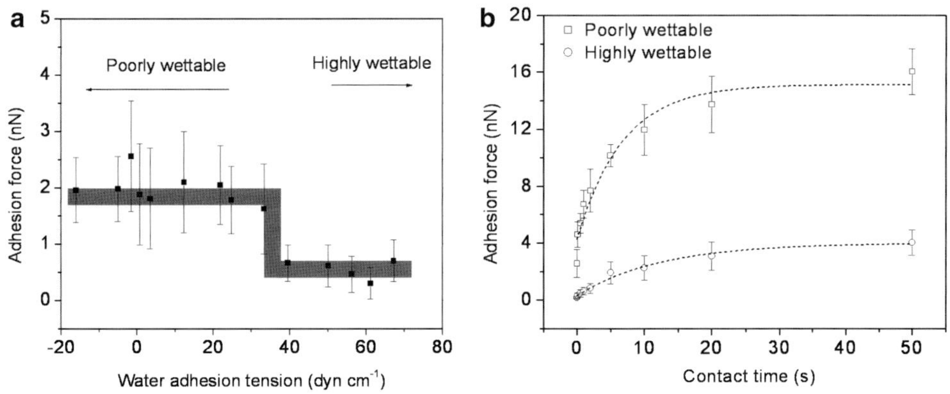

Figure 3.9. (**a**) Adhesion forces for human fibrinogen on LDPE surfaces as a function of water adhesion tension (τ); $\tau = \gamma\cos\theta$, $\gamma = 72.8$ dyn/cm for water. The *shaded area* is drawn to aid visualization of the results. (**b**) Adhesion forces between fibrinogen and material surfaces as a function of contact time. The curve is the fit of the exponential described in Eq. (3.4) to the experimental data. (Reprinted from [73]).

monolayer (SAM) surfaces. The results of these studies suggest that the range of contact angle of ~60–65° might be viewed as a criterion for distinguishing a material surface as either "protein adherent" or "protein nonadherent," consistent with other studies [66, 85]. More importantly, the presence of this type of step response in protein adhesion suggests that modifying material surfaces through subtle changes in wettability may not be an effective tool in affecting protein adhesion to surfaces unless that change yields a transition across the $\theta = 60$–65° region.

Early time points for time-dependent analysis of protein–surface interactions were investigated by bringing a protein into contact with a material surface for brief (<50 s) times and then assessing the effects of contact times on protein adhesion forces. In all cases investigated, the adhesion forces between protein and substrate increased with increasing contact time (Figure 3.9b), with smaller adhesion forces measured on highly wettable surfaces. Changes in adhesion forces with contact time were modeled by a simple exponential of the form:

$$F = F_e - F_0 \exp(-k_s t), \tag{3.4}$$

where F_e is the adhesion force at very long times (assumed to be equilibrium), F_0 is an empirical coefficient related to the initial interaction force, t is the contact time, and k_s is a rate constant. This rate constant was used to calculate an energy barrier for unfolding of the protein by using the Arrhenius equation [78]:

$$k_s = A(T)\exp(-E_a / kT), \tag{3.5}$$

where T is the absolute temperature, E_a is the activation energy for protein unfolding, and k is the Boltzmann constant. A prefactor, $A(T) = 10^7$–10^9/s was used [78]. This analysis yields 18.0–22.6 kT, and 18.6–23.2 kT for the activation energies of human fibrinogen on hydrophobic surfaces and hydrophilic surfaces, respectively [73].

3.3.3. Dynamic Force Microscopy Studies of Fibrinogen–Material Surface Interactions

AFM has been used to measure the strength of bonds between biological receptor molecules and their ligands [86–88]. The bond strengths measured under a variety of loading rates are governed by the barriers traversed in the energy landscape along the force-driven, bond dissociation pathway [89, 90]. Bell [91] was the first to suggest that the dissociation of adhesive receptor–ligand complexes is influenced by force. Using the Bell model, dissociation rates and energy barriers in biological molecule reactions can be quantified. Test systems include biotin and avidin [92, 93], P-selectin and P-selectin glycoprotein ligand 1 [94], aptamer and protein [95], and cell adhesion molecules [96, 97]. We have extended these measurements to probe protein–surface interactions and reveal the energy profiles in interactions of fibrinogen with both hydrophobic and hydrophilic surfaces. The results provide new insights into the effects of surface wettability on protein interaction with biomaterial substrates.

In investigating protein–surface interactions by AFM, protein is immobilized either on the AFM probe or on a material substrate. In the first method, the proteins on the probe undergo repeated compression and decompression cycles during multiple measurements, potentially leading to errors in the force measurement and analysis. Alternatively, proteins can be immobilized to the underlying substrate so that only one force measurement and protein perturbation is performed. Glass colloids (~4 μm in diameter) modified with SAMs to produce different surface chemistry and wettability were attached to AFM cantilevers. The "bond strengths" (in this case, adhesive forces) between proteins and surfaces were measured by AFM at different loading rates (r_f), where the loading rate (r_f) is defined as the product of the probe retraction velocity and the spring constant of the cantilever.

Figure 3.10 illustrates the trends for the average adhesive forces between fibrinogen and different model material surfaces as a function of the loading rate. A linear relationship between adhesion forces and the $\ln(r_f)$ was obtained for hydrophilic surfaces (bare glass and 3-aminopropyltrichlorosilane [APS]-coated glass) (Figures 3.10a and 3.10b), while two line segments with ascending slopes were observed on hydrophobic surfaces, (specifically, polystyrene and methyl-terminated n-butyltrichlorosilane [BTS]) (Figures 3.10c and 3.10d). The two distinct linear regions suggest that there are multiple energy barriers and multiple transition states present during interactions of fibrinogen with hydrophobic surfaces, while the single linear region suggests a single energy barrier during interactions of fibrinogen with hydrophilic surfaces. Dynamic force spectroscopy analysis was performed to fit the plots of adhesion force, F, against $\ln(r_f)$ by fitting the data to Eq. (3.6):

$$F = \frac{kT}{\chi_\beta}\ln(r_f) - \frac{kT}{\chi_\beta}\ln\left(\frac{k_{off}kT}{\chi_\beta}\right), \tag{3.6}$$

where k is the Boltzmann constant, T is the absolute temperature, k_{off} is the thermal off-rate at zero force, and χ_β is the effective distance between the bound and transition states along the direction of applied force that measures the width of the energy well that kinetically traps the interacting molecules in the bond state. The parameters, χ_β, k_{off}, and the binding lifetime $\tau\,(=1/k_{off})$, were derived and are shown in Table 3.1. These results show that the off-rate (k_{off}) for protein interactions with hydrophilic material surfaces lie in the range of 1.35–2.92/s,

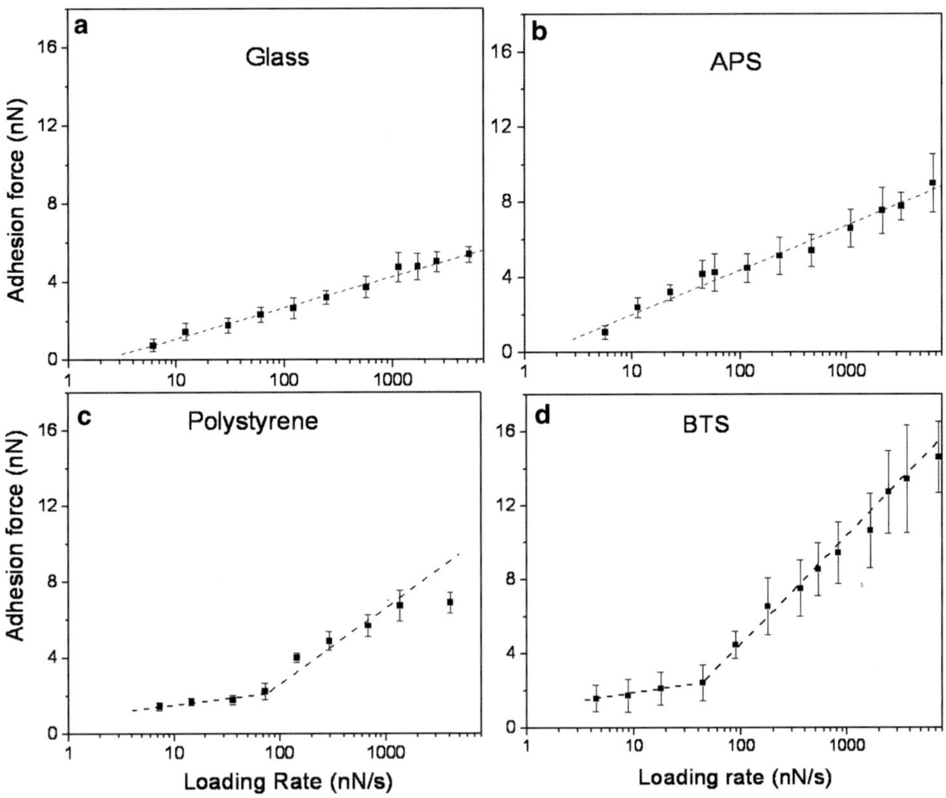

Figure 3.10. Adhesion forces of fibrinogen and colloid surfaces under different loading rates (residence time = 1 s). (**a**) Glass, (**b**) APS, (**c**) polystyrene, and (**d**) BTS. Multiple energy barriers are present between fibrinogen and hydrophobic surfaces while a single energy barrier is present between fibrinogen and hydrophilic surfaces.

Table 3.1. Bell model parameters for Eq. (3.6).

	Colloid surface	Loading rate (nN/s)	χ_β (Å)	k_{off} (s^{-1})	τ (s)
Highly wettable (hydrophilic)	Glass	6–5100	0.060	2.92	0.34
	APS	5–6500	0.040	1.35	0.74
Poorly wettable (hydrophobic)	BTS	4–100	0.11	0.19	5.12
		100–7400	0.02	5.50	0.18
	OTS	5–100	0.07	0.42	2.38
		100–3200	0.01	13.8	0.07

about three- to tenfold greater than those obtained for protein interactions with hydrophobic material surfaces at the lower loading rates. These results show that the bond lifetimes on the hydrophobic material surfaces are longer than they are on hydrophilic surfaces, and are consistent with the increased adhesion forces observed between proteins and hydrophobic material surfaces.

3.4. Recognition of Proteins on Material Surfaces by AFM

A critical step in using AFM to study protein adsorption at the molecular scale is the ability to detect specific proteins when numerous proteins simultaneously adsorb to the same material surface. Although molecular recognition of a specific protein might be feasible through topography alone on an ideally smooth surface, clinically used biomaterials have much rougher topographies; for this reason, it is difficult to detect proteins on these surfaces through standard imaging techniques, let alone to identify different proteins [98].

3.4.1. Immunological Recognition of Protein with Polyclonal Antibodies by AFM Force Mode

The ability to functionalize an AFM probe with a recognition molecule such as an antibody offers a unique method for AFM-based immunoassay detection [99–101]. This technique allows examination of multiple proteins at the molecular scale. These techniques may also be useful for studying protein–surface interactions, and with the appropriate choice of antibodies, studies of the conformation, orientation, and activity of adsorbed proteins may be an attainable goal [102].

Force-mode AFM imaging acquires both topographic and interaction force data simultaneously across the material surface. The force–distance curves can be analyzed so that the tip–surface interactions (e.g., adhesion force or rupture length) can be used to generate various interaction maps [103, 104]. We utilized techniques for adhesion mapping with a polyclonal antibody against fibrinogen coupled to the end of the AFM probe [84]. Figure 3.11 illustrates the technique using proteins patterned in a known distribution. BSA was patterned by microcontact printing on a mica surface, fibrinogen was then added to the fluid cell and allowed to adsorb into the nonpatterned areas. An AFM probe coupled with a polyclonal antibody against human fibrinogen was used to measure the topography and the forces across the substrate surface. The adhesive force measured at each point resulted in an adhesion map (Figure 3.11b) where the lighter regions having higher adhesive forces correspond closely

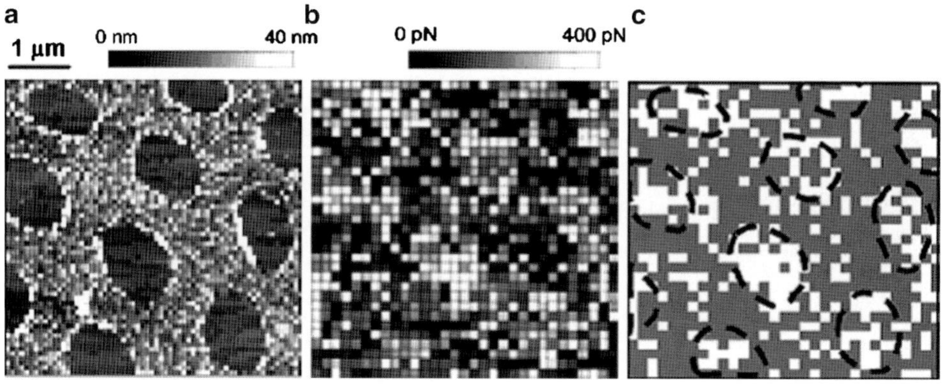

Figure 3.11. Images of microcontact printing patterned BSA on mica after backfilling with fibrinogen. (**a**) Topographic image of the pattern, (**b**) adhesion map showing the interaction strength, and (**c**) binary recognition map showing fibrinogen recognition as *white pixels* and nonspecific interactions with BSA as *black pixels*. (Reprinted from [84]).

with the holes in the height image, indicating the presence of fibrinogen. Control experiments were used to determine the nonspecific interaction forces between BSA and the antibody-modified probes so that a binary recognition image showing the distribution of specific and nonspecific interactions could be generated (Figure 3.11c).

3.4.2. Immunological Nanogold Labeling Technique

Limitations of the force-mode AFM imaging method are time requirements and relatively low resolution. Producing an adhesion map with resolution of 32 × 32 pixels as shown in Figure 3.11 may require between 30 and 60 min. This is a particularly important limitation when studying the postadsorptive transitions in fibrinogen structure and activity since the conformation of adsorbed fibrinogen is time dependent, as discussed in Sect. 3.3.1. An alternative approach is to use nanogold beads coupled to antibodies to characterize the distribution of proteins. Gold-bead labeling techniques are commonly used in electron microscopy [105, 106] and have become increasingly common in AFM studies in order to visualize small molecule binding to either individual proteins or on cells [107]. Hussain et al. [108] used nanogold beads to visualize individual ligand–receptor interactions in a physically relevant environment where the nanogold beads were conjugated to an RGD peptide. The peptide ligand bound to an integrin receptor was detected by AFM phase imaging with high resolution and by standard scanning techniques. These nanogold beads have also been used to identify proteins in dual-protein films [109]. In this case, proteins were patterned on PDMS surfaces and labeled using polyclonal antibodies against human fibrinogen that were coupled to 1.4 nm gold particles. The films were imaged by tapping mode AFM (Figure 3.12). Fibrinogen in the dual-protein layer was detected after incubation with the nanogold–antibody conjugate (Figure 3.12c).

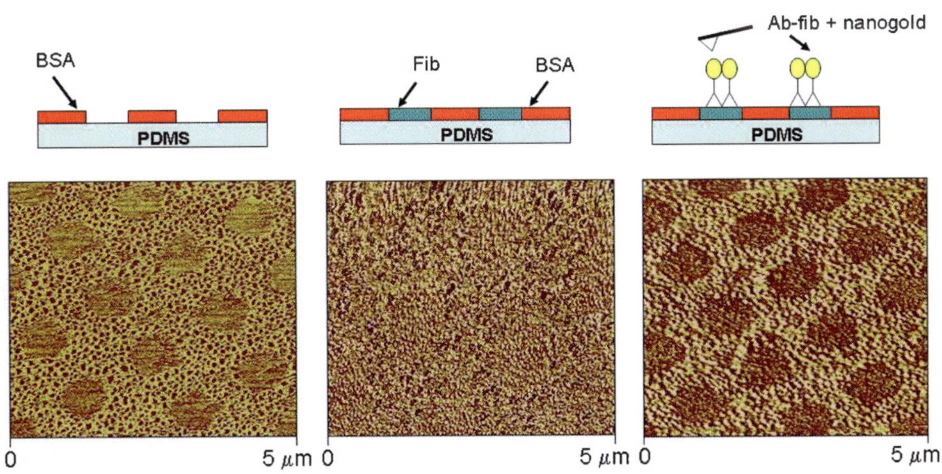

Figure 3.12. AFM tapping-mode phase images on PDMS substrates with (**a**) BSA pattern microcontact printed onto substrates, (**b**) after human fibrinogen (*Fib*) adsorption (1 mg/mL). Note that the microcontact printed pattern is not visible after backfilling of the holes. (**c**) Following nanogold–antibody incubation. *Dark regions* indicate the location of the conjugated nanogold binding to fibrinogen while the *bright regions* indicate lack of binding to BSA [109].

This AFM immunodetection technique is potentially applicable to complex multicomponent protein films adsorbed on clinically relevant polymers used in medical devices. The advantages of the nanogold labeling techniques are that images can be obtained at higher resolution and more rapidly than those obtained using force measurements.

3.5. Measuring the Functional Activity of Adsorbed Fibrinogen

Fibrinogen possesses three pairs of potential platelet-binding peptide sequences: two RGD sequences in each of the Aα chains and a dodecapeptide sequence in each of the γ chains. There is ample evidence that the γ-chain dodecapeptide sequence is the primary ligand for platelet binding to adsorbed fibrinogen [30, 31, 110]. Many investigators have demonstrated that the functional activity and availability of these dodecapeptide sequences in fibrinogen are related to the conformation and/or orientation of adsorbed fibrinogen. Furthermore, fibrinogen conformation and activity are time-dependent processes [72, 76, 111].

Monoclonal antibodies (mAbs) have been used to study the protein states and bioactivity in conjunction with platelet adhesion studies [112, 113]. Tunc et al. [114] used a mAb clone to determine the conformation changes in the carboxyl-terminal region of the γ chain of fibrinogen. Balasubramanian et al. [76] studied the effects of fibrinogen residence time on various biomaterials using a mAb directed against the C-terminal dodecapeptide of the γ chain (γ400–411).

The force mode of AFM provides an approach to study the fibrinogen activity (specifically, the availability of the dodecapeptide sequence in fibrinogen) as recognized by mAbs as a function of residence time [115]. A mAb that recognizes the last 20 amino acids of the γ chain of fibrinogen (γ392–411) was coupled to the AFM probe and was used to measure the interaction forces between mAb and adsorbed fibrinogen. Figure 3.13(a) illustrates the probability of a mAb against fibrinogen γ392–411 encountering the antigen in the platelet-binding region as a function of fibrinogen residence time after adsorption to a hydrophilic muscovite mice surface. The maximum likelihood of recognition occurred near 45 min, but decreased at longer adsorption times.

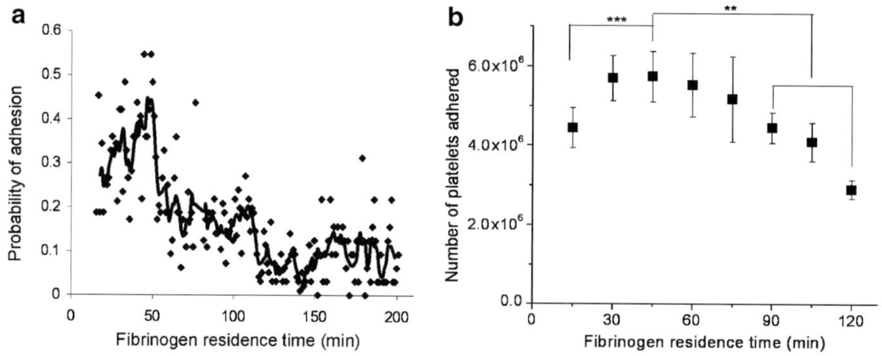

Figure 3.13. (**a**) Probability of interaction between a mAb against γ-chain 392–411 and its antigen on fibrinogen measured by AFM. The *solid line*, the average of 5 points, is included for visualization purposes. (**b**) Platelet adhesion data from multiple experiments ($n \geq 6$ for each time point) showing changes in platelet adhesion as a function of fibrinogen residence time on mica substrates. Figure 3.13(b) is reprinted from reference [115].

Time-dependent macroscale platelet adhesion measurements on the same muscovite mica substrates were carried out using a lactate dehydrogenase assay [116, 117]. The results illustrated in Figure 3.13(b) show that platelet adhesion reached a peak at ~45 min of fibrinogen residence time; this result correlated well with the molecular scale AFM results. Thus, the methodology described in this section combined with the high resolution of AFM offers a unique opportunity to assess the biological activity of single adsorbed protein molecules.

3.6. Measuring Protein Adsorption on PU Surfaces at the Molecular Scale

3.6.1. Microphase Separation Structure Affects Protein Adsorption

The microphase structure in PU biomaterials creates a surface microenvironment where chemical composition and physicochemical properties are distributed spatially at the molecular scale. Most studies of the role of PU and PUU microphase-separated structures on biological responses investigated the bulk properties of these materials, which may be substantially different from those of the surface microphase structures [118, 119]. AFM tapping mode with phase detection illustrates distribution of hard and soft domains at the surface of PU biomaterials [63], and provides an approach to simultaneously characterize the nature of adsorbed proteins on the material surface and the phase/chemical structure of the underlying substrate; for this reason, the relationships between material surface properties and biological response can be directly measured.

Direct visualization of single proteins adsorbed on PU surfaces is difficult due to the rough topography of these materials. Heights of the features on these materials are generally much larger than the size of the proteins of interest. An alternative approach to detect single-protein adsorption on these rough biomaterial surfaces is to add a label to the proteins [120]. Nanogold particles are suitable for labeling biological samples with the advantages of stability, being available in well-defined sizes, and being detectable by AFM. A commercial product consisting of BSA conjugated to 6-nm-diameter nanogold beads was used to follow protein adsorption onto PUU films. Figure 3.14 illustrates the height and phase images of a PUU film under aqueous condition following BSA–nanogold adsorption. The protein-conjugated nanogold beads are detectable on both soft and hard segment regions; quantitative measurements of the labeled proteins showed that there were 131 ± 36 proteins/μm^2 on the soft segment regions and 61 ± 20 proteins/μm^2 on the hard segment regions of PUU, suggesting that these proteins are more likely to adsorb to the more hydrophobic soft segment than to the more hydrophilic, polar hard segments of the material substrate.

As AFM probes can often be used for multiple imaging modes under aqueous conditions, different maps of tip–surface interactions can be obtained at the same location, allowing for direct correlation of protein interactions with microphase structures. A BSA-modified AFM probe was used to measure the adhesion force and local modulus of PUU films. Once the force data were collected, a phase image was recorded at the same location. Figure 3.15 displays the phase image, adhesion force map, and modulus map (determined from the compression region of the force curves) for a hydrated PUU film. The adhesion map illustrates the distribution of the measured adhesion forces between the BSA-modified probe and the PUU surface, where lighter regions represent large forces and darker regions indicate smaller adhesion forces. Note the strong correlation between the hard domains in the phase image (Figure 3.15a) and the low adhesion forces in the maps (Figure 3.15b). The white lines

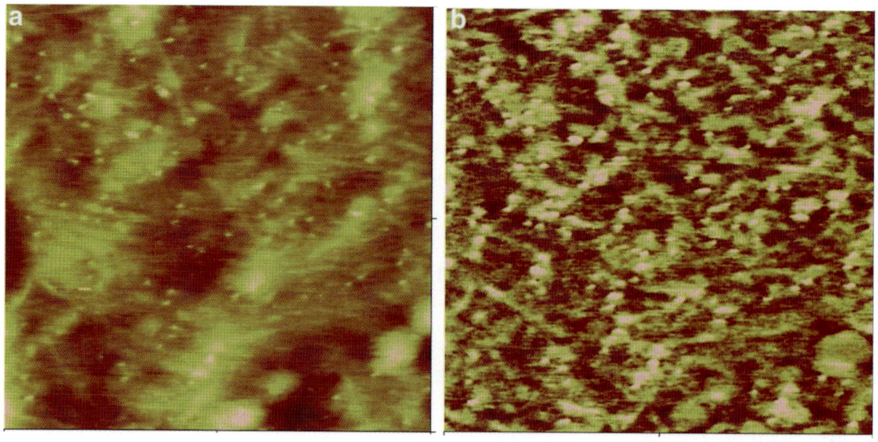

Figure 3.14. AFM images of BSA conjugated to 6-nm nanogold labels adsorbed on PUU and imaged under PBS buffer. Gold beads are seen in both height (**a**) and phase (**b**) images. Quantification revealed more than twofold increase in the number of labeled proteins per unit area on the soft segment regions compared with those obtained on the hard segment. Scale: height image, 50 nm; phase image 20°. Scan size: 500 × 500 nm^2.

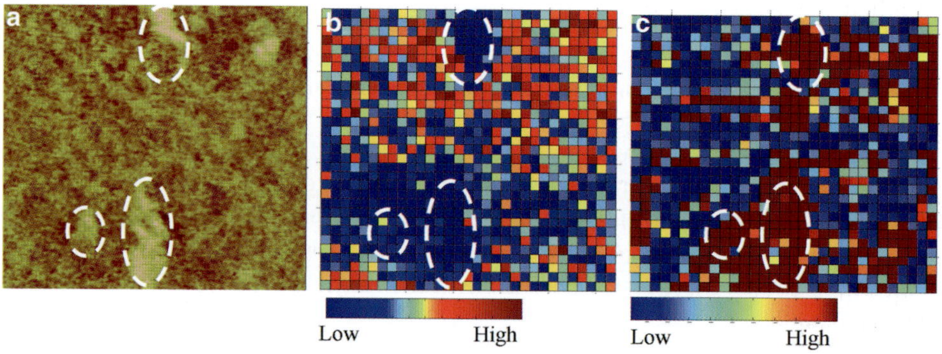

Figure 3.15. (**a**) AFM phase image, (**b**) adhesion map, and (**c**) modulus map of hydrated PUU obtained with a BSA-modified probe. Scan size: 500 × 500 nm^2.

denote the hard domain regions that were identified on the phase image and then superimposed on the adhesion map; these regions overlay predominantly onto low adhesion force areas. A secondary analysis using the modulus map generated from the compression region of the curves showed a similar distribution of hard and soft regions on the phase image and corroborated the correlation between hard segment domains and low adhesion forces (Figure 3.15c). Taken together, these results provide direct evidence of decreased protein adhesion forces to the hard segment regions in PUU.

3.6.2. Soft Segment Chemistry and Hard Segment Content of PUs Affect Fibrinogen Adsorption/Bioactivity and Platelet Adhesion

A variety of PUs have been developed and synthesized for meeting multiple requirements in medical device applications. Principle candidate PU biomaterials generally consist of similar hard segments, but have different soft segment chemistries, including polyether, aliphatic PC, and PDMS. These different soft segments produce different surface chemistries and affect material properties. The conventional polyether soft segments are sensitive to oxidative degradation [121]; poly(carbonate urethane) has superior oxidative biostability but is prone to enzymatic hydrolytic degradation [122]. The more recently developed PDMS-based PUs have good oxidative stability and good hemocompatibility [123, 124]. Although bulk studies have been undertaken on the phase separation structure and biocompatibility of these PUs, little is known about how the soft segment chemistry and microphase separation affects protein adsorption and platelet adhesion, particularly at the molecular scale.

A series of PU biomaterials with PC, PTMO, and PDMS soft segments were used to address the links between heterogeneous surface chemistries and fibrinogen adsorption/bioactivity. Blood platelet adhesion was measured so that the correlations between surface chemistry and biocompatibility could be addressed. Note that since pure PDMS-based PUs result in materials with poor mechanical properties, a second soft segment, poly(hexamethylene oxide), is generally added at up to 20% in order to improve the mechanical properties of the material. Addition of the second soft segment results in strong phase separation and formation of three phases characterized as soft, intermediate, and hard [56].

Protein adsorption was carried out by immersion of polymers in either mixed protein solutions consisting of BSA (40 mg/mL) and fibrinogen (5 mg/mL), or alternatively in plasma solutions. Fibrinogen adsorption was analyzed by an AFM probe modified with polyclonal antifibrinogen antibody. For the PDMS-based PUs, the amount of adsorbed fibrinogen recognized within a unit area generally increased with hard segment content except for PDMS-PU32.5 (Figure 3.16a). This trend was roughly consistent with platelet adhesion results measured by fluorescent microscopy (Figure 3.16c). However, larger amounts of fibrinogen were measured on the PC-PU40 and PTMO-PU40 surfaces despite the fact that fewer platelets adhered on these materials. These inconsistent trends between platelet adhesion and fibrinogen adsorption suggest the importance of fibrinogen bioactivity, specifically availability of appropriate platelet-binding sites. The functional activity of fibrinogen adsorbed on polymer surfaces was detected by coupling a mAb against fibrinogen γ392–411 to AFM probes. Although more fibrinogen was again detected by the polyclonal antibody on the PC-PU40 and PTMO-PU40 surfaces, fewer specific binding events were observed between the mAb and the proteins adsorbed on these materials (Figure 3.16b). Generally, the activity of fibrinogen (measured by the mAb) was consistent with platelet adhesion on the three types of polymers tested. These results suggest a correlation between platelet adhesion and availability of platelet-binding sites in the fibrinogen γ chain.

3.7. Summary

The microphase-separated structure of segmented PUs creates a unique surface microenvironment that is important in mediating protein adsorption, platelet adhesion, and thrombus formation on these materials. New and advanced AFM methods for the molecular-scale study of protein–surface interactions are important tools in understanding platelet adhesion and blood coagulation on biomaterial surfaces.

AFM is becoming an increasingly powerful tool for the characterization of biomaterials, including the study of microphase structure of block copolymers. Additionally, these techniques have led to advancements in the measurement of protein–surface interactions at

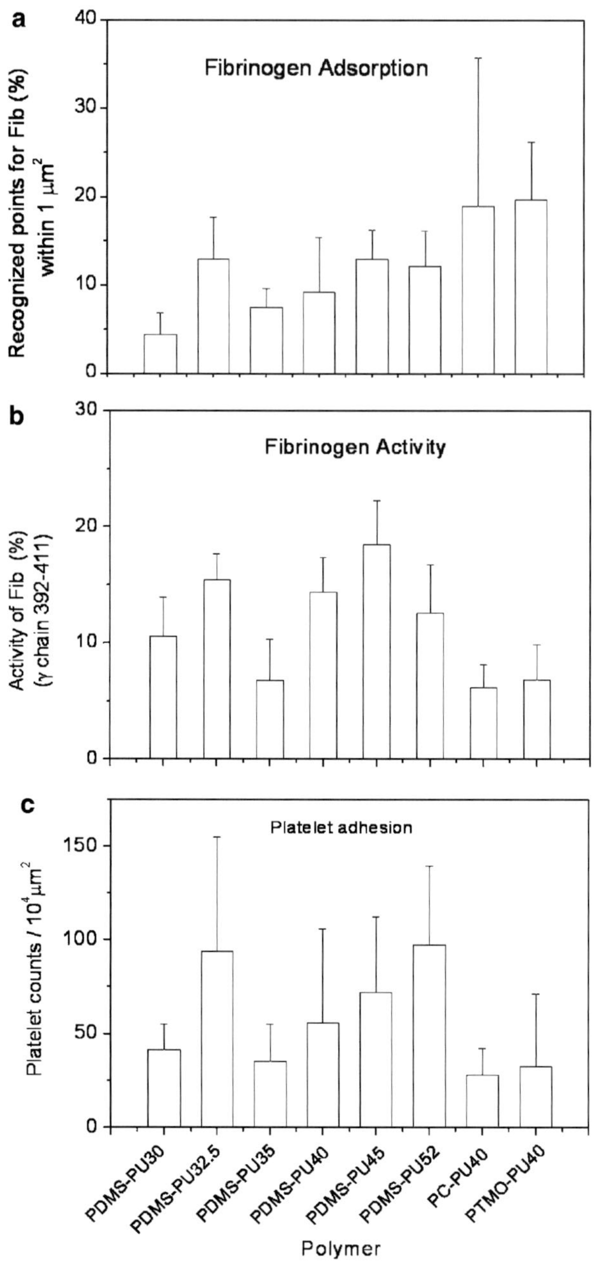

Figure 3.16. (**a**) Fibrinogen adsorbed on PUs as recognized by a polyclonal antibody; (**b**) adsorbed fibrinogen detected by a mAb against the fibrinogen γ-chain 392–411; (**c**) platelets adherent to PDMS-PU, PC-PU, and PTMO-PU surfaces. *Note*: The number following the type of PU is the percentage hard segment content.

the molecular scale. Combinations of these new techniques with traditional in vitro techniques for determining the biocompatibility of materials offer new insights in the fundamental relationships between PU microphase structures and biological responses.

Acknowledgments

The authors would like to thank Dr. James Runt, Dr. Jadwiga Weksler, and Dr. Ajay Padsalgikar for providing the PU biomaterials used in these studies. The authors would also like to thank Zachary Rice and Dr. James Garrett for technical contributions to the work.

References

1. Montdargent, B. and D. Letourneur, *Toward new biomaterials*. Infection Control and Hospital Epidemiology, 2000. 21(6):404–410.
2. Horbett, T.A., *Principles underlying the role of adsorbed plasma-proteins in blood interactions with foreign materials*. Cardiovascular Pathology, 1993. 2(3):S137–S148.
3. Gorbet, M.B. and M.V. Sefton, *Biomaterial-associated thrombosis: Roles of coagulation factors, complement, platelets and leukocytes*. Biomaterials, 2004. 25(26):5681–5703.
4. Bajpai, A.K., *Adsorption of fibrinogen onto macroporous, biocompatible sponges based on poly(2-hydroxyethyl methacrylate)*. Journal of Applied Polymer Science, 2006. 102(2):1341–1355.
5. Gray, J.J., *The interaction of proteins with solid surfaces*. Current Opinion in Structural Biology, 2004. 14(1): 110–115.
6. Hsu, S.H. and Y.C. Kao, *Biocompatibility of poly(carbonate urethane)s with various degrees of nanophase separation*. Macromolecular Bioscience, 2005. 5(3):246–253.
7. Castner, D.G. and B.D. Ratner, *Biomedical surface science: Foundations to frontiers*. Surface Science, 2002. 500(1–3):28–60.
8. Anderson, J.M., *Biological responses to materials*. Annual Review of Materials Research, 2001. 31:81–110.
9. Wang, Y.X., et al., *Effects of the chemical structure and the surface properties of polymeric biomaterials on their biocompatibility*. Pharmaceutical Research, 2004. 21(8):1362–1373.
10. Norde, W. and J. Lyklema, *Why proteins prefer interferences*. Journal of Biomaterials Science. Polymer Edition, 1991. 2(3):183–202.
11. Malmsten, M., *Formation of adsorbed protein layers*. Journal of Colloid and Interface Science, 1998. 207(2): 186–199.
12. Thevenot, P., W.J. Hu, and L.P. Tang, *Surface chemistry influences implant biocompatibility*. Current Topics in Medicinal Chemistry, 2008. 8(4):270–280.
13. Heynes, C.A. and W. Norde, *Globular proteins at solid/liquid interfaces*. Colloids and Surfaces B-Biointerfaces, 1994. 2:517–566.
14. Israelachvili, J. and H. Wennerstrom, *Role of hydration and water structure in biological and colloidal interactions*. Nature, 1996. 379(6562):219–225.
15. Kidoaki, S. and T. Matsuda, *Mechanistic aspects of protein/material interactions probed by atomic force microscopy*. Colloids and Surfaces B-Biointerfaces, 2002. 23(2–3):153–163.
16. Xu, L.C. and B.E. Logan, *Interaction forces between colloids and protein-coated surfaces measured using an atomic force microscope*. Environmental Science & Technology, 2005. 39(10):3592–3600.
17. Lubarsky, G.V., et al., *The influence of electrostatic forces on protein adsorption*. Colloids and Surfaces B-Biointerfaces, 2005. 44(1):56–63.
18. Vogler, E.A., *Structure and reactivity of water at biomaterial surfaces*. Advances in Colloid and Interface Science, 1998. 74:69–117.
19. Noh, H., *Volumetric interpretation of protein adsorption: Mass and energy balance for albumin adsorption to particulate adsorbents with incrementally increasing hydrophilicity*. Biomaterials, 2006. 27(34):5801–5812.
20. Noh, H. and E.A. Vogler, *Volumetric interpretation of protein adsorption: Partition coefficients, interphase volumes, and free energies of adsorption to hydrophobic surfaces*. Biomaterials, 2006. 27(34):5780–5793.
21. Amiji, M.M., K.R. Kamath, and K. Park, *Albumin-modified biomaterial surfaces for reduced thrombogenicity*, in *Encyclopedic Handbook of Biomaterials and Bioengineering*, D.L. Wise, Editor. 1995, New York: CRC.
22. Hantgan, R.R., C.W. Francis, and V.J. Mardner, *Fibrinogen structure and physiology*, in *Hemostasis and Thrombosis: Basic Principles and Clinical Practice*, R.W. Colman, et-al., Editors. 1994, Philadelphia:J.B. Lippincott, pp. 277–300.
23. Grunkemeier, J.M., *Platelet adhesion and procoagulant activity induced by contact with radiofrequency glow discharge polymers: Roles of adsorbed fibrinogen and vWF*. Journal of Biomedical Materials Research, 2000. 51(4):669–679.

24. Goodman, S.L., S.L. Cooper, and R.M. Albrecht, *Integrin receptors and platelet-adhesion to synthetic surfaces*. Journal of Biomedical Materials Research, 1993. 27(5):683–695.
25. Kerrigan, S.W., et al., *Molecular basis for* Staphylococcus aureus-*mediated platelet aggregate formation under arterial shear in vitro*. Arteriosclerosis Thrombosis and Vascular Biology, 2008. 28:335–340.
26. Walsh, E.J., et al., *Identification of the* Staphylococcus aureus *MSCRAMM clumping factor B (ClfB) binding site in the alpha C-domain of human fibrinogen*. Microbiology, 2008. 154:550–558.
27. Mitchell, J., A. Tristan, and T.J. Foster, *Characterization of the fibrinogen-binding surface protein Fbl of* Staphylococcus lugdunensis. Microbiology, 2004. 150:3831–3841.
28. Marchant, R.E., et al., Three dimensional structure of fibrinogen visualized under aqueous conditions. Thrombosis and Haemostasis, 1997:P3098.
29. Marchant, R.E., et al., *Three dimensional structure of human fibrinogen under aqueous conditions visualized by atomic force microscopy*. Thrombosis and Haemostasis, 1997. 77(6):1048–1051.
30. Farrell, D.H. and P. Thiagarajan, *Binding of recombinant fibrinogen mutants to platelets*. Journal of Biological Chemistry, 1994. 269(1):226–231.
31. Farrell, D.H., et al., *Role of fibrinogen alpha-chain and gamma-chain sites in platelet aggregation*. Proceedings of the National Academy of Sciences of the United States of America, 1992. 89(22):10729–10732.
32. Wu, Y.G., et al., *The role of adsorbed fibrinogen in platelet adhesion to polyurethane surfaces: A comparison of surface hydrophobicity, protein adsorption, monoclonal antibody binding, and platelet adhesion*. Journal of Biomedical Materials Research Part A, 2005. 74A(4):722–738.
33. Remijn, J.A., et al., *Reduced platelet adhesion in flowing blood to fibrinogen by alterations in segment gamma 316–322, part of the fibrin-specific region*. British Journal of Haematology, 2002. 117(3):650–657.
34. Agashe, M., et al., *Molecular simulation to characterize the adsorption behavior of a fibrinogen gamma-chain fragment*. Langmuir, 2005. 21(3):1103–1117.
35. Lamba, N.M.K., Woodhouse, K.A., Cooper S.L., *Polyurethanes in Biomedical Applications*. 1998, Boca Raton, FL:CRC.
36. Poksinski, M. and H. Arwin, *Total internal reflection ellipsometry: Ultrahigh sensitivity for protein adsorption on metal surfaces*. Optics Letters, 2007. 32(10):1308–1310.
37. Seitz, R., R. Brings, and R. Geiger, *Protein adsorption on solid-liquid interfaces monitored by laser-ellipsometry*. Applied Surface Science, 2005. 252(1):154–157.
38. Green, R.J., et al., *Surface plasmon resonance for real time in situ analysis of protein adsorption to polymer surfaces*. Biomaterials, 1997. 18(5):405–413.
39. Green, R.J., et al., *Competitive protein adsorption as observed by surface plasmon resonance*. Biomaterials, 1999. 20(4):385–391.
40. Welle, A., *Competitive plasma protein adsorption on modified polymer surfaces monitored by quartz crystal microbalance technique*. Journal of Biomaterials Science. Polymer Edition, 2004. 15(3):357–370.
41. Nonckreman, C.J., P.G. Rouxhet, and C.C. Dupont-Gillain, *Dual radiolabeling to study protein adsorption competition in relation with hemocompatibility*. Journal of Biomedical Materials Research Part A, 2007. 81A(4):791–802.
42. Hylton, D.M., S.W. Shalaby, and R.A. Latour, *Direct correlation between adsorption-induced changes in protein structure and platelet adhesion*. Journal of Biomedical Materials Research Part A, 2005. 73A(3):349–358.
43. Yokoyama, Y., et al., *Quantitative analysis of protein adsorption on a planar surface by Fourier transform infrared spectroscopy: Lysozyme adsorbed on hydrophobic silicon-containing polymer*. Journal of Colloid and Interface Science, 2003. 268(1):23–32.
44. Morin, C., et al., *Selective adsorption of protein on polymer surfaces studied by soft X-ray photoemission electron microscopy*. Journal of Electron Spectroscopy and Related Phenomena, 2004. 137:785–794.
45. Li, L., et al., *X-ray microscopy studies of protein adsorption on a phase-segregated polystyrene/polymethyl methacrylate surface*. 1. Concentration and exposure-time dependence for albumin adsorption. The Journal of Physical Chemistry. B, 2006. 110(33):16763–16773.
46. Tidwell, C.D., et al., *Static time-of-flight secondary ion mass spectrometry and x-ray photoelectron spectroscopy characterization of adsorbed albumin and fibronectin films*. Surface and Interface Analysis, 2001. 31(8):724–733.
47. Wagner, M.S., T.A. Horbett, and D.G. Castner, *Characterization of the structure of binary and ternary adsorbed protein films using electron spectroscopy for chemical analysis, time-of-flight secondary ion mass spectrometry, and radiolabeling*. Langmuir, 2003. 19(5):1708–1715.
48. Gettens, R.T.T., Z.J. Bai, and J.L. Gilbert, *Quantification of the kinetics and thermodynamics of protein adsorption using atomic force microscopy*. Journal of Biomedical Materials Research Part A, 2005. 72A(3): 246–257.

49. Muguruma, H., S. Miura, and N. Murata, *Adsorption of antibody protein onto plasma-polymerized film characterized by atomic force microscopy and quartz crystal microbalance*. IEICE Transactions on Electronics, 2007. E90C(3):649–651.
50. Schon, P., et al., *Nonspecific protein adsorption at the single molecule level studied by atomic force microscopy*. Langmuir, 2007. 23:9921–9923.
51. Nakamae, K., et al., *Microphase separation and surface properties of segmented polyurethane – Effect of hard segment content*. International Journal of Adhesion and Adhesives, 1996. 16(4):233–239.
52. Gisselfalt, K. and B. Helgee, *Effect of soft segment length and chain extender structure on phase separation and morphology in poly(urethane urea)s*. Macromolecular Materials and Engineering, 2003. 288(3):265–271.
53. Garrett, J.T., J.S. Lin, and J. Runt, *Influence of preparation conditions on microdomain formation in poly(urethane urea) block copolymers*. Macromolecules, 2002. 35(1):161–168.
54. Garrett, J.T., J. Runt, and J.S. Lin, *Microphase separation of segmented poly(urethane urea) block copolymers*. Macromolecules, 2000. 33(17):6353–6359.
55. Garrett, J.T., et al., *Phase separation of diamine chain-extended poly(urethane) copolymers: FTIR spectroscopy and phase transitions*. Polymer, 2003. 44(9):2711–2719.
56. Hernandez, R., et al., *Microstructural organization of three-phase polydimethylsiloxane-based segmented polyurethanes*. Macromolecules, 2007. 40(15):5441–5449.
57. Tingey, K.G. and J.D. Andrade, *Probing surface microheterogeneity of poly(Ether Urethanes) in an aqueous environment*. Langmuir, 1991. 7(11):2471–2478.
58. Lewis, K.B. and B.D. Ratner, *Observation of surface rearrangement of polymers using ESCA*. Journal of Colloid and Interface Science, 1993. 159(1):77–85.
59. Agnihotri, A., et al., *Atomic force microscopy visualization of poly(urethane urea) microphase rearrangements under aqueous environment*. Journal of Biomaterials Science. Polymer Edition, 2006. 17(1–2):227–238.
60. McLean, R.S. and B.B. Sauer, *Tapping-mode AFM studies using phase detection for resolution of nanophases in segmented polyurethanes and other block copolymers*. Macromolecules, 1997. 30(26):8314–8317.
61. Sheth, J.P., et al., *Probing the hard segment phase connectivity and percolation in model segmented poly(urethane urea) copolymers*. Macromolecules, 2005. 38(13):5681–5685.
62. Garrett, J.T., C.A. Siedlecki, and J. Runt, *Microdomain morphology of poly(urethane urea) multiblock copolymers*. Macromolecules, 2001. 34(20):7066–7070.
63. Xu, L.C., et al., *Characterization of surface microphase structures of poly(urethane urea) biomaterials by nanoscale indentation with AFM*. Journal of Biomaterials Science. Polymer Edition, 2007. 18(4):353–368.
64. Chizhik, S.A., et al., *Micromechanical properties of elastic polymeric materials as probed by scanning force microscopy*. Langmuir, 1998. 14(10):2606–2609.
65. Hansen, J.C., et al., *Effect of surface nanoscale topography on elastic modulus of individual osteoblastic cells as determined by atomic force microscopy*. Journal of Biomechanics, 2007. 40:2865–2871.
66. Vogler, E.A., *Water and the acute biological response to surfaces*. Journal of Biomaterials Science. Polymer Edition, 1999. 10(10):1015–1045.
67. Choee, J.H., et al., *Proliferation rate of fibroblast cells on polyethylene surfaces with wettability gradient*. Journal of Applied Polymer Science, 2004. 92(1):599–606.
68. Lee, J.H. and H.B. Lee, *Platelet adhesion onto wettability gradient surfaces in the absence and presence of plasma proteins*. Journal of Biomedical Materials Research, 1998. 41(2):304–311.
69. Faucheux, N., et al., *Self-assembled monolayers with different terminating groups as model substrates for cell adhesion studies*. Biomaterials, 2004. 25(14):2721–2730.
70. Sethuraman, A., et al., *Effect of surface wettability on the adhesion of proteins*. Langmuir, 2004. 20(18):7779–7788.
71. Dupont-Gillain, C.C., et al., *Use of AFM to probe the adsorption strength and time-dependent changes of albumin on self-assembled monolayers*. Journal of Biomedical Materials Research Part A, 2003. 67A(2):548–558.
72. Agnihotri, A. and C.A. Siedlecki, *Time-dependent conformational changes in fibrinogen measured by atomic force microscopy*. Langmuir, 2004. 20(20):8846–8852.
73. Xu, L.C. and C.A. Siedlecki, *Effects of surface wettability and contact time on protein adhesion to biomaterial surfaces*. Biomaterials, 2007. 28(22):3273–3283.
74. Weisel, J.W., et al., *A model for fibrinogen-domains and sequence*. Science, 1985. 230(4732):1388–1391.
75. Sit, P.S. and R.E. Marchant, *Surface-dependent conformations of human fibrinogen observed by atomic force microscopy under aqueous conditions*. Thrombosis and Haemostasis, 1999. 82(3):1053–1060.
76. Balasubramanian, V., et al., *Residence-time dependent changes in fibrinogen adsorbed to polymeric biomaterials*. Journal of Biomedical Materials Research, 1999. 44(3):253–260.
77. Balasubramanian, V. and S.M. Slack, *Effects of fibrinogen residence time and shear rate on the morphology and procoagulant activity of human platelets adherent to polymeric biomaterials*. ASAIO Journal, 2001. 47(4):354–360.

78. Santore, M.M. and C.F. Wertz, *Protein spreading kinetics at liquid-solid interfaces via an adsorption probe method*. Langmuir, 2005. 21(22):10172–10178.
79. Slack, S.M. and T.A. Horbett, *Changes in fibrinogen adsorbed to segmented polyurethanes and hydroxyethylmethacrylate-ethylmethacrylate copolymers*. Journal of Biomedical Materials Research, 1992. 26(12):1633–1649.
80. Wertz, C.F. and M.M. Santore, *Fibrinogen adsorption on hydrophilic and hydrophobic surfaces: Geometrical and energetic aspects of interfacial relaxations*. Langmuir, 2002. 18(3):706–715.
81. Wertz, C.F. and M.M. Santore, *Adsorption and relaxation kinetics of albumin and fibrinogen on hydrophobic surfaces: Single-species and competitive behavior*. Langmuir, 1999. 15(26):8884–8894.
82. Wertz, C.F. and M.M. Santore, *Effect of surface hydrophobicity on adsorption and relaxation kinetics of albumin and fibrinogen: Single-species and competitive behavior*. Langmuir, 2001. 17(10):3006–3016.
83. Chowdhury, P.B. and P.F. Luckham, *Probing recognition process between an antibody and an antigen using atomic force microscopy*. Colloids and Surfaces. A, Physicochemical and Engineering Aspects, 1998. 143(1):53–57.
84. Agnihotri, A. and C.A. Siedlecki, *Adhesion mode atomic force microscopy study of dual component protein films*. Ultramicroscopy, 2005. 102(4):257–268.
85. Yoon, R.H., D.H. Flinn, and Y.I. Rabinovich, *Hydrophobic interactions between dissimilar surfaces*. Journal of Colloid and Interface Science, 1997. 185(2):363–370.
86. Florin, E.L., V.T. Moy, and H.E. Gaub, *Adhesion forces between individual ligand-receptor pairs*. Science, 1994. 264(5157):415–417.
87. Moy, V.T., E.L. Florin, and H.E. Gaub, *Intermolecular forces and energies between ligands and receptors*. Science, 1994. 266(5183):257–259.
88. Hinterdorfer, P., et al., *Detection and localization of individual antibody-antigen recognition events by atomic force microscopy*. Proceedings of the National Academy of Sciences of the United States of America, 1996. 93(8):3477–3481.
89. Merkel, R., et al., *Energy landscapes of receptor-ligand bonds explored with dynamic force spectroscopy*. Nature, 1999. 397(6714):50–53.
90. Evans, E. and K. Ritchie, *Dynamic strength of molecular adhesion bonds*. Biophysical Journal, 1997. 72(4):1541–1555.
91. Bell, G.I., *Models for specific adhesion of cells to cells*. Science, 1978. 200(4342):618–627.
92. Lo, Y.S., Y.J. Zhu, and T.P. Beebe, *Loading-rate dependence of individual ligand-receptor bond-rupture forces studied by atomic force microscopy*. Langmuir, 2001. 17(12):3741–3748.
93. Izrailev, S., et al., *Molecular dynamics study of unbinding of the avidin-biotin complex*. Biophysical Journal, 1997. 72(4):1568–1581.
94. Lu, S.Q., et al., *Quantifying the effects of contact duration, loading rate, and approach velocity on P-selectin-PSGL-1 interactions using AFM*. Polymer, 2006. 47(7):2539–2547.
95. Yu, J.P., et al., *Energy landscape of aptamer/protein complexes studied by single-molecule force spectroscopy*. Chemistry, an Asian Journal, 2007. 2(2):284–289.
96. Marshall, B.T., et al., *Direct observation of catch bonds involving cell-adhesion molecules*. Nature, 2003. 423(6936):190–193.
97. Zhu, C., *Kinetics and mechanics of cell adhesion*. Journal of Biomechanics, 2000. 33(1):23–33.
98. Holland, N.B. and R.E. Marchant, *Individual plasma proteins detected on rough biomaterials by phase imaging AFM*. Journal of Biomedical Materials Research, 2000. 51(3):307–315.
99. Barattin, R., and N. Voyer, *Chemical modifications of AFM tips for the study of molecular recognition events*. Chemical Communications, 2008. (13):1513–1532.
100. Dufrene, Y.F. and P. Hinterdorfer, *Recent progress in AFM molecular recognition studies*. Pflugers Archiv:European Journal of Physiology, 2008. 456(1):237–245.
101. Dupres, V., C. Verbelen, and Y.F. Dufrene, *Probing molecular recognition sites on biosurfaces using AFM*. Biomaterials, 2007. 28(15):2393–2402.
102. Fotiadis, D., et al., *Imaging and manipulation of biological structures with the AFM*. Micron, 2002. 33(4):385–397.
103. Ludwig, M., W. Dettmann, and H.E. Gaub, *Atomic force microscope imaging contrast based on molecular recognition*. Biophysical Journal, 1997. 72(1):445–448.
104. Willemsen, O.H., et al., *Simultaneous height and adhesion imaging of antibody-antigen interactions by atomic force microscopy*. Biophysical Journal, 1998. 75(5):2220–2228.
105. Powell, R.D., C.M.R. Halsey, and J.F. Hainfeld, *Combined fluorescent and gold immunoprobes:Reagents and methods for correlative light and electron microscopy*. Microscopy Research and Technique, 1998. 42(1):2–12.

106. Montesano-Roditis, L., et al., *Cryo-electron microscopic localization of protein L7/L12 within the Escherichia coli 70 S ribosome by difference mapping and Nanogold labeling*. Journal of Biological Chemistry, 2001. 276(17):14117–14123.
107. Wiechmann, M., et al., *Nanoscale lines of supported nanogold particles and lysozyme-nanogold conjugates generated by atomic force microscopy in aqueous solution*. Surface and Interface Analysis, 2006. 38(6): 1004–1009.
108. Hussain, M.A., A. Agnihotri, and C.A. Siedlecki, *AFM imaging of ligand binding to platelet integrin alpha(IIb)beta(3) receptors reconstituted into planar lipid bilayers*. Langmuir, 2005. 21(15):6979–6986.
109. Soman, P., Z. Rice, and C.A. Siedlecki, *Immunological identification of fibrinogen in dual-component protein films by AFM imaging*. Micron, 2008. 39(7):832–842.
110. Tsai, W.B., *Variations in the ability of adsorbed fibrinogen to mediate platelet adhesion to polystyrene-based materials: A multivariate statistical analysis of antibody binding to the platelet binding sites of fibrinogen*. Journal of Biomedical Materials Research Part A, 2003. 67A(4):1255–1268.
111. Chiumiento, A., S. Lamponi, and R. Barbucci, *Role of fibrinogen conformation in platelet activation*. Biomacromolecules, 2007. 8(2):523–531.
112. Shiba, E., et al., *Antibody-detectable changes in fibrinogen adsorption affecting platelet activation on polymer surfaces*. American Journal of Physiology, 1991. 260(5):C965–C974.
113. Shiba, E., et al., *Detection by monoclonal antibodies of conformational change in fibrinogen adsorbed on artificial surfaces*. Thrombosis and Haemostasis, 1987. 58(1):33–33.
114. Tunc, S., et al., *In situ conformational analysis of fibrinogen adsorbed on Si surfaces*. Colloids and Surfaces B-Biointerfaces, 2005. 42(3–4):219–225.
115. Soman, P., Z. Rice, and C.A. Siedlecki, *Measuring the time-dependent functional activity of adsorbed fibrinogen by atomic force microscopy*. Langmuir, 2008. 24(16):8801–8806.
116. Tamada, Y., E.A. Kulik, and Y. Ikada, *Simple method for platelet counting*. Biomaterials, 1995. 16(3):259–261.
117. Vanickova, M., J. Suttnar, and J.E. Dyr, *The adhesion of blood platelets on fibrinogen surface: Comparison of two biochemical microplate assays*. Platelets, 2006. 17(7):470–476.
118. Groth, T., et al., *Protein adsorption, lymphocyte adhesion and platelet-adhesion activation on polyurethane ureas is related to hard segment content and composition*. Journal of Biomaterials Science. Polymer Edition, 1994. 6(6):497–510.
119. Huang, S.L., et al., *Microphase separated structure and protein adsorption of polyurethanes with butadiene soft segment*. European Polymer Journal, 2000. 36(2):285–294.
120. Goodman, S.L., et al., *Preferential adsorption of plasma-proteins onto apolar polyurethane microdomain*. Journal of Colloid and Interface Science, 1990. 139(2):561–570.
121. Christenson, E.M., J.M. Anderson, and A. Hiltner, *Oxidative mechanisms of poly(carbonate urethane) and poly(ether urethane) biodegradation: In vivo and in vitro correlations*. Journal of Biomedical Materials Research Part A, 2004. 70A(2):245–255.
122. Tang, Y.W., R.S. Labow, and J.P. Santerre, *Enzyme-induced biodegradation of polycarbonate-polyurethanes: Dependence on hard-segment chemistry*. Journal of Biomedical Materials Research, 2001. 57(4):597–611.
123. Ward, R., et al., *In vivo biostability of polysiloxane polyether polyurethanes: Resistance to biologic oxidation and stress cracking*. Journal of Biomedical Materials Research Part A, 2006. 77A(3):580–589.
124. Simmons, A., et al., *Long-term in vivo biostability of poly(dimethylsiloxane)/poly(hexamethylene oxide) mixed macrodiol-based polyurethane elastomers*. Biomaterials, 2004. 25(20):4887–4900.

4

Molecular Simulation of Protein–Surface Interactions

Robert A. Latour

Protein-surface interactions are fundamentally important in a broad range of applications in biomedical engineering and biotechnology. The adsorption behavior of a protein is governed by the complex set of interactions between the atoms making up the protein, surface, and surrounding solution. Molecular simulation provides a means of theoretically viewing, studying, and understanding these types of interactions at the atomic level. However, as with any application, molecular simulation methods must be properly developed and applied if protein-surface interactions are to be accurately portrayed. This chapter provides an overview of the basics of molecular simulation with a focus on the simulation of protein-surface interactions using molecular dynamics. Important issues such as force field transferability, solvation effects, and statistical sampling are addressed, as well as directions for further development. Molecular simulation methods have the potential to greatly enhance our ability to understand protein surface interactions, leading to improvements in biomaterials design to control cellular response and to improve the sensitivity of biosensors and other bioanalytical systems.

Abbreviations and Symbols

b	bond length
b_0	force field parameter: bond length at zero energy
CM	classical mechanics
E	potential energy
F	force
f_{ij}	force vector between atoms i and j
GB	generalized Born
K.E.	kinetic energy
k_B	Boltzmann's constant
k_b	force field parameter: bond stretching stiffness
k_u	stiffness constant for an umbrella-sampling simulation

R.A. Latour • Department of Bioengineering, Clemson University, Clemson, SC 29634, USA, LatourR@clemson.edu

D.A. Puleo and R. Bizios (eds.), *Biological Interactions on Materials Surfaces*,
DOI 10.1007/978-0-387-98161-1_4, © Springer Science+Business Media, LLC 2009

k_{φ}	force field parameter: dihedral angle rotational stiffness
k_{θ}	force field parameter: bond angle bending stiffness
L-J	Lennard–Jones
m	mass
MC	Monte Carlo
MD	molecular dynamics
MM	molecular mechanics
N	number of degrees of freedom
N_{c}	number of constraints on the degrees of freedom
NPT	constant number of atoms, pressure, and temperature
NVT	constant number of atoms, volume, and temperature
P	pressure
P-B	Poisson–Boltzmann
p_i	probability of system to be in state i
q	partial charge
Q	configurational partition function of the system
QM	quantum mechanics
REMD	replica-exchange molecular dynamics
r_{ij}	distance between atoms i and j
SAM	self-assembled monolayer
T	temperature
TIGER	temperature intervals with global energy reassignment
TIGER2	temperature intervals with global exchange of replicas-2
TIP3P	name of a type of three-site water model
U	internal energy
V	velocity
V	volume
x	coordinate position
ΔB_{ij}	biasing energy function between states i and j
Δt	time-step increment for a molecular dynamics simulation
δ	force field parameter: rotational dihedral shift
ε_{r}	relative dielectric constant
ε_{ij}	well-depth for Lennard–Jones interactions
ε_{0}	permittivity of free space
θ	bond angle
θ_{0}	force field parameter: bond angle at zero energy
λ	variable coordinate parameter for an umbrella-sampling simulation
λ_{h}	fixed coordinate parameter for an umbrella-sampling simulation
σ_{ij}	collision diameter between atoms i and j
φ	dihedral angle

4.1. Introduction

When a synthetic material either is placed in the body or is exposed to a protein-containing media, the surface of the material generally becomes rapidly coated with an irreversibly adsorbed layer of protein. The characteristics of the material surface (e.g., surface chemistry and topology) and the conditions of the solution to which it is exposed (e.g., types of proteins and their concentrations, solution temperature, pH, salt concentration, and flow conditions) determine the types and amounts of proteins that are adsorbed to the material surface, as well as their arrangement, orientation, and conformation. All of these factors together then determine the bioactive state of this adsorbed layer. The bioactivity of this layer in turn provides the cues that direct the cellular response to a biomaterial surface or determine the sensitivity of the surface if it is used as a biocatalyst, biosensor, or bioanalytical system for diagnostics or detection [1–9]. If we are to learn how to optimally design surfaces for these types of applications, it is essential that an understanding of the cause-and-effect relationships that govern protein-surface interactions first be achieved.

Protein adsorption processes occur as a result of coordinated associations between chemical functional groups presented by the biomaterials surface, the proteins, and the surrounding solvent, with influence from the topology of the surface at scales at, and below, the dimensions of the adsorbing proteins. As such, experimental methods are very limited at this time in terms of assessing the behavior of a system at this level of scale. For this reason, molecular simulation has become increasingly recognized as a valuable addition to research programs aimed at understanding and interpreting protein adsorption processes at the molecular level. While the molecular simulation of protein interactions with synthetic surfaces is a relatively new area of research in the biomaterials field, molecular simulation methods have been widely developed over the past three decades in other fields (such as biophysics and pharmacology) for the investigation of protein folding [10–12], protein-protein [13–15] and protein-cell membrane [16, 17] interactions, and drug design [18–20], and in materials science for the investigation of the atomic-level structure and behavior of materials [21–24]. Similar potential exists for the application of molecular simulation methods to help understand interactions between proteins and synthetic material surfaces. However, molecular simulation methods cannot just be borrowed from other applications, but must be carefully and specifically developed and validated for applications that are relevant to the biomaterials field [25]. Once properly developed and applied, molecular simulation methods hold promise to provide an excellent complement to experimental studies to understand and predict protein adsorption behavior and to proactively design biomaterials at the atomic level to control protein adsorption behavior in a manner that is simply not attainable by any other means at this time.

The objective of this chapter is to provide an introduction to the topic of molecular simulation with a specific focus on protein-surface interactions. This will be addressed by first presenting an overview of some of the fundamentals regarding protein structure, folding, and adsorption (Sect. 4.2), followed by a summary of the basic types of methods that are used to conduct molecular simulations, with a focus on molecular dynamics (MD) (Sect. 4.3). Some of the key issues that should be considered when conducting a molecular simulation of protein-surface interactions are addressed, namely the suitability of the force field that is used (in the section "The Empirical Force Field Equation"), the representation of solvation effects (Sect. 4.3.3), and statistical sampling (Sect. 4.3.4). The final sections in this chapter will then address future directions for development that are particularly relevant to the simulation of protein-surface interactions (Sect. 4.4), followed by concluding remarks (Sect. 4.5).

4.2. Fundamentals of Protein Structure and Protein–Surface Interactions

Proteins are complex biomolecules that are generally composed of four levels of structural organization, which are termed primary, secondary, tertiary, and quaternary structures [26, 27]. The primary structure of a protein details the specific amino acid sequence that makes up the continuous polypeptide chain of the protein, with the individual amino acids along the chain referred to as "amino acid residues." DNA codes for 20 different types of L-amino acids, which have the general structure of -(NH-CHR-CO)-, where R represents the side group that differentiates between the different amino acid types. Amino acids are classified based on the chemical functionality of these R groups as nonpolar, polar, and positively or negatively charged. The primary sequence of a protein is defined as beginning with the amino group of the first amino acid in the chain and ending with the carbonyl group of the last amino acid in the chain. At pH 7.4, the beginning amino group will typically be protonated to form a positively charged end, referred to as the N terminus, while the opposite end is oxidized to form a negatively charged carboxylate group, referred to as the C terminus. The secondary structure of a protein defines how the polypeptide chain wraps with itself to form three basic units of structure: α-helices, β-sheets, and random loops. The α-helices typically form a right-handed helical structure with 3.6 amino acids per complete helical rotation, with the NH and CO functional groups of the main chain forming stabilizing hydrogen bonds with the portions of the polypeptide chain along the helical axis and with the R-groups extending in a radial direction outward from the long axis of the helix. The β-sheets elements are formed from either parallel or antiparallel chain segments of the polypeptide chain, with the NH and CO groups of the main chain forming hydrogen bonds that stabilize the association of these chain segments with one another. In this case, the R-groups of each amino acid extend out essentially perpendicularly above and below the sheet. Random loop segments of the polypeptide chain represent the amino acid residues in the polypeptide chain that lie in between the α-helix and β-sheet forming elements of the chain. The tertiary structure of the protein is determined by the manner in which the secondary structures organize with one another and is primarily stabilized by interactions between the R-groups that extend out from the secondary structures. Because proteins typically reside in an aqueous environment, formation of the tertiary structure is primarily driven by the tendency of the nonpolar amino acid side groups to complex together to reduce their exposure to the surrounding aqueous environment. This structure is also stabilized by hydrogen bonding, salt-bridge formation (i.e., associations between oppositely charged side groups), and covalently bonded disulfide bridges formed between cysteine amino acid residues. Finally, the quaternary structure of a protein is formed by the association of multiple polypeptide chains, each with their own tertiary structure; these interactions result from the same types of interactions that stabilize the tertiary structure of each individual polypeptide chain. The combination of the chemical functionality provided by the primary structure of the protein and the geometric structure provided by the secondary through quaternary organization of the polypeptide chain(s) in turn create the bioactive sites within a protein that enable it to perform its designated biological functions.

The final folded structure of a protein is primarily driven by thermodynamics [27–30], with each type of protein adapting its own uniquely folded structure on the basis of achieving its lowest free-energy state within its physiological environment. This is conceptualized by a "folding funnel" [27–30], such as that depicted in Figure 4.1a, which represents the free-energy state of a given protein as a function of all of the possible configurations in which the polypeptide chain making up the protein can arrange itself. As depicted in Figure 4.1a, there are numerous local low free-energy minima that the protein can become trapped in as it folds into its native structure, with the depth of each well effectively representing the stability of that localized state. As illustrated in Figure 4.1a, there is only one global free-energy minimum conformation, which represents the native structure of the protein.

Conceptually, when a protein adsorbs to a material surface, the atoms making up the surface impose additional external forces on the protein, which perturb the free-energy state of the combined protein-surface system, likely resulting in a shift in the shape of the folding funnel to a different global low free-energy state. This behavior is depicted in Figure 4.1b.

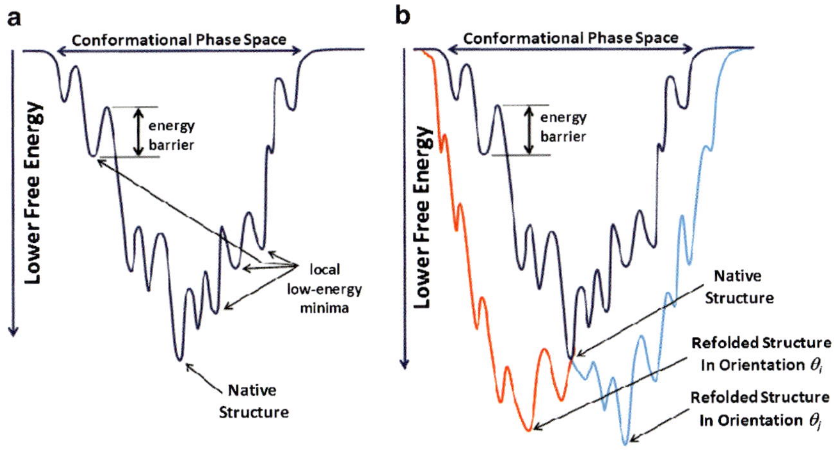

Figure 4.1. (**a**) Illustration of the protein-folding funnel for a protein in aqueous solution. The horizontal axis represents the conformational phase space and the vertical axis represents the free energy of the protein. The global free-energy minimum position represents the native structure of the protein. (**b**) Conceptual illustration of the way that protein-surface interactions shift the protein-folding funnel, resulting in a new low free-energy state of the system. The *dark blue* folding funnel represents the aqueous solution behavior, as shown in (**a**), with the *red* and *light blue* folding funnels depicting how different orientations of the protein on the surface will shift the folding funnel to different free-energy minima. (Adapted from [28]).

Based on this concept, the change in the conformation of a protein following adsorption may be most properly understood as surface-induced protein refolding as opposed to unfolding. Furthermore, because the external forces applied by a material surface on a protein will act in an anisotropic manner on the protein (i.e., the surface will not surround the protein, but will apply external forces from one side of the protein), it can be expected that the protein will refold into different states depending on the orientation of the protein on the material surface. Therefore, two different factors will be important for determining the adsorbed conformation of a protein: the first factor is the orientation of the protein on the material surface, which influences how the surface will interact with the protein, and the second is how the folding funnel of the protein will be perturbed for each orientation, thus determining how the protein conformation will shift after it adsorbs in a given orientation. The dependence of protein refolding behavior on the adsorbed orientation of the protein then further complicates the behavior of the system based on the kinetics of protein reorientation vs. refolding once a protein adsorbs to a material surface. If the process of protein reorientation on a material surface is much faster than surface-induced refolding, then it can be expected that the adsorbed proteins will generally exhibit the same orientation and subsequent refolding behavior, thus resulting in a fairly uniform layer of restructured proteins on the material surface. However, if the opposite is true, adsorbed proteins will tend to undergo surface-induced refolding before reorientation can occur, resulting in proteins being adsorbed in a much more heterogeneous manner [31]. With this perspective, if the horizontal axis of Figure 4.1a were changed to adsorbed orientation, then Figure 4.1a could equally well represent the relationship between adsorption free energy and adsorbed orientation, with one specific orientation resulting in an overall global free-energy minimum for the system. Again, the energy barriers separating the different free-energy states of the protein after it refolds from a given initial adsorbed orientation may be so large that transitions between these refolded states may be extremely slow. This depiction of the thermodynamic behavior of the system is consistent with the experimentally observed behavior that an adsorbed layer of protein may continue to change its conformation on a material surface over a time period of hours to days following initial adsorption on that surface [32–35].

As described in the preceding paragraphs, while protein adsorption is of critical importance in a wide-range of technologies, the adsorption behavior of proteins is extremely complex and is dependent on the collective interactions between the functional groups making up the protein, the material surface that the protein is interacting with, and the molecules making up the solvent (i.e., water and salt ions). These complexities make it very difficult to investigate protein adsorption behavior using experimental methods alone and make molecular simulation such a powerful tool to be used to investigate how all of these factors combine together to determine the bioactive state of the adsorbed protein layer on a biomaterials surface.

4.3. Molecular Simulation Methods

4.3.1. Overview

Molecular simulation methods can basically be separated into two distinct categories [36]: (1) quantum mechanics (QM) methods that treat electrons as the fundamental elements of the system and use approximations to Schrödinger's equation to calculate the behavior and the properties of the molecules in the system, and (2) classical mechanics (CM) methods that treat individual atoms or groups of atoms as the fundamental units of the system and use empirical force fields to calculate the behavior and the properties of the system. While QM methods are generally very accurate and may require no empirical data about the molecular system being analyzed, they are much more computationally rigorous and, therefore, are generally restricted to small set of atoms (i.e., a couple dozen atoms), thus making them unsuitable for the simulation of protein-surface interactions. CM methods, on the other hand, are much less computationally demanding and can be used to conduct simulations containing hundreds of thousands of atoms, or even much larger systems if united-atom methods are used, thus making them much better suited for the simulation of protein-surface interactions. This reduction in computational complexity, however, comes at a cost of the simulations being dependent on the accuracy of the empirical force field parameters that are used to determine how the atoms or groups of atoms in the system interact with one another during the simulation. More comments will be made about force field parameters in the section "The Empirical Force Field Equation."

CM simulation methods can be further subdivided into three basic types [36]: (1) molecular mechanics (MM), (2) Monte Carlo (MC), and (3) MD methods, each of which have their own advantages and limitations. MM methods are used to calculate the potential energy of a molecular structure based on the coordinates and bonded state of the atoms in the system. This method is most commonly used to calculate the geometry or organization of a molecule or system of molecules that will minimize the potential energy of the molecular system, referred to as "energy minimization" or "geometry optimization." Because this calculation does not include thermal energy effects, it should be realized that the resultant geometries represent conditions under zero Kelvin. Although at first glance this aspect may not seem to be of much practical use, MM methods are almost always used at the front end of an MC or MD simulation, especially when representing a condensed-phase system, in order to relax oddly stretched, bent, or twisted covalent bonds in the system and to avoid energetically unfavorable overlaps between atoms. The effect of temperature on the system is then introduced by the subsequent MC or MD simulation.

MC and MD methods are similar to MM methods in that they may use the same empirical force field for the calculations; however, the MC and MD methods incorporate temperature in the simulation, which provides the very valuable capability of calculating thermodynamic properties of the system, such as changes in free energy. MC methods are particularly used to efficiently search the conformational phase space (or coordinate degrees of freedom) of a molecular system by applying various algorithms to randomly move the positions of atoms within a defined move set (e.g., atom translation or dihedral rotation around a covalent bond) to efficiently explore the conformational phase space of the system. Conformational phase space simply represents all of the different ways that the atoms can be arranged within a given system of molecules. Because the move algorithms applied are not time dependent, MC simulations are primarily used to efficiently calculate the thermodynamic properties (e.g., changes in free energy) of a system but do not provide direct information regarding the rate, or the kinetics, of molecular processes. In contrast, MD simulations use Newton's equations of motion to predict the behavior of the molecules in the system over time in a manner that reflects both the thermodynamic and kinetic behavior of the system. The decision of whether a MC or a MD method would be the best choice for the simulation of protein-surface interactions will depend on the system at hand and the properties that are desired. For example, if it is desired to conduct a simulation to address which face of a protein will be most strongly attracted to a given surface, thus addressing the issue of the initial adsorbed protein orientation, and if the solvent is being represented using a mean-field implicit solvation algorithm (a topic that is addressed in Sect. 4.3.3), a MC method would probably be most efficient. In this case, the protein could be represented as a rigid molecule that is rotated and translated over a designated set of incremental moves to rapidly generate a converged set of states that depict how the potential energy of the system changes as a function of the protein orientation on the material surface. An example of this type of approach is illustrated in the paper by Zhou et al. [37]. An MD simulation in this case would not be nearly as efficient because of the relatively slow kinetics involved for a protein rotating on a material surface. Alternatively, if this same simulation was conducted in explicitly represented water (i.e., individual molecules of the solvent represented in the simulation, see Sect. 4.3.3), or if conformational changes in the protein structure were to be investigated as it adsorbs to a material surface, MC moves would be drastically limited because any large change in the position of the atoms of the protein would cause it to overlap with neighboring atoms, leading to unacceptably high values of potential energy. In this case, MC and MD methods would both be expected to provide about the same general degree of efficiency for sampling the various states of the system, with the MD method providing additional data regarding the predicted time course of the simulation [36]. For this reason (along with other reasons), MD methods are much more commonly used to simulate the behavior of large biomolecular systems and, therefore, this chapter will primarily focus on MD methods and their application to simulate the adsorption behavior of a protein.

4.3.2. Molecular Simulation of Protein–Surface Interactions

An MD simulation uses an empirical force field equation to calculate the forces acting on the atoms represented in the molecular system as a function of their spatial configuration and state of bonding and then applies Newton's equations of motion to calculate how the atoms in the system will move as a result of the forces acting upon them. Additional equations, based on molecular thermodynamics and statistical mechanics, are then also used in the calculations to control and determine thermodynamic properties of the system, such as temperature and pressure. In this section, each of these components of an MD simulation will be addressed with specific focus on their application for the simulation of protein-surface interactions.

The Empirical Force Field Equation

The empirical force field is an equation that is used in an MM, MC, or MD simulation to calculate the potential energy (E) of the system as a summation of the individual energetic contributions from each pair of atoms in the system (Figure 4.2). This is called a "force field equation" because, when differentiated with respect to the coordinate positions of the atoms (x_i), it yields an expression that provides the force acting on each atom, i.e., $F_i = -\mathrm{d}E/\mathrm{d}x_i$; these forces determine how the atoms will move for the next MD step of the simulation.

The general form of the force field equation that is most widely used for biomolecular simulations (e.g., proteins, carbohydrates, phospholipids, RNA, and DNA) is shown in Eq. (4.1), which is referred to as a "Class I force field" [36].Class I Force Field Equation:

$$E = \sum_{\text{bonds}} k_\text{b}(b-b_0)^2 + \sum_{\text{angle}} k_\theta(\theta-\theta_0)^2 + \sum_{\text{dihedrals}} k_\varphi[1+\cos(n\varphi-\delta)] + \sum_{\text{vdW}} 4\varepsilon_{ij}\left[\left(\frac{\sigma_{ij}}{r_{ij}}\right)^{12} - \left(\frac{\sigma_{ij}}{r_{ij}}\right)^{6}\right] + \sum_{\text{Coulomb}} \frac{q_i q_j}{4\pi\varepsilon_0\varepsilon_r r_{ij}} \tag{4.1}$$

The first three terms on the right-hand side of Eq. (4.1) represent the potential energy of the covalently bonded atoms in the system. These terms represent covalent bond stretching (b–b_0), bond bending (θ–θ_0), and dihedral bond rotation (ϕ). For these terms, the parameters

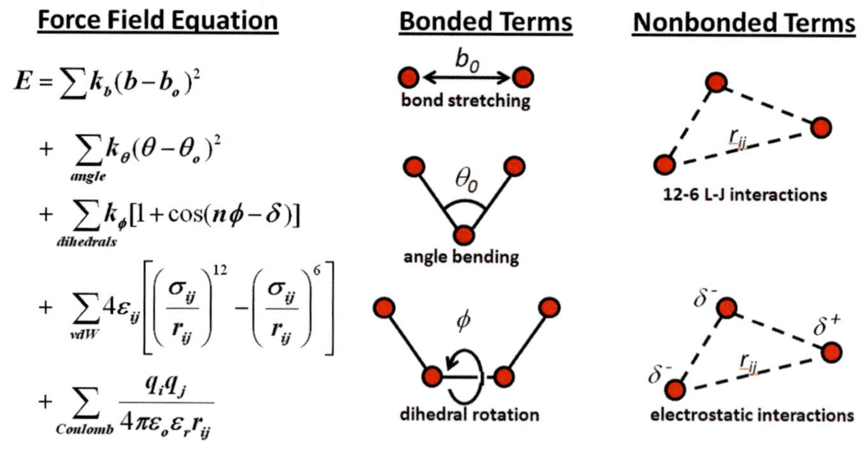

Figure 4.2. Illustration of the various types of bonded and nonbonded contributions to the potential energy of the molecular system as accounted for in the force field equation.

b, θ, and ϕ represent the bond length, bond angle, and dihedral angle, respectively, of the bonded state at a given time point, relative to a reference position (designated by force field parameters b_0, θ_0, and ϕ_0, with $\phi_0 = 0$); the ks represent additional empirical force field parameters that reflect the stiffness of the system. The last two terms of the force field equation (Eq. 4.1) represent the contributions of the nonbonded interactions to the potential energy of the system. The first of these two represent Lennard-Jones (L-J) 12-6 interactions, which are considered as pairwise interactions for all nonbonded atoms (i.e., atoms separated by more than two covalent bonds along a covalently bonded chain or between two atoms in different molecules in the system), where ε_{ij} and σ_{ij} represent the L-J well-depth and the collision diameter (i.e., the distance between atoms when the interaction energy between them equals zero) between the designated pairs of atoms, respectively, for atoms i and j. The second nonbonded term represents the energy contributions from the electrostatic interactions between pairs of atoms, which are represented by Coulomb's law, where q_i and q_j represent the partial charges centered on atoms i and j, ε_0 is the relative permittivity of free space, and ε_r is the relative dielectric constant; ε_r is typically set to 1.0 or about 79 depending whether an explicit or implicit solvation method is being used, respectively (this aspect is further addressed in Sect. 4.3.3).

A force field for a given molecular simulation is thus described by both the form of the force field equation (e.g., Eq. 4.1) and the values of the empirical force field parameters, which are assigned for each designated atom type of the system. For example, a C-C covalent bond will have one set of defined b_0 and k_b values, while a C-O bond will have a different set of b_0 and k_b values, etc. The L-J parameters, ε_{ij} and σ_{ij}, are handled slightly differently, with each individual atom type given an ε and σ parameter value. These values are then combined for a given atom-atom pair interaction using Lorentz-Berthelot combination rules [36], in which ε_{ij} and σ_{ij} are derived using either the geometric or arithmetic mean of the two values (depending on the rules of the type of force field used). The values of these empirical parameters for both the bonded and nonbonded terms are generally determined either using QM calculations of small molecules containing similar functional groups with parameters adjusted to optimize the force field to match the QM-predicted behavior, or by adjusting the parameters to match experimentally determined properties, such as density, heat of vaporization, and vibrational spectra. If parameterized properly, an empirical force field can provide a very accurate representation for a given molecular system. However, if not parameterized properly, the predicted behavior of the system can be very far from reality.

One of the difficulties involved in force field parameterization is that an appropriate set of force field parameters for one set of atoms in one type of molecular environment may not be appropriate for the same set of atoms in a different type of molecular environment; this situation is referred to as the "issue of force field transferability." Force field transferability is particularly a problem for the fixed partial charges assigned to each atom type in an empirical force field (i.e., q_i), because, in reality, the partial charges of the atoms in a molecule will shift substantially depending on their polarizability and the distribution of charges in their immediate surroundings. As an excellent example of this problem, van Gunsteren and co-workers [38] showed that two different sets of partial charge parameters were needed to accurately represent the behavior of the same set of molecules depending on whether they were represented in their pure condensed liquid state compared with when they were represented as individual molecules in aqueous solution. As a potential solution to this problem, a substantial amount of research is currently underway in the computational chemistry community for the development of polarizable force field methods that are capable of adjusting the partial charges of each atom in the simulation in response to the local environment [39–41].

Currently, however, these methods are too computationally expensive to be considered of practical use [40] and fixed-charged empirical force fields are still the only methods available to handle the large types of systems that must be addressed for the simulation of protein-surface interactions. Given this situation, it is important to use a validated set of parameters for a given molecular simulation.

The question of force field transferability raises a particularly important issue for the use of empirical force field-based methods for the simulation of protein-surface interactions. Generally, a force field is validated for a given molecular system by using the force field to conduct simulations for the calculation of experimentally measurable properties of the system (e.g., density, structure, vibrational frequencies, heat of evaporation, and free energy of solvation). By comparing calculated values to experimentally measured data, the validity of the force field to accurately represent the behavior of the given molecular system can then be assessed. If substantial errors are found in the calculated values, the experimental data then also provide a means whereby the force field parameters can be adjusted until suitable agreement is found, thereby validating the force field for the designated type of molecular system under consideration. The difficulty that this approach presents for the simulation of protein-surface interactions is that presently there are relatively few experimental data available that can be used to determine whether the results from a molecular simulation of protein adsorption behavior are correct or not. Without such data, it is impossible to determine whether a simulation realistically portrays the molecular behavior of the system. As one approach to address this deficiency, we have been conducting carefully matched experimental and simulation studies to measure and calculate the free energy of adsorption for a broad series of peptide-surface systems using a custom-designed host-guest peptide model and a series of alkanethiol self-assembled monolayer (SAM) surfaces with surface functionality set to represent common functional groups present in polymeric biomaterials [42–44]. This method provides a means of assessing the validity of a given force field for the simulation of protein-surface interactions. Similar efforts are also needed to develop experimental methods that can be used to quantitatively measure the orientation, conformation, and bioactive state of adsorbed proteins, from which the accuracy of simulations of protein adsorption behavior, using a given force field, can be further assessed.

Newton's Laws of Motion and System Control for an MD Simulation

An MD simulation uses the force field equation combined with the numerical integration of Newton's equations of motion over defined time-step increments to determine the movement of each of the atoms represented in a molecular simulation [36, 45]. This process is graphically presented in Figure 4.3. As indicated in this figure, the MD cycle starts with some defined initial starting coordinates for all of the atoms of the system and a set of initial velocities (Figure 4.3, steps 1 to 3). As described in the section "The Empirical Force Field Equation," the force field equation provides an analytical expression to calculate the potential energy (E) of the system as a function of the designated force field parameters and the coordinates of the positions of the atoms in the system (Figure 4.3, step 4). When differentiated with respect to x, y, and z, this expression then provides the x, y, and z components, respectively, of the force vector acting on each of the atoms in the system (Figure 4.3, step 5), and, when divided by atomic mass, the acceleration of each atom from Newton's second law, $a_n = F_n/m$, is determined (Figure 4.3, step 6). The acceleration of each of the atoms is then numerically integrated over a designated time step (Δt) to calculate an incremental change in velocity (Figure 4.3, step 7), which, when added with the velocity at the end of the

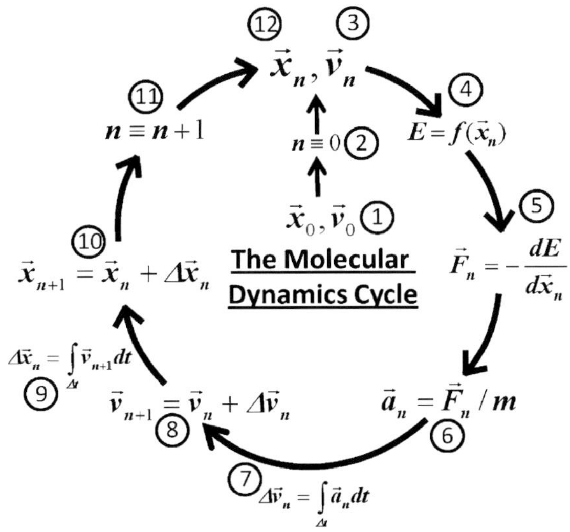

Figure 4.3. Illustration of the cycle that is followed during a molecular dynamics simulation in order to calculate the motions of the atoms in the system. The process begins with a set of initial values that define the coordinate position (x_0) and velocity (v_0) of each atom in the system (step 1). The cycle counter (n) is then set to zero (step 2) and the positions (x_n) and velocities (v_n) of the atoms are updated with these values (step 3). The positions of the atoms are plugged into Eq. (4.1) from the section "The Empirical Force Field Equation," and into the differentiated form of this equation to calculate the potential energy of the system (step 4) and the force vector (F_n) (step 5) acting on each atom, respectively. The acceleration of each atom (a_n) is next calculated from Newton's second law (step 6). The acceleration of each atom (taken as a constant over the time interval, Δt) is integrated over the time interval to calculate the change in velocity (step 7), which is then added to the previous velocity (v_n) to calculate the new velocity of each atom, v_{n+1} (step 8). The new velocity (taken as a constant over the time interval, Δt) is integrated over the time interval to calculate the change in position (step 9), which is then added to the previous position (x_n) to calculate the new position of each atom, x_{n+1} (step 10). The cycle counter is updated (step 11), and the new positions and velocities (step 12) are then used to begin the next cycle.

previous cycle, provides the new velocity of each atom (Figure 4.3, step 8). The velocities are then numerically integrated over the time step Δt to provide the incremental change in coordinate position (Figure 4.3, step 9), which is added to the position from the previous cycle to update the new positions of all of the atoms of the system (Figure 4.3, step 10). The coordinate positions and velocities of all of the atoms are then updated (Figure 4.3, steps 11 and 12) for use in the next cycle. This process completes one full round of the MD cycle, which describes the behavior of all atoms in the system after the new incremental time period of Δt. These new positions are then reinputted into the force field equation to calculate the new force vectors acting on each atom to begin the next MD cycle. This cyclic process is continued, ideally until the system converges to a steady state, from which thermodynamic and kinetic properties of system behavior can be determined. One of the critical issues in this cycle is that the time increment, Δt, must be sufficiently small to smoothly follow the fastest motions of the system, which are the vibrational motions of covalent bond stretches. A time step of $\Delta t = 1.0$ fs is typically used for an all-atom MD simulation without any constraints, or if covalent bond lengths for hydrogen atoms are held fixed (commonly done using an algorithm called "Shake" [46] or "Rattle" [47]), a longer time step of $\Delta t = 2.0$ fs can be used. While this adjustment may seem like a minor issue, the use of a time step of 2.0 vs. 1.0 fs will reduce the wall-clock time required to perform a given simulation by a factor of two (e.g., a month-long simulation can be completed in 2 weeks).

While the integration cycle represents the "engine" of an MD simulation, there are several other factors that must be included to control the simulation in order to accurately represent a real molecular system. For molecular interactions within a condensed-phase system, such as in aqueous solution using explicitly represented water (i.e., water molecules atomically represented in the simulation, see Sect. 4.3.3), three-dimensional (3D) periodic boundary conditions are used in which the "real space" part of the simulation represents a unit "cell" that is repeated in each direction to create an infinite system of "images," thus representing a continuous condensed-phase system, as depicted in Figure 4.4. Because of the infinite periodicity of this type of representation, it is important that the set of atoms represented in the real-space system (e.g., the center box highlighted in red in Figure 4.4) have a net charge of zero. The use of periodic boundary conditions creates the situation where an atom or molecule that moves through one of the boundaries of the real-space system reenters the system from the adjacent image through the opposite boundary, as illustrated in Figure 4.4. This technique enables a constant number of atoms to be maintained in the real-space system of the simulation and allows the molecules to freely move around as if they really were in an infinitely large condensed-phase system. The use of periodic boundary conditions

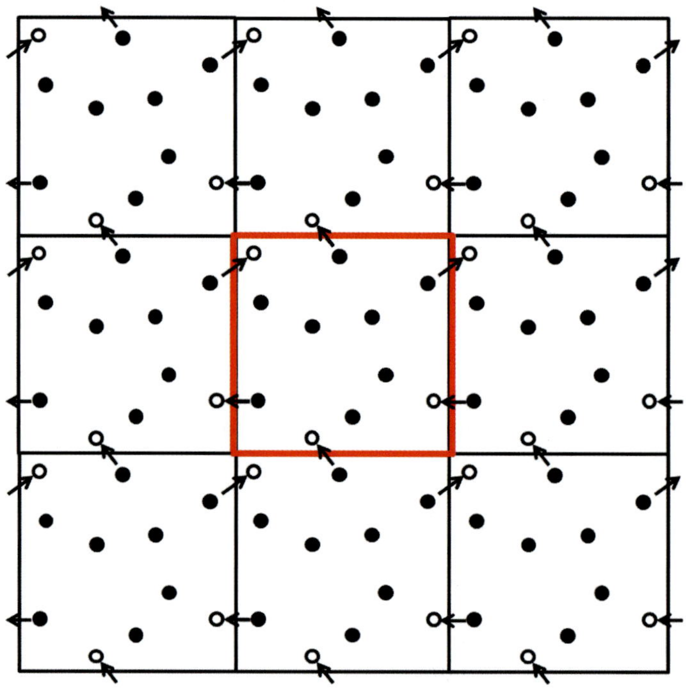

Figure 4.4. Illustration depicting how periodic boundary conditions are used to represent an infinite, condensed-phase system. The real-space system is highlighted in *red* with images of this same system reproduced as unit cells surrounding the real-space cell. When an atom passes through one side of the real-space cell, the same atom will reenter the real-space cell through the opposite side from the adjacent imaged cell. The *solid black circles* represent the atoms in the system, with the open circles representing the same atoms after they cross a boundary into the adjacent cell.

then also enables thermodynamic properties to be determined from the behavior of the molecules contained within the real-space system, such as internal energy (U), temperature (T), and pressure (P), which can be calculated from molecular thermodynamics principles using the following equations (Eq. 4.2) [36]:

$$U = \langle E \rangle = \frac{1}{M}\sum_{i=1}^{M} E_i$$

$$T = \frac{2}{(3N - N_c)k_B}\langle K.E. \rangle = \frac{< m_i v_i^2 >}{(3N - N_c)k_B}$$

$$P = \frac{1}{V}\left[Nk_B T - \frac{1}{3}\sum_{i=1}^{N}\sum_{j=i+1}^{N} r_{ij} f_{ij} \right], \tag{4.2}$$

where M represents the number of times the potential energy of the system (E) is sampled during the MD simulation, N is the number of atoms in the system (each with three degrees of freedom), and N_c is the number of constraints applied to the system (which reduces the total number of degrees of freedom in the system). $<K.E.>$ is the ensemble average of the kinetic energy of the system, which equals $<m_i v_i^2>$, where m_i and v_i are the mass and velocity, respectively, of each atom in the system, k_B is Boltzmann's constant, V is the volume of the real-space system that is represented, and r_{ij} and f_{ij} are the radial distance and the force vector, respectively, between each *i-j* pair of atoms in the system. The ability to calculate and monitor the temperature and pressure of the system enables a simulation to be performed under either NVT or NPT conditions, which represent a constant number of atoms (N) and constant volume (V), temperature (T), and pressure (P) conditions.

While the use of a periodic system for an MD simulation is a very useful and efficient technique to represent a condensed-phase system, it also causes a problem with respect to how the nonbonded interactions of the force field equation (i.e., the last two terms in Eq. 4.1) are calculated. Ideally the nonbonded interactions for the condensed-phase system would represent the summation of L-J and electrostatic interactions between each of the atoms in the real-space system with each of the surrounding atoms, including the imaged atoms that extend out to infinity. This, of course, is not possible, and thus simplifying assumptions must be made for the calculation of these terms. The most widely accepted practice is to use a cutoff method for the L-J interactions in which the L-J interactions for a given atom are only calculated for atoms within a defined cutoff distance from the atom in question; a correction is then added to account for the remaining atoms beyond the cutoff. This method is usually implemented using a "switching" or "shifting" function at the cutoff to smoothly transition L-J interactions to zero [48]. This smooth transitioning is necessary to avoid artifacts that can be induced if the energy vs. distance function is reduced to zero too rapidly, in which case the resulting force, which is the gradient of the energy function, would be calculated as an unnaturally large value, thus causing artifacts in the motions of the atoms. For example, a 10-12 switching function would mean that L-J interactions would be calculated for all atoms within 10 Å of the atom in question, with L-J interactions then smoothly being decreased to zero for atoms between 10 and 12 Å away, with interactions beyond 12 Å being ignored. Because L-J interactions are relatively short ranged (i.e., they decrease as a function of r_{ij}^{-6}; see Eq. 4.1), this truncation results in negligible error. When using this procedure with periodic boundary conditions, it is important that the minimum dimension of the periodic cell be at least twice as long as the outer range of the cutoff distance. This condition ensures that an atom in the real-space system that is represented will not feel a force from its own image in the neighboring cells; such a situation will introduce undesirable correlations in the motion of the atoms in the simulated system. In contrast to the short-ranged L-J interactions, electrostatic interactions are much more long ranged and decrease on the order of r_{ij}^{-1} (see Eq. 4.1). Thus, use of a cutoff for electrostatic interactions is generally not advised, as it will tend to introduce a substantial amount of error in the simulation.

Instead, Ewald methods, such as particle mesh Ewald [49, 50], which closely approximate electrostatic interactions in an efficient manner including all of the images surrounding the real-space system, are recommended. While this representation does mean that atoms will effectively feel forces from their own images in the neighboring periodic "cells," errors introduced from this condition are minimal and much lower than those that are introduced by the truncation of these types of interactions. While the use of 3D periodic boundary conditions with particle mesh Ewald represents a very realistic situation for the simulation of either a peptide or a protein in aqueous solution, it does raise concerns regarding its applicability for the simulation of a protein interacting with a material surface. In this case, a model system that represented an adsorbing surface at the bottom of the real-space unit "cell" would then also represent an image of the surface immediately above the real-space "cell" as well, as depicted in Figure 4.5. Thus, a peptide or protein in solution would just as easily interact with the "bottom" of the surface that was imaged at the top of the real-space "cell" as it would with the top of the adsorbent surface that is represented at the bottom of the real-space "cell." Under this circumstance, the ideal situation would be to use two-dimensional (2D) periodicity and a 2D Ewald method where the surface is periodically imaged only along the sides of the real-space system, but not above and below it. Unfortunately, 2D Ewald methods have not yet been adequately developed to represent this type of molecular system, and 3D periodicity with 3D particle mesh Ewald is still the best way of representing this type of system. To avoid the situation where a protein will be able to interact with the bottom of the surface via the top image "cell," a constraint can be added to the MD simulation that prevents the protein from approaching within about 20 Å from the top of the real-space unit "cell." While long-range electrostatic interactions are theoretically still a concern, especially when representing a material surface with a high charge density, in practice we have found that when representing an aqueous solution with 140 mM Na^+ and Cl^- (i.e., representing physiological conditions [51]), charged molecules in solution that are separated more than 20 Å from the top image of the surface are sufficiently far away that artifacts in the simulated behavior do not occur to a noticeable degree.

4.3.3. Representation of Solvation Effects for Protein–Surface Interactions

The process of protein adsorption to a material surface is often simply represented as a chemical reaction in the form of:

$$P + S \rightarrow P.S, \tag{4.3}$$

where P, S, and $P{\cdot}S$ represent the protein, the surface, and the protein-surface complex after the protein adsorbs, respectively. Equation 4.3, however, is actually very misleading in that it omits the critically important interactions of the molecules of the solvent during this

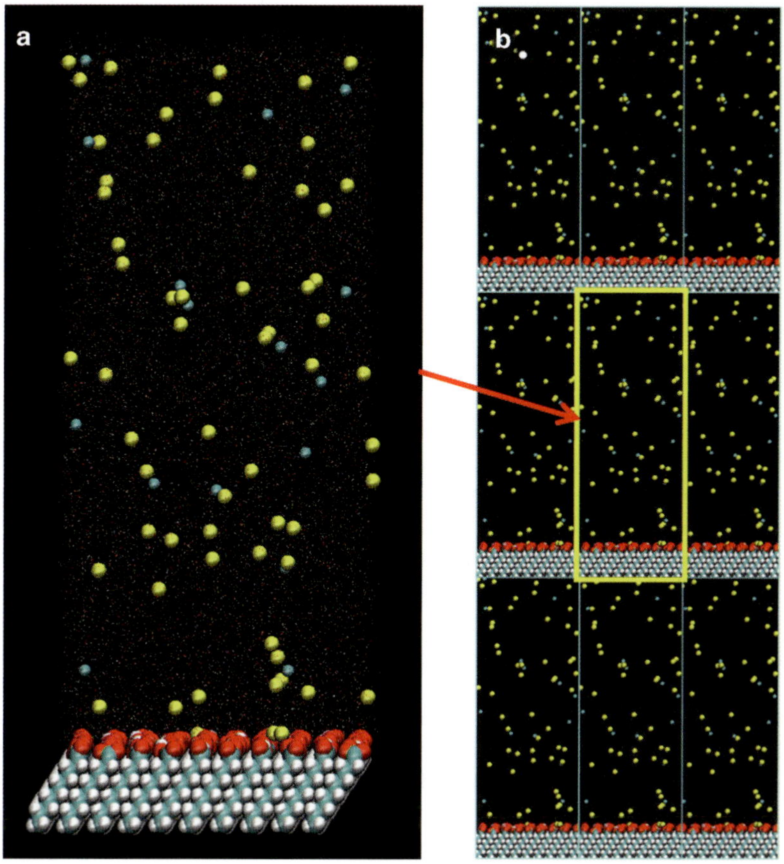

Figure 4.5. **(a)** Real-space system showing an aqueous solution made up of explicitly represented water molecules (*small red* and *white dots*) and Na^+ and Cl^- ions (*yellow* and *blue spheres*, respectively) over a self-assembled monolayer (SAM) surface functionalized with negatively charged COOH groups. **(b)** Illustration showing the periodic system that is represented in the simulation to reveal how the use of periodic boundary conditions results in the SAM surface being represented both above and below the real-space surface. A protein in solution will therefore interact with both the top of the SAM surface in the real-space cell and the bottom of the SAM surface that is imaged above the real-space cell. A constraining function can be used to keep the protein sufficiently far from the bottom of the imaged surface so that the presence of this imaged surface has negligible influence on the behavior of the protein during the simulation.

process. A much more appropriate description of the protein adsorption process on a material surface is represented by Eq. (4.4):

$$P{\cdot}m\mathrm{H_2O} + S{\cdot}n\mathrm{H_2O} \rightarrow P{\cdot}S + \left(\frac{m+n}{2}\right)\mathrm{H_2O{\cdot}H_2O}, \tag{4.4}$$

in which case m and n represent the number of water molecules that are complexed with the protein and the material surface that must be consequently dissociated in order for the interaction between the protein and the surface to take place; these water molecules then subsequently interact with one another and with the other water molecules in the bulk solvent [1]. As indicated by this expression, the process of a protein thus adsorbing to a material surface does not just involve interactions between the atoms of the protein and atoms of the surface, but rather involves competitive interactions between the water molecules with the atoms making up the protein and surface vs. interactions between the atoms of the protein with those of the surface and between the water molecules with one another. This process is further complicated when the protein and/or material surface includes charged functional groups, in which case the counter-ions in solution also take part in the adsorption process as very influential components of the overall system. As indicated in Eq. (4.4), water molecules thus are not passive elements in the system, but very active components that are just as involved in the protein adsorption process as are the protein and the adsorbing surface themselves. Because of this situation, the proper representation of solvation effects is essential if a simulation is to be carried out in a manner that realistically represents the interaction of a protein with a material surface.

The most direct way of representing solvation effects in an MD simulation of protein-surface interactions is to include water molecules and salt ions (e.g., 140 mM Na^+ and Cl^- to represent physiological conditions [51] plus additional counter-ions for overall charge neutrality) as explicitly represented molecules in the simulation. Many different water models, and their associated force field parameters, have been developed over the years and they generally provide an excellent representation of solvation effects at around 298 K. Some of the most popular water models are SPC [52, 53], SPC/E [53], TIP3P [53, 54], TIP4P [55, 56], TIP4P/Ew [57, 58], and TIP5P [59]. Decisions regarding which water model should be used for a given simulation of protein-surface interactions should primarily be made based on selecting the water model that is recommended by the force field developers to go with the force field used to represent the protein. For example, the CHARMM force field was parameterized with the CHARMM version of TIP3P water [48]. Unfortunately, the explicit representation of water molecules greatly increases the number of atoms that must be represented in a molecular simulation; this situation then directly translates into the need for much greater computing capacity and/or much longer wall-clock time that is required to perform a given simulation.

To get around the high computational cost of using explicit solvation, several methods have been developed to represent solvation effects using mean-field continuum approximations, which are referred to as "implicit solvation methods." The most simple of these methods involves merely changing the relative dielectric constant (ε_r) in the Coulomb's term of the force field (see Eq. 4.1) from a value of 1.0, which is used in explicit solvation simulations, to a value of about 79 [36], which is the dielectric constant of bulk water at about 298 K. A similar technique involves setting $\varepsilon_r = r_{ij}$ (i.e., the distance between atoms i and j), which provides a situation where electrostatic interactions are very strong when atoms are close to one another, but die down proportional to r_{ij}^{-1} as the separation distance increases; this approach is intended to account for the influence of intervening water molecules that would dampen the strength of electrostatic interactions. This procedure is known as the "distant-dependent dielectric method" [36]. These methods were developed as very crude approaches to represent the dampening of electrostatic effects for protein-folding simulations. For this type of process, however, the behavior of the protein is largely mediated by the covalently bonded state of the protein. The structural stability provided by the bonded state of the protein enables this type of approximation to be considered, even though its use in this case is still not recommended because of the possible artifacts in the simulated behavior that it may cause [36, 60]. On the other hand, the simulation of protein-surface interactions is completely dominated by nonbonded interactions between the atoms of the protein, material surface, and solvent, with the solvent molecules playing a critical role in the process. Under these conditions, the use of the relative dielectric constant alone (either as a constant or in the distance-dependent form) to represent solvation effects is strongly advised against because this procedure will completely neglect the competitive interaction of water for the atoms of the protein and the material surface through both electrostatic and L-J interactions; the result will be to substantially overestimate the strength of adsorption, particularly for polar functional groups. For example, Sun and Latour [61] have shown that the use of a distance-dependent dielectric method will cause polar hydrophilic groups of the amino acid residues of a protein to be attracted to polar groups on a material surface nearly as strongly as they would in a vacuum; in contrast, in aqueous solution, the attraction between these same functional groups can be expected to be relatively weak due to their capability to interact just as strongly with water.

A much more sophisticated and accurate form of implicitly representing solvation effects due to electrostatic interactions is provided through the use of the Poisson-Boltzmann (P-B) equation to calculate the free energy of molecular solvation [62, 63]. Because this method is computationally too rigorous to be used in MD simulations, its use is primarily limited to MM simulations for the calculation of the solvation free energy contribution of molecules in a given conformation. A much less computationally expensive approximation of the P-B equation, known as the "generalized Born (GB) equation," has been developed and widely applied as an implicit solvation method for MD simulations of peptide and protein folding with a reasonable degree of accuracy [64–68]. One of the well-established problems with GB methods, however, is that they tend to overemphasize the strength of hydrogen bond interactions [69]. While this approach may actually be advantageous for simulations seeking to fold a polypeptide chain because of the importance in hydrogen bonding for the formation of secondary and tertiary protein structures, this same tendency will again tend to greatly misrepresent protein adsorption processes. This situation occurs because the overemphasis of hydrogen-bonding interactions in this case will lead to overly strong interactions between the hydrogen-bondable functional groups of a protein and a material surface while neglecting the competitive nature of water to form hydrogen bonds to these same functional groups.

Finally, as noted in the previous paragraph, P-B and GB methods only address the electrostatic component of solvation effects while completely neglecting hydrophobic effects. To address this deficiency when using either a P-B or GB method to represent hydrophobic interactions, an additional term is required in the force field equation. This term is generally expressed in the form of a surface tension per unit area parameter that is then multiplied by the surface area of the molecule in solution. However, because of the complexities involved in discriminating between surface areas of functional groups with hydrophobic vs. hydrophilic character, this surface energy term is generally applied to the total surface area of the molecules in the simulation. Once again, while this approach may be suitable for protein-folding applications, for which these methods have been primarily developed, their suitability for use in protein adsorption simulations is questionable at best. This issue is addressed in a paper by Sun et al. [70] in which simulations were performed for peptide adsorption to a hydrophobic material surface with results from two different types of implicit GB models compared with results using explicit water (TIP3P). These simulations showed that substantially different peptide adsorption behavior is predicted between each of the GB methods and the explicit solvent case; the GB methods tend to over predict the strength of hydrophobic interactions.

Based on the points discussed in the preceding paragraphs, at this time there is really no basis to support the use of any of the implicit solvation methods for the simulation of protein-surface interactions [25]. All of the currently available methods fall far short of providing an accurate representation of protein interactions with a material surface. Use of any of these methods will simply result in the prediction of protein adsorption behavior that has little to do with the real molecular behavior of these types of systems. Thus, simulations of protein-surface interactions should be performed with an appropriately selected explicit water model along with the desired concentration of counter- and co-ions in the system. This being stated, the use of implicit solvation methods is still highly desirable because they greatly reduce the computational requirements that are needed to perform a given simulation. There is thus a need for the development of new implicit solvation methods that are specifically designed, tuned, and validated for use with protein adsorption simulations.

4.3.4. Statistical Sampling Considerations for Protein-Surface Interactions

Statistical sampling is just as important in a molecular simulation as it is in an experimental study. In an experiment, it is necessary to test a sufficient number of independent samples of a system in order to generate an estimate of the mean and variance of the population, from which statistical analyses can be performed to determine whether changes in the measured behavior (caused by different treatments and conditions) can be considered to be statistically different. In an MD simulation, it is necessary to run the simulation over a sufficient length of time in order to obtain converged ensemble-average estimates of the properties of the system. This condition requires that the MD simulation be performed ergodically, which means that the simulation should be run sufficiently long to enable it to visit all of the available microstates of the system, such that the calculated time-averaged properties accurately represent the macroscopic properties of the system that would be measured experimentally [45]. The results of such a simulation can then be taken as being equivalent to one sample for an experimental system. In practice, this situation can be problematic for an MD simulation since it is often very difficult to determine whether or not a simulation has run sufficiently long to explore all of the pertinent microstates of the system. This determination is particularly difficult for systems that have a “rugged” energy landscape, in which case the relatively high-energy barriers that separate local low-energy states may trap the simulation in localized regions of the conformational phase space and prevent proper sampling of the system, as illustrated in Figure 4.6. Under such circumstances, a given property of the system, such as either the average potential energy or the orientation of a protein, may appear to converge to a stable value, but in fact a much longer simulation would reveal that this property only represents the conditions for a localized state or set of states, while the properly converged property has a substantially different value.

Unfortunately, protein folding and adsorption-induced refolding behavior, as illustrated in Figure 4.1, presents a situation where the microstates of the system are separated by these types of high-energy barriers. The presence of these types of energy barriers results in a situation where a conventional MD simulation is generally not able to adequately sample system behavior within time frames that are practically accessible with current-day computational resources. To address this issue, several different types of advanced sampling algorithms have been developed that enable energy barriers to be readily crossed, thus enabling an MD simulation to escape from these types of local low-energy wells, explore the relevant phase space of the system, and generate ensemble-averaged properties that can be confidently compared with experimental results. Estimates of the mean and variance of the calculated properties can then be obtained by either running separate simulations of the same system with different starting conditions to provide multiple independent “samples” of the system, or a single long simulation can be broken up into uncorrelated blocks of time with each block then representing an independent, converged representation of the system; this approach is referred to as “block averaging.”

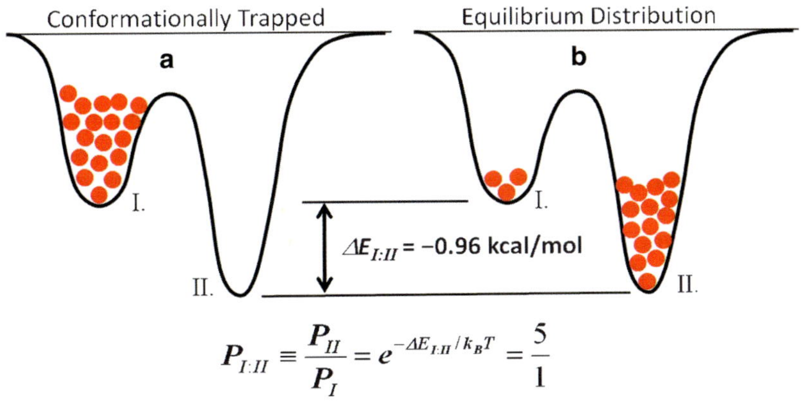

Figure 4.6. **(a)** Illustration depicting a sampling problem that often occurs during a conventional MD simulation in which the molecular system is "stuck" in a local low-energy well (conformational state I), which in this case is shown as being 0.96 kcal/mol higher in potential energy than conformational state II. The *red spheres* represent the states of the system that are sampled during the MD simulation. This approach results in improper sampling of the system with subsequent errors in the calculation of the thermodynamic properties of the system. **(b)** The proper Boltzmann-weighted equilibrium distribution of the system at 298 K that would be sampled if the MD simulation were run infinitely long (see Eq. 4.6).

The basic principles behind most advanced sampling algorithms can be understood based on the statistical mechanics relationships that describe the probability (p_i) of a given state for a given molecular system and the relative probability of sampling a different state, p_j, relative to p_i (i.e., p_j/p_i). These relationships can be described by the following Eqs. (4.5) and (4.6) [71]:

$$p_i = \frac{\mathrm{e}^{-E_i/k_\mathrm{B}T}}{Q}\ ;\ \text{with } Q = \sum_i \mathrm{e}^{-E_i/k_\mathrm{B}T} \tag{4.5}$$

and

$$p_{ij} \equiv \frac{p_j}{p_i} = \mathrm{e}^{-\Delta E_{ij}/k_\mathrm{B}T}, \tag{4.6}$$

where E_i is the potential energy of the system for state "*i*", k_B is the Boltzmann constant, T is absolute temperature, Q is the configurational partition function of the system, and ΔE_{ij} is the difference in potential energy between states "*i*" and "*j*." Given these relationships, the probability of the system being in a given energy state can be adjusted by altering the value of the group of parameters in the exponential function (i.e., $E_i/k_\mathrm{B}T$), either by introducing a biasing energy function into the force field equation to influence the energy state of the system (E_i) or by adjusting the temperature of the system (T).

The introduction of a biasing energy function (ΔB_{ij}) can be done by adding a user-defined term into the force field equation that controls how the atoms interact with one another during the simulation in a manner that lowers the energy difference that separate the various states of the system that one wishes to sample during the simulation [72]. The introduction of a biasing function can be expressed mathematically by Eq. (4.7) [43]:

$$\overline{p}_{ij} \equiv \left(\frac{p_j}{p_i}\right) e^{-\Delta B_{ij}/k_B T} = e^{-(\Delta E_{ij} + \Delta B_{ij})/k_B T}, \tag{4.7}$$

where $\overline{p}_{ij}$ represents the biased probability distribution that results from the addition of the biased energy function. As is clear from this relationship, if B_{ij} is set to exactly cancel the potential energy difference between the two states (i.e., $B_{ij} = -E_{ij}$), then the energy difference separating them will be zero, creating a situation where there is equal probability of being in either state, thus eliminating the sampling problem. After the simulation is completed, the biased probability ratio that is obtained from the trajectory of the simulation can then be corrected to remove the applied bias to obtain the unbiased probability distribution by inverting Eq. (4.7) to obtain Eq. (4.8):

$$p_{ij} \equiv \frac{p_j}{p_i} = \overline{p}_{ij} e^{\Delta B_{ij}/k_B T}. \tag{4.8}$$

This method is highly effective if the coordinate position and the depth of a given local low-energy well are known such that a biasing energy function can be appropriately determined and applied in the simulation. If these parameters are not known a priori, they can be determined by running preliminary MD simulations to assess where the system tends to become trapped, and then adaptively adding a biasing function until the sampling problem is removed [43, 72–75].

The use of a biased-energy function is most widely used to force sampling over a single-designated system coordinate, such as either the dihedral rotation about a bond in a peptide chain [72, 76, 77] or the separation distance between a solute and a material surface [43]. For these types of applications, an umbrella-sampling technique [72, 78] is typically used in the form of Eq. (4.9):

$$\Delta B = k_u (\lambda - \lambda_u)^2, \tag{4.9}$$

which is also referred to as a "restraining potential," where k_u is the stiffness constant, λ_u is the coordinate parameter of interest, which is set at a designated coordinate value, and λ is the sampled value of the coordinate at a given time point in the simulation. In this form, the biasing energy function penalizes the system in a quadratically increasing manner as it deviates from the designated position, λ_h. A series of independent parallel simulations, referred to as "windows," can then be carried out with the value of λ_h incrementally varied over the full range of interest; these increments are set so that the neighboring sampled populations overlap one another. The resulting overlapping sampled distributions from all of these simulations can then be combined using the weighted histogram analysis method (WHAM) [79, 80], which serves to remove the applied bias from the sampling distribution and to generate an unbiased probability distribution of states over the full range of the parameter of interest (i.e., over the full range of λ).

While a biasing energy function in an MD simulation can be very helpful to overcome many types of sampling problems, its use is primarily restricted to a single degree of freedom, such as either protein-surface separation distance or the dihedral rotation of a peptide chain about a designated covalent bond. However, for cases where there are numerous different types of energy barriers that must be overcome, such as the case of simulating protein adsorption behavior, temperature is a much more versatile parameter to use since it operates on all degrees of freedom at the same time, thus enabling all energy barriers to be overcome by the addition of thermal energy to the molecular system without requiring a priori knowledge of where the energy barriers are.

One of the most widely used methods to accomplish this type of sampling is known as "replica-exchange molecular dynamics" (REMD) [81–83], which is based on a method previously developed for MC simulations called "parallel tempering" [84]. With this method, independent MD simulations are run in parallel for a series of replicas of a given molecular system, with each replica being run at an increasingly higher temperature level above the baseline temperature of interest, e.g., 298 K. After a short user-designated time period (e.g., 250 steps of MD), a statistical mechanics-based exchange algorithm (similar to that used in a Metropolis MC exchange process [36]) is applied to compare the potential energy levels between replicas at neighboring temperature levels. If the exchange is accepted, then the temperatures between the pair of replicas are swapped. If it is not, the replicas remain at their prior temperature levels for another sampling cycle, following which, the swapping decision process is repeated again. A diagram depicting this process is shown in Figure 4.7. By this method, a replica that is trapped in a local low-energy well that actually represents a relative high-energy state of the system tends to be exchanged upward in temperature; this procedure provides additional thermal energy to help a given replica escape from the local energy well and explore other states of the system. Similarly, replicas that happen to be in relatively low-energy states tend to be exchanged downward in temperature, thus increasing their

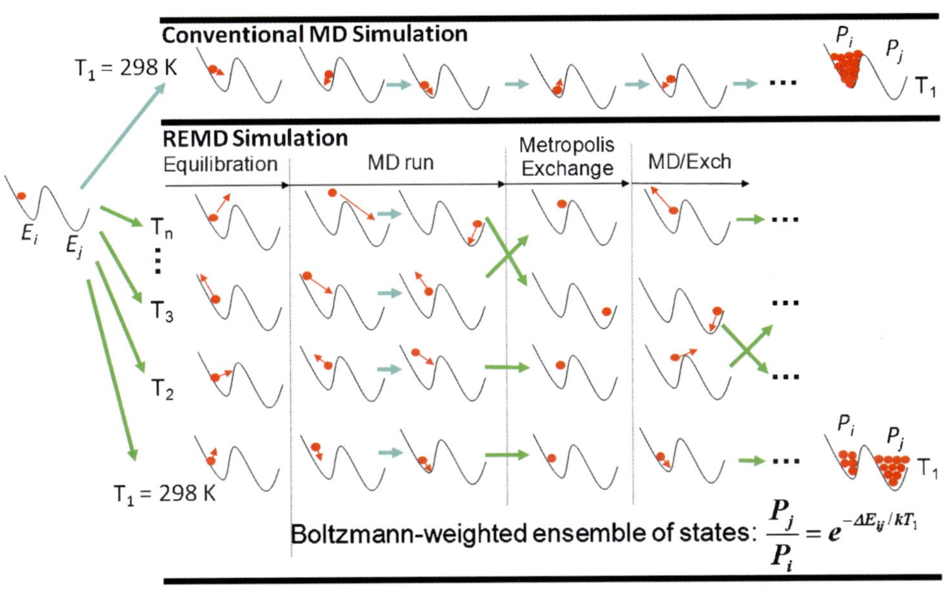

Figure 4.7. Schematic comparison of a conventional MD simulation compared with an REMD simulation for a simple model system containing two potential energy wells separated by an energy barrier. As in Figure 4.6, the conventional MD simulation is shown to be trapped in a local low-energy well during the entire simulation, resulting in a poorly sampled system. The REMD simulation uses thermal energy to enable the system to readily cross the energy barrier. The Metropolis-type exchange procedure used to swap temperature levels enables the development of a Boltzmann-weighted ensemble of states being sampled at the baseline temperature, thus representing a properly sampled, equilibrated system with ensemble-average properties that should be comparable to experimental measurements. (Adapted from [25]).

probability of ending up being sampled at the baseline temperature. The effect of this somewhat complicated process is to greatly enhance the sampling of the conformational phase space of a molecular system and, in the end, generate a Boltzmann-weighted ensemble of states at each temperature level. The resulting sampled distribution is no longer a time sequence as the one generated by a conventional MD simulation, but rather it represents an equilibrated system of states from which ensemble-averaged thermodynamic properties of the system can be calculated; the calculated values should be comparable to experimental measurements of the same system under equilibrium conditions. While this is an extremely useful computational technique, it can require that a very large number of replicas be used in order to span the temperature range that is necessary to be used to enhance energy-barrier crossing, and it still requires that the simulations be run for a sufficiently long time in order to obtain a properly converged ensemble of states at the baseline temperature of interest. The combination of these two requirements thus means that a very large amount of computing resources is often needed in order to use this method. In fact, for a very large molecular system, as is required for the simulation of protein adsorption behavior with explicit solvation, the use of a conventional REMD method may require that hundreds of computer processors be run in parallel, thus potentially making it prohibitively expensive to use.

To address these limitations, many different versions of this method have been, and are being, developed to improve computational efficiency [85–90]. In this respect, methods that vary the potential energy function instead of temperature are referred to as "Hamiltonian replica exchange methods" [91, 92]. To address the limitations of conventional REMD simulations for application to a large system simulation, such as would be needed to simulate protein adsorption behavior, we have recently developed a new advanced sampling method, which is called "Temperature Intervals with Global Energy Reassignment" (TIGER) [89, 90]; a revised method, currently under development, is called "TIGER2." These methods effectively uncouple the number of replicas that must be used to span a given temperature range from the size of the simulation, thus enabling a REMD-like simulation to be conducted with however, many processors that a given user has on hand to perform the simulation.

When performing simulations of protein-surface interactions, it is generally necessary to combine both energy-biasing and temperature-based advanced sampling algorithms to control the behavior of the molecular system. While elevated temperature is very useful to overcome the multiple energy barriers that are needed to be surmounted to sample the different conformational states of the protein on the material surface, elevated temperature also tends to cause the protein to desorb and diffuse away from that surface. To overcome this type of problem, a constraining potential can be used in the simulation, like that described in Eq. (4.9), to cause the protein to be maintained close to the surface. In this case, the restraining potential would only be turned on if the center of gravity of the protein moved further than a designated distance away from the material surface; the restraining potential then applies a force to the protein to push it back toward the material surface. In addition, if the surface being modeled is a crystalline polymer, elevated temperature will tend to cause the polymer to melt and transition into an amorphous state; this transition represents an undesirable event. In such a case, a second thermostat may be used during the simulation to maintain the atoms of the material surface at a desirable temperature level (e.g., 298 K), while the temperature of the protein and aqueous solution is allowed to be elevated to overcome the activation energy barriers of the system [93]; such an approach is necessary to explore how interactions of the protein with the material surface influence the folded state of the protein.

4.4. Future Directions

New methods and algorithms in the computational chemistry field are continually being developed and refined to improve current molecular simulation capabilities. For example, as mentioned in the section "The Empirical Force Field Equation," there is a substantial effort currently underway for the development of polarizable empirical force fields with force field parameterization that is able to adapt to the local surroundings of an atom during the simulation [39–41]; this approach should then greatly enhance the transferability of a force field to different molecular environments. This capability would be particularly useful for protein adsorption simulations to get around issues related to the appropriateness of a given force field to accurately represent this type of molecular system. Another area of ongoing research is the development of empirical force field methods that provide for bond breaking and forming capabilities [94–96]. This capability would enable the protonation state of functional groups to shift as a function of their environment; this result can have a substantial effect on protein adsorption behavior as well as enabling simulations to be conducted to represent the hydrolysis of hydrolytically degradable polymers (such as poly[glycolic acid] and poly[lactic acid]), and the subsequent influence of these chemical processes on protein adsorption behavior. Unfortunately, at this time, neither one of these types of methods is computationally efficient enough to be used for MD simulations of protein adsorption behavior; their implementation will have to wait until computational methods and computational power increases well beyond current capabilities.

One of the most promising areas of new development that will most directly impact the ability to predict protein-surface interactions, and potentially even cellular response, is the development of multiscale modeling methods [97–99]. These methods involve the development and application of coarse-graining techniques to combine groups of atoms, such as those making up either a single amino acid residue or a monomer unit of a polymer chain, into a single particle with force field parameterization adjusted to capture the general molecular behavior of that group of atoms. Conceptually, coarse graining can then be further applied in a sequential manner to join length scales all the way from the atomic to the macroscopic scale, with time scales of the simulation similarly extending from nanoseconds to seconds and longer. However, as with all of the other methods described in this chapter, these types of methods can only be realized if the biomaterials field itself works to develop and validate them for biomaterials-related applications.

4.5. Concluding Remarks

In this chapter a general overview of the topic of molecular simulation was presented with a focus on the use of MD methods for the simulation of protein-surface interactions. Emphasis was placed on addressing some of the key issues that must be considered when conducting simulations of peptide and protein adsorption behavior; namely, the validity of force field parameterization for the system being considered, the accurate representation of solvation effects, and statistical sampling to obtain converged results that are representative of an equilibrated system.

When these methods are properly developed and applied, they have the potential to provide the biomaterials field with an extremely valuable tool for the design of surfaces at the atomic level that will enable the adsorption behavior and subsequent bioactivity of adsorbed proteins to be manipulated and controlled in an optimal manner. This capability should directly translate to the development of biomaterial systems with improved performance for a broad range of applications in biomedical engineering and biotechnology. This situation, however, will only be realized if the biomaterials field will recognize this potential and promote further development of methods specific for simulation of protein-surface interactions.

Acknowledgments

I thank my colleague Dr. Steven Stuart, Department of Chemistry, Clemson University, for numerous helpful discussions over the years regarding molecular simulation and statistical thermodynamics, Mr. Galen Collier, a current doctoral student in my group for assistance with Fig. 4.5, and Dr. Feng Wang, a former doctoral student in my group, for helpful discussions regarding the protein-folding funnel and protein adsorption. I also thank the NIH and NSF for providing funding support for my research program on the development of molecular simulation methods to simulate protein-surface interactions: NIH R01 EB006163, R01 GM074511, the NJ Center for Biomaterials RESBIO (NIH, P41 EB001046), and the Center for Advanced Engineering Fibers and Films (CAEFF, NSF-ERC, EPS-0296165).

References

1. Latour RA (2008) Biomaterials: Protein-surface interactions. In: The Encyclopedia of Biomaterials and Bioengineering, 2nd Ed. Informa Healthcare, New York, NY.
2. Castner DG, Ratner BD (2002) Biomedical surface science: Foundations to frontiers. Surf. Sci. 500:28–60.
3. Hlady V, Buijs J (1996) Protein adsorption on solid surfaces. Curr. Opin. Biotechnol. 7:72–77.
4. Tsai WB, Grunkemeier JM, et al. (2002) Platelet adhesion to polystyrene-based surfaces preadsorbed with plasmas selectively depleted in fibrinogen, fibronectin, vitronectin, or von Willebrand's factor. J. Biomed. Mater. Res. 60:348–359.
5. Dee KC, Puleo DA, et al. (2002) Protein-Surface Interactions, Chapter 3. Wiley, Hoboken, NJ.
6. Geelhood SJ, Horbett TA, et al. (2007) Passivating protein coatings for implantable glucose sensors:Evaluation of protein retention. J. Biomed. Mater. Res. B 81B:251–260.
7. Kusnezow W, Hoheisel JD (2002) Antibody microarrays: promises and problems. Biotechniques Suppl. 14–23.
8. Schüler C, Carusa F (2000) Preparation of enzyme multilayers on colloids for biocatalysis. Macromol. Rapid Comm. 21:750–753.
9. Yu AM, Liang ZJ, et al. (2005) Enzyme multilayer-modified porous membranes as biocatalysts. Chem. Mater. 17:171–175.
10. Beck DAC, Daggett V (2004) Methods for molecular dynamics simulations of protein folding/unfolding in solution. Methods 34:112–120.
11. Brooks III CL (1998) Simulations of protein folding and unfolding. Curr. Opin. Struct. Biol. 8:222–226.
12. Gnanakaran S, Nymeyer H, et al. (2003) Peptide folding simulations. Curr. Opin. Struct. Biol. 13:168–174.
13. Wang W, Donini O, et al. (2001) Biomolecular simulations: Recent developments in force fields, simulations of enzyme catalysis, protein-ligand, protein-protein, and protein-nucleic acid noncovalent interactions. Annu. Rev. Biophys. Biomol. Struct. 30:211–243.
14. Ehrlich LP, Nilges M, et al. (2005) The impact of protein flexibility on protein-protein docking. Proteins 58: 126–133.
15. Chandrasekaran V, Ambati J, et al. (2007) Molecular docking and analysis of interactions between vascular endothelial growth factor (VEGF) and SPARC protein. J. Mol. Graph. Model. 26:775–782.
16. Hyvonen MT, Oorni K, et al. (2001) Changes in a phospholipid bilayer induced by the hydrolysis of a phospholipase A(2) enzyme: A molecular dynamics simulation study. Biophys. J. 80:565–578.
17. Bond PJ, Sansom MSP (2006) Insertion and assembly of membrane proteins via simulation. J. Am. Chem. Soc. 128:2697–2704.

18. Muegge I (2003) Selection criteria for drug–like compounds. Med. Res. Rev. 23:302–321.
19. Bernard D, Coop A, et al. (2005) Conformationally sampled pharmacophore for peptidic delta opioid ligands. J. Med. Chem. 48:7773–7780.
20. Chen HF (2008) Computational study of the binding mode of epidermal growth factor receptor kinase inhibitors. Chem. Biol. Drug Des. 71:434–446.
21. Nick B, Suter UW (2001) Solubility of water in polymers – Atomistic simulations. Comput. Theor. Polym. Sci. 11:49–55.
22. Pan R, Liu XK, et al. (2007) Molecular simulation on structure–property relationship of polyimides with methylene spacing groups in biphenyl side chain. Comp. Mater. Sci. 39:887–895.
23. Tarmyshov KB, Muller–Plathe F (2007) The interface between platinum(111) and poly(vinyl alcohol) melt: A molecular dynamics study. Soft Mater. 5:135–154.
24. Zhang J, Liang Y, et al. (2007) Study of the molecular weight dependence of glass transition temperature for amorphous poly(l–lactide) by molecular dynamics simulation. Polymer 28:4900–4905.
25. Latour RA (2008) Molecular simulation of protein–surface interactions: Benefits, problems, solutions, and future directions. Biointerphases 3:FC2–FC12.
26. Brandon C, Tooze J (1999) Introduction to Protein Structure, 2nd Ed. Garland, New York, NY.
27. Voet D, Voet JG, et al. (2002) Fundamentals of Biochemistry. Wiley, New York, NY.
28. Bryngelson JD, Onuchic JN, et al. (1995) Funnels, pathways, and the energy landscape of protein–folding – A synthesis. Proteins 21:167–195.
29. Onuchic JN, Wolynes PG, et al. (1995) Toward an outline of the topography of a realistic protein–folding funnel. Proc. Natl. Acad. Sci. USA. 92:3626–3630.
30. Wolynes PG, Luthey–Schulten Z, et al. (1996) Fast folding experiments and the topography of protein folding energy landscapes. Chem. Biol. 3:425–432.
31. Agashe M, Raut V, et al. (2005) Molecular simulation to characterize the adsorption behavior of a fibrinogen gamma–chain fragment. Langmuir 21:1103–1117.
32. Lee C–S, Belfort G (1989) Changing activity of ribonuclease A during adsorption: A molecular explanation. Biophysics 86:8392–8396.
33. Lenk T, Horbett T, et al. (1991) Infrared spectroscopic studies of time–dependent changes in fibrinogen adsorbed to polyurethanes. Langmuir 7:1755–1764.
34. Chinn JA, Posso SE, et al. (1992) Postadsorptive transitions in fibrinogen adsorbed to polyurethanes – Changes in antibody–binding and sodium dodecyl–sulfate elutability. J. Biomed. Mater. Res. 26:757–778.
35. Agnihortri A, Siedlecki CA (2004) Time–dependent conformational changes in fibrinogen measured by atomic force micrscopy. Langmuir 20:8846–8852.
36. Leach AR (1996) Molecular Modelling. Principles and Applications. Pearson Education, Harlow, UK.
37. Zhou J, Zheng J, et al. (2004) Molecular simulation studies of the orientation and conformation of cytochrome c adsorbed on self–assembled monolayers. J. Phys. Chem. B 108:17418–17424.
38. Oostenbrink C, Villa A, et al. (2004) A biomolecular force field based on the free enthalpy of hydration and solvation: The GROMOS force–field parameter sets 53A5 and 53A6. J. Comput. Chem. 25:1656–1676.
39. Rick SW, Stuart SJ (2002) Potentials and algorithms for incorporating polarizability in computer simulations. In: Reviews in Computational Chemistry, Wiley, New York, NY.
40. Kaminski GA, Stern HA et al. (2002) Development of a polarizable force field for proteins via ab initio quantum chemistry: First generation model and gas phase tests. J. Comput. Chem. 23:1515–1531.
41. Warshel A, Kato M et al. (2007) Polarizable force fields: History, test cases, and prospects. J. Chem. Theory Comput. 3:2034–2045.
42. Wei Y, Latour RA (2008) Determination of the adsorption free energy for peptide–surface interactions by SPR spectroscopy. Langmuir 24:6721–6729.
43. Wang F, Stuart SJ, et al. (2008) Calculation of adsorption free energy for solute–surface interactions using biased replica–exchange molecular dynamics. Biointerphases 3:9–18.
44. Raut VP, Agashe M, et al. (2005) Molecular dynamics simulations of peptide–surface interactions. Langmuir 21:1629–1639.
45. Frenkel D, Smit B (1996) Understanding Molecular Simulation. Academic, New York, NY.
46. Ryckaert JP, Ciccotti G, et al. (1977) Numerical integration of the cartesian equation of motion of a system with constraints: molecular dynamics of n–alkanes. J. Comput. Phys. 23:327–341.
47. Andersen HC (1983) Rattle: A "velocity" version of the Shake algorithm for molecular dynamics calculations. J. Comput. Phys. 52:24–34.
48. MacKerell AD, Bashford D, et al. (1998) All–atom empirical potential for molecular modeling and dynamics studies of proteins. J. Phys. Chem. B 102:3586–3616.

49. Darden T, York D, et al. (1993) Particle mesh Ewald: An N–log(N) method for Ewald sums in large systems. J. Chem. Phys. 98:10089–10092.
50. Essmann U, Perera L, et al. (1995) A smooth particle mesh Ewald method. J. Chem. Phys. 103:8577–8593.
51. West JB (1985) Physiology of the Body Fluids, Chapter 26. Williams, Baltimore, MD.
52. Glattli A, Daura X, et al. (2002) Derivation of an improved simple point charge model for liquid water: SPC/A and SPC/L. J. Chem. Phys. 116:9811–9828.
53. Mark P, Nilsson L (2001) Structure and dynamics of the TIP3P, SPC, and SPC/E water models at 298 K. J. Phys. Chem. A 105:9954–9960.
54. Jorgensen WL, Chandrasekhar J, et al. (1983) Comparison of simple potential functions for simulating liquid water. J. Chem. Phys. 79:926–935.
55. Jorgensen WL, Madura JD (1985) Temperature and size dependence for Monte–Carlo simulations of TIP4P water. Mol. Phys. 56:1381–1392.
56. Jorgensen WL, Jenson C (1998) Temperature dependence of TIP3P, SPC, and TIP4P water from NPT Monte Carlo simulations: Seeking temperatures of maximum density. J. Comput. Chem. 19:1179–1186.
57. Horn HW, Swope WC, et al. (2004) Development of an improved four–site water model for biomolecular simulations: TIP4P–Ew. J. Chem. Phys. 120:9665–9678.
58. Horn HW, Swope WC, et al. (2005) Characterization of the TIP4P–Ew water model: Vapor pressure and boiling point. J. Chem. Phys. 123:1–12.
59. Mahoney MW, Jorgensen WL (2001) Diffusion constant of the TIP5P model of liquid water. J. Chem. Phys. 114:363–366.
60. Schaefer M, Bartels C, et al. (1999) Solution conformations of structured peptides: Continuum electrostatics versus distance–dependent dielectric functions. Theor. Chem. Acc. 101:194–204.
61. Sun Y, Latour RA (2006) Comparison of implicit solvent models for the simulation of protein–surface interactions. J. Comp. Chem. 27:1908–1922.
62. Sharp KA, Honig B (1990) Calculating total electrostatic energies with the nonlinear Poisson–Boltzmann equation. J. Phys. Chem. 94:7684–7692.
63. Bertonati C, Honig B, et al. (2007) Poisson–Boltzmann calculations of nonspecific salt effects on protein–protein binding free energies. Biophys. J. 92:1891–1899.
64. Still WC, Tempczyk A, et al. (1990) Semianalytical treatment of solvation for molecular mechanics and dynamics. J. Am. Chem. Soc. 112:6127–6129.
65. Dominy BN, Brooks III CL (1999) Development of a generalized Born model parameterization for proteins and nucleic acids. J. Phys. Chem. B 103:3765–3773.
66. Bashford D, Case DA (2000) Generalized Born models of macromolecular solvation effects. Annu. Rev. Phys. Chem. 51:129–152.
67. Feig M, Brooks III CL (2004) Recent advances in the development and application of implicit solvent models in biomolecule simulations. Curr. Opin. Struct. Biol. 14:217–224.
68. Feig M, Onufriev A, et al. (2004) Performance comparison of generalized born and Poisson methods in the calculation of electrostatic solvation energies for protein structures. J. Comput. Chem. 25:265–284.
69. Formaneck MS, Cui Q (2006) The use of a generalized born model for the analysis of protein conformational transitions: A comparative study with explicit solvent simulations for chemotaxis Y protein (CheY). J. Comput. Chem. 27:1923–1943.
70. Sun Y, Dominy BN, et al. (2007) Comparison of solvation–effect methods for the simulation of peptide interactions with a hydrophobic surface. J. Comput. Chem. 28:1883–1892.
71. McQuarrie DA (1976) The Canonical Ensemble, Chapter 2. In: Statistical Thermodynamics, Harper, New York, NY.
72. Beutler TC, van Gunsteren WF (1994) The computation of a potential of mean force – Choice of the biasing potential in the umbrella sampling technique. J. Chem. Phys. 100:1492–1497.
73. Friedman RA, Mezei M (1995) The potentials of mean force of sodium–chloride and sodium dimethylphosphate in water – An application of adaptive umbrella sampling. J. Chem. Phys. 102:419–426.
74. Bartels C, Karplus M (1998) Probability distributions for complex systems: Adaptive umbrella sampling of the potential energy. J. Phys. Chem. B 102:865–880.
75. Bartels C, Schaefer M, et al. (1999) Adaptive umbrella sampling of the potential energy: Modified updating procedure of the umbrella potential and application to peptide folding. Theor. Chem. Acc. 101:62–66.
76. Depaepe JM, Ryckaert JP, et al. (1993) Sampling of molecular–conformations by molecular–dynamics techniques. Mol. Phys. 79:515–522.
77. Souaille M, Roux B (2001) Extension to the weighted histogram analysis method: Combining umbrella sampling with free energy calculations. Comput. Phys. Commun. 135:40–57.
78. Harvey SC, Prabhakaran M (1987) Umbrella sampling – Avoiding possible artifacts and statistical biases. J. Phys. Chem. 91:4799–4801.

79. Kumar S, Bouzida D, et al. (1992) The weighted histogram analysis method for free–energy calculations of biomolecules. 1. The Method. J. Comput. Chem. 13:1011–1021.
80. Kumar S, Rosenberg JM, et al. (1995) Multidimensional free–energy calculations using the weighted histogram analysis method. J. Comput. Chem. 16:1339–1350.
81. Sugita Y, Okamoto Y (1999) Replica–exchange molecular dynamics method for protein folding. Chem. Phys. Lett. 314:141–151.
82. Garcia AE, Sanbonmatsu KY (2001) Exploring the energy landscape of a beta hairpin in explicit solvent. Proteins 42:345–354.
83. Gallicchio E, Andrec M, et al. (2005) Temperature weighted histogram analysis method, replica exchange, and transition paths. J. Phys. Chem. B. 109:6722–6731.
84. Hansmann UHE (1997) Parallel tempering algorithm for conformational studies of biological molecules. Chem. Phys. Lett. 281:140–150.
85. Mitsutake A, Sugita Y, et al. (2001) Generalized–ensemble algorithms for molecular simulations of biopolymers. Biopolymers 60:96–123.
86. Okamoto Y (2004) Generalized–ensemble algorithms: Enhanced sampling techniques for Monte Carlo and molecular dynamics simulations. J. Mol. Graph. Model.22: 425–439.
87. Okur A, Wickstrom L, et al. (2006) Improved efficiency of replica exchange simulations through use of a hybrid explicit/implicit solvation model. J. Chem. Theory Comput. 2:420–433.
88. Okur A, Roe DR, et al. (2007) Improving convergence of replica–exchange simulations through coupling to a high–temperature structure reservoir. J. Chem. Theory Comput. 3:557–568.
89. Li XF, O'Brien CP, et al. (2007) An improved replica–exchange sampling method: Temperature intervals with global energy reassignment. J. Chem. Phys. 127:1–10.
90. Li XF, Stuart SJ, Latour RA (2009) TIGER2: An improved algorithm for temperature intervals with global exchange of replicas, J. Chem. Phys., in press.
91. Fukunishi H, Watanabe O, et al. (2002) On the Hamiltonian replica exchange method for efficient sampling of biomolecular systems: Application to protein structure prediction. J. Chem. Phys. 116:9058–9067.
92. Affentranger R, Tavernelli I, et al. (2006) A novel Hamiltonian replica exchange MD protocol to enhance protein conformational space sampling. J. Chem. Theory Comput. 2:217–228.
93. O'Brien CP, Stuart SJ, et al. (2008) Modeling of peptide adsorption interactions with a poly(lactic acid) surface. Langmuir 24:14115–14124.
94. Brenner DW, Shenderova OA, et al. (2002) A second–generation reactive empirical bond order (REBO) potential energy expression for hydrocarbons. J. Phys. Condens. Matter 14:783–802.
95. Ni B, Lee KH, et al. (2004) A reactive empirical bond order (REBO) potential for hydrocarbon–oxygen interactions. J. Phys. Condens. Matter 16:7261–7275.
96. Liu A, Stuart SJ (2008) Empirical bond–order potential for hydrocarbons: Adaptive treatment of van der Waals interactions. J. Comp. Chem. 29:601–611.
97. Chu JW, Izveko S, et al. (2006) The multiscale challenge for biomolecular systems: Coarse–grained modeling. Mol. Simul. 32:211–218.
98. de Pablo JJ, Curtin WA (2007) Multiscale modeling in advanced materials research: Challenges, novel methods, and emerging applications. MRS Bull. 32:905–911.
99. Zhou J, Thorpe IF, et al. (2007) Coarse–grained peptide modeling using a systematic multiscale approach. Biophys. J. 92:4289–4303

5

Biomolecule–Nanomaterial Interactions: Effect on Biomolecule Structure, Function, and Stability

Ravindra C. Pangule, Shyam Sundhar Bale, Dhiral A. Shah, Amit Joshi, Prashanth Asuri, Jonathan S. Dordick, and Ravi S. Kane

We have characterized the influence of protein–carbon nanotube interactions on protein structure and function using various techniques such as Fourier transform infrared spectroscopy, circular dichroism spectroscopy, and atomic force microscopy. This structure-based analysis revealed that different proteins interact with nanotubes differentially, consistent with the observed biological activity data. Furthermore, the high degree of surface curvature of the nanoscale support was found to play an important role in stabilizing proteins under denaturing conditions. Along with these fundamental studies, various applications of such highly active and stable nanotube–protein conjugates have been pursued, which include self-cleaning nanobiocomposite films, interfacial biocatalysis in a biphasic medium, and synthesis of nanotube–nanoparticle hybrids, among others.

Abbreviations

ADH	alcohol dehydrogenase
AFM	atomic force microscopy
AGP	α1-acid glycoprotein
AOT	Aerosol-OT
BSA	bovine serum albumin
CALB	*Candida antarctica* lipase B
CD	circular dichroism
CNT	carbon nanotube
CT	α-chymotrypsin
d-SBP	deglycosylated soybean peroxidase
EDC	*N*-ethyl-*N′*-(3-dimethylaminopropyl) carbodiimide hydrochloride
EDX	energy dispersive X-ray
FAM	carboxyfluorescein

R.C. Pangule, S.S. Bale, D.A. Shah, A. Joshi, P. Asuri, J.S. Dordick, and R.S. Kane • Department of Chemical and Biological Engineering, Rensselaer Polytechnic Institute, Troy, NY 12180, USA

D.A. Puleo and R. Bizios (eds.), *Biological Interactions on Materials Surfaces*,
DOI 10.1007/978-0-387-98161-1_5,

FT-IR	Fourier transform infrared
HOPG	highly ordered pyrolytic graphite
HRP	horseradish peroxidase
HSA	human serum albumin
Lys	lysozyme
MJL	*Mucor javanicus* lipase
MWNT	multiwalled nanotube
NaDDBS	sodium dodecylbenzene sulfonate
NHS	*N*-hydroxysuccinimide
NIR	near infrared
PAGE	polyacrylamide gel electrophoresis
PLL	poly-L-lysine
PMMA	poly(methyl methacrylate)
PMSF	phenylmethansulfonyl fluoride
ROS	reactive oxygen species
SAM	self-assembled monolayer
SBP	soybean peroxidase
SC	subtilisin Carlsberg
SWNT	single-walled nanotube
TEM	transmission electron microscopy
TRY	trypsin
UV	ultraviolet

5.1. Introduction

There has been considerable progress in the synthesis of nanomaterials with unique optical, electrical, and physicochemical properties [1–4]. Recently, there has been an increasing interest in realizing their wide-ranging applications in solid-state nanoelectronics, nanocomposites, nanolithography, chemical and biomolecule sensing, biomedical imaging and diagnostics, and targeted delivery for cancer therapy [5–11]. The applications of such biomolecule-functionalized nanomaterials have been described in recent reviews [12–15]. For example, Kane and coworkers [12, 13] reviewed recent advances in understanding and applying protein–nanomaterial interactions in nanomaterial assembly, molecular imaging, targeted cancer therapy, and biomolecule delivery. Katz et al. [14] provided a broad overview of the synthesis of biomolecule–nanoparticle hybrid systems and their application in the generation of two-dimensional (2D) and three-dimensional (3D) architectures in solutions and on surfaces. Lacerda et al. [15] summarized recent work on the application of carbon nanotubes in nucleic acid delivery via in vitro and in vivo models. While nanomaterials, in general, offer numerous opportunities in materials, electronics, biotechnology, and medical research, there has been particular interest in understanding and controlling the interactions of carbon nanotubes with biomolecules (peptides, proteins, and nucleic acids [e.g., DNA and RNA]) and cells, and their ultimate use for biosensing, designing functional surfaces and coatings, biomolecule delivery, and targeted cancer therapy [7, 8, 16–22].

The use of carbon nanotubes in biomaterials for scaffolds or implants is another rapidly growing area of research. While some studies have provided evidence that carbon-based nanomaterials, e.g., single-walled nanotubes (SWNTs), multiwalled nanotubes (MWNTs), and fullerenes, exhibit low biocompatibility when dispersed in cell culture [23], there are some examples where carbon nanomaterial-based biomaterials have been found to be compatible with neural cells, osteoblasts, and fibroblasts [24–27]. It has recently been shown that the adsorption, conformation, and bioactivity of specific proteins such as vitronectin and fibronectin play an important role in facilitating the adhesion and proliferation of osteoblasts onto carbon nanofibers, nanophase alumina, and nanophase ceramics, thereby improving the efficacy of corresponding orthopedic and dental implants [27–30]. Clearly, protein–nanomaterial interactions are of great importance in designing such novel biomaterials and realizing their applications. Moreover, interactions between nanomaterials and other biomolecules such as DNA and enzymes are also important as the function of such hybrids is largely governed by the conformation of the biomolecules on the nanomaterial.

In this chapter, we review our own research on designing, characterizing, and utilizing nanomaterial–biomolecule conjugates. We have considered carbon nanotubes and proteins as model nanomaterials and biomolecules, respectively, because of their growing importance in the various aforementioned applications.

5.2. Structure and Function of Proteins on Carbon Nanotubes

In one of the earliest reports on protein–nanotube interactions, Karajanagi et al. [17] studied the structure and function of proteins adsorbed onto SWNTs, using enzymatic activity as a highly sensitive probe of protein function. This work represented the first detailed evaluation of both the structure and the function of proteins adsorbed onto carbon nanotubes. Two structurally and functionally distinct enzymes, soybean peroxidase (SBP) and α–chymotrypsin (CT), were adsorbed onto SWNTs (Figure 5.1a) to assess the influence of the hydrophobic, nanoscale environment on protein structure and function. Activity measurements for the SWNT–SBP conjugates indicated that SBP retained ~30% of its native activity when loaded onto nanotubes at 50% of maximal surface coverage, while CT retained no greater than 1% of its native activity. These results suggest that attachment onto SWNTs affects the structure of the two enzymes in a fundamentally different manner [17].

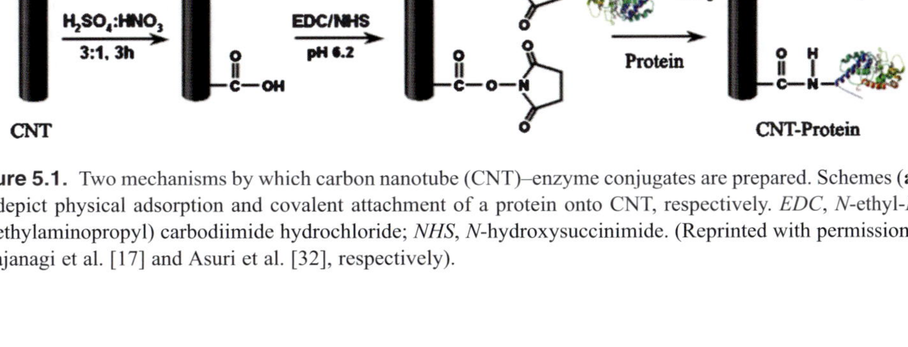

Figure 5.1. Two mechanisms by which carbon nanotube (CNT)–enzyme conjugates are prepared. Schemes (**a**) and (**b**) depict physical adsorption and covalent attachment of a protein onto CNT, respectively. *EDC*, *N*-ethyl-*N′*-(3-dimethylaminopropyl) carbodiimide hydrochloride; *NHS*, *N*-hydroxysuccinimide. (Reprinted with permission from Karajanagi et al. [17] and Asuri et al. [32], respectively).

Next, transmission-mode Fourier transform infrared (FT-IR) spectroscopy and atomic force microscopy (AFM) were used to characterize the changes in secondary structure of the proteins and overall structural perturbations, respectively. The FT-IR spectra were analyzed in the amide I region, between 1,600 and 1,700 cm^{-1}; analysis of this region has been previously used to obtain a quantitative estimation of the secondary structural features of proteins [31]. As shown in Table 5.1, both SBP (minor) and CT (major) undergo changes in secondary structure (13% and 44% overall change, respectively). Figure 5.2 shows AFM images of SBP and CT adsorbed onto SWNTs. In both images, regions of the nanotube not covered by enzyme had a height of ~4 nm, indicating the presence of bundles of SWNTs. In the case of SWNT–SBP conjugates, distinct globular structures were observed, as shown in Figure 5.2a; line scans suggested that the height of these structures ranged from 4 to 6 nm, which matched very well with the molecular dimensions of SBP ($6.1 \times 3.5 \times 4.0$ nm^3). On the other hand, the AFM image of SWNT–CT conjugates shows a layer of CT on the SWNT surface. The height of the layer varied between 2 and 20 nm, which is far different from the molecular size

Table 5.1. Secondary structure of SBP and CT (solution phase and adsorbed onto SWNTs) as determined by FT-IR spectroscopy. (Reprinted with permission from Karajanagi et al. [17].)

Sample	% α-Helix	% β-Sheet
Solution SBP	36.1 ± 1.2	25.1 ± 2.5
SBP adsorbed onto SWNTs	27.9 ± 4.1	20.6 ± 6.9
Solution CT	13.6 ± 3.5	50.0 ± 2.4
CT adsorbed onto SWNTs	31.5 ± 2.9	23.5 ± 5.1

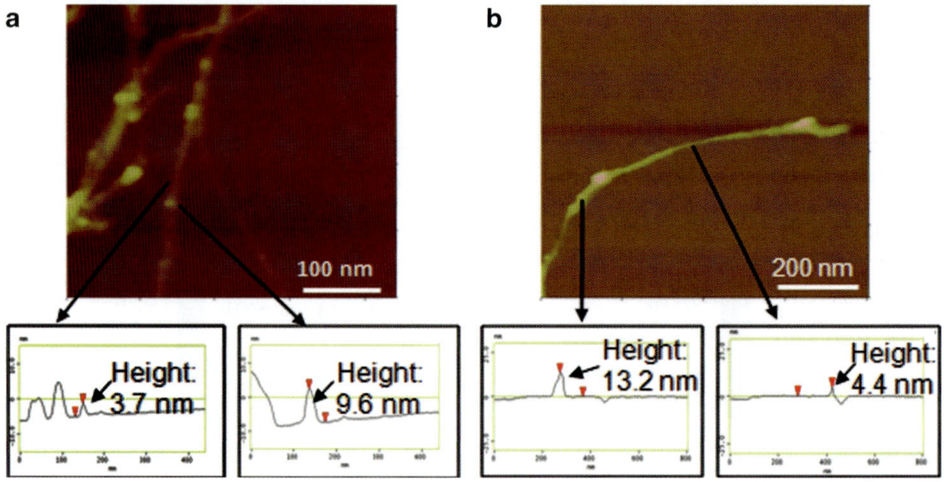

Figure 5.2. AFM images of SBP and CT adsorbed onto SWNTs at half-maximal surface coverage. The globular structures seen on SWNTs in image (**a**) represent SBP molecules. Image (**b**) shows a nonuniform layer of CT on SWNT. (Reprinted with permission from Karajanagi et al. [17]).

of CT (5.1 × 4.0 × 4.0 nm^3), suggesting that significant unfolding and aggregation of CT occurred on the nanotubes [17].

Collectively, these results suggest that retention of protein structure and function are related, and that the extent of retention of protein structure and activity on carbon nanotubes is protein specific.

In addition to water-insoluble conjugates formed by non-covalently attaching proteins onto the nanotubes, we have also designed active and water-soluble nanotube–protein conjugates in which the protein is covalently attached to the nanotube (Figure 5.1b). We first synthesized water-soluble MWNT–SBP conjugates by a covalent attachment method [32]. MWNT–SBP conjugates were prepared with an enzyme loading of at least 0.15 mg/mg of MWNT; the resulting conjugates were active (see Table 5.3a) and stable in solution for as long as a week. The conjugates were tested for their activity with *p*-cresol as the substrate, which followed conventional Michaelis–Menten kinetics; the kinetic parameters obtained are shown in Table 5.2. To further characterize the nanotube–enzyme conjugates, we employed Hammett analysis, which can be used to probe local structural perturbations in the active site of the enzyme. Kinetic parameters for native SBP and MWNT–SBP conjugates were determined by activity measurements against a range of phenolic substrates with varying electronic parameters. Hammett coefficients (ρ), in each case, were then determined using these kinetic parameters in Eq. (5.1):

$$\log\left(\frac{V_{max}}{K_M}\right) = \sigma\rho + \text{constant}, \tag{5.1}$$

where σ is the substrate electronic parameter (whose values vary between −0.24 and +0.24 for electron donating [*p*-OC_2H_5, *p*-CH_3] and electron withdrawing [*p*-CH_2OH, *p*-Cl] phenolic substrates), V_{max} is the maximum reaction rate, and K_M represents the Michaelis constant. We obtained nearly identical values of Hammett coefficients for native SBP and MWNT–SBP conjugates (−1.6 ± 0.1 and −1.5 ± 0.2, respectively), indicating that SBP retains its intrinsic active site structure in MWNT–SBP conjugates.

We also recently reported the design and characterization of water-soluble conjugates of SWNTs with enzymes [33]. We were able to obtain high protein loadings (up to 1.3 mg/mg of SWNT; see Table 5.3b). The combination of high solubility and high enzyme loading enabled a detailed spectroscopic characterization of protein structure in the conjugates, which complements the kinetic analysis shown in Table 5.3b.

Figure 5.3 shows far-UV circular dichroism (CD) spectra for native horseradish peroxidase (HRP), covalently attached SWNT–HRP conjugates, and oxidized SWNTs alone. Quantification of the α-helix content by measuring the mean residue ellipticity at 222 nm indicated that HRP attached to SWNTs retained 68% of its native α-helix content (Table 5.3b); consistent with the significant retention of HRP activity [33].

Table 5.2. Kinetic parameters for native SBP and MWNT–SBP conjugates. (Reprinted with permission from Asuri et al. [32].)

	V_{max} (mM mg^{-1} s^{-1})	K_M (mM)	V_{max}/K_M (mg^{-1} s^{-1})
Native SBP	45.1	4.9	9.2
MWNT–SBP	24.8	7.2	3.4

Table 5.3. Loading values for various enzymes when attached covalently to MWNTs and SWNTs.

Table 5.3a

Enzyme	Loading (mg enzyme/ mg MWNT)	Relative activity (%)
SBP	0.172	55 ± 5
SC	0.168	54 ± 4
CALB	0.203	58 ± 7

SBP soybean peroxidase, *SC* subtilisin Carlsberg, *CALB Candida antarctica* lipase B

Table 5.3b

Enzyme	Loading (mg enzyme/mg SWNT)	Relative activity (%)	Relative structure (%)
HRP	1.3	49 ± 5	68 ± 4
SC	1.3	53 ± 4	76 ± 3
Lys	1.2	43 ± 6	63 ± 3

HRP horseradish peroxidase, *SC* subtilisin Carlsberg, *Lys* lysozyme

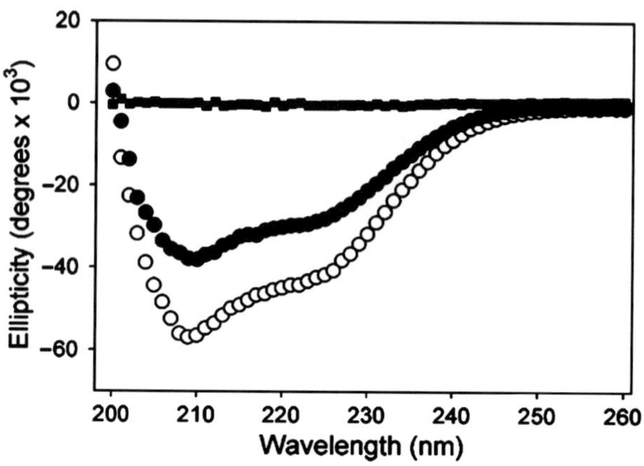

Figure 5.3. Far-UV CD spectra of native HRP (*empty circles*), HRP–SWNT conjugates (*filled circles*), and oxidized SWNTs alone (control; *filled squares*). (Reprinted with permission from Asuri et al. [33]).

5.3. Enhanced Protein Stability on Nanomaterials

The results described in Sect. 5.2 indicate that proteins can retain a significant fraction of their native activity when attached to nanoscale supports; the extent of such retention depends both on the protein and on the physicochemical properties of the nanomaterial. However, we have recently found that nanoscale supports can also *enhance* protein stability in harsh environments to a significantly greater extent than on macroscopic supports [34, 35]. This finding may be useful for preparing conjugates for incorporation into biomaterials for implants, preparation of functional coatings, and designing drug delivery systems.

As shown in Figure 5.4, SBP adsorbed onto SWNTs was significantly more stable under harsh denaturing conditions (e.g., 95°C or 100% methanol) than native SBP or SBP adsorbed onto conventional flat supports such as graphite flakes. SBP was ca. two times more stable on SWNTs than on relatively flat supports, under both conditions (Figure 5.4a, b).

We hypothesized that the enhanced stability of SBP resulted from a decrease in protein–protein interactions, one of the important factors that induce protein denaturation under harsh conditions. As depicted in Figure 5.5, a highly curved support, such as SWNTs,

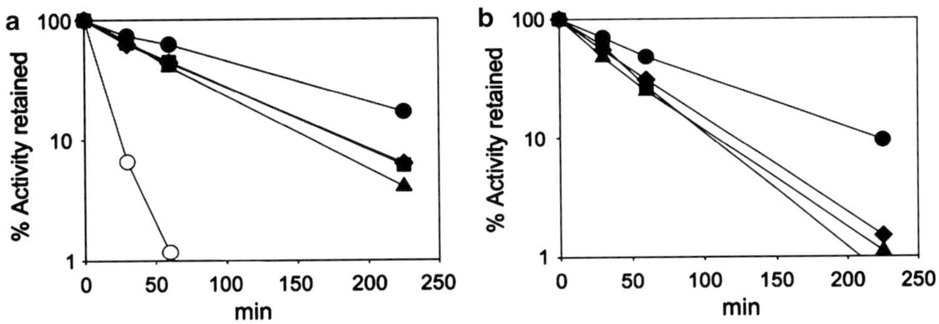

Figure 5.4. Time-dependent deactivation of native SBP (*open circles*) and SBP on various supports – SWNTs (*filled circles*), highly ordered pyrolytic graphite (HOPG) (*diamonds*), self-assembled monolayer (SAM) of undecanethiolate on gold (*squares*), and graphite flakes (*triangles*) (**a**) at 95°C and (**b**) in 100% methanol. The activities were normalized relative to the initial activity (activity at $t = 0$ min). Figure 5.4b does not contain percent activity data for native SBP as the enzyme showed no activity in 100% methanol. (Reprinted with permission from Asuri et al. [34]).

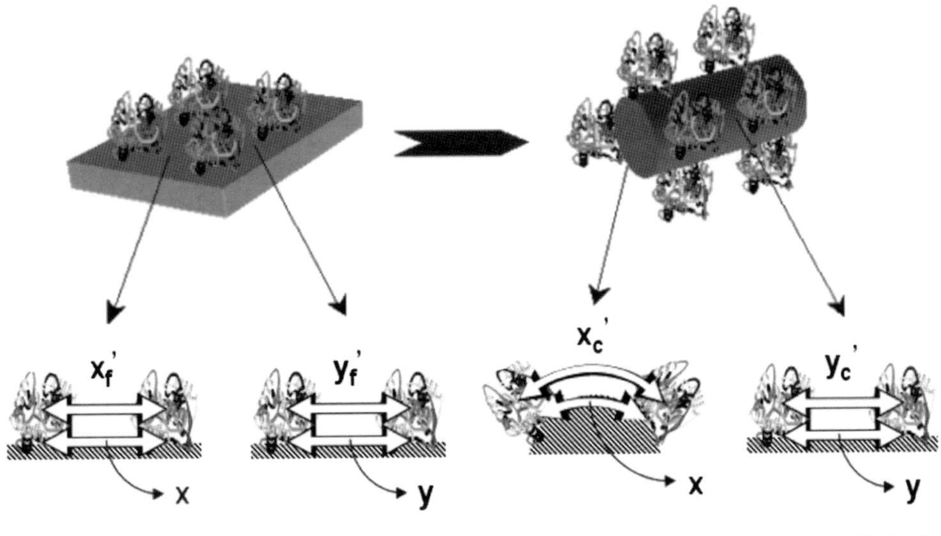

Figure 5.5. Schematic depicting the reduced lateral interactions between enzyme molecules on a cylindrical support, namely SWNT, compared with that observed on a flat graphite support. (Reprinted with permission from Asuri et al. [34]).

suppresses unfavorable lateral protein–protein interactions by keeping adjacent enzyme molecules separated. Specifically, a geometric model was developed wherein the center-to-center distance between neighboring molecules, measured along the circumference of the particle, is given by Eq. (5.2):

$$x_c^{'} = \frac{(R+r)}{R}x, \tag{5.2}$$

where R is the radius of the nanomaterial, r represents the average radius of an enzyme molecule, and x signifies the distance between adjacent proteins (measured along the protein–substrate interface). As depicted in Figure 5.5, for the same value of x, the center-to-center distance between adjacent molecules increases on highly curved cylindrical nanomaterials. As a result of this increased separation distance, there may be reduction in the interaction, which could in turn lead to the decrease in protein deactivation. In other words, this curvature effect could contribute to the greater protein stability on nanosupports when compared with that on a flat support.

To support the hypothesis that the observed stability was indeed a result of reduced protein–protein interactions, deactivation constants for SBP adsorbed onto SWNTs and graphite flakes were determined at different fractional surface coverage values on these supports. At low surface coverage where protein molecules are spaced far apart, protein–protein interactions would be nearly absent, resulting in similar rates of deactivation, irrespective of the support used for protein immobilization (as shown in Figure 5.6). On the other hand, at higher surface coverage, protein molecules are closer together, thereby leading to a greater likelihood of unfavorable protein–protein interactions that would enhance the rate of protein deactivation. Importantly, at higher surface coverage, SBP on SWNTs showed less protein deactivation compared with SBP on graphite flakes (Figure 5.6), thus demonstrating that surface *chemistry* is less critical than surface curvature in affecting protein stability [34].

To gain additional mechanistic insights, we introduced a dimensionless parameter ε to capture the effect of both surface curvature and surface coverage on protein–protein interactions. We determined the value of ε by taking into account the geometric mean of center-to-center distances between adjacent proteins measured along the two orthogonal axes (S). This

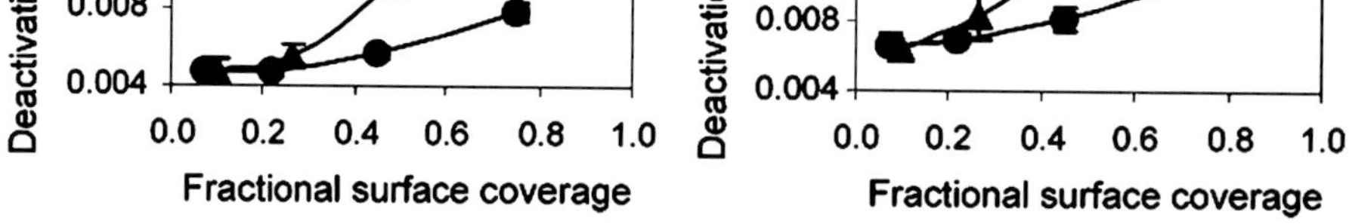

Figure 5.6. Deactivation constants for SBP adsorbed onto SWNTs (*circles*) and graphite flakes (*triangles*) as a function of surface coverage (**a**) at 95°C and (**b**) in 100% methanol. The deactivation constant is defined as the first-order exponential decay constant (k) in the equation $A/A_0 = \exp(-kt)$, where A and A_0 are the enzymatic activities at time $t = t$ and $t = 0$, respectively; k is determined from the slope of a straight-line fit through the plot of $\ln(A/A_0)$ versus time. (Reprinted with permission from Asuri et al. [34]).

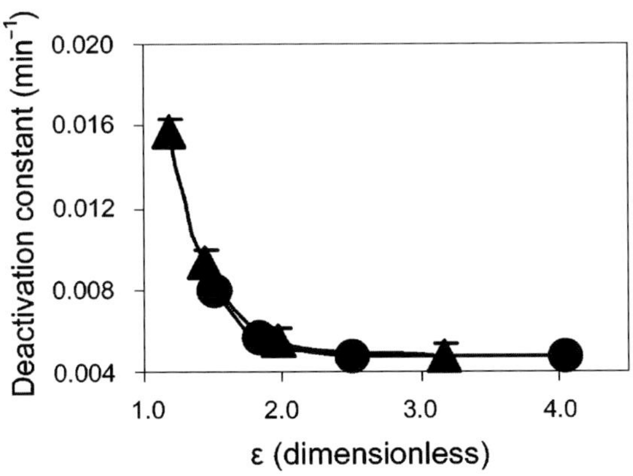

Figure 5.7. Influence of average center-to-center distance between adsorbed proteins on deactivation rate. Deactivation constants for SBP adsorbed onto SWNTs (*circles*) and graphite flakes (*triangles*) at 95°C plotted as a function of the dimensionless variable ε. (Reprinted with permission from Asuri et al. [34]).

mean value (S) in the case of a flat and a cylindrical support is given by $(x_f'y_f')^{1/2}$ and $(x_c'y_c')^{1/2}$, respectively, where x_f' and y_f' are center-to-center distances between neighboring molecules measured along the x and y axes, respectively, on a flat support, and x_c' and y_c' are the distances corresponding to protein molecules on a cylindrical support. Clearly, S is greater in the case of a cylindrical support compared with its flat counterpart as a result of surface curvature (Figure 5.5). The dimensionless parameter ε is then determined from the equation: $\varepsilon = S/S_m$, where S_m is the value of S calculated on a flat support at maximum surface coverage.

In Figure 5.7, the deactivation rate constants are plotted as a function of ε that now takes into account both surface coverage and surface curvature. The observed collapse of the data onto a single curve (Figure 5.7) provides strong evidence supporting our hypothesis. While a cylindrical support is curved in one dimension, spherical particles are curved in all dimensions, which should lead to even greater enhancement in protein stability due to further suppression of protein–protein interactions. Indeed, our recent experimental results support this hypothesis (data not shown) [35].

5.4. Functional Materials

In addition to an improved understanding of protein–nanotube conjugate structure, function, and stability, we have demonstrated various applications of such highly active and stable conjugates, which include self-cleaning nanobiocomposite films, interfacial biocatalysis in biphasic media, and nanomaterials assembly, among others.

5.4.1. Polymer–Nanotube–Enzyme Composites for Antifouling Applications

As discussed in Sects. 5.2 and 5.3, nanotube–enzyme conjugates were found to be active and highly stable. In this section, we show that nanotubes serve as ideal supports for incorporating enzymes into polymeric matrices due to their large surface area that leads to large functional enzyme loadings and to their high aspect ratio, which aids in the retention of enzyme–nanotube conjugates in the matrix. We recently described the generation of highly stable and active polymer–nanotube–enzyme composites that prevent fouling of material surfaces by proteins [36]. These nanocomposite films contained proteases that carried out proteolytic degradation of adsorbed proteins thereby preventing nonspecific protein adsorption. We demonstrated the ability of these nanocomposites to prevent fouling by two proteins, specifically human serum albumin (HSA) and fibrinogen, that are present in blood plasma and readily adsorb onto most surfaces. Composites incorporating conjugates of nanotubes with proteases such as subtilisin Carlsberg (SC) and trypsin (TRY) showed significant reduction in protein fouling (as shown in Figure 5.8). Further experiments were conducted using dextranase to test the films for hydrolysis of glucan-based polymers that are the major constituents of biofilms (data not shown).

Such antifouling nanocomposite films may ultimately find use in preventing the fouling of a variety of material surfaces ranging from membranes (for downstream bioprocessing and water purification) to medical instruments and implantable materials.

5.4.2. Nanotube-Assisted Protein Deactivation

A number of nanomaterials, including nanoshells and carbon nanotubes, strongly absorb near-infrared (NIR) light (700–1,100 nm) [22, 37, 38]. Since biological systems are

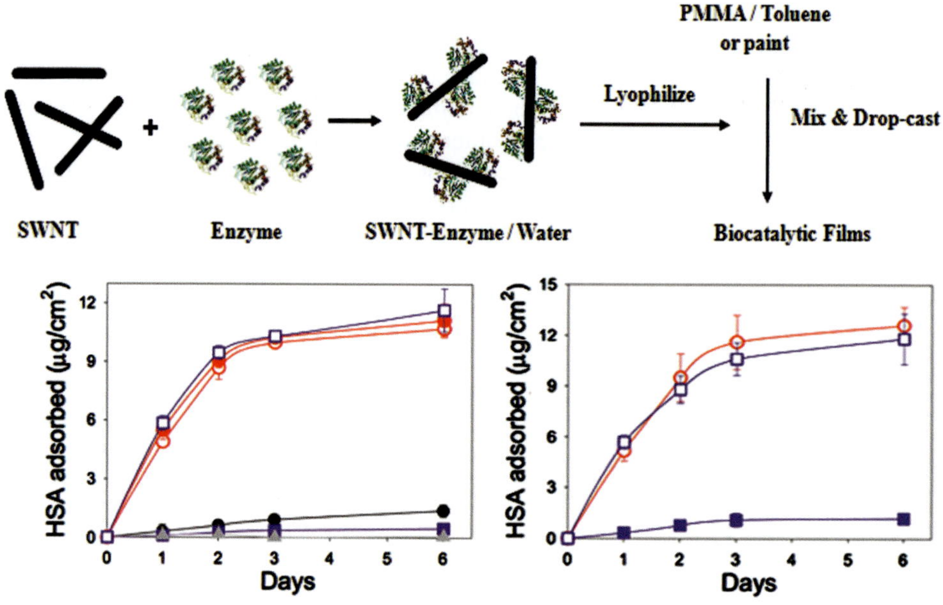

Figure 5.8. Antifouling properties of polymer–nanotube–enzyme composite films. (**a**) Schematic of preparation of biocatalytic films. (**b**) Amount of HSA adsorbed onto poly(methyl methacrylate) (PMMA) films (control; *red open circles*), PMMA–SWNT–SC films (*black-filled circles*), PMMA–SWNT–SC–TRY films (*blue-filled squares*), SWNT–SC–TRY buckypaper (*grey-filled triangles*), PMMA–SWNT–SBP (control films containing nonproteolytic enzyme; *red-filled circles*), and PMMA–SWNT–SC–TRY–PMSF films (control films containing protease conjugates preincubated with a serine protease inhibitor, phenylmethansulfonyl fluoride [PMSF]; *blue open squares*). (**c**) Amount of HSA adsorbed onto paint films (control; *red open circle*), paint–SWNT–SC–TRY films (*blue-filled squares*), and paint–SWNT–SC–TRY–PMSF films (control samples containing protease conjugates preincubated with PMSF, *blue open squares*). (Reprinted with permission from Asuri et al. [36]).

relatively transparent in this range of wavelengths, these nanomaterials have been utilized for NIR thermal therapy of tumors wherein the localized generation of heat, resulting from NIR irradiation of nanomaterials, causes cell death [22, 38]. We decided to test whether the ability of carbon nanotubes to absorb NIR radiation could also be used to provide local control over the activity of adsorbed proteins [39].

We adsorbed the enzyme alcohol dehydrogenase (ADH) from aqueous buffer onto SWNTs. ADH remained catalytically active on SWNTs, with specific activity ca. 70% relative to that of native ADH in aqueous buffer. To test the influence of NIR irradiation on enzyme activity, a solution of the SWNT–ADH conjugates in phosphate buffer was transferred to a quartz cuvette and exposed to radiation from a continuous wave laser at room temperature. Aliquots were removed periodically, and the enzymatic activity was determined. As shown in Figure 5.9, NIR irradiation resulted in a significant loss of enzymatic activity of SWNT–ADH conjugates. In contrast, NIR irradiation of a solution of ADH alone did not result in a loss of enzymatic activity.

Interestingly, this NIR-induced deactivation of ADH was inhibited in the presence of free radical quenchers, thus suggesting that the mechanism of protein deactivation was photochemical, i.e., mediated by reactive oxygen species (ROS) generated upon irradiation of SWNTs. We also confirmed that this phenomenon was broadly applicable by demonstrating the ability to deactivate protease SC, another protein different from ADH in function, and by using MWNTs instead of SWNTs [39].

This unique property of carbon nanotubes enables the nanotube-assisted protein deactivation to be exploited in a variety of contexts. As a first demonstration, we functionalized MWNTs with a peptide that binds to the heptameric subunit of the anthrax toxin. On exposing a solution containing a mixture of the polyvalent, functionalized MWNTs, the anthrax toxin heptamer, and ADH (an unrelated protein) to NIR radiation, we were able to selectively destroy the anthrax toxin heptamer [39]. This selective polyvalent recognition and destruction strategy may be broadly applicable for the neutralization of bacterial and viral pathogens and toxins.

Our next demonstration exploited the ability of carbon nanotubes to form ultrathin, flexible, and transparent films [40]. Hydroxyl radicals and superoxide anions are known to cause protein fragmentation and amino acid modification [41]. We therefore reasoned that

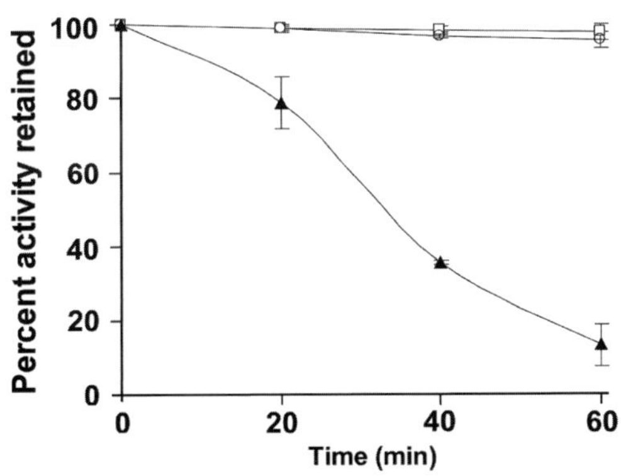

Figure 5.9. Retention of activity of SWNT–ADH conjugates (*filled triangles*) and native ADH (*open squares*) with NIR irradiation at room temperature, and of SWNT–ADH (*open circles*) conjugates at 60°C without NIR irradiation. (Reprinted with permission from Joshi et al. [39]).

these transparent films may self-clean (e.g., eliminate protein fouling) on exposure to either NIR or even visible radiation. To that end, we fabricated nanotube films that were optically transparent and exposed them to a solution of a model protein, bovine serum albumin (BSA), followed by exposure to either NIR or visible radiation. Exposure of these films to radiation eliminated protein fouling as confirmed by several characterization techniques [39].

The ability to generate ultrathin, transparent coatings that self-clean on exposure to visible light would be particularly advantageous in a variety of both indoor and outdoor applications.

5.4.3. Protein-Mediated Formation of Nanotube–Nanoparticle Hybrid Materials

Nanotube–protein conjugates can also mediate the formation of hybrid nanotube–nanoparticle conjugates [42]. Furthermore, this process does not require any functionalization of the nanotube surface and preserves its properties, which may be important for applications of these materials. Along these lines, we prepared MWNT conjugates of a variety of polypeptides and proteins, specifically poly-L-lysine (PLL), BSA, α1-acid glycoprotein (AGP), and SBP, and tested their ability to mediate the formation of silver nanoparticles. The conjugates were exposed to a solution of silver nitrate for 24 h in the dark, washed, and treated with sodium borohydride, a reducing agent. Characterization by transmission electron microscopy (TEM) and energy dispersive X-ray (EDX) spectroscopy indicated that PLL and BSA mediated the formation of silver nanoparticles when adsorbed onto nanotubes, whereas glycosylated proteins such as SBP and AGP mediated the formation of silver nanoparticles only upon deglycosylation (Figure 5.10). Moreover, when silver nanoparticle conjugates were formed using conjugates of nanotubes with deglycosylated SBP (d-SBP), the enzyme retained significant activity, making such a mechanism an attractive approach for preparing hybrid multifunctional nanomaterials.

5.4.4. Nanotube-Directed Interfacial Biocatalysis

Production of various pharmaceuticals and fine chemicals involve biphasic biocatalytic reactions of water-insoluble compounds. The rate of phase transfer biocatalysis, however, is limited by mass transfer of the substrate from the organic phase into the aqueous phase. Therefore, interfacial adsorption of enzymes is desired to carry out biotransformations at the aqueous–organic interface. However, interfacial adsorption of enzymes is one of the main sources of protein inactivation, which can be reduced by decreasing either the enzyme concentration or the interfacial area, both of which cause reduction in the rate of biocatalytic reactions and product yield.

Nanomaterials such as SWNTs can be directed to aqueous–organic interfaces with the aid of surfactants. SWNTs have a high intrinsic surface area without any intraparticle diffusional limitations. As a result, the nanotube-mediated interfacial assembly of enzymes can be very advantageous in directing greater amounts of enzymes from the bulk aqueous phase to the interface and increasing the stability of enzymes against inactivation [43].

Asuri et al. [43] prepared an aqueous dispersion of purified SWNTs, as per a previously described protocol [17]. When contacted with an equal volume of hexane (or isooctane, $CHCl_3$, or CH_2Cl_2) containing a 2-mM concentration of the anionic surfactant Aerosol-OT (AOT), the SWNTs were transferred from the aqueous phase (Figure 5.11a) to the interface (Figure 5.11b). Thus, nanotubes can serve as carriers to transport proteins to an aqueous–

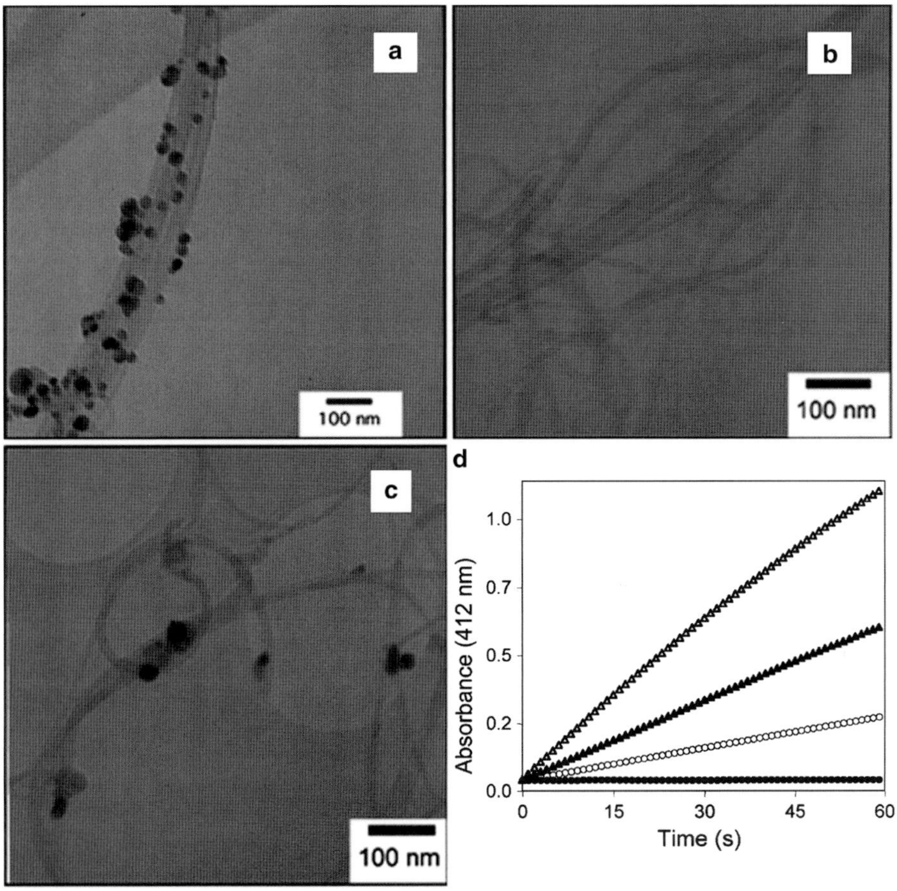

Figure 5.10. TEM showing the ability of proteins to selectively mediate the formation of silver nanoparticles. TEM images (**a**) and (**c**) show silver nanoparticles formation on nanotubes mediated by PLL and deglycosylated SBP (d-SBP), respectively. Image (**b**) corresponds to MWNT–SBP indicating the inability of the glycosylated protein to mediate nanoparticle formation. The absorbance versus time plot in panel (**d**) presents the enzymatic activity of d-SBP (*open triangles*), MWNT–d-SBP (*filled triangles*), MWNT–d-SBP–nanoparticle hybrids (*open circles*), and solution without enzyme (*filled circles*). (Reprinted with permission from Bale et al. [42]).

organic interface. For example, using a model hydrophobic peroxidase substrate, *p*-cresol (soluble in the organic phase), and the hydrophilic H_2O_2 (soluble in the aqueous phase), the SWNT–SBP conjugates assembled at the hexane–water interface and showed catalytic activity. The initial rate of reaction was at least three orders of magnitude higher than that observed with identical enzyme concentrations for either native SBP or SWNT–SBP in the aqueous phase of a biphasic system in the absence of AOT, or for native SBP in the aqueous phase in the presence of AOT. In addition to enhancing the rate of interfacial biotransformations, the nanotubes also enhance enzyme stability at higher temperatures and facilitate long-term storage at the interface.

Thus, the confluence of the three properties afforded by nanomaterials like SWNTs, specifically their high surface area, ability to assemble at the aqueous–organic interface, and absence of intraparticle diffusional limitations, can be employed to assemble SWNT–enzyme conjugates at the aqueous–organic interface, and facilitate interfacial biotransformations.

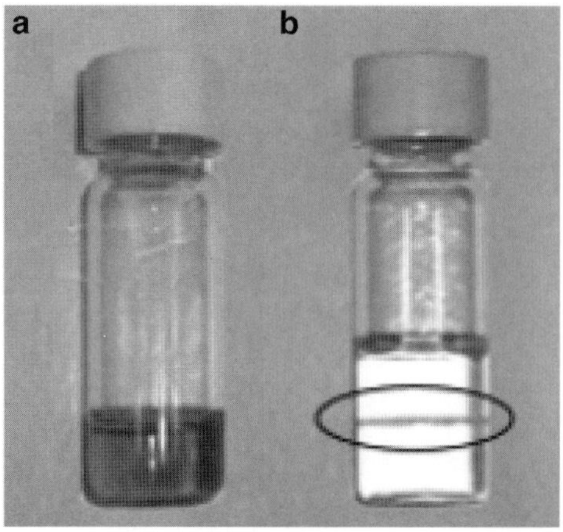

Figure 5.11. Photographs demonstrating the assembly of SWNTs at an interface: (**a**) suspension of SWNTs in water and (**b**) interfacial assembly of SWNTs on addition of a solution of AOT in hexane (2 mM) to the aqueous dispersion of SWNTs. (Reprinted with permission from Asuri et al. [43]).

5.4.5. Solubilization of SWNTs Using Proteins

Water-insoluble SWNTs could be solubilized by ultrasonicating a mixture of proteins and nanotubes [44]. The proteins essentially act as pseudosurfactants that facilitate solubilization of the intrinsically hydrophobic carbon nanotubes in water (Figure 5.12a). The ability to achieve dispersion at the individual nanotube level was confirmed by using multiple complementary techniques such as ultraviolet–visible (UV-Vis) spectroscopy, Raman spectroscopy, and AFM. Figure 5.12b shows UV-Vis spectra of aqueous dispersions of SWNTs prepared using three different methods. The spectrum for aqueous solution of pristine SWNTs without any dispersing agent was featureless (data not shown) indicating the presence of SWNT aggregates. Conversely, sharp and well-resolved peaks were observed for SWNTs solubilized using the proteins BSA and *Mucor javanicus* lipase (MJL); such spectra are characteristic of solutions containing SWNTs in debundled and individually dispersed form (Figure 5.12b). As a positive control, we prepared a uniform dispersion of SWNTs using the surfactant sodium dodecylbenzene sulfonate (NaDDBS), which showed a similar UV-Vis spectrum (Figure 5.12b). Further analysis using Raman spectroscopy was consistent with UV-Vis spectroscopy results, confirming the solubilization of nanotubes (data not shown). SWNT–BSA conjugates were further characterized by AFM, which also revealed the presence of individually dispersed SWNTs (Figure 5.12c).

Formulations containing such solubilized nanotubes may be useful for applications in biomedical engineering, such as the processing of nanocomposite biomaterials. In addition, the wide variety of functional groups present on adsorbed proteins could be used as orthogonal reactive handles for further functionalization of the carbon nanotubes. Furthermore, biorecognition properties of proteins may also be used in preparing self-assembled nanostructures and materials composed of SWNTs.

5.4.6. DNA Degradation by MWNT–DNAzyme Hybrids

In nature, certain small single-stranded DNA fragments are known to possess catalytic activity, e.g., endonuclease-type activity, and hence are called "DNAzymes" [45]. Conjugating such

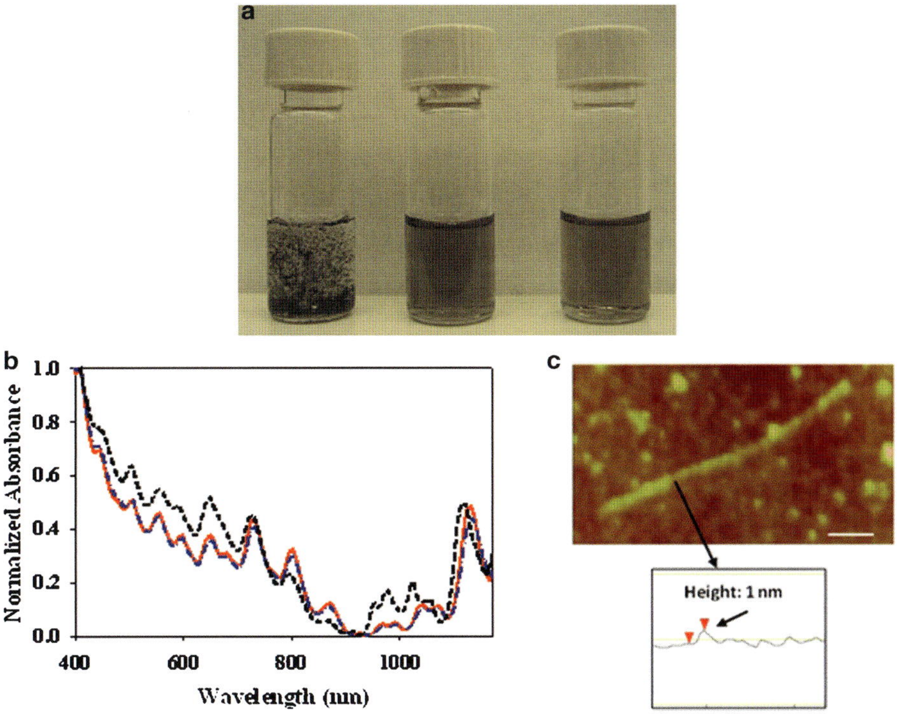

Figure 5.12. (**a**) Photographs of SWNT dispersed (*left* to *right*) in water without protein, water with BSA, and water with MJL. (**b**) UV-Vis absorption spectra of SWNTs dispersed in water using NaDDBS (*black, dashed line*), BSA (*red, solid line*), and MJL (*blue, dash–dot line*) (**c**) Height image of SWNT–BSA. The scale bar represents 100 nm. Line scan revealed that the diameter of the SWNT is ~1 nm. (Reprinted with permission from Karajanagi et al. [44]).

DNAzymes with nanomaterials has resulted in the development of biosensors to detect metal ions and nucleic acids as well as strategies for directing nanoparticle assembly [46, 47]. We have studied the generation and characterization of MWNT–DNAzyme conjugates, which might provide novel opportunities as templates for nanoscale assembly and as vectors for gene delivery [48].

Yim et al. [48] covalently attached streptavidin to acid-treated MWNTs using *N*-ethyl-*N′*-(3-dimethylaminopropyl) carbodiimide hydrochloride (EDC)–*N*-hydroxysuccinimide (NHS) chemistry, followed by the binding of biotinylated DNAzyme to yield MWNT–DNAzyme conjugates that were soluble in aqueous buffer. The DNAzyme used was a 38-mer, which gets activated in the presence of Pb^{2+} ions; the substrate for this DNAzyme consisted of a 20-mer labeled with 6-carboxyfluorescein (FAM) at its 5′ end, also incorporating a ribonucleotide. The activated DNAzyme cleaved the substrate at the ribonucleotide position to give two double-stranded DNA fragments.

The MWNT–DNAzyme conjugates followed Michaelis–Menten kinetics when the substrate concentration was higher than that of DNAzyme (Figure 5.13). The conjugates showed multiple turnover kinetics in more than 400 catalytic turnovers of the DNAzyme. There was essentially no deactivation of DNAzyme on the MWNT conjugates; moreover, there was no need to heat up and cool the MWNT–DNAzyme conjugate to hybridize and then cleave substrate DNA [48].

The catalytic activity and selectivity of MWNT–DNAzyme conjugates may be useful for the directed assembly of nanomaterials into functional 3D architectures and for the development of therapeutics that rely on the RNA cleaving ability of DNAzymes and the delivery capability of nanotubes [49].

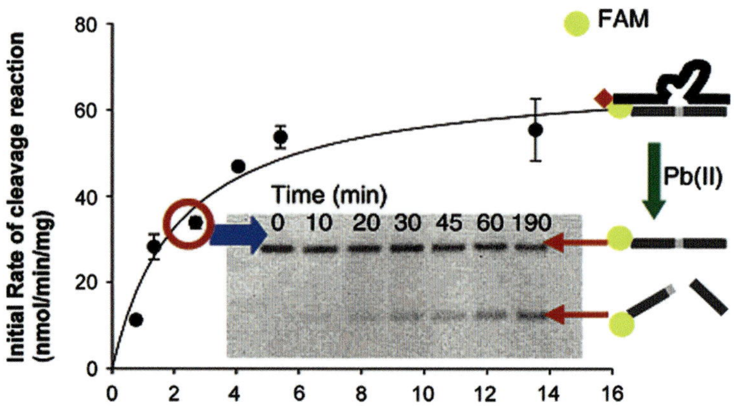

Figure 5.13. Catalytic activity of MWNT–DNAzyme conjugates. The line represents a nonlinear fit of the Michaelis–Menten expression to the data. The inset shows analysis of the extent of conversion of the fluorescently labeled substrate DNA by polyacrylamide gel electrophoresis (PAGE), with the *upper band* representing uncleaved DNA and the *lower band* representing cleaved fragments. (Reprinted with permission from Yim et al. [48]).

5.5. Conclusion and Future Directions

While we have shown the ability to characterize protein structure and function in protein–nanomaterial conjugates, it is also useful to characterize the orientation of proteins on the nanomaterial surface. Measurements of hydrogen–deuterium exchange by solid-state nuclear magnetic resonance spectroscopy and liquid chromatography mass spectrometry show promise in investigating the structure and orientation of proteins on nanomaterials [50, 51]. Furthermore, conjugates prepared with nanoparticles varying in size and surface properties could also help to elucidate the relationship between the physicochemical properties of the nanomaterial and the structure and function of attached proteins. Collectively, these studies will provide control over the orientation and function of proteins attached to nanomaterials thereby helping in designing novel and optimal hybrids for various applications.

Because biomolecule–nanomaterials conjugates possess significant potential as key components of biosensors, gene delivery systems, nanomaterial assemblies, etc., we continue to explore different areas where these functional composites may be of use. One interesting opportunity involves the design of enzyme-based self-decontaminating antimicrobial coatings, which could be used to avoid the fouling of bioimplants and to thwart the transmission and proliferation of pathogens in hospital settings. It is known that the interaction between the surface of a material and the proteins and polysaccharides present on microbial surfaces facilitate microbial adhesion and proliferation, which in turn leads to biofouling. We have already demonstrated the antiprotein-fouling properties of protease-containing nanobiocomposite films [36]; we believe that polymer coatings containing a suite of enzymes could be used to prepare antimicrobial films. Furthermore, the ROS-generating capabilities of nanotubes could also be used to design coatings that are effective not only against proteins but also against other contaminants including organic compounds, bacteria, and viruses. The design of such novel materials is underway in our laboratories.

References

1. Dabbousi BO, RodriguezViejo J, Mikulec FV, Heine JR, Mattoussi H, Ober R, et al. (CdSe)ZnS core-shell quantum dots: Synthesis and characterization of a size series of highly luminescent nanocrystallites. J Phys Chem B 1997;101(46):9463–9475.
2. Strano MS, Dyke CA, Usrey ML, Barone PW, Allen MJ, Shan HW, et al. Electronic structure control of single-walled carbon nanotube functionalization. Science 2003;301(5639):1519–1522.
3. Peng XG, Manna L, Yang WD, Wickham J, Scher E, Kadavanich A, et al. Shape control of CdSe nanocrystals. Nature 2000;404(6773):59–61.
4. Hong R, Fischer NO, Verma A, Goodman CM, Emrick T, Rotello VM. Control of protein structure and function through surface recognition by tailored nanoparticle scaffolds. J Am Chem Soc 2004;126(3):739–743.
5. Ajayan PM. Nanotubes from carbon. Chem Rev 1999;99(7):1787–1799.
6. Zhao YL, Hu LB, Stoddart JF, Gruner G. Pyrenecyclodextrin-decorated single-walled carbon nanotube field-effect transistors as chemical sensors. Adv Mater 2008;20(10):1910–1915.
7. Besteman K, Lee JO, Wiertz FGM, Heering HA, Dekker C. Enzyme-coated carbon nanotubes as single-molecule biosensors. Nano Lett 2003;3(6):727–730.
8. Yan YM, Yehezkeli O, Willner I. Integrated, electrically contacted NAD(P)(+)-dependent enzyme – carbon nanotube electrodes for biosensors and biofuel cell applications. Chem Eur J 2007;13(36): 10168–10175.
9. Medintz IL, Uyeda HT, Goldman ER, Mattoussi H. Quantum dot bioconjugates for imaging, labelling and sensing. Nat Mater 2005;4(6):435–446.
10. Michalet X, Pinaud FF, Bentolila LA, Tsay JM, Doose S, Li JJ, et al. Quantum dots for live cells, in vivo imaging, and diagnostics. Science 2005;307(5709):538–544.
11. Brannon-Peppas L, Blanchette JO. Nanoparticle and targeted systems for cancer therapy. Adv Drug Del Rev 2004;56(11):1649–1659.
12. Kane RS, Stroock AD. Nanobiotechnology: Protein-nanomaterial interactions. Biotechnol Progr 2007;23(2):316–319.
13. Asuri P, Bale SS, Karajanagi SS, Kane RS. The protein-nanomaterial interface. Curr Opin Biotechnol 2006;17(6):562–568.
14. Katz E, Willner I. Integrated nanoparticle-biomolecule hybrid systems: Synthesis, properties, and applications. Angew Chem Int Ed Engl 2004;43(45):6042–6108.
15. Lacerda L, Bianco A, Prato M, Kostarelos K. Carbon nanotube cell translocation and delivery of nucleic acids in vitro and in vivo. J Mat Chem 2008;18(1):17–22.
16. Kostarelos K, Lacerda L, Pastorin G, Wu W, Wieckowski S, Luangsivilay J, et al. Cellular uptake of functionalized carbon nanotubes is independent of functional group and cell type. Nat Nanotechnol 2007;2(2):108–113.
17. Karajanagi SS, Vertegel AA, Kane RS, Dordick JS. Structure and function of enzymes adsorbed onto single-walled carbon nanotubes. Langmuir 2004;20(26):11594–11599.
18. Singh R, Pantarotto D, McCarthy D, Chaloin O, Hoebeke J, Partidos CD, et al. Binding and condensation of plasmid DNA onto functionalized carbon nanotubes: Toward the construction of nanotube-based gene delivery vectors. J Am Chem Soc 2005;127(12):4388–4396.
19. Rege K, Viswanathan G, Zhu GY, Vijayaraghavan A, Ajayan PM, Dordick JS. In vitro transcription and protein translation from carbon nanotube-DNA assemblies. Small 2006;2(6):718–722.
20. Pantarotto D, Singh R, McCarthy D, Erhardt M, Briand JP, Prato M, et al. Functionalized carbon nanotubes for plasmid DNA gene delivery. Angew Chem Int Ed Engl 2004;43(39):5242–5246.
21. Pantarotto D, Briand JP, Prato M, Bianco A. Translocation of bioactive peptides across cell membranes by carbon nanotubes. Chem Commun 2004 (1):16–17.
22. Kam NWS, O'Connell M, Wisdom JA, Dai HJ. Carbon nanotubes as multifunctional biological transporters and near-infrared agents for selective cancer cell destruction. Proc Natl Acad Sci USA 2005;102(33):11600–11605.
23. Jia G, Wang HF, Yan L, Wang X, Pei RJ, Yan T, et al. Cytotoxicity of carbon nanomaterials: Single-wall nanotube, multi-wall nanotube, and fullerene. Environ Sci Technol 2005;39(5):1378–1383.
24. MacDonald RA, Laurenzi BF, Viswanathan G, Ajayan PM, Stegemann JP. Collagen-carbon nanotube composite materials as scaffolds in tissue engineering. J Biomed Mater Res A 2005;74A(3):489–496.
25. Correa-Duarte MA, Wagner N, Rojas-Chapana J, Morsczeck C, Thie M, Giersig M. Fabrication and biocompatibility of carbon nanotube-based 3D networks as scaffolds for cell seeding and growth. Nano Lett 2004;4(11):2233–2236.
26. Hu H, Ni YC, Montana V, Haddon RC, Parpura V. Chemically functionalized carbon nanotubes as substrates for neuronal growth. Nano Lett 2004;4(3):507–511.

27. Price RL, Waid MC, Haberstroh KM, Webster TJ. Selective bone cell adhesion on formulations containing carbon nanofibers. Biomaterials 2003;24(11):1877–1887.
28. Webster TJ, Ergun C, Doremus RH, Siegel RW, Bizios R. Specific proteins mediate enhanced osteoblast adhesion on nanophase ceramics. J Biomed Mater Res 2000;51(3):475–483.
29. Webster TJ, Schadler LS, Siegel RW, Bizios R. Mechanisms of enhanced osteoblast adhesion on nanophase alumina involve vitronectin. Tissue Eng 2001;7(3):291–301.
30. Price RL, Haberstroh KM, Webster TJ. Improved osteoblast viability in the presence of smaller nanometre dimensioned carbon fibres. Nanotechnology 2004;15(8):892–900.
31. Vedantham G, Sparks HG, Sane SU, Tzannis S, Przybycien TM. A holistic approach for protein secondary structure estimation from infrared spectra in H2O solutions. Anal Biochem 2000;285(1):33–49.
32. Asuri P, Karajanagi SS, Sellitto E, Kim DY, Kane RS, Dordick JS. Water-soluble carbon nanotube-enzyme conjugates as functional biocatalytic formulations. Biotechnol Bioeng 2006;95(5):804–811.
33. Asuri P, Bale SS, Pangule RC, Shah DA, Kane RS, Dordick JS. Structure, function, and stability of enzymes covalently attached to single-walled carbon nanotubes. Langmuir 2007;23(24):12318–12321.
34. Asuri P, Karajanagi SS, Yang HC, Yim TJ, Kane RS, Dordick JS. Increasing protein stability through control of the nanoscale environment. Langmuir 2006;22(13):5833–5836.
35. Asuri P, Karajanagi SS, Vertegel AA, Dordick JS, Kane RS. Enhanced stability of enzymes adsorbed onto nanoparticles. J Nanosci Nanotechnol 2007;7(4–5):1675–1678.
36. Asuri P, Karajanagi SS, Kane RS, Dordick JS. Polymer-nanotube-enzyme composites as active antifouling films. Small 2007;3(1):50–53.
37. Liu Z, Cai WB, He LN, Nakayama N, Chen K, Sun XM, et al. In vivo biodistribution and highly efficient tumour targeting of carbon nanotubes in mice. Nat Nanotechnol 2007;2(1):47–52.
38. Hirsch LR, Stafford RJ, Bankson JA, Sershen SR, Rivera B, Price RE, et al. Nanoshell-mediated near-infrared thermal therapy of tumors under magnetic resonance guidance. Proc Natl Acad Sci USA 2003;100(23):13549–13554.
39. Joshi A, Punyani S, Bale SS, Yang HC, Borca-Tasciuc T, Kane RS. Nanotube-assisted protein deactivation. Nat Nanotechnol 2008;3(1):41–45.
40. Wu ZC, Chen ZH, Du X, Logan JM, Sippel J, Nikolou M, et al. Transparent, conductive carbon nanotube films. Science 2004;305(5688):1273–1276.
41. Davies KJ. Protein damage and degradation by oxygen radicals. I. General aspects. 1987; 262(20):9895–9901.
42. Bale SS, Asuri P, Karajanagi SS, Dordick JS, Kane RS. Protein-directed formation of silver nanoparticles on carbon nanotubes. Adv Mater 2007;19(20):3167–3170.
43. Asuri P, Karajanagi SS, Dordick JS, Kane RS. Directed assembly of carbon nanotubes at liquid-liquid interfaces: Nanoscale conveyors for interfacial biocatalysis. J Am Chem Soc 2006;128(4):1046–1047.
44. Karajanagi SS, Yang HC, Asuri P, Sellitto E, Dordick JS, Kane RS. Protein-assisted solubilization of single-walled carbon nanotubes. Langmuir 2006;22(4):1392–1395.
45. Li YF, Breaker RR. Deoxyribozymes: New players in the ancient game of biocatalysis. Curr Opin Struct Biol 1999;9(3):315–323.
46. Liu JW, Lu Y. A colorimetric lead biosensor using DNAzyme-directed assembly of gold nanoparticles. J Am Chem Soc 2003;125(22):6642–6643.
47. Sando S, Sasaki T, Kanatani K, Aoyama Y. Amplified nucleic acid sensing using programmed self-cleaving DNAzyme. J Am Chem Soc 2003;125(51):15720–15721.
48. Yim TJ, Liu JW, Lu Y, Kane RS, Dordick JS. Highly active and stable DNAzyme – Carbon nanotube hybrids. J Am Chem Soc 2005;127(35):12200–12201.
49. Santoro SW, Joyce GF. A general purpose RNA-cleaving DNA enzyme. Proc Natl Acad Sci USA 1997;94(9):4262–4266.
50. Cotten M, Fu R, Cross TA. Solid-state NMR and hydrogen-deuterium exchange in a bilayer-solubilized peptide: Structural and mechanistic implications. Biophys J 1999;76(3):1179–1189.
51. Hoofnagle AN, Resing KA, Ahn NG. Protein analysis by hydrogen exchange mass spectrometry. Ann Rev Biophys Biomol Struct 2003;32:1–25.

6

Phage Display as a Strategy for Designing Organic/Inorganic Biomaterials

Sharon Segvich and David H. Kohn

To extend and optimize the performance of biomaterials, better control of biofunctionality is needed. In this chapter, we focus on the integration of peptides into biomaterials as a strategy for providing a biomaterial with greater ability to control subsequent protein, cell, and tissue responses. The focus of this chapter is on phage display, a high-throughput selection technique used to identify peptides that have preferential affinity to a specific material or cell type. The use of phage display provides a genetic engineering platform for designing new materials at the nanoscale. The basics of the phage display technique are presented, and postprocessing approaches to analyze the combinatorial data derived from phage display are discussed. Specific examples of the use of phage display with calcium phosphate biomaterials are presented, as are examples from the use of phage display to define amino acid sequences that preferentially bind to specific cell types. Data from multiple phage pannings can be used to create dual-functioning peptides that serve as linkers between the organic and inorganic worlds.

Abbreviations

BLM	bone-like mineral
CN	carbon nanotubes
DNT	2,4-dinitrotoluene
ECM	extracellular matrix
ELISA	enzyme-linked immunosorbent assay
FASTA	DNA and protein alignment software
PEG	polyethylene glycol
PPyCl	chlorine-doped polypyrrole
RELIC	REceptor Ligand Contacts

S. Segvich • Department of Biomedical Engineering, University of Michigan, Ann Arbor, MI, USA
D.H. Kohn • Departments of Biologic and Material Sciences and Biomedical Engineering, University of Michigan, Ann Arbor, MI, USA

D.A. Puleo and R. Bizios (eds.), *Biological Interactions on Materials Surfaces*,
DOI 10.1007/978-0-387-98161-1_6,

6.1. Introduction: Biomaterials Development and the Need for More Robust Approaches to Control Protein, Cell, and Tissue Responses

Historically, most biomaterials have had their origins in other fields and only secondarily found utility in medicine and dentistry. For example, high-strength alloys used in joint replacements were first developed in the aerospace field, and acrylic bone cement was developed in the paint industry. Although many such materials have functioned well, few interact with their surrounding host environment or promote integration with host tissue in an intelligent and proactive fashion. The desire to implement more biological approaches toward biomaterials design such that materials can provide instructions has led to an expansion and paradigm shift in the field of biomaterials in the last two decades. Many biomaterials are now being rationally designed to interact with the biological milieu they will encounter in vivo and, in some cases, facilitate tissue regeneration.

Biomaterial systems that can promote tissue regeneration should satisfy the following design requirements [1, 2]: (1) biocompatibility; (2) conductivity for attachment and proliferation of committed cells or their progenitors, and production of new, functional extracellular matrix (ECM); (3) ability to incorporate inductive factors to direct and enhance new tissue growth; (4) support of vascular ingrowth for transport of oxygen and biomolecules; (5) mechanical integrity to support loads at the implant site; (6) controlled and predictable degradation into nontoxic species that are easily metabolized or excreted; and (7) simple and inexpensive processing into irregular three-dimensional (3D) shapes of sufficient volume to fill clinically relevant tissue defects. Integration of these criteria into a single material presents design challenges that require more biomimetic complexity than many of the current simplified ECM mimics can provide.

The first generation of biomaterials that mimicked structural and/or functional aspects of ECMs and satisfied a subset of the design requirements listed earlier in this section included both organic and inorganic biomaterials: copolymers of poly(lactic–glycolic acid) [3], collagen [4], polyphosphazenes [5], polyurethanes [6], polycaprolactone [7], polyethylene glycol (PEG) [8], poly(propylene fumarate) [9], starch-based materials [10], alginate [11], silk [12], bioactive glasses and glass ceramics [13, 14], calcium–phosphate ceramics [15–17], and composites of calcium–phosphates and collagen [4] or synthetic polymers [18–21]. Varying the material properties of a biomaterial, such as composition, topology, and crystallinity can lead to a significant variation in a number of cell functions in vitro, including cell attachment, cell proliferation, RNA transcription, and protein synthesis [1, 13,14, 22–28]. The material properties of a biomaterial can also significantly affect cell differentiation, the rate and amount of tissue formation, and the duration and magnitude of inflammatory responses in vivo [1, 16, 22, 29, 30].

To extend the performance of biomaterials beyond the capabilities of what these first-generation materials can provide and to incorporate more of the above design criteria into a single material, better control of biofunctionality is needed. The specific microenvironment that interacts with the material must be considered in the design process, such that in vivo functionality and tissue remodeling can be maintained in the long-term.

Biomaterial modification can take on different levels of complexity, resulting in increasing levels of physiological replication and functionality. One approach is to capitalize on the instructive cues inherent in natural ECMs; in fact, variations of ECM molecules serve as one basis for formulating biomaterials [31, 32]. Surface and bulk chemical modifications of synthetic materials can also help control cell–material interactions and enhance tissue integration. These chemical modifications include changes in hydrophilicity and surface functionalization with charged groups [33, 34], supramolecular self-assembly [35–38], and development of materials that bind and release soluble factors [39–43]. Incorporation of insoluble ligands and cell recognition sequences from peptides, as well as attachment of larger proteins, are also key strategies to impart communication between a material and cells [33, 44–47]. Strategies based on physical, rather than chemical cues, include reproduction of the nanoscale topology of natural ECMs [31, 32, 48] and superposition of mechanical cues [49].

In this chapter, we focus on the integration of peptides into biomaterials as a strategy for providing a biomaterial with greater ability to control protein, cell, and tissue responses. Starting with a discussion of the rationale for utilizing peptides over larger proteins and the physiological importance of using peptides as a component of a biomaterial, we next focus on the high-throughput technique of phage display as a tool to discover peptides that have preferential affinity to specific materials. Following a general presentation of the phage display technique and postprocessing approaches to analyze the combinatorial data derived from phage display, we present specific examples of the use of phage display with calcium phosphate biomaterials. We next discuss the analogous use of phage display to define amino acid sequences that preferentially bind to progenitor cells. Finally, we discuss how data from multiple phage pannings can be used to create dual-functioning peptides that serve as linkers between the organic and inorganic worlds. The objective of this chapter is to discuss the use of the phage display technique as a means of linking cells to biomaterials in a more rational manner via the discovery of linker peptides.

Table 6.1. List of peptides discussed in this chapter and their proven functions.

Peptide sequence	Derived from	Proven function	Reference
RGD	Multiple proteins	Relevant to bone engineering, RGD enhances cell adhesion and differentiation into bone, cartilage, neural, and endothelial tissue	[53–59]
FHRRIKA	Heparin-binding domain	Increases osteoblast adhesion and mineralization	[60]
KRSR	Heparin-binding domain	Increases osteoblast adhesion and mineralization	[53]
YIGSR	Laminin	Increases human foreskin fibroblast adhesion	[66]
IKVAV	Laminin	Increases neurite extension	[61]
REDV	Fibronectin	Increases endothelial cell adhesion	[67]
KHIFSDDSSE	Neural cell adhesion molecules	Increases astrocyte adhesion	[68]
VPGIG	Elastin	Increases stiffness of synthetic matrices	[64]
SVSVGMKPSPRP	Phage display	High selectivity toward hydroxyapatite and tooth enamel	[117]
VTKHLNQISQSY	Phage display	High selectivity toward hydroxyapatite and bone-like mineral (BLM)	[47]

(continued)

Table 6.1. (continued)

Peptide sequence	Derived from	Proven function	Reference
CRKRLDRNC	Phage display	Interacts with IL-4 receptor on endothelial cells, macrophages, and smooth muscle cells	[131]
THRTSTLDYFVI	Phage display	High selectivity toward polypyrrole that increases adhesion of PC12 cells	[81]
STFTKSP	Phage display	Homes to primitive hematopoietic progenitor cells in bone marrow	[98]
WYRGRL	Phage display	Binds to collagen IIα1	[93]
ASSLINA	Phage display	Binds to both skeletal and cardiac tissue	[97]
CAGALCY	Phage display	Targets brain tissue	[94]

6.2. Peptide–Biomaterial Interactions

Functionalizing a surface via either protein or peptide adsorption or attachment can increase binding of specific cell receptors. Incorporation of proteins or their subsequences into the backbone of a polymer can control processes, such as cell differentiation and matrix synthesis. Proteins, growth factors, and peptides have been either ionically or covalently attached to biomaterial surfaces to promote cell adhesion, and, ultimately, the amount of tissue regenerated [33, 40, 47, 50]. While several proteins enhance cell adhesion, proteins are challenging to isolate and prone to degradation [51]. Proteins can also change conformation or orientation because they possess sections with varying hydrophobicities that modulate cellular functions other than adhesion. On the other hand, peptides can mimic the same response as a protein while being smaller, cheaper, and less susceptible to degradation. Peptides may, therefore, have a greater potential for controlling initial biological response to a material, because they can contain specific target amino acid sequences and can permit control of hydrophilic properties through specific sequence design [52].

In addition to its structural role, the ECM contains adhesive ligands, such as fibronectin, vitronectin, and, laminin that direct cell function. Identification of peptide sequences within proteins that are responsible for cell adhesion led to the development of peptide-functionalized biomaterials [44]. Incorporation of peptide motifs containing sequences, such as the arginine–glycine–aspartic acid (RGD)-based sequences, that are recognized by integrin receptors on cell membranes is now a common strategy to enhance the biological functionality of substrates [33, 45, 46]. Materials with appropriate concentrations of RGD-containing sequences can enhance cell adhesion and direct differentiation into cells of the bone [53–55], cartilage [56, 57], neural [58], and endothelial [59] tissues.

Using recombinant DNA technology, synthetic proteins can be designed to mimic specific ECM constituents (Table 6.1). In addition to the ubiquitous RGD sequence, sequences derived from the heparin-binding domain, such as FHRRIKA [60] and KRSR [53], increase osteoblast adhesion and mineralization. Peptide sequences that mimic sections of collagen [50] and of the noncollagenous proteins laminin [61], bone sialoprotein [62], osteopontin [63], statherin [63], elastin [64], and osteonectin [65] also increase cell adhesion, proliferation, and lineage-specific progenitor cell differentiation. A noninclusive list of synthetic sequences used in tissue engineering and their functions includes: YIGSR, derived from laminin, increases human fibroblast adhesion [66]; IKVAV, derived from laminin, increases neurite extension [61]; REDV, derived from fibronectin, increases endothelial cell adhesion [67]; KHIFSDDSSE, derived from neural cell adhesion molecules, increases astrocyte adhesion [68]; and VPGIG, derived from elastin, increases stiffness of synthetic matrices [64].

In addition to using recombinant technologies to synthesize peptide sequences found within proteins known to promote a specific biological function, domains within a protein can be deleted to investigate the effect of targeted sequence deletions on the function of the protein. Subsequently, sequences deemed to control a specific function could be synthesized for integration with a biomaterial. The focus of this chapter is, however, on another discovery technique, phage display, a high-throughput selection technique in which a bacteriophage library expressing combinations of either linear or cyclic peptide inserts is used to identify amino acid sequences that have high affinity to either a substrate or a cell type. The use of phage display provides a genetic engineering platform for designing new materials at the nanoscale.

6.3. Phage Display as a Selection Technique

Typically, phage display technologies introduce a combinatorial library (on the order of 10^9 sequences) of 7-mer or 12-mer peptide sequences to a molecule, ligand, or material. The phage display technique has been utilized to identify amino acid sequences that recognize specific substrates, serving as a strategy to create biological linkers to bridge biomolecules and synthetic materials at the nanoscale. Targeted inorganic substrates include $BaTiO_3$ for electronic applications [69], SiO_2 [70], TiO_2 [70], aluminum [71], steel [71], semiconductors [72–74], platinum [75], and silver [76, 77]. Organic substrates include carbon [78,79], helical wrapping of DNA [80], the electrically conductive chlorine-doped polypyrrole (PPyCl) [81], plastics [82], and poly(methyl methacrylate) [83]. The use of phage display is not limited to solid surfaces. To detect volatile organic compounds, phage libraries have been used to identify peptide recognition motifs for 2,4,6-trinitrotoluene (TNT) and 2,4-dinitrotoluene (DNT) and to develop gas-binding assays [84].

In addition to creating functional materials via the adhesion of phage motifs to a synthetic material, M13 phage can be assembled into a material itself. The M13 phage is a nanofiber-like virus that can self-align to form higher-order structures [74, 85–87]. Phage fibers can also be fabricated via wet spinning and electrospinning techniques [88]. The fiber-like organization and specific recognition motifs also allow this phage to coassemble with other molecules. Ordered phage-based materials can be constructed by controlling the concentration of virus and ionic strength of the solution in which the virus is suspended, as well as exogenous forces [86–88].

In the biomedical sector, phage display has been used to identify biological species that selectively bind to polystyrene (*Saccharomyces cerevisiae*; [89]), poly(hydroxybutyrate) (immunoglobulin variable regions of human antibodies; [90]), catheters (*Staphylococcus aureus*; [91]), and carbon nanotubes (CN) for drug delivery [92]. Peptides that target specific tissues, including cartilage [93], brain microvasculature [94], kidney tubules [95], breast vasculature [96], muscle [97], bone marrow [98], cell/organ targets in vivo [99], and malignant cell types [100, 101] have also been identified. In vivo tissue screening via phage display allows targeted delivery of drugs [102] and imaging agents [103]. Furthermore, phage displays have identified sequences with high affinity to cell lines cultured in vitro [104, 105], enzymes and their inhibitors, DNA, proteins, and specific tissues [52, 106, 107].

Although phage display has been mostly used in identifying enzyme substrates and inhibitors, DNA and protein-binding peptides, tissue-specific peptides, and receptors [52, 106, 107], the principle of the technique can be applied to identifying peptides that have high affinity and specificity to biomaterials.

Introduction of phage display libraries to the research fields of cancer, tissue engineering, and molecular biology has proven fruitful since its debut in 1990 [108]. The concept behind phage display is to create an oligonucleotide insertion mutation on the gene of a virus allowing a library of sequences to be expressed on the exterior protein coat of the phage [109]. After several rounds of panning, or expanding and reintroducing the sequences that adhere, consensus sequences emerge and are identified by DNA sequencing (Figure 6.1). The peptide sequences physically presented on the exterior of the bacteriophage (or phage) coats are the result of a genetic modification within the virion DNA encapsulated within the phage.

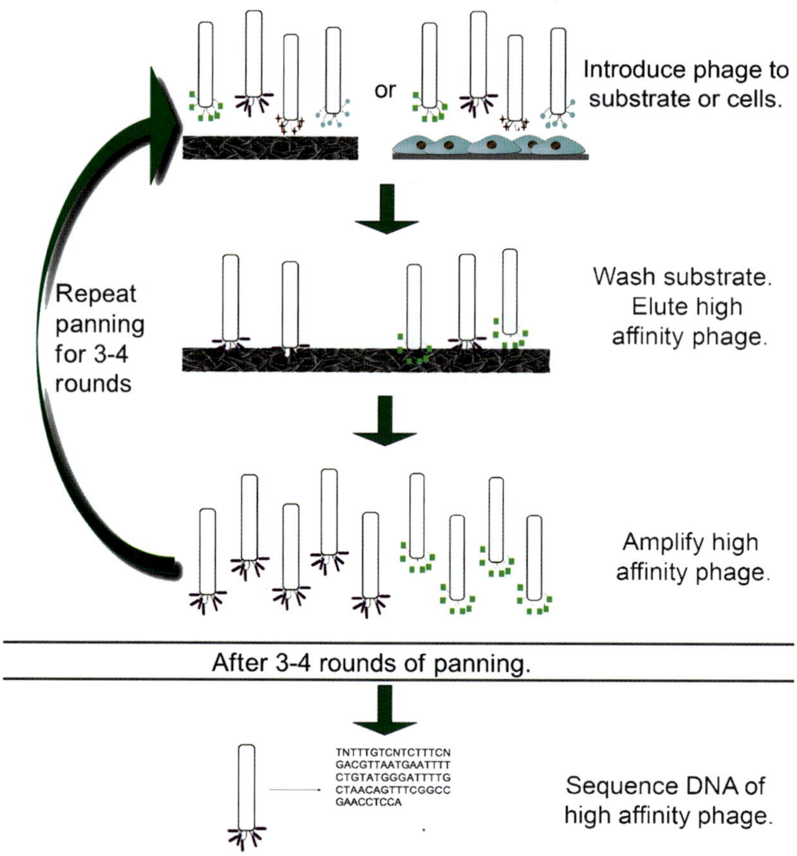

Figure 6.1. Schematic showing the phage display panning technique in vitro on either a substrate or a cell population. One round of panning includes introducing the parent library (~10^9) to the substrate or cell population, washing off the phages that do not adhere, and eluting the phages that have high affinity. Multiple rounds of panning can yield consensus sequences identified via DNA sequencing.

This provides the user with a link between the physically presented peptide and the encoded DNA within the virion, allowing simultaneous testing of a vast number of possible peptide sequences (e.g., 10^9 sequences). This simple biopanning approach allows researchers to investigate either stabilized molecules or substrates in vitro in addition to organ and tissue targeting in vivo. Once the phages display library is introduced in vitro, multiple rounds of rigorous washing discard nonadherent phages, allowing highly adherent phages to be eluted and amplified, constituting one round of phage panning. Multiple rounds of panning, typically between three and five rounds, can elucidate phages that physically bind to either the substrate or biological constituent of interest. DNA sequencing of the adherent phages reveals the peptide sequence presented on the phage protein coat. This sequence information is often used to synthesize a peptide, and its affinity to either the material or biological target is verified via peptide adsorption and/or immunohistochemical assays.

If the phage display library is introduced in vivo, targeted organs are often harvested, the bound phages are carefully eluted and amplified, and the amplified phages are reintroduced in vivo, constituting one round of in vivo panning. Similar steps as in in vitro studies are then taken to achieve peptide sequences that home to the targeted tissue.

The biological vehicle that is oftentimes used in phage display combinatorial libraries is the filamentous M13 bacteriophages. M13 bacteriophages are a well understood and characterized strain with three main structural protein regions on the virion, pIII, pVIII, and pVI, which have been manipulated to display random peptides [109]. The most commonly used region is the pIII, which allows presentation of approximately five copies of the randomly generated peptide expressed in either a linear or cyclic manner. In commercial phage display libraries (e.g., from New England Biolabs), the random peptide is fused to the pIII protein coat and is expressed at the N terminus of the pIII, typically with a Gly–Gly–Gly–Ser spacer (Figure 6.2). Commercial libraries contain ~10^9 sequences amplified such that ~55 copies are presented in 10 μL of the parent library. The bacterial host used for M13 bacteriophage is typically a strain of *Escherichia coli.*

The technique of biopanning is not limited to combinatorial peptide libraries, as other polypeptide libraries such as immunological protein and cDNA libraries have been also utilized. In addition, the means of peptide display is not limited to bacterial display in eukaryotic

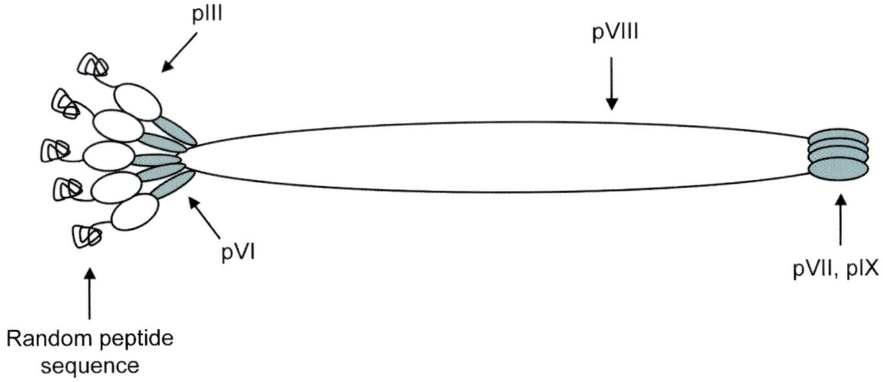

Figure 6.2. Schematic of a bacteriophage identifying the protein coat regions encoded in the virion DNA. The phage display library expresses the same random peptide sequence in three to five copies at the pIII end of the bacteriophage. A parent library can have on the order of 10^9 random sequences.

viruses, since yeast, mammalian cells, ribosomes, messenger RNA (mRNA), and DNA display techniques have also been investigated [110].

Phage display combinatorial libraries offer potential in isolating binding peptide sequences from a vast library, making otherwise unfathomable experiments possible [106, 107, 111]. However, as with any biological system, limitations intrinsic to the technique do exist. Phage libraries can express an unequal representation of each amino acid within the library, but are not able to include every peptide combination possible, and oftentimes omit important post-translational modifications such as phosphorylation of serine residues in the parent library. Nonetheless, the array of peptides presented in these libraries provides the opportunity to identify sequences specific to biological molecules, tissues, as well as synthetic and natural materials. It is possible with phage display techniques to identify high adsorbing, nonmodified sequences that could achieve even higher affinity after synthetic phosphorylation.

6.3.1. Computational Analysis Tools

Because the phage display experiment involves a biological system that can produce biased results from broth growth conditions and/or ease of sequence fabrication by the bacteriophages, computational tools have been developed to supplement experimentation in order to further analyze phage display data. An example of such a tool is REceptor Ligand Contacts (RELIC), a database designed to advance functional genomics through identification of small molecule-binding regions on proteins (http://relic.bio.anl.gov/programs.aspx). This publicly accessible database allows the user to utilize a statistically based approach in analyzing peptide populations, searching for common motifs, and comparing homologous regions with protein sequences [112]. RELIC is broken into five main categories: translation, characterization of peptide populations, peptide motif identification, comparison of peptide population to known sequence structures, and analysis of either single or multiple FASTA (the DNA and protein sequence alignment software) sequences. Translation involves taking DNA sequences of individual phage plaques and translating them to protein code. Although the program is written for New England Biolabs' libraries, other libraries can be inserted if the start and end vector sequences are known. Characterization of the peptide population provides multiple programs that can provide analysis of amino acid frequency with position within a given peptide population, diversity of a peptide population, and the likelihood of observing sequences by chance based on amino acid composition and position for a given peptide population. Both continuous and discontinuous peptide motifs allowing conservative substitutions can be identified using the peptide motif identification programs. To better link the identified sequences with a protein–ligand interaction, peptides within a population that could be in contact with proteins listed in the Protein Data Base are listed and visually displayed. Sequence alignments between peptides and a given protein sequence list (from text only, no coordinates) can also be generated using the analysis of either single or multiple FASTA sequences. Depending on the aim of the study, one or many of the programs can facilitate data analysis.

Other computational tools exist for evaluating parent libraries and they can be used to estimate diversity [113]. Another computational approach uses genetic algorithms to assign sequence similarity scores for individual sequences in a population of peptides [114]. Computational programs are also available to facilitate the design and construction of randomized unbiased libraries [113]. With increased phage display research, we can only hope to mirror the evolution of a variety of computational tools now available in molecular biology to help us advance our abilities of identifying significant peptide sequences [115].

6.3.2. Characterization Techniques

Computational analysis tools are available to aid in the analysis of data sets resulting from phage display experiments; however, characterization techniques to verify that identified peptides are indeed specific binders include analyses ranging from biological assays to computer modeling. The purpose of this section is to give an overview of the techniques researchers are utilizing to support their phage display findings.

The enzyme-linked immunosorbent assay (ELISA), is a common technique used to determine the binding affinity of the isolated and amplified phage binders to both substrate materials and cell sources [47, 116]. High background noise on some substances, such as apatite-based materials, has been reported as one limitation of the ELISA [117]. Another common technique used to visually display the binding of either phage or synthetic peptides uses microscopy to image fluorescently tagged phage or peptides [81, 93, 95, 117]. While this technique can provide qualitative analysis, it is important to couple this technique with a quantitative measure. If a cell source or tissue is being panned against, immunohistochemistry methods can be employed to visually detect phage binding [96–98, 101, 104, 118]. Another powerful instrument commonly used in imaging nanoscale features, the atomic force microscope, is used to determine peptide–substrate binding affinities [81, 116]. Less common techniques include quartz crystal microbalance [70, 90, 117], surface plasmon resonance [90], and circular dichroism [76], which are used to detect phage binding, binding of antibodies to a given substrate, and secondary structure before and after binding to particles, respectively.

Synthetic peptide fabrication is a logical next step after identifying highly specific phage to either a material or cell of interest, and thus, the adsorption behavior of the peptide should also be thoroughly investigated. Being able to control the amount of peptide that adsorbs to the surface of a substrate is imperative when linking peptide concentration to subsequent cellular function. Controlling the amount of peptide on a surface can be as simple as covalently linking the peptide to a functionalized polymer, but mainly relies on peptide adsorption in the case of ceramic materials that do not have modifiable surface chemistries. Fluorescently tagged or untagged peptide adsorbed to a substrate can be quantified using an adsorption assay that detects the peptide via UV spectrophotometry [47]. Additionally, a fluorescein isothiocyanate tag is a common tag that can be added during or after peptide synthesis. Radiolabeling tyrosine is an alternative to fluorescently tagging a peptide that can also provide quantitative results [119]. Less commonly used peptide quantification methods include x-ray photoelectron spectroscopy and amino acid analysis [53, 120]. Once adsorbed peptide quantities are established, cell functions can be tested using various cell attachment assays [99, 121–123].

Synthetic peptide can also be useful in performing competitive binding studies between either phage and peptides or untagged and tagged peptides, to provide evidence that peptide binding to material, cell, or tissue is being established [93, 95, 101, 104, 117, 124]. Finally, computational modeling of identified peptide sequences on a substrate of interest, with or without water, is another tool useful in determining the utility of the phage data sets [47, 94, 125, 126].

6.4. Phage Display on Apatite-Based Mineral

A complete set of sequences known to selectively adhere to each of the variety of biomaterials used in bone tissue engineering would provide the opportunity to more optimally functionalize the material surfaces. An appropriately functionalized material surface could minimize inflammation, deliver biological molecules, and direct tissue growth when implanted in vivo. To achieve this, combinatorial phage display technology has been utilized to elucidate specific sequences or sequence patterns that adhere to biomaterials of interest in orthopedic and dental applications.

Since the osteoconductivity of a bone implant material is improved if the material is either made or coated with an apatite-based material [25, 127], it is of interest to design peptide sequences with preferential adsorption to apatite. Compared with polymer-based materials, apatite-based materials are less amenable to surface modification required for covalent molecular attachment; for this reason, these materials must rely on their inherent material properties to achieve consistent, ionic adhesion of peptides. Post-translational modifications and poly-acidic peptide strings are two strategies that have been used to encourage peptide adhesion to apatite-based materials [62, 65]. Phage display technology can be utilized to identify 12-mer peptide sequences with preferential binding abilities to apatite-based materials, including bone-like mineral (BLM) (Figure 6.3). It has also been suggested that 7-mer peptide sequences have been identified on hydroxyapatite substrates, but limited information on either the substrate preparation or pertinent properties is reported [128, 129]. Another study identified the peptide SVSVGMKPSPRP as having high selectivity toward hydroxyapatite and tooth enamel [117].

The limited number of known peptides that selectively adhere to apatite led the pursuit of sequences that preferentially bind to BLM. Since BLM is a synthetic form of carbonated apatite fabricated at benign physiologic conditions (pH = 6.8; 37°C), it is capable of biomolecular incorporation and positively influences cell spreading [24, 39, 43], making it an

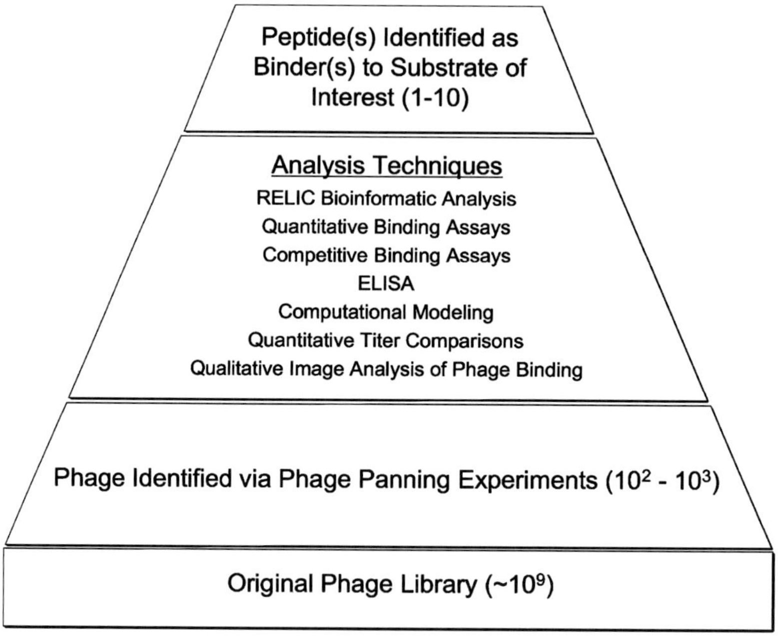

Figure 6.3. Representation of the steps utilized to identify peptide sequences with preferential adsorption to substrates via phage display. The analysis techniques listed are used to identify a small number of peptides from the hundreds of phage sequences identified after multiple rounds of phage panning.

important apatite-based substrate to investigate. Specifically, compared with bone sialoprotein-derived peptide E_7PRGDT, the sequence VTKHLNQISQSY was identified as having increased ability to bind to BLM and hydroxyapatite [47].

Understanding the conformation of peptides and proteins adsorbed to a material surface is imperative to control the presentation of cell adhesion sequences. Therefore, links between surface adsorption density, free peptide in solution, and peptide conformation must be established. For example, an optimal peptide surface density of $\rho \geq 0.62$ pmol/cm^2 has been reported for a bone sialoprotein-derived peptide covalently attached to quartz [54]. As another example, adsorption of proteins onto hydroxyapatite occurs preferentially on certain mineral faces, specifically the (100) planes [130]. Effects of peptide density and peptide conformation on subsequent cell adhesion have not been studied on mineral surfaces; such information is, however, critical in presenting concentrations and appropriate sequences that will promote cellular responses and functions pertinent to new tissue formation.

Phage display techniques can also be utilized to identify peptide sequences that bind to mineral deposits formed in vivo, such as in atherosclerotic plaques. Identifying peptides that home to calcified plaques in cardiovascular tissue can prove effective in delivering target drugs to the diseased areas. The peptide CRKRLDRNC has been reported as an atherosclerotic plaque-homing peptide that interacts with the interleukin-4 (IL-4) receptor on endothelial cells, macrophages, and smooth muscle cells [131]. Attachment of nanoparticles acting as both drug delivery carriers and imaging probes could advance in vivo imaging of atherosclerotic lesions, as well as have therapeutic benefits. Other pathological mineral systems known to develop in vivo that could potentially benefit from this approach include kidney stones, gallstones, and dental calculus.

Phage display experiments are not limited to ceramics and have been used to identify peptide sequences on polymer substrates. An advantage of most polymer substrates is their modifiable surface chemistry that can be altered to covalently link drugs or therapeutic molecules. However, the polymer polypyrrole, mainly used in neural tissue engineering, does not have a surface group capable of modification; as a result, polypyrrole is a good candidate for surface modification via peptides. A 12-mer phage display library run on polypyrrole isolated the peptide, THRTSTLDYFVI, as a positive binder to polypyrrole that increased adhesion of PC12 cells when the identified peptide was conjugated with an RGD sequence [81].

6.5. Phage Display on Cells and the Role of Dual-Functioning Peptides

In order to improve guided tissue formation using peptides, the peptides identified to selectively bind to a material of interest need to be functionalized with a cell-instructive sequence that will promote appropriate cell function for the desired cell lineage needed. This approach encourages the idea of a dual-functioning peptide where one portion of the peptide attaches to the material and the other portion controls cell function, particularly cell adhesion. The usefulness of phage display in cell adhesion has been proven through the isolation of integrin receptor sequences that bind to RGD [99]. Understanding the steps involved in cell adhesion is imperative when trying to modulate function of anchorage-dependent cells using peptides. Cell adhesion progresses in four steps: cell attachment, cell spreading, production of organized actin filaments, and focal adhesion formation [132]. The importance of cell adhesion has been demonstrated in a variety of tissues including the differentiation of myoblasts [133], keratinocytes [134], mammary epithelium [135], and osteoblasts [136]. Since cell

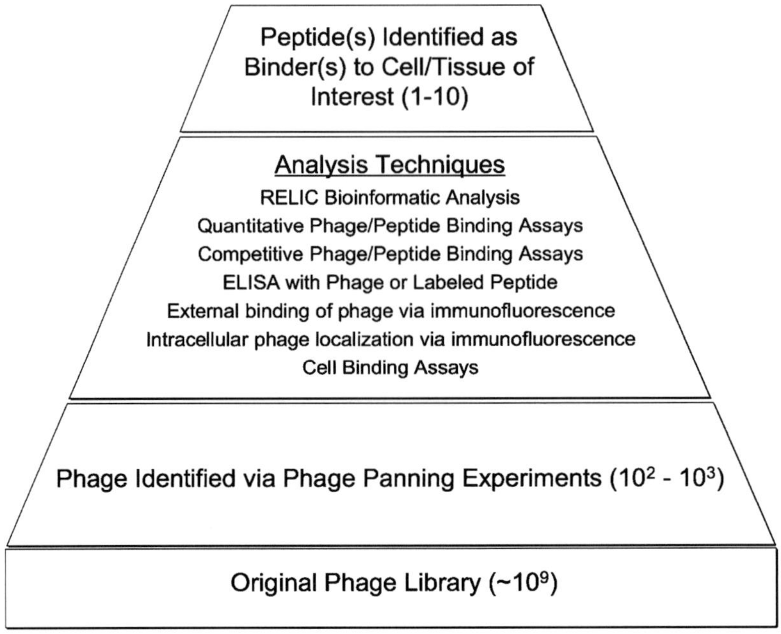

Figure 6.4. Representation of the steps utilized to identify peptide sequences with preferential adsorption to cell populations via phage display. The analysis techniques listed are used to identify a small number of peptides from the hundreds of phage sequences identified after multiple rounds of phage panning.

attachment is the first step of adhesion, dual-functioning peptides can aim to improve attachment of cells to the biomaterial. In the process of bone and other tissue repair, establishing cell attachment is essential, as subsequent phenotypic differentiation cannot occur if anchorage-dependent cells do not attach. Before a dual-functioning peptide can be constructed, both ends of the peptide need to be identified; phage display techniques, therefore, can be utilized on cells and tissues to identify appropriate peptide sequences (Figure 6.4).

The cell sequence can be identified either by panning against an in vitro population of cells, or by injecting the phage library in vivo and harvesting target organs. For example, the 7-mer peptide, STFTKSP, was identified via in vivo phage display panning to possess the specific ability to home to primitive hematopoietic progenitor cells in the bone marrow [98]. For cartilage, the peptide sequence, WYRGRL, was found to bind to collagen IIα1 after biopanning against denuded cartilage. Functionalizing such peptides could provide a therapeutic approach in targeting avascular tissue [93]. In another example, phage display was able to identify the peptide, ASSLINA, as a muscle-specific sequence by panning against murine C2C12 myotubes both in vitro and in vivo. This muscle specific sequence binds to both skeletal and cardiac tissue [97]. Phage display was also used to isolate the cyclic sequence, CAGALCY, that targets brain tissue [94]; most importantly, this study demonstrated that this sequence inhibits platelet adhesion to the brain microvasculature.

Ongoing work to identify peptide sequences that adhere to clonally derived human bone marrow stromal cells could benefit dual function peptide design for guided bone regeneration. Such an identified sequence can be attached to an apatite-binding sequence, providing an effective peptide that can increase initial cell attachment, the imperative first step in anchorage-dependent cell viability and function.

Identification of specific substrate-binding moieties has led to the development of dual-functioning peptides for a diverse range of applications, including digital printing. For this purpose, peptide elements are binding domains for a pigment (cellulose) and for a printing surface (carbon black). Incorporation of a hydrophilic linker in between the binding domains isolates the two-functional elements, allowing dual binding activity [78]. Another use of dual peptides is in drug delivery. By combining peptide motifs that have affinity to the surfaces of CN with peptides that can target specific organs, the CN can be functionalized, leading to CN drug carriers [92]. For cytophobic coatings, amphiphilic macromolecules containing a polystyrene-adherent peptide domain and a cell-repellent PEG domain have been designed [116]. Moreover, a dual-functioning peptide was formulated by linking a phage, with high binding capability, to the electrically conductive polymer PPyCl, to a cell-adhesive sequence to promote cell attachment to PPyCl [81].

6.6. Advancing Phage Display in Biomaterials Research – Summary

Developing multifunctioning biomolecules is necessary in biomaterials design if we are to better mimic the ability of nature to fabricate intricate tissues in situations that require either repair or restoration. In this chapter, we present phage display as a lesser-used discovery technique that can potentially identify useful peptide sequences to be used in drug targeting, establishing a union between two substrates, and increasing initial cell attachment on a biomaterial prior to implantation. Furthermore, identification of such peptides could lead to a better understanding of the behavior of biomolecules, including proteins, during tissue growth, disease, and repair. In order to capitalize on the powerful phage display technique, multiple bioinformatics, experimental, and analytical tools in addition to characterization techniques should be implemented in tandem with phage display.

Acknowledgments

Parts of the authors' work discussed in this chapter were supported by the National Institutes of Heath: R01 DE 013380, R01 DE 015411, and T32 DE07057.

References

1. Kohn DH. "Bioceramics," Chapter 15, In: Biomedical Engineering and Design Handbook, Volume 1, M. Kutz, Ed., McGraw-Hill, New York, 2009.
2. Yaszemski MJ, Payne RG, Hayes WC, Langer R, Mikos AG. Evolution of bone transplantation: molecular, cellular and tissue strategies to engineer human bone. Biomaterials 1996;17(2):175–185.
3. Ishaug-Riley SL, Crane GM, Gurlek A, Miller MJ, Yasko AW, Yaszemski MJ, Ectopic bone formation by marrow stromal osteoblast transplantation using poly(DL-lactic-co-glycolic acid) foams implanted into the rat mesentery. Journal of Biomedical Materials Research 1997;36(1):1–8.
4. Krebsbach PH, Kuznetsov SA, Satomura K, Emmons RV, Rowe DW, Robey PG. Bone formation in vivo: comparison of osteogenesis by transplanted mouse and human marrow stromal fibroblasts. Transplantation 1997;63(8):1059–1069.
5. Laurencin CT, ElAmin SF, Ibim SE, Willoughby DA, Attawia M, Allcock HR, A highly porous 3-dimensional polyphosphazene polymer matrix for skeletal tissue regeneration. Journal of Biomedical Materials Research 1996;30(2):133–138.

6. Gorna K, Gogolewski S. Preparation, degradation, and calcification of biodegradable polyurethane foams for bone graft substitutes. Journal of Biomedical Materials Research Part A 2003;67A(3):813–827.
7. Li WJ, Tuli R, Huang X, Laquerriere P, Tuan RS. Multilineage differentiation of human mesenchymal stem cells in a three-dimensional nanofibrous scaffold. Biomaterials 2005;26(25):5158–5166.
8. Nuttelman CR, Tripodi MC, Anseth KS. In vitro osteogenic differentiation of human mesenchymal stem cells photoencapsulated in PEG hydrogels. Journal of Biomedical Materials Research Part A 2004;68A(4):773–782.
9. Payne RG, McGonigle JS, Yaszemski MJ, Yasko AW, Mikos AG. Development of an injectable, in situ crosslinkable, degradable polymeric carrier for osteogenic cell populations. Part 3. Proliferation and differentiation of encapsulated marrow stromal osteoblasts cultured on crosslinking poly(propylene fumarate). Biomaterials 2002;23(22):4381–4387.
10. Marques AP, Cruz HR, Coutinho OP, Reis RL. Effect of starch-based biomaterials on the in vitro proliferation and viability of osteoblast-like cells. Journal of Materials Science. Materials in Medicine 2005;16(9):833–842.
11. Alsberg E, Anderson KW, Albeiruti A, Franceschi RT, Mooney DJ. Cell-interactive alginate hydrogels for bone tissue engineering. Journal of Dental Research 2001;80(11):2025–2029.
12. Kim HJ, Kim UJ, Vunjak-Novakovic G, Min BH, Kaplan DL. Influence of macroporous protein scaffolds on bone tissue engineering from bone marrow stem cells. Biomaterials 2005;26(21):4442–4452.
13. Ducheyne P, el-Ghannam A, Shapiro I. Effect of bioactive glass templates on osteoblast proliferation and in vitro synthesis of bone-like tissue. Journal of Cellular Biochemistry 1994;56(2):162–167.
14. El-Ghannam A, Ducheyne P, Shapiro IM. Porous bioactive glass and hydroxyapatite ceramic affect bone cell function in vitro along different time lines. Journal of Biomedical Materials Research 1997;36(2):167–180.
15. Yoshikawa T, Ohgushi H, Tamai S. Immediate bone forming capability of prefabricated osteogenic hydroxyapatite. Journal of Biomedical Materials Research 1996;32(3):481–492.
16. Ohgushi H, Okumura M, Tamai S, Shors EC, Caplan AI. Marrow cell induced osteogenesis in porous hydroxyapatite and tricalcium phosphate – a comparative histomorphometric study of ectopic bone-formation. Journal of Biomedical Materials Research 1990;24(12):1563–1570.
17. Kruyt MC, Dhert WJ, Yuan H, Wilson CE, van Blitterswijk CA, Verbout AJ, Bone tissue engineering in a critical size defect compared to ectopic implantations in the goat. Journal of Orthopaedic Research 2004;22(3):544–551.
18. Murphy WL, Kohn DH, Mooney DJ. Growth of continuous bonelike mineral within porous poly(lactide-co-glycolide) scaffolds in vitro. Journal of Biomedical Materials Research 2000;50(1):50–58.
19. Shin K, Jayasuriya AC, Kohn DH. Effect of ionic activity products on the structure and composition of mineral self assembled on three-dimensional poly(lactide-co-glycolide) scaffolds. Journal of Biomedical Materials Research Part A 2007;83(4):1076–1086.
20. Thomson RC, Yaszemski MJ, Powers JM, Mikos AG. Hydroxyapatite fiber reinforced poly(alpha-hydroxy ester) foams for bone regeneration. Biomaterials 1998;19(21):1935–1943.
21. Kretlow JD, Mikos AG. Review: mineralization of synthetic polymer scaffolds for bone tissue engineering. Tissue Engineering 2007;13(5):927–938.
22. James K, Levene H, Parsons JR, Kohn J. Small changes in polymer chemistry have a large effect on the bone-implant interface: evaluation of a series of degradable tyrosine-derived polycarbonates in bone defects. Biomaterials 1999;20(23–24):2203–2212.
23. Rouahi M, Champion E, Hardouin P, Anselme K. Quantitative kinetic analysis of gene expression during human osteoblastic adhesion on orthopaedic materials. Biomaterials 2006;27(14):2829–2844.
24. Leonova EV, Pennington KE, Krebsbach PH, Kohn DH. Substrate mineralization stimulates focal adhesion contact redistribution and cell motility of bone marrow stromal cells. Journal of Biomedical Materials Research Part A 2006;79(2):263–270.
25. Kohn DH, Shin K, Hong SI, Jayasuriya AC, Leonova EV, Rossello RA, et al. Self-assembled mineral scaffolds as model systems for biomineralization and tissue engineering. Proceedings of the Eighth International Conference on the Chemistry and Biology of Mineralized Tissues 2005:216–219.
26. Puleo DA, Holleran LA, Doremus RH, Bizios R. Osteoblast responses to orthopedic implant materials in-vitro. Journal of Biomedical Materials Research 1991;25(6):711–723.
27. Zreiqat H, Evans P, Howlett CR. Effect of surface chemical modification of bioceramic on phenotype of human bone-derived cells. Journal of Biomedical Materials Research 1999;44(4):389–396.
28. Simon CG, Jr, Eidelman N, Kennedy SB, Sehgal A, Khatri CA, Washburn NR. Combinatorial screening of cell proliferation on poly(L-lactic acid)/poly(D,L-lactic acid) blends. Biomaterials 2005;26(34):6906–6915.
29. Krebsbach PH, Kuznetsov SA, Bianco P, Robey PG. Bone marrow stromal cells: characterization and clinical application. Critical Reviews in Oral Biology and Medicine 1999;10(2):165–181.

30. Hartman EHM, Vehof JWM, Spauwen PHM, Jansen JA. Ectopic bone formation in rats: the importance of the carrier. Biomaterials 2005;26(14):1829–1835.
31. Buttafoco L, Kolkman NG, Engbers-Buijtenhuijs P, Poot AA, Dijkstra PJ, Vermes I, Electrospinning of collagen and elastin for tissue engineering applications. Biomaterials 2006;27(5):724–734.
32. Ji Y, Ghosh K, Li B, Sokolov JC, Clark RA, Rafailovich MH. Dual-syringe reactive electrospinning of cross-linked hyaluronic acid hydrogel nanofibers for tissue engineering applications. Macromolecular Bioscience 2006;6(10):811–817.
33. Lutolf MP, Hubbell JA. Synthetic biomaterials as instructive extracellular microenvironments for morphogenesis in tissue engineering. Nature Biotechnology 2005;23(1):47–55.
34. Stevens MM. Biomaterials for bone tissue engineering. Materials Today 2008;11(5):18–25.
35. Zhang S. Fabrication of novel biomaterials through molecular self-assembly. Nature Biotechnology 2003;21(10):1171–1178.
36. Menger FM. Supramolecular chemistry and self-assembly. Proceedings of the National Academy of Sciences of the United States of America 2002;99(8):4818–4822.
37. Kisiday J, Jin M, Kurz B, Hung H, Semino C, Zhang S, Self-assembling peptide hydrogel fosters chondrocyte extracellular matrix production and cell division: implications for cartilage tissue repair. Proceedings of the National Academy of Sciences of the United States of America 2002;99(15):9996–10001.
38. Niece KL, Hartgerink JD, Donners JJ, Stupp SI. Self-assembly combining two bioactive peptide-amphiphile molecules into nanofibers by electrostatic attraction. Journal of American Chemical Society 2003;125(24):7146–7147.
39. Segvich S, Smith HC, Luong LN, Kohn DH. Uniform deposition of protein incorporated mineral layer on three-dimensional porous polymer scaffolds. Journal of Biomedical Materials Research Part B. Applied Biomaterials 2008;84(2):340–349.
40. Whang K, Tsai DC, Nam EK, Aitken M, Sprague SM, Patel PK, Ectopic bone formation via rhBMP-2 delivery from porous bioabsorbable polymer scaffolds. Journal of Biomedical Materials Research 1998;42(4):491–499.
41. Shea LD, Smiley E, Bonadio J, Mooney DJ. DNA delivery from polymer matrices for tissue engineering. Nature Biotechnology 1999;17(6):551–554.
42. Murphy WL, Peters MC, Kohn DH, Mooney DJ. Sustained release of vascular endothelial growth factor from mineralized poly(lactide-co-glycolide) scaffolds for tissue engineering. Biomaterials 2000;21(24):2521–2527.
43. Luong LN, Hong SI, Patel RJ, Outslay ME, Kohn DH. Spatial control of protein within biomimetically nucleated mineral. Biomaterials 2006;27(7):1175–1186.
44. Massia SP, Hubbell JA. Covalent surface immobilization of Arg-Gly-Asp- and Tyr-Ile-Gly-Ser-Arg-containing peptides to obtain well-defined cell-adhesive substrates. Analytical Biochemistry 1990;187(2):292–301.
45. Shin H, Jo S, Mikos AG. Biomimetic materials for tissue engineering. Biomaterials 2003;24(24):4353–4364.
46. Garcia AJ. Get a grip: integrins in cell-biomaterial interactions. Biomaterials 2005;26(36):7525–7529.
47. Segvich S, Biswas S, Becker U, Kohn D. Identification of peptides with targeted adhesion to bone-like mineral via phage display and computational modeling. Cells, Tissues, Organs 2009;189(1–4):245–251.
48. Stevens MM, George JH. Exploring and engineering the cell surface interface. Science 2005;310(5751):1135–1138.
49. Guilak F, Butler D, Goldstein SA, Mooney D (eds). *Functional Tissue Engineering*. New York: Springer, 2003.
50. Reyes CD, Petrie TA, Burns KL, Schwartz Z, Garcia AJ. Biomolecular surface coating to enhance orthopaedic tissue healing and integration. Biomaterials 2007;28(21):3228–3235.
51. Hersel U, Dahmen C, Kessler H. RGD modified polymers: biomaterials for stimulated cell adhesion and beyond. Biomaterials 2003;24(24):4385–4415.
52. Ladner RC, Sato AK, Gorzelany J, de Souza M. Phage display-derived peptides as therapeutic alternatives to antibodies. Drug Discovery Today 2004;9(12):525–529.
53. Dee KC, Andersen TT, Bizios R. Design and function of novel osteoblast-adhesive peptides for chemical modification of biomaterials. Journal of Biomedical Materials Research 1998;40(3):371–377.
54. Rezania A, Healy KE. The effect of peptide surface density on mineralization of a matrix deposited by osteogenic cells. Journal of Biomedical Materials Research 2000;52(4):595–600.
55. Itoh D, Yoneda S, Kuroda S, Kondo H, Umezawa A, Ohya K, Enhancement of osteogenesis on hydroxyapatite surface coated with synthetic peptide (EEEEEEEPRGDT) in vitro. Journal of Biomedical Materials Research 2002;62(2):292–298.
56. Hwang NS, Varghese S, Zhang Z, Elisseeff J. Chondrogenic differentiation of human embryonic stem cell-derived cells in arginine-glycine-aspartate-modified hydrogels. Tissue Engineering 2006;12(9):2695–2706.
57. Salinas CN, Cole BB, Kasko AM, Anseth KS. Chondrogenic differentiation potential of human mesenchymal stem cells photoencapsulated within poly(ethylene glycol)-arginine-glycine-aspartic acid-serine thiol-methacrylate mixed-mode networks. Tissue Engineering 2007;13(5):1025–1034.

58. Gunn JW, Turner SD, Mann BK. Adhesive and mechanical properties of hydrogels influence neurite extension. Journal of Biomedical Materials Research Part A 2005;72A(1):91–97.
59. Patel S, Tsang J, Harbers GM, Healy KE, Li S. Regulation of endothelial cell function by GRGDSP peptide grafted on interpenetrating polymers. Journal of Biomedical Materials Research Part A 2007;83(2):423–433.
60. Rezania A, Healy KE. Biomimetic peptide surfaces that regulate adhesion, spreading, cytoskeletal organization, and mineralization of the matrix deposited by osteoblast-like cells. Biotechnology Progress 1999;15(1):19–32.
61. Schense JC, Hubbell JA. Three-dimensional migration of neurites is mediated by adhesion site density and affinity. Journal of Biological Chemistry 2000;275(10):6813–6818.
62. Fujisawa R, Mizuno M, Nodasaka Y, Kuboki Y. Attachment of osteoblastic cells to hydroxyapatite crystals by a synthetic peptide (Glu7-Pro-Arg-Gly-Asp-Thr) containing two functional sequences of bone sialoprotein. Matrix Biology 1997;16(1):21–28.
63. Gilbert M, Shaw WJ, Long JR, Nelson K, Drobny GP, Giachelli CM, Chimeric peptides of statherin and osteopontin that bind hydroxyapatite and mediate cell adhesion. Journal of Biological Chemistry 2000;275(21):16213–16218.
64. Panitch A, Yamaoka T, Fournier MJ, Mason TL, Tirrell DA. Design and biosynthesis of elastin-like artificial extracellular matrix proteins containing periodically spaced fibronectin CS5 domains. Macromolecules 1999;32(5):1701–1703.
65. Fujisawa R, Wada Y, Nodasaka Y, Kuboki Y. Acidic amino acid-rich sequences as binding sites of osteonectin to hydroxyapatite crystals. Biochimica et Biophysica Acta 1996;1292(1):53–60.
66. Massia SP, Hubbell JA. Covalent surface immobilization of Arg-Gly-Asp- and Tyr-Ile-Gly-Ser-Arg-containing peptides to obtain well-defined cell-adhesive substrates. Analytical Biochemistry 1990;187(2):292–301.
67. Hubbell JA, Massia SP, Desai NP, Drumheller PD. Endothelial cell-selective materials for tissue engineering in the vascular graft via a new receptor. Biotechnology 1991;9(6):568–572.
68. Kam L, Shain W, Turner JN, Bizios R. Selective adhesion of astrocytes to surfaces modified with immobilized peptides. Biomaterials 2002;23(2):511–515.
69. Ahmad G, Dickerson MB, Cai Y, Jones SE, Ernst EM, Vernon JP, Rapid bioenabled formation of ferroelectric BaTiO3 at room temperature from an aqueous salt solution at near neutral pH. Journal of American Chemical Soceity 2008;130(1):4–5.
70. Chen H, Su X, Neoh KG, Choe WS. QCM-D analysis of binding mechanism of phage particles displaying a constrained heptapeptide with specific affinity to SiO2 and TiO2. Analytical Chemistry 2006;78(14):4872–4879.
71. Zuo R, Ornek D, Wood TK. Aluminum- and mild steel-binding peptides from phage display. Applied Microbiology & Biotechnology 2005;68(4):505–509.
72. Mao C, Flynn CE, Hayhurst A, Sweeney R, Qi J, Georgiou G, Viral assembly of oriented quantum dot nanowires. Proceedings of the National Academy of Sciences of the United States of America 2003;100(12):6946–6951.
73. Lee SW, Mao C, Flynn CE, Belcher AM. Ordering of quantum dots using genetically engineered viruses. Science 2002;296(5569):892–895.
74. Whaley SR, English DS, Hu EL, Barbara PF, Belcher AM. Selection of peptides with semiconductor binding specificity for directed nanocrystal assembly. Nature 2000;405(6787):665–668.
75. Seker UO, Wilson B, Dincer S, Kim IW, Oren EE, Evans JS, Adsorption behavior of linear and cyclic genetically engineered platinum binding peptides. Langmuir 2007;23(15):7895–7900.
76. Zhang X, Chen J, Yang P, Yang W. Biomimetic synthesis silver crystallite by peptide AYSSGAPPMPPF immobilized on PET film in vitro. Journal of Inorganic Biochemistry 2005;99(8):1692–1697.
77. Naik RR, Stringer SJ, Agarwal G, Jones SE, Stone MO. Biomimetic synthesis and patterning of silver nanoparticles. Nature Materials 2002;1(3):169–172.
78. Qi M, O'Brien JP, Yang J. A recombinant triblock protein polymer with dispersant and binding properties for digital printing. Biopolymers 2008;90(1):28–36.
79. Wang S, Humphreys ES, Chung SY, Delduco DF, Lustig SR, Wang H, Peptides with selective affinity for carbon nanotubes. Nature Materials 2003;2(3):196–200.
80. Gigliotti B, Sakizzie B, Bethune DS, Shelby RM, Cha JN. Sequence-independent helical wrapping of single-walled carbon nanotubes by long genomic DNA. Nano Letters 2006;6(2):159–164.
81. Sanghvi AB, Miller KP, Belcher AM, Schmidt CE. Biomaterials functionalization using a novel peptide that selectively binds to a conducting polymer. Nature Materials 2005;4(6):496–502.
82. Adey NB, Mataragnon AH, Rider JE, Carter JM, Kay BK. Characterization of phage that bind plastic from phage-displayed random peptide libraries. Gene 1995;156(1):27–31.

83. Serizawa T, Sawada T, Matsuno H, Matsubara T, Sato T. A peptide motif recognizing a polymer stereoregularity. Journal of American Chemical Society 2005;127(40):13780–13781.
84. Jaworski JW, Raorane D, Huh JH, Majumdar A, Lee SW. Evolutionary screening of biomimetic coatings for selective detection of explosives. Langmuir 2008;24(9):4938–4943.
85. Mao C, Solis DJ, Reiss BD, Kottmann ST, Sweeney RY, Hayhurst A, Virus-based toolkit for the directed synthesis of magnetic and semiconducting nanowires. Science 2004;303(5655):213–217.
86. Flynn CE, Lee S, Peelle BR, Belcher AM. Viruses as vehicles for growth, organization and assembly of materials. Acta Materialia 2003;51(19):5867–5880.
87. Merzlyak A, Lee SW. Phage as templates for hybrid materials and mediators for nanomaterial synthesis. Current Opinion in Chemical Biology 2006;10(3):246–252.
88. Lee S, Belcher AM. Virus-based fabrication of micro- and nanofibers using electrospinning. Nano Letters 2004;4(3):387–390.
89. Mortensen HD, Dupont K, Jespersen L, Willats WG, Arneborg N. Identification of amino acids involved in the Flo11p-mediated adhesion of *Saccharomyces cerevisiae* to a polystyrene surface using phage display with competitive elution. Journal of Application Microbiology 2007;103(4):1041–1047.
90. Watanabe H, Tsumoto K, Taguchi S, Yamashita K, Doi Y, Nishimiya Y, A human antibody fragment with high affinity for biodegradable polymer film. Bioconjugate Chemistry 2007;18(3):645–651.
91. Bjerketorp J, Rosander A, Nilsson M, Jacobsson K, Frykberg L. Sorting a *Staphylococcus aureus* phage display library against ex vivo biomaterial. Journal of Medical Microbiology 2004;53(Pt 10):945–951.
92. Shiba K. Functionalization of carbon nanomaterials by evolutionary molecular engineering: potential application in drug delivery systems. Journal of Drug Targeting 2006;14(7):512–518.
93. Rothenfluh DA, Bermudez H, O'Neil CP, Hubbell JA. Biofunctional polymer nanoparticles for intra-articular targeting and retention in cartilage. Nature Materials 2008;7(3):248–254.
94. Fan X, Venegas R, Fey R, van der Heyde H, Bernard MA, Lazarides E, An in vivo approach to structure activity relationship analysis of peptide ligands. Pharmaceutical Research 2007;24(5):868–879.
95. Odermatt A, Audige A, Frick C, Vogt B, Frey BM, Frey FJ, Identification of receptor ligands by screening phage-display peptide libraries ex vivo on microdissected kidney tubules. Journal of the American Society of Nephrology 2001;12(2):308–316.
96. Essler M, Ruoslahti E. Molecular specialization of breast vasculature: a breast-homing phage-displayed peptide binds to aminopeptidase P in breast vasculature. Proceedings of the National Academy of Sciences of the United States of America 2002;99(4):2252–2257.
97. Samoylova TI, Smith BF. Elucidation of muscle-binding peptides by phage display screening. Muscle and Nerve 1999;22(4):460–466.
98. Nowakowski GS, Dooner MS, Valinski HM, Mihaliak AM, Quesenberry PJ, Becker PS. A specific heptapeptide from a phage display peptide library homes to bone marrow and binds to primitive hematopoietic stem cells. Stem Cells 2004;22(6):1030–1038.
99. Pasqualini R, Koivunen E, Ruoslahti E. A peptide isolated from phage display libraries is a structural and functional mimic of an RGD-binding site on integrins. The Journal of Cell Biology 1995;130(5):1189–1196.
100. Samoylova TI, Petrenko VA, Morrison NE, Globa LP, Baker HJ, Cox NR. Phage probes for malignant glial cells. Molecular Cancer Therapeutics 2003;2(11):1129–1137.
101. Morita Y, Mamiya K, Yamamura S, Tamiya E. Selection and properties for the recognition of P19 embryonic carcinoma stem cells. Biotechnology Progress 2006;22(4):974–978.
102. Wu M, Sherwin T, Brown WL, Stockley PG. Delivery of antisense oligonucleotides to leukemia cells by RNA bacteriophage capsids. Nanomedicine 2005;1(1):67–76.
103. Frenkel D, Solomon B. Filamentous phage as vector-mediated antibody delivery to the brain. roceedings of the National Academy of Sciences of the United States of America 2002;99(8):5675–5679.
104. Samoylova TI, Ahmed BY, Vodyanoy V, Morrison NE, Samoylov AM, Globa LP, Targeting peptides for microglia identified via phage display. Journal of Neuroimmunology 2002;127(1–2):13–21.
105. Zhang J, Spring H, Schwab M. Neuroblastoma tumor cell-binding peptides identified through random peptide phage display. Cancer Letters 2001;171(2):153–164.
106. Uchiyama F, Tanaka Y, Minari Y, Toku N. Designing scaffolds of peptides for phage display libraries. Journal of Bioscience and Bioengineering 2005;99(5):448–456.
107. Marks C, Marks JD. Phage libraries – a new route to clinically useful antibodies. The New England Journal of Medicine 1996;335(10):730–733.
108. Scott JK, Smith GP. Searching for peptide ligands with an epitope library. Science 1990;249(4967):386–390.
109. Rodi DJ, Makowski L. Phage-display technology – finding a needle in a vast molecular haystack. Current Opinion in Biotechnology 1999;10(1):87–93.

110. Sergeeva A, Kolonin MG, Molldrem JJ, Pasqualini R, Arap W. Display technologies: application for the discovery of drug and gene delivery agents. Advanced Drug Delivery Reviews 2006;58(15):1622–1654.
111. Scott JK, Craig L. Random peptide libraries. Current Opinion in Biotechnology 1994;5(1):40–48.
112. Mandava S, Makowski L, Devarapalli S, Uzubell J, Rodi DJ. RELIC – a bioinformatics server for combinatorial peptide analysis and identification of protein-ligand interaction sites. Proteomics 2004;4(5):1439–1460.
113. Patrick WM, Firth AE. Strategies and computational tools for improving randomized protein libraries. Biomolecular Engineering 2005;22(4):105–112.
114. Tamerler C, Sarikaya M. Molecular biomimetics: utilizing nature's molecular ways in practical engineering. Acta Biomaterialia 2007;3(3):289–299.
115. Tompa M, Li N, Bailey TL, Church GM, De Moor B, Eskin E, Assessing computational tools for the discovery of transcription factor binding sites. Nature Biotechnology 2005;23(1):137–144.
116. Kenan DJ, Walsh EB, Meyers SR, O'Toole GA, Carruthers EG, Lee WK, Peptide-PEG amphiphiles as cytophobic coatings for mammalian and bacterial cells. Chemistry and Biology 2006;13(7):695–700.
117. Roy MD, Stanley SK, Amis EJ, Becker ML. Identification of a highly specific hydroxyapatite-binding peptide using phage display. Advanced Materials 2008;20(10):1830–1836.
118. Lu H, Jin D, Kapila YL. Application of laser capture microdissection to phage display peptide library screening. Oral Surgery, Oral Medicine, Oral Pathology, Oral Radiology, and Endodontics 2004;98(6):692–697.
119. Hern DL, Hubbell JA. Incorporation of adhesion peptides into nonadhesive hydrogels useful for tissue resurfacing. Journal of Biomedical Materials Research 1998;39(2):266–276.
120. Rezania A, Thomas CH, Branger AB, Waters CM, Healy KE. The detachment strength and morphology of bone cells contacting materials modified with a peptide sequence found within bone sialoprotein. Journal of Biomedical Materials Research 1997;37(1):9–19.
121. Garcia AJ, Gallant ND. Stick and grip: measurement systems and quantitative analyses of integrin-mediated cell adhesion strength. Cell Biochemistry and Biophysics 2003;39(1):61–73.
122. Koivunen E, Gay DA, Ruoslahti E. Selection of peptides binding to the alpha 5 beta 1 integrin from phage display library. Journal of Biological Chemistry 1993;268(27):20205–20210.
123. Okochi M, Nomura S, Kaga C, Honda H. Peptide array-based screening of human mesenchymal stem cell-adhesive peptides derived from fibronectin type III domain. Biochemical and Biophysical Research Communications 2008;371(1):85–89.
124. Sheu TJ, Schwarz EM, O'Keefe RJ, Rosier RN, Puzas JE. Use of a phage display technique to identify potential osteoblast binding sites within osteoclast lacunae. Journal of Bone and Mineral Research 2002;17(5): 915–922.
125. Ehrlich GK, Bailon P. Identification of peptides that bind to the constant region of a humanized IgG(1) monoclonal antibody using phage display. Journal of Molecular Recognition 1998;11(1–6):121–125.
126. Kantarci N, Tamerler C, Sarikaya M, Haliloglu T, Doruker P. Molecular dynamics simulations on constraint metal binding peptides. Polymer 2005;46(12):4307–4313.
127. Abe Y, Kokubo T, Yamamuro T. Apatite coating on ceramics, metals and polymers utilizing a biological process. Journal of Materials Science. Materials In Medicine 1990;1(4):233–238.
128. Sarikaya M, Tamerler C, Jen AK, Schulten K, Baneyx F. Molecular biomimetics: nanotechnology through biology. Nature Materials 2003;2(9):577–585.
129. Tamerler C, Dincer S, Heidel D, Zareie MH, Sarikaya M. Biomimetic multifunctional molecular coatings using engineered proteins. Progress in Organic Coatings 2003;47(3–4):267–274.
130. Fujisawa R, Kuboki Y. Preferential adsorption of dentin and bone acidic proteins on the (100) face of hydroxyapatite crystals. Biochimica et Biophysica Acta 1991;1075(1):56–60.
131. Park K, Hong HY, Moon HJ, Lee BH, Kim IS, Kwon IC, A new atherosclerotic lesion probe based on hydrophobically modified chitosan nanoparticles functionalized by the atherosclerotic plaque targeted peptides. Journal of Controlled Release 2008;128(3):217–223.
132. LeBaron RG, Athanasiou KA. Extracellular matrix cell adhesion peptides: functional applications in orthopedic materials. Tissue Engineering 2000;6(2):85–103.
133. Menko AS, Boettiger D. Occupation of the extracellular matrix receptor, integrin, is a control point for myogenic differentiation. Cell 1987;51(1):51–57.
134. Adams JC, Watt FM. Changes in keratinocyte adhesion during terminal differentiation: reduction in fibronectin binding precedes alpha 5 beta 1 integrin loss from the cell surface. Cell 1990;63(2):425–435.
135. Streuli CH, Bailey N, Bissell MJ. Control of mammary epithelial differentiation: basement membrane induces tissue-specific gene expression in the absence of cell-cell interaction and morphological polarity. The Journal of Cell Biology 1991;115(5):1383–1395.
136. Damsky CH. Extracellular matrix-integrin interactions in osteoblast function and tissue remodeling. Bone 1999;25(1):95–96.

7

Extracellular Matrix-Derived Ligands for Selective Integrin Binding to Control Cell Function

Timothy A. Petrie and Andrés J. García

Integrins are key cell surface receptors that function as the primary bridge between the intracellular and extracellular matrix environments and play a critical role in modulating adhesive cell–material interactions. Over the past decade, several biomimetic material surface modification strategies to actively engage integrins have been employed, including the use of robust integrin-binding sequences (i.e., RGD) and native matrix proteins, with varying levels of success in vivo. This chapter details these aforementioned ligand strategies as well as more recent, next-generation biomaterial surface strategies geared toward specifically engineering integrin selectivity. Building on our expanding knowledge of the functional roles of particular integrin subunit combinations, including $\alpha_5\beta_1$, $\alpha_4\beta_1$, $\alpha_2\beta_1$, and $\alpha_v\beta_3$, in multiple cell and tissue types, these integrin-specific biointerfaces incorporate unique protein fragments, engineered oligopeptides, multivalent ligands, and exclusive integrin-binding sequences to direct selective integrin binding and activation. This chapter highlights these major integrin-specific ligands and details the varying successes that these designs have achieved as implant biomaterial coatings, therapeutics, in vitro adhesive substrates, and templates for matrix assembly. Collectively, these biomolecular strategies have contributed to the rational engineering of bioactive materials to achieve desired tissue responses.

Abbreviations

BSP	Bone sialoprotein
COL	Collagen
DGEA	Aspartic acid–glycine–glutamic acid–alanine
ECM	Extracellular matrix
ERK	Extracellular signal-related kinase

T.A. Petrie, A.J. García • Woodruff School of Mechanical Engineering, Petit Institute for Bioengineering and Bioscience, Georgia Institute of Technology, Atlanta, GA 30332

D.A. Puleo and R. Bizios (eds.), *Biological Interactions on Materials Surfaces*,
DOI 10.1007/978-0-387-98161-1_7,

FAK	Focal adhesion kinase
FN	Fibronectin
$FNIII_{7-10}$	Fibronectin fragment encompassing the seventh through tenth type III repeats
GAG	Glycosaminoglycan
GDP	Glow discharge plasma
GFOGER	Glycine–phenylalanine–hydroxyproline–glycine–glutamic acid–arginine
GRGDSP	Glycine–arginine–glycine–aspartic acid–serine–proline
HA	Hydroxyapatite
MAPK	Mitogen-activated protein kinase
PHSRN	Proline–histidine–serine–arginine–asparagine
PLL-g-PEG	Poly(L-lysine)–poly(ethylene) glycol
REDV	Arginine–glutamic acid–aspartic acid–valine
RGD	Arginine–glycine–aspartic acid
SAMs	Self-assembled monolayers

7.1. Extracellular Matrix: Composition and Role

The interactions of cells with their extracellular matrix (ECM) are critically involved in mediating the development, organization, and repair of numerous tissues. At the cell level, cell–matrix interactions can directly modulate cell morphology, survival, proliferation, and differentiation in multiple cell systems [1, 2].

The ECM is primarily composed of a complex meshwork of fibrous proteins (fiber-forming elements) surrounded by space-filling molecules such as glycosaminoglycans (GAGs)—as well as mineral deposits in tissues such as bone. Many of these components are secreted by resident cells and primarily provide support and anchorage, but also serve important roles as tissue boundaries and in the modulation of intracellular and intercellular communication [3]. Moreover, the ECM is not only made, but also organized and remodeled, by the cells within it. Formation and remodeling of this ECM network is vital for a number of tissue responses such as cell migration during tissue formation and repair, growth, wound healing, and fibrotic responses [4]. Many of the fibrous proteins that make up the ECM, including collagen, fibronectin (FN), elastin, and laminin, have both structural and adhesive functions, and are secreted in specific tissues by specialized cells, including fibroblasts (present in most connective tissue), chondroblasts (cartilage), and osteoblasts (bone).

The collagens constitute a family of abundant ECM molecules that contribute significantly to the integrity and mechanical properties of tissues such as bone, skin, cartilage, and tendon [5]. Elastin is the dominant ECM protein in elastic fibers, a major component of arteries, lungs, and skin [3]. Laminin is the major noncollagenous component of the basal lamina, playing a critical role as the "glue" that holds many body structures together [6]. Fibronectin (FN) is a complex, multifunctional protein with multiple domains, each with specific binding sites for other matrix macromolecules and for receptors on the surface of cells. FN therefore contributes to both organizing the matrix and helping cells attach to it [4, 7]. These essential roles of the ECM in regulating cell and tissue structure and function are illustrated schematically in Figure 7.1.

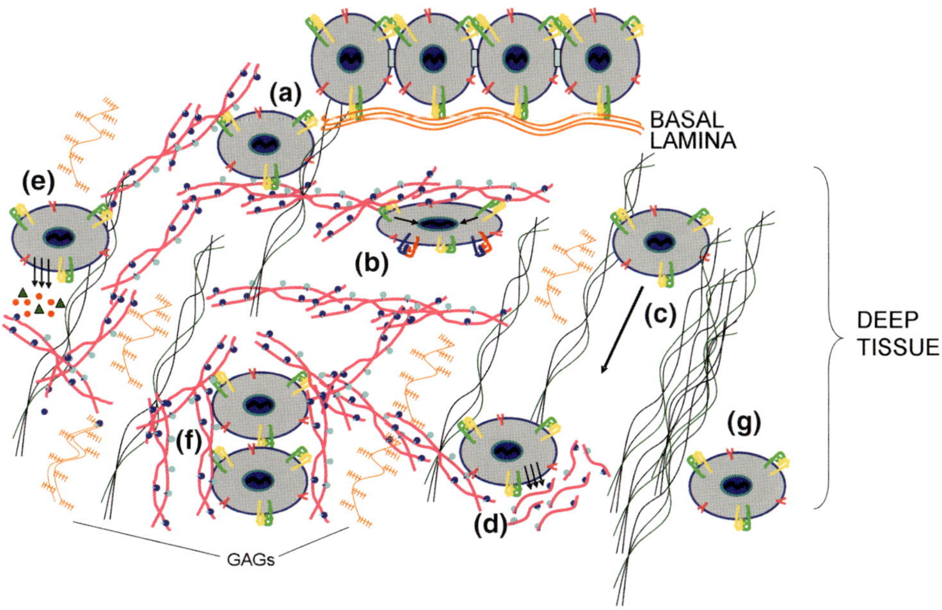

Figure 7.1. General key roles of the extracellular matrix (ECM) in regulating cellular response in tissue. (**a**) ECM interactions provide structure and anchorage for cells residing in tissue and modulate; (**b**) cell phenotype and differentiation; (**c**) migration of cells to specific spatial locales during development, repair, and growth; (**d**) matrix secretion, assembly, and remodeling of the extracellular environment; (**e**) secretion of local, soluble tissue-specific factors for degradation, repair, and migration through tissue; (**f**) intercellular communication; and (**g**) boundaries from one tissue to another, or within specialized regions of one tissue. *GAGs* glycosaminoglycans.

7.2. Cell–ECM Adhesive Interactions: Integrins as Pivotal Linkers

Most of these aforementioned structural proteins also play fundamental roles in promoting cell adhesion and mediating intracellular signals critical to tissue function. In particular, cell adhesion to the ECM is essential for controlling such complex biological processes as embryonic development, wound healing, immune responses, and tissue organization and repair [8, 9]. For example, cell–adhesive interactions with specific ECM proteins, such as FN and type I collagen, regulate bone cell survival, cell cycle progression, differentiation, and matrix formation; at the tissue level, these interactions can play crucial roles in bone remodeling, maintenance, and formation [10, 11]. Moreover, ECM-mediated cell adhesion plays critical roles in animal development, as evidenced by the embryonic lethality of mice that have genetic deletions for particular ECM ligands and receptors [8, 12]. Proteins such as FN are important in guiding cell movements during organism development, including migration of mesodermal cells at different embryonic stages [13]. These ECM components regulate cell behavior by interacting with cell surface receptors to activate particular intracellular signaling pathways, resulting in tissue-specific alterations in cell spreading, migration, cell–cell communication, and differentiation [14–16].

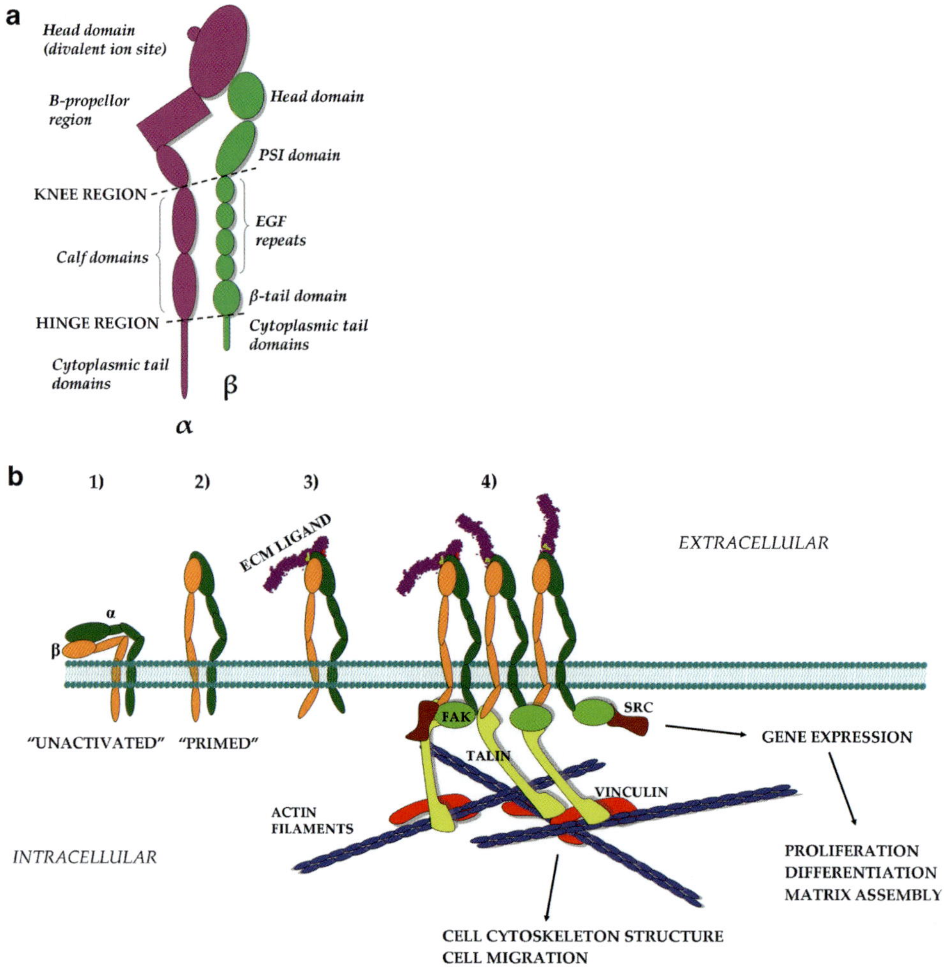

Figure 7.2. Integrins link the extracellular and intracellular environments. (**a**) Integrin subunit structure. Each subunit (α and β) noncovalently associates with each other and contains a head (ligand binding), knee (genuflection), and hinge (at the cell membrane) region. (**b**) Primary integrin activation and binding steps. (1) "low affinity" ligand binding unactivated integrin state demonstrating bending at knee regions, (2) "high affinity" ligand binding activated state as a result of interaction with intracellular (talin) and/or extracellular (divalent ions, ligands) factors, (3) integrin clustering into (4) focal adhesions, which also consist of large intracellular scaffolds of structural and signaling components, important in regulating functions such as cell adhesion, cell shape, migration, proliferation, and differentiation. *FAK* focal adhesion kinase; *EGF* endothelial growth factor; *PSI* plexin–semaphorin–integrin.

Although there are many types of cell receptors that facilitate cell–ECM interactions, cells recognize and adhere to ECM ligands primarily via integrins, a widely expressed class of cell surface receptors. Integrins are transmembrane, heterodimeric proteins consisting of a particular combination of two noncovalently associated subunits (α and β), of which 24 specific combinations have been currently identified in humans [17] (Figure 7.2a). Integrins interact with the ECM through their extracellular domains, and are linked to cytoskeletal elements and signaling molecules through their intracellular domains, functioning as the primary bridge between the ECM environment and the cell. Different α and β subunit combinations have the capacity to bind to one or more ECM ligands, whereas many ECM proteins can act as ligands for more than one integrin. Moreover, a wide array of integrins can be expressed, typically in tissue- and development-specific patterns. For example, osteoblasts and osteoprogenitor cells express multiple integrins, including $\alpha_1\beta_1$, $\alpha_2\beta_1$, $\alpha_3\beta_1$, $\alpha_4\beta_1$, $\alpha_5\beta_1$, $\alpha_6\beta_1$, $\alpha_8\beta_1$, $\alpha_v\beta_3$, and $\alpha_v\beta_5$, that can vary in degree of expression with the stage of the osteoblast and bind to numerous ECM components [18–20].

Most integrins recognize specific binding sites, typically small peptide sequences, such as the ubiquitous arginine–glycine–aspartic acid (RGD) motif present in a variety of ECM proteins [21]. Integrin binding to ligands is a dynamic and highly regulated process that requires "activation" of the integrin and mechanical coupling to the ligand. This activation changes the conformation of the extracellular portion of the receptor into a more "high-affinity" structural orientation for ligand binding, although divalent metal ions are also typically needed for functional binding. After integrin–ligand binding, integrins associate with the actin cytoskeleton and begin to aggregate together in clusters forming focal adhesions, distinct structures that form intracellular scaffolds of structural and signaling molecules [1, 22, 23]. Focal adhesions mediate stable adhesion by providing structural links between the ECM and cell cytoskeleton to regulate force-associated cell functions such as adhesion, spreading, morphology, and migration. These adhesive structures also activate discrete signaling pathways (mitogen-activated protein kinase [MAPK], JNK, extracellular signal-related kinase [ERK]) through accumulation and activation of signaling mediators, such as focal adhesion kinase (FAK) and Src, that can ultimately regulate transcription factor activity and direct major cell functions such as migration, proliferation, and differentiation [1, 24] (Figure 7.2b). For example, $\alpha_2\beta_1$-mediated cell attachment to type I collagen stimulates the tyrosine phosphorylation of FAK and, subsequently, the activation of ERK, a MAPK that has been implicated in the control of osteoblast-specific gene expression and matrix mineralization [25, 26].

Specialized integrin–ECM ligand interactions can regulate many cell functions, including survival, proliferation, motility, morphology, and differentiation in a cell/tissue-specific and time-dependent manner. Numerous studies using antibodies that block specific integrins have underscored the importance of the FN interaction with β_1 integrins in regulating osteoblast, chondrocyte, myoblast (muscle cells), neural, and human mesenchymal stem cell survival, proliferation, gene expression, and cell fate [18, 27–29]. In bone cells, blocking antibodies against collagen-specific $\alpha_2\beta_1$ integrin impede the expression of osteoblast-specific genes, such as osteocalcin, and inhibit calcification and formation of a mineralized matrix [30, 31]. Engagement of distinct integrins will often direct particular cell responses that are specific to different tissues. While $\alpha_v\beta_3$ binding is pro-proliferative and aids in migration of osteoblasts, liver, and even cancer cells, activation of this same integrin will impede bone mineralization and osteoblast differentiation; these functions can be rescued by binding of $\alpha_5\beta_1$ [32–35]. Leukocyte-specific β_2, including the Mac-1 receptor $\alpha_M\beta_2$, regulates macrophage adhesion to FN, as well as interactions with complement macromolecules and various domains of fibrinogen that may become exposed on different biomaterial surfaces [14, 36, 37]. A compendium of integrin subunit combinations and their demonstrated individual high-affinity recognition to various ECM proteins and their binding motifs is presented in Figure 7.3. These studies highlight the importance of specific integrin–ECM interactions to control tissue-specific cell function.

a

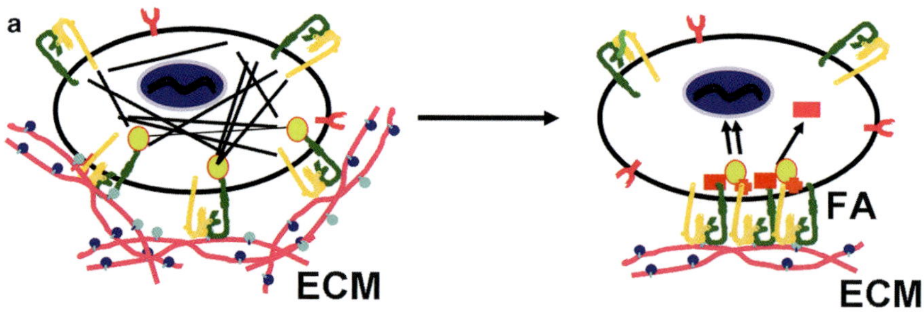

b

ECM Ligand	Integrin	Binding Motif
FN	$\alpha_4\beta_1$	IIICS (REDV, EILDV)
	$\alpha_5\beta_1$	RGD + PHSRN
	$\alpha_{II}\beta_3$	RGD
	$\alpha_v\beta_3$	RGD
COL-I	$\alpha_2\beta_1$	RGD, DGEA
COL-IV	$\alpha_1\beta_1$	CNBr a1(IV)2
VITRONECTIN	$\alpha_{II}\beta_3$	RGD
	$\alpha_v\beta_3$	RGD
FIBRINOGEN	$\alpha_{II}\beta_3$	RGD, KQAGD
	$\alpha_v\beta_3$	RGD
	$\alpha_M\beta_2$	P1,P2
LN	$\alpha_1\beta_1$	P1, E1-4
	$\alpha_3\beta_1$	GD6 peptide, E3
	$\alpha_6\beta_1$	E8
VON WILLENBRAND FACTOR	$\alpha_{II}\beta_3$	RGD
	$\alpha_v\beta_3$	RGD
OSTEOPONTIN	$\alpha_4\beta_1$	IDAPS
	$\alpha_v\beta_3$	RGD
THROMBOSPONDIN	$\alpha_3\beta_1$	TSP-768
	$\alpha_v\beta_3$	RGD
BONE SIALOPROTEIN	$\alpha_v\beta_3$	RGD
TENASCIN	$\alpha_v\beta_3$	RGD
FACTOR X	$\alpha_M\beta_2$	

Figure 7.3. Integrin–matrix interactions are finely tuned and regulated, and are essential for subsequent cell functional pathways. (**a**) Focal adhesions (FA) are initiated when integrins bound to ECM or ECM-derived ligands are activated and subsequently "pulled" by cytoskeletal elements together. (**b**) List of selected native ECM proteins and associated binding epitopes noted for specific recognition of individual integrin subunit combinations.

7.3. Engineering Biomaterial Surface Properties for Integrin Binding

Given the central roles of integrins in cellular processes, ongoing biomaterials research has focused on identifying key material surface characteristics that modulate the type, strength, and degree of ECM–integrin binding. Cell adhesion to ECM components or ECM-derived ligands on engineered surfaces governs host responses to implanted devices, integration of biomedical prostheses and tissue-engineered constructs, and the performance of biotechnological supports [38]. More specifically, integrins promote this robust cell adhesion to biomaterials by binding adhesive ligands on the material that have been (a) adsorbed from solution, (b) deposited by cells onto the surface, or (c) specifically engineered onto a surface, or a combination of these mechanisms [39]. Several material surface properties have been identified that modulate the type, amount, strength and conformation of proteins adsorbed or immobilized on the surface, which, in turn, directly modulate the level and specificity of engaged integrins. A summary of these different adsorption states and their impact on integrin binding is illustrated in Figure 7.4. For example, following protein adsorption, various protein unfolding events may occur, either (a) exposing "cryptic" binding sites that promote engagement of more selective integrins or (b) "stretching" integrin-binding epitopes that may negatively affect integrin recognition and levels of binding [40, 41].

Physiochemical properties of material surfaces that influence the nature of integrin binding include substrate composition, surface energy, surface charge, surface chemistry, and topography. The underlying chemical composition of a biomaterial substrate alone may regulate cell type-specific integrin expression. Osteoblasts on titanium alloys express α_2, α_3, α_4, α_6, α_v, β_1, and β_3 integrin subunits; however, on CoCrMo alloys, these same cells do not always express α_3, α_6, or β_3 [42]. Osteosarcoma and bone marrow stromal cells cultured on hydroxyapatite (HA), titanium, and CaP-coated titanium surfaces display differential trends in integrin expression [43, 44]. It has been reported that HA-coated surfaces promote more

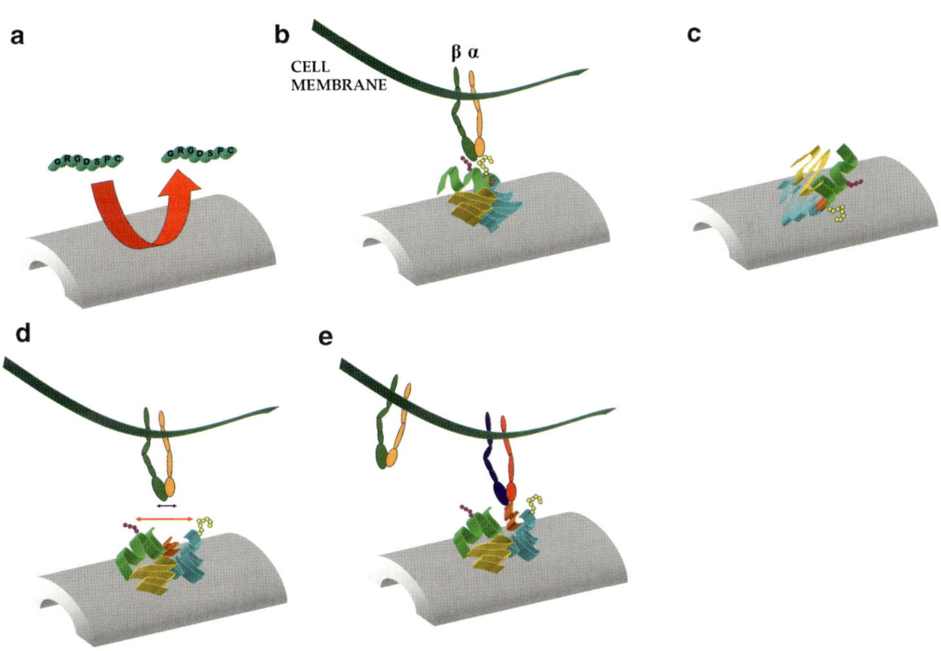

Figure 7.4. Different modes of adsorption to biomaterial surfaces for integrin binding ligands. (**a**) Short oligopeptides typically do not physisorb well to metal and polymeric surfaces. (**b**) Favorable adsorption of larger ligands presenting adhesive epitopes in an accessible orientation promote integrin binding. (**c**) Unfavorable adsorption of ligands presenting adhesive epitopes in a nonaccessible orientation reduce integrin binding. (**d**) Protein unfolding post-adsorption may "stretch" integrin-binding epitopes, resulting in suboptimal or no integrin binding. (**e**) Protein unfolding may also expose "cryptic" epitopes that may promote interaction with different integrins.

robust adsorption of serum proteins compared with titanium, and that these proteins are in distinct conformations on each surface [45]. Surface energy and charge, which are closely related, also may elicit differential integrin expression profiles. Materials that are hydrophobic, including poly(methyl methacrylate) and polystyrene, typically exhibit higher levels of total protein adsorption compared with more hydrophilic surfaces, such as ceramics and metallic materials [3]. More general biomaterial surface strategies to utilize unique charged and higher surface energy substrates have been used to take advantage of common electrostatic interactions that occur in many biomaterial interfacial events in physiological conditions. Titanium implants have been engineered via glow discharge plasma (GDP) technology to exhibit altered charge density and increased surface energy [46]. It is hypothesized that this treatment enables favorable surface energy for binding serum proteins and growth factors that increase integrin expression for enhanced osteoblast differentiation compared with untreated uncharged surfaces.

Tailoring the surface chemistry of material and scaffold surfaces has emerged as another strategy to regulate protein adsorption and conformation in order to better control the biological activity of adsorbed adhesive proteins [47]. Using self-assembled monolayers (SAMs) of functionalized alkanethiols on gold to present well-defined chemistries, biomaterial-dependent differences in the total amount and conformation of adsorbed FN have been identified [28, 48]. These differences in FN conformation in turn affect integrin receptor binding [49–51]. Osteoblast-like cells bound selectively with $\alpha_5\beta_1$ to OH- and NH_2-presenting SAMs, but with both $\alpha_5\beta_1$ and $\alpha_v\beta_3$ to COOH-presenting surfaces [47]. This regulation of $\alpha_5\beta_1$ vs. $\alpha_v\beta_3$ binding was found to directly modulate osteoblast adhesion, signaling, proliferation, and differentiation, as well as to influence myoblast proliferation and myogenic differentiation [29, 52]. Aside from demonstrating surface chemistry-dependent effects on integrin specificity, these data also underscore the importance of engaging specific integrins to achieve a directed cellular response.

Nano-scale and micro-scale surface roughness also regulates integrin-mediated cell interactions. Human osteoblasts and bone marrow stromal cells exhibit distinct expression profiles of α_3, α_5, and α_6 when cultured on either rough or smooth titanium substrates, ultimately influencing relative levels of adhesion and tissue-specific gene and protein expression [43]. Furthermore, MG63 human osteoblast-like cells exhibit an increase in α_2, α_3, α_5, β_1, and β_3 integrin subunits on micro-scale rougher surfaces compared with smooth titanium, resulting in enhanced integrin-mediated spreading and adhesion [53]. Micro-scale surface roughness modulates $\alpha_5\beta_1$ binding, correlating to a boost in FAK activation, local growth factor production, and osteoblast-specific bone markers [54–56].

7.4. Modulating Cellular Response to Biomaterial Surfaces Through ECM-Mimetic Surface Modification Strategies

7.4.1. General ECM Surface Modifications to Regulate Integrin-Mediated Cell Function

Cell adhesion to ECM components or ECM-derived ligands on engineered surfaces governs the host response to implanted devices, the degree of integration of biomedical prostheses and tissue-engineered constructs, and the performance of biotechnological supports [38]. While typically easy to implement, most physiochemical surface strategies alone lack the ability to consistently generate selective integrin binding and direct robust integrin-mediated

a

ECM Ligand	Ligand	Reference	Targeted Integrin	Demonstrated Effect on Cell Function
ADSORBED	FN	[18,27-29,50-54,59]	$\alpha_v\beta_3$, $\alpha_5\beta_1$	adhesion, migration, proliferation, differentiation
	COL-1	[30,60,61]	$\alpha_2\beta_1$	adhesion, migration, proliferation, differentiation
	COL-IV	[58]	$\alpha_1\beta_1$,$\alpha_2\beta_1$,$\alpha_3\beta_1$	adhesion
	BSP	[33 62,63]	$\alpha_v\beta_3$	adhesion, osteoconductivity
	VN	[19,64]	$\alpha_v\beta_3$, $\alpha_v\beta_5$	adhesion
SHORT OLIGOPEPTIDE	EILDV, REDV	[69]	$\alpha_4\beta_1$	adhesion
	IDAPS, KQAGD	[72,73]	$\alpha_{IIb}\beta_3$, $\alpha_4\beta_1$	adhesion
	RGD	[21,74-76,79,80,84]	$\alpha_v\beta_3$	adhesion, proliferation, migration, differentiation
	DGEA	[71,74]	$\alpha_2\beta_1$	adhesion, differentiation
	GFOGER	[26,31,113,114]	$\alpha_2\beta_1$	adhesion, differentiation
	KGGGAHEEICTTNEGVM (FN13)	[134]		matrix assembly, proliferation
MULTI-MOTIF PEPTIDE/AMPHIPHILE	RGD-PHSRN amphiphile	[99-104]	$\alpha_v\beta_3$	adhesion, differentiation
	RGD-PHSRN phospholipid	[100]	$\alpha_v\beta_3$, $\alpha_5\beta_1$	adhesion
	RGD with elastin-like repeats	[77]	$\alpha_v\beta_3$	adhesion
	GFOGER-SPARC	[122-124]	$\alpha_2\beta_1$	adhesion, proliferation
	RGD-amphiphile	[81]	$\alpha_3\beta_1$	adhesion, migration
ENGINEERED ECM-MIMETIC	FNIII $_{10}$	[107]	$\alpha_v\beta_3$	adhesion
	FNIII $_{9\text{-}10}$	[130]	$\alpha_5\beta_1$	adhesion, differentiation
	FNIII$_{7\text{-}10}$	[87,106,109-111]	$\alpha_5\beta_1$	adhesion, profliferation, differentiation
	V25/CS-1	[108]	$\alpha_4\beta_1$	adhesion, matrix assembly
	GFOGER (triple-helical)	[31,70, 115-121]	$\alpha_2\beta_1$	adhesion, profliferation, differentiation

b

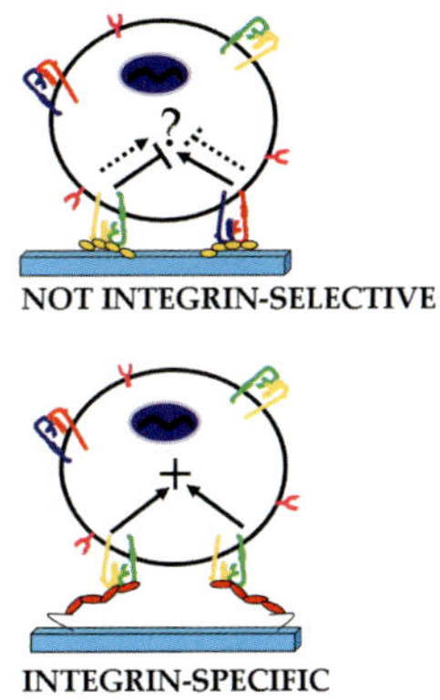

Figure 7.5. Surfaces mimicking the natural ECM to direct integrin specificity and modulate cell response. (**a**) Major ligand surface modification schemes utilizing full and modified ECM ligands to design integrin-selective interfaces. (**b**) Substrates that do not discriminate between integrin binding and activation often may produce antagonistic signaling pathways. Integrin-specific surfaces facilitate single integrin-mediated adhesive and signaling pathways to produce a more directed and robust cell response.

cell responses, especially for in vivo applications. Advances in understanding the role of specific ECM biomolecules in regulating integrin expression, cell adhesion, differentiation, and tissue remodeling have led to the investigation of targeted biochemical methods of surface modification. Recent biomimetic strategies have focused on the immobilization of selected ECM matrix components, including native structural proteins, peptide sequences, or synthetic derivatives based on matrix molecules [57]. The goal of these surface strategies is to modulate integrin–matrix interactions and promote general integrin binding to direct a specific cell/tissue response. Many of these ECM-derived surface ligand modifications and their integrin-specific effect on cellular functions will be discussed in further detail in the following sections of this chapter; a summary is given in Figure 7.5.

The coating of various biomaterial substrates with full-length ECM proteins modulates integrin-mediated cell function to varying degrees. For instance, FN coatings on polystyrene increase osteoblast expression of α_3 and α_5 and reduce expression of α_2, α_6, and α_v subunits compared with uncoated polystyrene [42]. Stromal cell, calvarial osteoblast, and osteoblast-like cells exhibit enhanced β_1-mediated adhesion, proliferation, and differentiation on FN-coated and collagen type I-coated surfaces compared with uncoated glass, polystyrene, or titanium surfaces [58]. It has been postulated that plasma FN-coated implants enhanced early bone formation due to a chemotactic attraction of osteoprogenitors near the ECM-coated implant surface [59]. Although there are conflicting data regarding the functional effects of collagen coating on polystyrene and titanium for various cell types, there is evidence of cell type-specific distinct integrin expression profiles on collagen vs. non-collagen-coated surfaces [60, 61]. Osteoblasts upregulate α_3 expression over α_6 on collagen type I-coated polystyrene, whereas collagen type IV coatings mediate greater $\alpha_{1,}$ $\alpha_{2,}$ $\alpha_{3,}$ and β_1 expression and activation [42]. Other bone-specific matrix proteins, including bone sialoprotein (BSP) and osteopontin, have also been explored as bioactive surface coatings to control β_3-mediated binding and signaling for human osteoblasts and stromal cells. BSP is an acidic, noncollagenous glycoprotein abundantly expressed in mineralized tissues. BSP bound to collagen type I was found to orient in a bioactive orientation that augments β_3 binding and subsequent osteoblast adhesion and differentiation over unmodified interfaces [62]. Implants coated with BSP were found to be fairly osteoconductive yet supported only minimal functional integration and mechanical fixation [63]. Vitronectin is another ECM protein that dominates serum-exposed adsorption to polystyrene [39]. Interestingly, coating vitronectin on polystyrene enhances mouse osteoblast expression of $\alpha_v\beta_3$, but also has been shown to weaken adhesion of murine cells compared with uncoated polystyrene [19]. However, for human osteoblasts, adhesion to vitronectin was found to be primarily $\alpha_v\beta_5$-mediated and more robust than adhesion to uncoated surfaces [64]. These studies highlight cell type-dependent effects on integrin binding and cell function that surface modifications using natural ECM proteins may elicit.

Although promoting cell-specific functions, the functionalization of biomaterials with natural matrix proteins has several drawbacks. Since many full-length matrix proteins contain multiple integrin and other biomolecule binding sites, it is likely that these proteins may induce significant binding of more than one integrin, and hence, do not truly support an "integrin-specific" surface. Moreover, immunogenic and pathogen transmission concerns across species limit widespread use of these ligands as bioactive coatings for implants in humans. Furthermore, processing issues including scale-up difficulties and long-term stability may reduce the widespread application of these matrix functionalized surfaces.

7.4.2. Small Biomimetic Peptide Surface Strategies

Synthetic biomimetic strategies to promote cell adhesion have primarily focused on presenting short bioadhesive motifs derived from ECM components onto biomaterial or implant surfaces [65]. A number of adhesion motifs from ECM components have been identified that can be subsequently incorporated in a small synthetic peptide. These motifs consist of a short linear sequence of amino acids, typically no more than a dozen amino acids in length, that one or more integrin receptors can recognize and bind [66]. These short sequences can be easily incorporated into adhesive oligopeptides that can be strategically immobilized in numerous ways on a biomaterial support to promote integrin-mediated adhesive and functional responses. These immobilization schemes include covalent tethering directly onto a biomaterial substrate, using plasma treatment to increase the functional groups on the substrate, or tethering onto polymer coatings functionalized on an underlying substrate. Nonfouling supports (polyethylene glycol or alginate) are often used to better isolate the activity of these motifs by utilizing a background with minimal nonspecific protein adsorption [67].

Many of these adhesive sequences are specifically associated with ECM components from a particular tissue environment. For example, BSP- and collagen-derived sequences have been identified that remain functional only with bone-associated cell types. Multiple integrin-binding sequences have also been identified on single ECM proteins. For instance, the EILDV and REDV sequences on FN mediate $\alpha_4\beta_1$-binding of endothelial cells, while the RGD and PHSRN sequences, when presented together correctly, promote $\alpha_5\beta_1$-mediated osteogenic differentiation [68, 69]. The adhesion motifs DGEA and GFOGER from collagen type I both mediate $\alpha_2\beta_1$-mediated function—although in separate cell systems, i.e., neurite migration for DGEA and osteoblast differentiation for GFOGER [70, 71]. Other ECM-specific integrin-binding sequences identified include the IDAPS sequence in osteopontin, which promotes $\alpha_4\beta_1$-mediated adhesion of lymphocytes, and the KQAGD sequence in fibrinogen, which controls $\alpha_{IIb}\beta_3$-adhesion of osteoclasts [72, 73]. Hydrogels functionalized with adhesive peptides incorporating IKLLI, IKVAV, LRE, PDSGR, RGD, or YIGSR, and the collagen type I sequence, DGEA, differentially regulated pancreatic cell survival [74]. Many of these synthetic oligopeptide-based strategies have been successfully used in vivo for tissue repair and regeneration. For example, GFOGER-coated implants promote extensive bone formation and implant integration in orthopedic applications [31].

The most common oligopeptide strategy to control integrin binding relies on the surface presentation of the arginine–glycine–aspartic acid (RGD) adhesive sequence, which mediates cell attachment to several matrix proteins, including FN, vitronectin, osteopontin, and BSP [75]. Since the RGD sequence is a ubiquitous adhesive motif in most ECM components, biomaterial substrates and tissue-engineered constructs coated with RGD peptides have been employed to control in vitro adhesion, migration, and differentiation in numerous cellular systems, including neural, endothelial, and bone cells [76]. For example, synthetic ECM proteins incorporating the GRGDSP cell-binding domain as well as elastin-like repeats (for more robust mechanical strength) enhance adhesion and spreading of endothelial cells over similar adhesive peptides [77]. The density and presentation of RGD peptides can have a profound effect on their overall influence on cell function for certain cell types. The degree of osteoblast proliferation, bone-specific gene expression, and differentiation varies directly with RGD ligand density [78]. However, relatively high RGD densities can also negatively affect neurite extension and outgrowth in three-dimensional (3-D) gel scaffolds [79]. Although RGD-immobilized peptides typically mediate osteoblast cell function through the $\alpha_v\beta_3$ integrin, altering the structural presentation of this motif can mediate other cell type functions via non-α_v and/or β_3 integrins [80]. For example, immobilized synthetic alkyl amphiphiles of RGD promote $\alpha_3\beta_1$-mediated adhesion and migration of melanoma cells [81]. Grafting implant materials with either cyclic or oligomeric peptides presenting multiple RGD arms increases human osteoblast adhesion and spreading in a primarily $\alpha_v\beta_5$-dependent mechanism, depending on alterations to the RGD structure and the exact flanking sequence surrounding the tripeptide [82, 83]. Although it has been shown that RGD presented in cyclic, or "constrained," conformations exhibit enhanced integrin affinity to many α_v or β_3 integrins compared with linear RGD, due to the conformation of RGD and its flanking regions, these peptides still retain minimal $\alpha_5\beta_1$ selectivity [80, 84].

Nonetheless, while many studies have demonstrated that RGD-functionalized materials support integrin-mediated adhesion, proliferation, and differentiation in vitro, mounting evidence suggests that this biomimetic surface strategy does not enhance biomedical implant integration in more rigorous animal models. A number of studies have concluded that RGD-immobilized titanium implant coatings do not improve peri-implant new bone formation and yield only marginal increases in implant integration and mechanical fixation [85–89]; in contrast, very few studies demonstrate a significant in vivo enhancement in osseointegration [90, 91]. Furthermore, when RGD peptides were immobilized onto hydroxyapatite discs in conjunction with serum proteins such as FN and vitronectin, there was a detrimental effect on mesenchymal stem cell survival as well as new bone formation [85]. Even the benefits of this biomimetic surface treatment in vitro are unclear. Osteoblasts cultured on RGD functionalized poly(L-lysine)–poly(ethylene) glycol (PLL-g-PEG) surfaces exhibit reduced bone-specific markers and osteogenic differentiation compared with unmodified surfaces [92]. Even RGD peptides of varying flanking sequences tethered on SAMs all displayed substandard adhesive capacity compared with full-length FN surfaces [93]. Although the stability, relative cost-effectiveness, ease of immobilization, and reduced immunogenicity are all advantages of using these small adhesive motifs, several limitations reduce their efficacy for more directed tissue repair and regeneration. First, the biological activity of these peptides is substantially lower than that of the whole protein due to the absence of modulatory integrin-binding domains. Second, this lack of essential modulatory domains limits surface selectivity for integrins. Because RGD is recognized by a large number of integrins in numerous cell types, this lack of integrin specificity may result in nondiscriminatory attachment of cells to the RGD-coated surfaces. For example, in order to achieve high affinity $\alpha_5\beta_1$-binding, both the RGD sequence in the tenth type III repeat of FN as well as its synergy site, the PHSRN sequence in the ninth type III repeat, are required to be presented together [94]. Since $\alpha_5\beta_1$ is especially critical for induction of various osteoblastic signaling pathways and activities, this lack of integrin binding may be detrimental for orthopedic applications. Third, these small peptides may lack the ability to bind specific receptors due to conformational differences among adhesive sequences compared with the native ECM ligand. As a consequence of these collective limitations, newer generations of bio-inspired surfaces have focused on more effectively mimicking the structure and make-up of natural ECM integrin-binding epitopes.

7.4.3. Multiple-Motif Integrin-Specific Ligands

An advanced biomimetic surface strategy that has been recently explored to obtain more directed cell/tissue response is the engineering of surfaces to engage specific integrin receptors among RGD-binding integrins [10]. Recent studies suggest that the marginal healing responses of RGD-functionalized implants might arise from the lack of selectivity of this adhesive ligand for specific integrin receptors [31, 95]. Since activation and signaling by more than one type of integrin receptor on the same cell may induce antagonistic cellular responses, as evidenced by the aforementioned effect of $\alpha_5\beta_1$ vs. $\alpha_v\beta_3$ binding on osteoblast function, a more selective integrin binding surface may achieve a more controlled cellular response [52]. Unfortunately, full-length matrix proteins possess multiple integrin binding sites, whereas short RGD peptides can also bind multiple integrins without a high degree of specificity. Therefore, several recent approaches have synthesized ligands presenting this RGD motif with additional modulatory domains to selectively target particular integrin receptors, while excluding other nonessential biological domains.

Although cyclic-RGD peptides improve ligand specificity for several integrins, including $\alpha_v\beta_3$ and $\alpha_3\beta_1$, the ability of these constrained peptides to achieve levels of high affinity $\alpha_5\beta_1$ binding comparable to full-length FN is limited. As previously mentioned, RGD and its PHSRN synergy sequence in FN individually contribute little to high affinity $\alpha_5\beta_1$ binding, but, when presented together, promote $\alpha_5\beta_1$-mediated cell adhesion [94]. The synergistic effects of the RGD and PHSRN sites is strongly dependent on the molecular structure as slight alterations in the nano-scale spacing (30–40 nm), relative angle, and flanking sequences between these two sites result in significant losses in biological activity and integrin binding behavior [76, 96, 97]. Because of this tight dependence of receptor binding on the structural context of the ligand, the ability to mimic the integrin binding and functional abilities of the native ligand is especially challenging with single RGD peptides. FN-mimetic peptide-amphiphiles and multi-motifed peptides presenting both the RGD and PHSRN site have been engineered to address some of these structural limitations and to study at different molecular levels the ligand binding properties of $\alpha_5\beta_1$ [98]. Surfaces of PEG molecules attached to phospholipids presenting the GRGDSP and PHSRN sites were used to demonstrate that $\alpha_5\beta_1$ binding to peptide-amphiphiles is dependent on membrane composition, temperature, and density [99]. Bioartificial membranes constructed from GRDGSP and PHSRN have led to insights on binding/unbinding events with $\alpha_5\beta_1$ down to a single-molecule level. The results of research with these FN-mimetic peptide-amphiphiles suggest that accessibility and relative spatial orientation is crucial for FN-like robust $\alpha_5\beta_1$ binding [100]. An RGD–PHSRN amphiphile that presented the RGD motif in structurally distinct variations exhibited enhanced $\alpha_v\beta_3$ specificity when the RGD was looped rather than linear [101]. Another RGD–PHSRN amphiphile consisting of the two adhesive motifs, a spacer, and a linker (designed to mimic the distance and hydrophobicity of the two motifs in FN) was reported to enhance endothelial cell adhesion over FN, although a direct comparison of the two surfaces on an equimolar basis was not performed [102, 103]. Moreover, another similar RGD–PHSRN amphiphile that was connected by a linker recapitulating the native spacing of FN enhanced osteoblast differentiation over surfaces immobilized with RGD alone [104]. Linear and cyclic lipid-linked RGD ligands also seem to enhance integrin-mediated adhesion, even at low surface densities [105].

Although several studies have indicated that FN-mimetic peptide-amphiphiles enhance integrin binding and cell function over RGD alone, the relative biological activity of these second-generation peptides, compared with the native protein and even simple RGD peptides, remains poorly characterized. Recently, a study in which both RGD and an RGD–PSRHN peptide, mimicking the natural spacing of FN, were tethered on a mixed SAM at equimolar levels revealed no significant differences in adhesion or integrin binding specificity between the two ligands [106]. Comprehensive analyses, including antibody blocking, varying ligand surface densities, and signaling evaluations are necessary to more fully establish the adhesive and functional potential of these engineered interfaces. Hence, although these adhesive interfaces do promote integrin-mediated adhesion, mimicking the full biological activity of the native protein (FN) using these synthetic short adhesive motifs remains challenging.

7.4.4. FN-Derived Highly Selective Integrin Ligands

As an alternative strategy, the engineering of high molecular weight ligands recapitulating the primary, secondary, and tertiary structure of the native protein to reconstitute full biological activity and convey integrin binding specificity has been pursued. In particular, recombinant fragments of FN incorporating either just the tenth type III repeat or both the tenth type III repeat (RGD) and the ninth type III repeat (PHSRN) have been recently engineered. These fragments present the aforementioned adhesive motifs and flanking sequences in the correct structural conformation and spatial orientation as native FN. These recombinant FN fragments were prepared using DNA recombinant technology, which affords flexibility in designing unique ligand characteristics to study structure/function, control immobilization and orientation, and enhance bioactivity. Using fragments of FN also limits antigenicity (compared with full-length FN) and eliminates other modulatory domains, which may either interfere with specific integrin binding or elicit undesired inflammatory responses. FN10, a FN fragment comprising the tenth type III FN repeat and engineered to present a modified RGDWXE motif, exhibited more robust binding to human $\alpha_v\beta_3$ compared with normal RGD-containing FN10 fragments [107]. This finding highlights the utility of using recombinant protein technology to correlate structural adhesive motif characteristics with receptor binding activity and cell function. Specific binding to $\alpha_4\beta_1$ has been linked to an alternatively spliced V (residues 1–25 of the IIICS) region of FN distant from the central cell-binding domain (V25/CS-1) [108]. FNIII(del1–7), a recombinant FN ligand that incorporates this V25/CS-1 region and the tetrapeptide Arg–Glu–Asp–Val (REDV) motif (recognized by $\alpha_4\beta_1$), display $\alpha_4\beta_1$-mediated function, including adhesion and matrix assembly [108].

A larger recombinant FN fragment ($FNIII_{7-10}$) presenting the seventh through tenth type III repeats of FN was functionalized on nonadhesive supports to effectively mimic the $\alpha_5\beta_1$-specific binding region of FN. These biomimetic surfaces supported $\alpha_5\beta_1$-mediated osteoblast adhesion and spreading at levels comparable to full-length FN [109]. This fragment was tethered onto model, nonfouling mixed SAM surfaces in order to compare the adhesive activity and integrin binding specificity on an equimolar basis with two RGD-based linear ligands: specifically, (1) a linear GRDGSPC peptide, and (2) the RGD–PHSRN peptide that recapitulates the normal spacing of these two motifs in FN. Importantly, biointerfaces presenting $FNIII_{7-10}$ exhibited significantly higher adhesion strength, FAK activation, and proliferation rate than substrates presenting an equivalent molar density of either RGD or RGD–PHSRN [106]. Moreover, $FNIII_{7-10}$-functionalized surfaces displayed specificity for $\alpha_5\beta_1$ integrin, while mixed SAMs presenting either RGD or RGD–PHSRN were mediated by $\alpha_v\beta_3$. For controlled in vivo analysis, a clinically relevant, robust polymer coating was grafted onto clinical-grade titanium (the gold standard for orthopedic implant materials), which afforded controlled ligand tethering and direct comparison of $FNIII_{7-10}$ to RGD in vitro and in vivo (Figure 7.6a) [110]. In vitro evaluation of these biointerfaces was performed

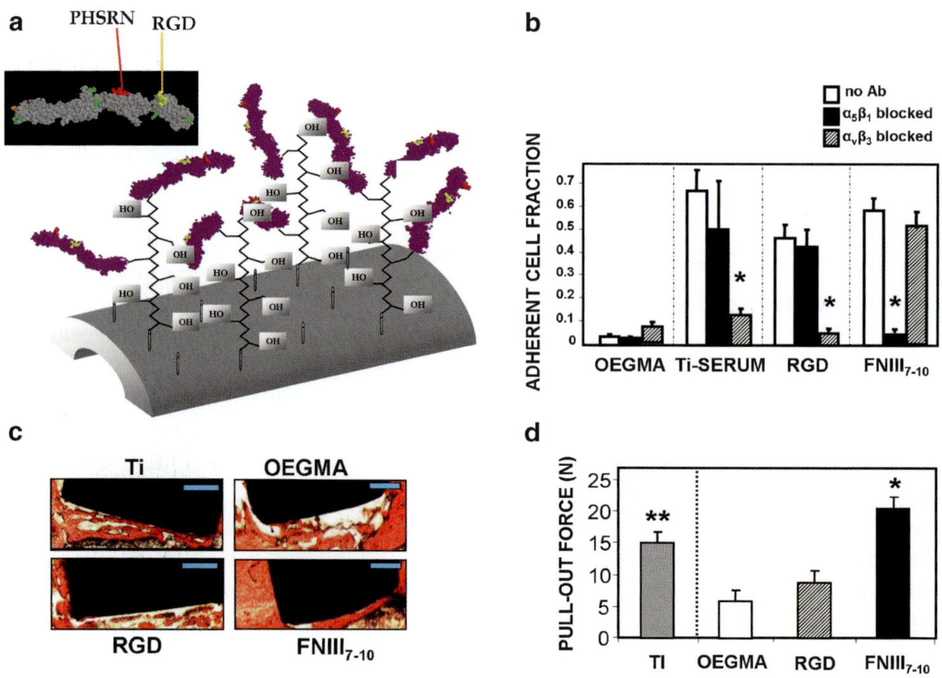

Figure 7.6. $FNIII_{7-10}$-functionalized implant coatings direct integrin specificity in vivo to enhance implant osseointegration. (**a**) $FNIII_{7-10}$, containing the RGD and PHSRN epitopes presented in the correct structural conformation and orientation for high-affinity $\alpha_5\beta_1$ binding, was tethered onto stable, nonfouling poly(OEGMA) brushes grafted onto clinical titanium substrates. (**b**) Integrin-subunit-blocking antibodies demonstrate that adhesive interactions on $FNIII_{7-10}$ surfaces are mediated mainly by $\alpha_5\beta_1$, whereas RGD-functionalized surfaces were primarily $\alpha_v\beta_3$ mediated* vs. no antibody control (Ti-serum: $p < 0.01$; RGD: $p < 0.01$; $FNIII_{7-10}$: $p < 0.005$). $FNIII_{7-10}$-functionalized brush implant coatings enhanced (**c**) bone–implant contact area (bone = *orange*, implant = *black*; scale bar is 0.5 mm) and (**d**) functional integration and mechanical fixation as measured by pull-out testing compared with RGD-functionalized, unmodified brushes and on clinical-grade titanium. *$FNIII_{7-10}$ vs. RGD ($p < 0.009$), Ti ($p < 0.05$), OEGMA ($p < 0.001$), **Ti vs. RGD ($p < 0.01$).

using rat primary bone marrow stromal cells; this heterogeneous cell population contains osteoprogenitors and is relevant to human clinical cell-based applications. $FNIII_{7-10}$-tethered brushes conferred $\alpha_5\beta_1$ binding specificity and enhanced cell adhesion, signaling, osteoblast-specific gene expression, and matrix mineralization compared with equimolar RGD-tethered brushes (Figure 7.6b). Importantly, integrin-specific $FNIII_{7-10}$-functionalized implants significantly enhanced in vivo implant integration and mechanical fixation compared with the current clinical standard (unmodified Ti) and RGD-based surface treatments in a rat tibial model (Figure 7.6c, d) [87]. Remarkably, simple dip-coating of $FNIII_{7-10}$ onto titanium supports also promoted enhanced in vitro osteoblastic signaling, gene expression, and differentiation over adsorbed plasma FN and RGD peptides. This simple, point-of-application coating of $FNIII_{7-10}$ enhanced implant integration and fixation almost threefold compared with unmodified Ti and 80% over plasma FN-coated Ti implants [111]. Importantly, this facile, single-step implant coating procedure can be conducted under physiological conditions immediately prior to implantation, reducing chance of infection, cytotoxicity concerns from other immobilization schemes, and biomaterial surface variability. Collectively, these studies suggest that conferring integrin specificity on bioadhesive interfaces may be a clinically applicable surface strategy to obtain a directed tissue healing response, and, moreover, may be a facile approach for more robust tissue integration of other biomedical devices.

7.4.5. Collagen-Mimetic Integrin-Specific Ligands

Although much surface-engineering work has centered around RGD-binding receptors, other integrin receptors play critical roles in cell and tissue responses. For example, $\alpha_1\beta_1$ and $\alpha_2\beta_1$ integrins are the major collagen-binding integrins, with $\alpha_2\beta_1$ dominating osteoblast integrin-mediated adhesion to type I collagen, the major ECM constituent of bone [5]. The $\alpha_2\beta_1$–collagen interaction promotes osteoblast-specific signaling, gene expression, and differentiation, even in multipotent bone marrow stromal cells [30, 112]. Adhesion sequences have been identified that regulate collagen–integrin interactions, including the collagen-binding motif DGEA and the hexapeptide sequence GFOGER (residues 502–507 of the alpha1[I] chain in type I collagen) [71, 113]. However, linear peptides using these sequences lack the full binding specificity for $\alpha_2\beta_1$ in vitro compared with full-length collagen in multiple cell types [114]. Not surprisingly, molecular binding studies found that full $\alpha_2\beta_1$ recognition and high affinity binding of this GFOGER sequence is critically dependent on the tertiary structure of collagen, which resembles a triple helix [115, 116]. To engineer a stable, biomimetic ligand that reconstitutes this crucial tertiary collagen structure, a 4-kDa peptide was engineered that incorporates this $\alpha_2\beta_1$-specific GFOGER sequence surrounded by GPP triplet repeats that provide structural motifs for the formation of a stable, right-handed triple-helical structure mimicking native collagen [117]. When GFOGER was adsorbed on a polystyrene surface, this peptide specifically targeted the $\alpha_2\beta_1$ integrin receptor on osteoblasts, and supported FAK activation, osteoblast-specific gene expression, and matrix mineralization at levels comparable to those obtained on surfaces with adsorbed collagen type I [70]. Integrin-subunit blocking-antibody studies confirmed $\alpha_2\beta_1$-mediated adhesion for GFOGER coatings on orthopedic-relevant titanium substrates. In addition, studies using rat bone marrow stromal cells verified that GFOGER-coated titanium surfaces induced higher levels of osteoblast-specific gene expression, enzyme activity, and matrix calcification compared with serum-exposed titanium substrates [31]. Notably, GFOGER peptide coatings also significantly improved peri-implant new bone formation, implant integration, and functional osseointegration compared with the current clinical standard, unmodified titanium [31].

Given the ubiquitous role of collagen in other tissue types, these studies may point toward a significant clinical role for this GFOGER biomolecular strategy to control even nonorthopedic implant–tissue responses.

In fact, similar triple-helical GFOGER peptides have indeed been employed fairly successfully in other tissue systems to promote in vitro $\alpha_2\beta_1$-mediated adhesion and function [118]. Peptides incorporating slight alterations in the flanking sequence to GFOGER, resulting in variable differences in helix stability, mediated liver cell and mouse fibroblast adhesion and spreading to varying degrees [119–121]. The collagen-mimetic peptides, which demonstrated greater relative stability than other peptides, afforded higher levels of adhesion for these cell types. Mixed ligand surfaces presenting this triple-helical GFOGER peptide with other ECM-derived adhesion motifs on the same surface have also been explored in various tissue systems. A triple-helical peptide utilizing the collagen residues 496–507 (incorporating the GFOGER motif as well as six other amino acids) presented with a second ligand, an endothelial proliferating motif from the ECM protein SPARC, mediated $\alpha_2\beta_1$-dependent activation of endothelial cells, although presentation with the second ligand had no significant effect compared with the GFOGER peptide alone [122, 123]. Interestingly, compared with single ligands, incorporation of this triple-helical GFOGER peptide in an amphiphile, which also incorporated this SPARC motif as well as a pseudolipid (monoalkyl hydrocarbon chain) for flexible interaction, showed the greatest activity in the activation and adhesion of endothelial cells [124]. These conflicting results may be explained by differences in ligand density as well as presentation of the two motifs to modulate optimal cell interactions.

Finally, nonfouling surfaces presenting both this $\alpha_2\beta_1$-specific triple-helical GFOGER peptide as well as the $\alpha_5\beta_1$-specific $FNIII_{7-10}$ were engineered in order to evaluate the effects of integrin crosstalk on adhesive responses. Mixed ligand surfaces synergistically enhanced cell adhesion strength, focal adhesion assembly, FAK activation, and proliferation rate compared with single FN and COL-I mimetic ligand surfaces [125]. Given the crucial and overlapping roles multiple integrins play in various cell signaling and functional pathways, engineering interfaces to present multiple integrin-specific ligands may be a successful strategy to achieve more robust and/or complex, directed cell/tissue responses. Overall, biointerfaces presenting engineered ligands that recapitulate the primary, secondary, and tertiary structure of the native matrix protein, including FN and type I collagen, have exhibited great potential in directing specific adhesive, signaling, and differentiation in targeted cell populations. The improved integrin selectivity of these surfaces points to a major clinical role of this biomolecular engineering strategy for biomedical and biotechnological devices.

7.4.6. FN-Derived Integrin Ligands to Direct Matrix Assembly

A key property of the ECM environment is its tightly regulated, supramolecular 3-D structure that provides positional and environmental information essential for tissue function. In particular, the fibrillar architecture of FN is a major ECM network that modulates cell cycle progression, migration, differentiation, and even assembly of other matrix proteins in tissues at all stages of development [126–128]. FN matrix assembly is a multistep process that involves integrin interactions with specific regions of FN that, in coordination with the actin cytoskeleton, expose "self-association" sites on FN (initiation) and mediate subsequent fibril assembly steps [129]. In addition, these specific integrin–FN interactions regulate cell-mediated FN secretion that is integral in normal matrix assembly and remodeling [128]. Unfortunately, although cells will organize assembled matrices on biomaterials, most synthetic materials do not directly recapitulate this essential function of the ECM, limiting the use of these materials in tissue engineering and regenerative medicine. Development of FN-derived ligands that interact with specific integrin receptors have afforded more precise control over FN matrix assembly on biointerfaces. FN10, a recombinant fragment of FN incorporating the adhesive RGD motif, is an essential requirement for initiation of $\alpha_5\beta_1$-mediated FN matrix assembly. Presentation of FN10 with FN9, incorporating the synergy site PHSRN of $\alpha_5\beta_1$, augmented assembly over FN10 alone [130]. Interestingly, when mice were mutated so that the RGD motif became RGE on all cellular FN, the FN matrix assembly was not compromised, although embryonic development was stunted [131]. Studies using other recombinant fragments of FN containing various deletions of type III repeats verified that the RGD sequence is essential for the initiation step, but fibrils can form independently of the $FNIII_{1-7}$ modules [132]. It is noteworthy that deletion of the first through seventh type III repeats did alter the rate of FN assembly, suggesting a role for nonessential epitopes such as these to modulate different stages of matrix formation. Although the RGD-dependent $\alpha_5\beta_1$-mediated assembly pathway is perhaps the most studied, other integrins and FN regions have also been identified that are actively involved in matrix assembly. FNIII(del1–7), a recombinant FN ligand that incorporates the V25/CS-1 region and the tetrapeptide Arg–Glu–Asp–Val (REDV) motif recognized by $\alpha_4\beta_1$, was found to promote $\alpha_4\beta_1$-mediated FN matrix assembly, although the actin filament morphology was altered compared with $\alpha_5\beta_1$-mediated assembly, and the cell proliferation rate was unaffected [133].

A 13-amino acid sequence localized between the type II2 and I7 repeats of FN, the self-assembly domain, was found to induce the assembly of fibrillar ECM; this sequence also exhibits collagen binding activity in tumorigenic cells [126]. Applying this knowledge, a synthetic oligopeptide FN13 (KGGGAHEEICTTNEGVM) incorporating this sequence was covalently tethered to a nonfouling mixed SAM surface to nucleate the assembly of fibrillar FN matrix networks [134]. A specific surface threshold of tethered FN13 not only nucleated robust FN fibril matrices, but also served as a template for collagen fibril assembly, along with increasing osteoblast cell proliferation rate over scrambled FN13, RGD-tethered, and unmodified surfaces. This synthetic matrix-nucleating ligand offers the distinct advantage of spatial control over matrix assembly, although reproduction of these results in a 3-D system may be more challenging. These synthetic matrix assembly-directing ligands may be of practical use in bottom-up organ building and other longer-term in vivo biotechnological applications, as well as fundamental studies on cell–matrix interactions.

7.5. Advanced ECM-Mimetic Surface Strategies: Multivalent, Clustered Integrin Ligands

After integrin binding to a ligand, clustering of integrins facilitates assembly of multiple cytoplasmic regulatory and structural proteins into important supramolecular structures, termed "focal adhesions." These structures activate various intracellular signaling pathways that regulate gene and protein expression, migration, and differentiation and mediate strong adhesive forces (Figure 7.7) [135]. Multivalent integrin-binding ECM ligands have been developed to promote integrin aggregation by presenting nano-scale clusters of adhesion ligands, in particular, RGD peptide-based motifs. These ligands have been immobilized on either two-dimensional (2-D) surfaces or beads to study the effect of integrin aggregation/clustering on cell responses. Key multi-clustered integrin ligands and the underlying bases of ligand clustering are shown schematically in Figure 7.7. For example, the adhesion ligand GRGDSPK was presented in nano-scale synthetic polymer clusters of varying size on

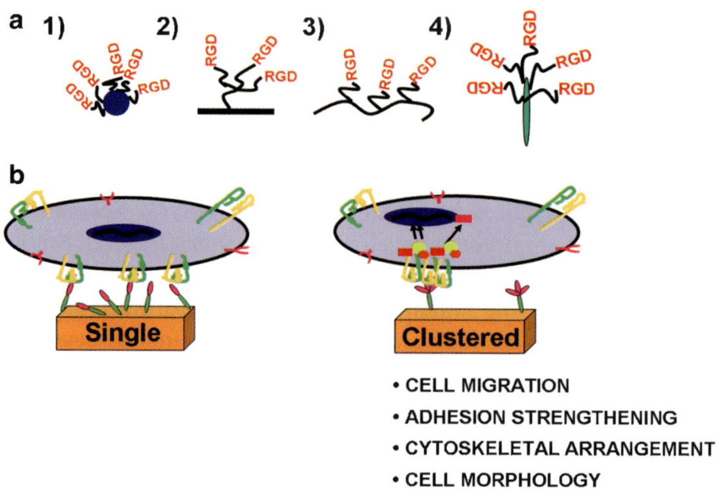

Figure 7.7. Integrin-specific ligand clustering may augment integrin-mediated cell functions. (**a**) Schematic of various clustered ligand schemes that have been developed recently, including: (1) nano-spaced ligands covalently tethered to micro-scale beads, (2) ligands functionalized on "star polymer" systems on substrate surfaces, (3) ligands tethered to the backbone of functional polymer gels and coatings in either two or three dimensions, and (4) "multi-headed" ligands that present individual integrin ligands with nano-scale spacing between them. (**b**) Various cell functions that have been demonstrated to be modulated by ligand clustering are also listed.

nonfouling substrates [136]. At higher cluster sizes (>3.6 ligands/cluster) an adhesion strengthening response was observed in which greater fractions of cells remained adherent as the detachment force was increased up to the force threshold. A similar polymeric RGD-dendrimer system correlated larger cluster sizes with an increase in cell migration behavior [137]. Alpha-helical coiled coil peptides fused with an RGD-containing fragment were more than 100-fold more efficient than linear RGD peptides in blocking the $\alpha_v\beta_3$-mediated adhesion and, once immobilized on a material surface, significantly more efficient in promoting cell adhesion and spreading [138]. Taking advantage of the robust recognition of $\alpha_v\beta_3$ for the RGD sequence, template-assembled cyclopeptides of RGD were used to successfully target in vitro $\alpha_v\beta_3$-expressing cells for drug delivery and detection applications [139, 140]. Possible explanations for the augmented efficacy of $\alpha_v\beta_3$ selectivity and related adhesive activities that these multivalent RGD ligands exhibit over single RGD peptides may be the relatively close RGD spacing, higher localized density, and constrained tripeptide and flanking sequence conformations. More rigorous studies must be conducted to ascertain the effect of these multimeric RGD ligands on integrin clustering and higher-order cell functions compared with native matrix proteins and single ECM-mimetic ligands.

However, given the lack of integrin selectivity of RGD peptides for one specific integrin, such as $\alpha_5\beta_1$, multimeric ligands presenting more integrin-specific components may be useful in examining functional effects of single integrin clustering. Moreover, these clustered ligand schemes rely mainly on "statistical averages" for valency values, rather than absolute ligand cluster sizes. Multimers presenting 1, 2, 3, and 5 heads of $\text{FNIII}_{7\text{–}10}$, an $\alpha_5\beta_1$-specific FN-mimetic ligand, have been developed to investigate the role of $\alpha_5\beta_1$ clustering on integrin–cytoskeleton interactions in fibroblasts [141]. Beads of these $\text{FNIII}_{7\text{–}10}$ multimers localized to actin filaments; the higher valency ($N > 3$) multimer beads translocated along the cell periphery on actin highways, indicating a link between ligand clustering and cytoskeletal pathways. The overall efficacy of these multimeric integrin-specific ligands in coordinating a variety of cell adhesive responses allude to the potential of these next-generation ligands for use on biointerfaces for more controlled cell responses and for therapeutic detection applications.

7.6. Summary

As we continue to accumulate greater understanding of how integrins regulate robust cell functions in distinct pathways and, in turn, how biointerfaces regulate integrin-mediated activities, new biomaterial surface modification strategies geared toward engineering integrin selectivity have been recently explored to take advantage of integrin function. Since integrins direct signaling pathways, cytoskeletal arrangements, and phenotype, and are major mediators of how cells detect and respond to their outside environment, it has become apparent that the better we control material–integrin interactions, the more directed cell and tissue response we might be able to achieve. These integrin-specific strategies utilize a multidisciplinary approach, incorporating synthetic and organic chemistry, material science, knowledge of biological pathways, and genetic engineering, to design ligands functionalized on biointerfaces that direct selective integrin binding and activation.

Although by no means inclusive, this chapter sought to highlight the major ligand engineering designs that implement this integrin-specific strategy, and also to detail the varying levels of success that these designs have achieved. Promising future approaches include targeting spatial, temporal, and multi-integrin components of integrin binding and signaling. These strategies include engineering multi-clustered integrin ligands, mixed integrin ligands geared to synergistically utilize multiple integrins for a directed functional effect, and, alternatively, reversible ligand presentation schemes to control temporal integrin activation.

References

1. Giancotti FG, Ruoslahti E. Integrin signaling. Science 1999;285:1028–1032.
2. Schwartz MA, Assoian RK. Integrins and cell proliferation: regulation of cyclin-dependent kinases via cytoplasmic signaling pathways. J Cell Sci 2001;114:2553–2560.
3. Temenoff JS, Mikos AG. *Biomaterials: The Intersection of Biology and Materials Science*. Upper Saddle River, NJ: Pearson Prentice-Hall, 2007, p. 1.
4. Reichardt LF. Introduction: extracellular matrix molecules. In: Kreis T, Vale R, eds. *Guidebook to the Extracellular Matrix, Anchor, and Adhesion Proteins*. Oxford Press: Oxford, 2nd edition, 1999, pp. 335–344.
5. Prockop DJ, Kivirikko KI. Collagens: molecular biology, diseases, and potentials for therapy. Annu Rev Biochem 1995;64:403–434.
6. Reddi AH. Role of morphogenetic proteins in skeletal tissue engineering and regeneration. Nat Biotechnol 1998;16:247–252.
7. Hynes RO. Molecular biology of fibronectin. Annu Rev Cell Biol 1985;1:67–90.
8. De Arcangelis A, Georges-Labouesse E. Integrin and ECM functions: roles in vertebrate development. Trends Genet 2000;16:389–395.
9. Danen EH, Sonnenberg A. Integrins in regulation of tissue development and function. J Pathol 2003;201:632–641.
10. Garcia AJ, Reyes CD. Bio-adhesive surfaces to promote osteoblast differentiation and bone formation. J Dent Res 2005;84:407–413.
11. Khan Y, Yaszemski MJ, Mikos AG, Laurencin CT. Tissue engineering of bone: material and matrix considerations. J Bone Joint Surg Am 2008;90(Suppl 1):36–42.
12. Hynes RO. Integrins: bidirectional, allosteric signaling machines. Cell 2002;110:673–687.

13. Alberts B, Johnson A, Lewis J, Raff M, Roberts K, Walter P. *Molecular Biology of the Cell*. Garland Science: New York, 4th edition, 2007, Chapter 21. AU8
14. Anderson JM. Biological responses to materials. Annu Rev Mater Res 2001;31:81–110.
15. Hench LL, Polak JM. Third-generation biomedical materials. Science 2002;295:1014–1017.
16. Vreeland WN, Barron AE. Functional materials for microscale genomic and proteomic analyses. Curr Opin Biotechnol 2002;13:87–94.
17. Siebers MC, ter Brugge PJ, Walboomers XF, Jansen JA. Integrins as linker proteins between osteoblasts and bone replacing materials. A critical review. Biomaterials 2005;26:137–146.
18. Moursi AM, Globus RK, Damsky CH. Interactions between integrin receptors and fibronectin are required for calvarial osteoblast differentiation in vitro. J Cell Sci 1997;110:2187–2196.
19. Gronthos S, Simmons PJ, Graves SE, Robey PG. Integrin-mediated interactions between human bone marrow stromal precursor cells and the extracellular matrix. Bone 2001;28:174–181.
20. Gronowicz G, McCarthy MB. Response of human osteoblasts to implant materials: integrin-mediated adhesion. J Orthop Res 1996;14:878–887.
21. Ruoslahti E, Pierschbacher MD. New perspectives in cell adhesion: RGD and integrins. Science 1987;238:491–497.
22. Geiger B, Bershadsky A, Pankov R, Yamada KM. Transmembrane crosstalk between the extracellular matrix and the cytoskeleton. Nat Rev Mol Cell Biol 2001;2:793–805.
23. Petit V, Thiery JP. Focal adhesions: structure and dynamics. Biol Cell 2000;92:477–494.
24. Anselme K. Osteoblast adhesion on biomaterials. Biomaterials 2000;21:667–681.
25. Takeuchi Y, Suzawa M, Kikuchi T, Nishida E, Fujita T, Matsumoto T. Differentiation and transforming growth factor-beta receptor down-regulation by collagen-alpha2beta1 integrin interaction is mediated by focal adhesion kinase and its downstream signals in murine osteoblastic cells. J Biol Chem 1997;272:29309–29316.
26. Nykvist P, Tasanen K, Viitasalo T, Kapyla J, Jokinen J, Bruckner-Tuderman L, Heino J. The cell adhesion domain of type XVII collagen promotes integrin-mediated cell spreading by a novel mechanism. J Biol Chem 2001;276:38673–38679.
27. Moursi AM, Damsky CH, Lull J, Zimmerman D, Doty SB, Aota S, Globus RK. Fibronectin regulates calvarial osteoblast differentiation. J Cell Sci 1996;109:1369–1380.
28. Lan MA, Gersbach CA, Michael KE, Keselowsky BG, Garcia AJ. Myoblast proliferation and differentiation on fibronectin-coated self assembled monolayers presenting different surface chemistries. Biomaterials 2005;26:4523–4531.
29. Stephansson SN, Byers BA, García AJ. Enhanced expression of the osteoblastic phenotype on substrates that modulate fibronectin conformation and integrin receptor binding. Biomaterials 2002;23:2527–2534.
30. Mizuno M, Fujisawa R, Kuboki Y. Type I collagen-induced osteoblastic differentiation of bone-marrow cells mediated by collagen-alpha2beta1 integrin interaction. J Cell Physiol 2000;184:207–213.
31. Reyes CD, Petrie TA, Burns KL, Schwartz Z, Garcia AJ. Biomolecular surface coating to enhance orthopaedic tissue healing and integration. Biomaterials 2007;28:3228–3235.
32. Cheng SL, Lai CF, Blystone SD, Avioli LV. Bone mineralization and osteoblast differentiation are negatively modulated by integrin alpha(v)beta3. J Bone Miner Res 2001;16:277–288.
33. Sung V, Stubbs JT, III, Fisher L, Aaron AD, Thompson EW. Bone sialoprotein supports breast cancer cell adhesion proliferation and migration through differential usage of the alpha(v)beta3 and alpha(v)beta5 integrins. J Cell Physiol 1998;176:482–494.
34. Niu JX, Zhang WJ, Ye LY, Wu LQ, Zhu GJ, Yang ZH, Grau GE, Lou JN. The role of adhesion molecules, alpha v beta 3, alpha v beta 5 and their ligands in the tumor cell and endothelial cell adhesion. Eur J Cancer Prev 2007;16:517–527.
35. Zhou X, Murphy FR, Gehdu N, Zhang J, Iredale JP, Benyon RC. Engagement of alphavbeta3 integrin regulates proliferation and apoptosis of hepatic stellate cells. J Biol Chem 2004;279:23996–24006.
36. Tang L, Ugarova TP, Plow EF, Eaton JW. Molecular determinants of acute inflammatory responses to biomaterials. J Clin Invest 1996;97:1329–1334.
37. Flick MJ, Du X, Witte DP, Jirouskova M, Soloviev DA, Busuttil SJ, Plow EF, Degen JL. Leukocyte engagement of fibrin(ogen) via the integrin receptor alphaMbeta2/Mac-1 is critical for host inflammatory response in vivo. J Clin Invest 2004;113:1596–1606.
38. Lutolf MP, Hubbell JA. Synthetic biomaterials as instructive extracellular microenvironments for morphogenesis in tissue engineering. Nat Biotechnol 2005;23:47–55.
39. Garcia AJ. Interfaces to Control Cell-Biomaterial Adhesive Interactions. Polymers for Regenerative Medicine 2006;203:171–190.
40. Vogel V, Thomas WE, Craig DW, Krammer A, Baneyx G. Structural insights into the mechanical regulation of molecular recognition sites. Trends Biotechnol 2001;19:416–423.

41. Xu LC, Siedlecki CA. Effects of surface wettability and contact time on protein adhesion to biomaterial surfaces. Biomaterials 2007;28:3273–3283.
42. Sinha RK, Tuan RS. Regulation of human osteoblast integrin expression by orthopedic implant materials. Bone 1996;18:451 457.
43. ter Brugge PJ, Torensma R, de Ruijter JE, Figdor CG, Jansen JA. Modulation of integrin expression on rat bone marrow cells by substrates with different surface characteristics. Tissue Eng 2002;8:615–626.
44. Ter Brugge PJ, Dieudonne S, Jansen JA. Initial interaction of U2OS cells with noncoated and calcium phosphate coated titanium substrates. J Biomed Mater Res 2002;61:399–407.
45. Kilpadi KL, Chang PL, Bellis SL. Hydroxylapatite binds more serum proteins, purified integrins, and osteoblast precursor cells than titanium or steel. J Biomed Mater Res 2001;57:258–267.
46. Yamamoto H, Shibata Y, Miyazaki T. Anode glow discharge plasma treatment of titanium plates facilitates adsorption of extracellular matrix proteins to the plates. J Dent Res 2005;84:668–671.
47. Brodbeck WG, Shive MS, Colton E, Nakayama Y, Matsuda T, Anderson JM. Influence of biomaterial surface chemistry on the apoptosis of adherent cells. J Biomed Mater Res 2001;55:661–668.
48. Keselowsky BG, Collard DM, Garcia AJ. Surface chemistry modulates fibronectin conformation and directs integrin binding and specificity to control cell adhesion. J Biomed Mater Res A 2003;66:247–259.
49. Keselowsky BG, Collard DM, Garcia AJ. Surface chemistry modulates focal adhesion composition and signaling through changes in integrin binding. Biomaterials 2004;25:5947–5954.
50. Michael KE, Vernakar VN, Keselowsky BG, Meredith JC, Latour RA, Garcia AJ. Adsorption-induced conformational changes in fibronectin due to interactions with well-defined surface chemistries. Langmuir 2003;19:8033–8040.
51. Lee MH, Ducheyne P, Lynch L, Boettiger D, Composto RJ. Effect of biomaterial surface properties on fibronectin-alpha5beta1 integrin interaction and cellular attachment. Biomaterials 2006;27:1907–1916.
52. Keselowsky BG, Collard DM, Garcia AJ. Integrin binding specificity regulates biomaterial surface chemistry effects on cell differentiation. Proc Natl Acad Sci U S A 2005;102:5953–5957.
53. Lange R, Luthen F, Beck U, Rychly J, Baumann A, Nebe B. Cell-extracellular matrix interaction and physico-chemical characteristics of titanium surfaces depend on the roughness of the material. Biomol Eng 2002;19:255–261.
54. Keselowsky BG, Wang L, Schwartz Z, Garcia AJ, Boyan BD. Integrin alpha(5) controls osteoblastic proliferation and differentiation responses to titanium substrates presenting different roughness characteristics in a roughness independent manner. J Biomed Mater Res A 2006;80A:700–710.
55. Lincks J, Boyan BD, Blanchard CR, Lohmann CH, Liu Y, Cochran DL, Dean DD, Schwartz Z. Response of MG63 osteoblast-like cells to titanium and titanium alloy is dependent on surface roughness and composition. Biomaterials 1998;19:2219–2232.
56. Anselme K, Bigerelle M, Noel B, Dufresne E, Judas D, Iost A, Hardouin P. Qualitative and quantitative study of human osteoblast adhesion on materials with various surface roughnesses. J Biomed Mater Res 2000;49:155–166.
57. Rezania A, Healy KE. Biomimetic peptide surfaces that regulate adhesion, spreading, cytoskeletal organization, and mineralization of the matrix deposited by osteoblast-like cells. Biotechnol Prog 1999;15:19–32.
58. Gronthos S, Stewart K, Graves SE, Hay S, Simmons PJ. Integrin expression and function on human osteoblast-like cells. J Bone Miner Res 1997;12:1189–1197.
59. Jimbo R, Sawase T, Shibata Y, Hirata K, Hishikawa Y, Tanaka Y, Bessho K, Ikeda T, Atsuta M. Enhanced osseointegration by the chemotactic activity of plasma fibronectin for cellular fibronectin positive cells. Biomaterials 2007;28:3469–3477.
60. Geissler U, Hempel U, Wolf C, Scharnweber D, Worch H, Wenzel K. Collagen type I-coating of Ti6Al4V promotes adhesion of osteoblasts. J Biomed Mater Res 2000;51:752–760.
61. Bierbaum S, Hempel U, Geissler U, Hanke T, Scharnweber D, Wenzel KW, Worch H. Modification of Ti6AL4V surfaces using collagen I, III, and fibronectin. II. Influence on osteoblast responses. J Biomed Mater Res A 2003;67:431–438.
62. Karadag A, Fisher LW. Bone sialoprotein enhances migration of bone marrow stromal cells through matrices by bridging MMP-2 to alpha(v)beta3-integrin. J Bone Miner Res 2006;21:1627–1636.
63. O'Toole GC, Salih E, Gallagher C, FitzPatrick D, O'Higgins N, O'Rourke SK. Bone sialoprotein-coated femoral implants are osteoinductive but mechanically compromised. J Orthop Res 2004;22:641–646.
64. Cheng SL, Lai CF, Fausto A, Chellaiah M, Feng X, McHugh KP, Teitelbaum SL, Civitelli R, Hruska KA, Ross FP, Avioli LV. Regulation of alphaVbeta3 and alphaVbeta5 integrins by dexamethasone in normal human osteoblastic cells. J Cell Biochem 2000;77:265–276.
65. Langer R, Tirrell DA. Designing materials for biology and medicine. Nature 2004;428:487–492.

66. Humphries MJ. Peptide recognition motifs involved in the binding of integrins to their ligands. Kidney Int 1992;41:645–649.
67. Merrill EW. Poly(ethylene oxide) and blood contact: a chronicle of one laboratory. In: Harris JM, ed. Poly(*Ethylene Glycol*) Chemistry: *Biotechnology and Biomedical Applications*. Plenum Press: New York, 1992, pp. 199–220.
68. Garcia AJ, Vega MD, Boettiger D. Modulation of cell proliferation and differentiation through substrate-dependent changes in fibronectin conformation. Mol Biol Cell 1999;10:785–798.
69. Hodde J, Record R, Tullius R, Badylak S. Fibronectin peptides mediate HMEC adhesion to porcine-derived extracellular matrix. Biomaterials 2002;23:1841–1848.
70. Reyes CD, Garcia AJ. Alpha2beta1 integrin-specific collagen-mimetic surfaces supporting osteoblastic differentiation. J Biomed Mater Res 2004;69A:591–600.
71. Gilbert M, Giachelli CM, Stayton PS. Biomimetic peptides that engage specific integrin-dependent signaling pathways and bind to calcium phosphate surfaces. J Biomed Mater Res 2003;67A:69–77.
72. Sanchez-Aparicio P, Dominguez-Jimenez C, Garcia-Pardo A. Activation of the alpha 4 beta 1 integrin through the beta 1 subunit induces recognition of the RGDS sequence in fibronectin. J Cell Biol 1994;126:271–279.
73. Horton MA, Nesbit MA, Helfrich MH. Interaction of osteopontin with osteoclast integrins. Ann N Y Acad Sci 1995;760:190–200.
74. Weber LM, Hayda KN, Haskins K, Anseth KS. The effects of cell-matrix interactions on encapsulated beta-cell function within hydrogels functionalized with matrix-derived adhesive peptides. Biomaterials 2007;28: 3004–3011.
75. Shin H, Jo S, Mikos AG. Biomimetic materials for tissue engineering. Biomaterials 2003;24: 4353–4364.
76. Hersel U, Dahmen C, Kessler H. RGD modified polymers: biomaterials for stimulated cell adhesion and beyond. Biomaterials 2003;24:4385–4415.
77. Liu JC, Heilshorn SC, Tirrell DA. Comparative cell response to artificial extracellular matrix proteins containing the RGD and CS5 cell-binding domains. Biomacromolecules 2004;5:497–504.
78. Harbers GM, Healy KE. The effect of ligand type and density on osteoblast adhesion, proliferation, and matrix mineralization. J Biomed Mater Res A 2005;75:855–869.
79. Schense JC, Hubbell JA. Three-dimensional migration of neurites is mediated by adhesion site density and affinity. J Biol Chem 2000;275:6813–6818.
80. Takagi J. Structural basis for ligand recognition by RGD (Arg-Gly-Asp)-dependent integrins. Biochem Soc Trans 2004;32:403–406.
81. Pakalns T, Haverstick KL, Fields GB, McCarthy JB, Mooradian DL, Tirrell M. Cellular recognition of synthetic peptide amphiphiles in self-assembled monolayer films. Biomaterials 1999;20:2265–2279.
82. Dahmen C, Auernheimer J, Meyer A, Enderle A, Goodman SL, Kessler H. Improving implant materials by coating with nonpeptidic, highly specific integrin ligands. Angew Chem Int Ed Engl 2004;43:6649–6652.
83. Lieb E, Hacker M, Tessmar J, Kunz-Schughart LA, Fiedler J, Dahmen C, Hersel U, Kessler H, Schulz MB, Gopferich A. Mediating specific cell adhesion to low-adhesive diblock copolymers by instant modification with cyclic RGD peptides. Biomaterials 2005;26:2333–2341.
84. Marchi-Artzner V, Lorz B, Hellerer U, Kantlehner M, Kessler H, Sackmann E. Selective adhesion of endothelial cells to artificial membranes with a synthetic RGD-lipopeptide. Chemistry 2001;7:1095–1101.
85. Hennessy KM, Clem WC, Phipps MC, Sawyer AA, Shaikh FM, Bellis SL. The effect of RGD peptides on osseointegration of hydroxyapatite biomaterials. Biomaterials 2008;29:3075–3083.
86. Barber TA, Ho JE, De Ranieri A, Virdi AS, Sumner DR, Healy KE. Peri-implant bone formation and implant integration strength of peptide-modified p(AAM-co-EG/AAC) interpenetrating polymer network-coated titanium implants. J Biomed Mater Res A 2007;80:306–320.
87. Petrie TA, Raynor JE, Reyes CD, Burns KL, Collard DM, Garcia AJ. The effect of integrin-specific bioactive coatings on tissue healing and implant osseointegration. Biomaterials 2008;29:2849–2857.
88. Schliephake H, Scharnweber D, Dard M, Rossler S, Sewing A, Meyer J, Hoogestraat D. Effect of RGD peptide coating of titanium implants on periimplant bone formation in the alveolar crest. An experimental pilot study in dogs. Clin Oral Implants Res 2002;13:312–319.
89. Ferris DM, Moodie GD, Dimond PM, Gioranni CW, Ehrlich MG, Valentini RF. RGD-coated titanium implants stimulate increased bone formation in vivo. Biomaterials 1999;20:2323–2331.
90. Elmengaard B, Bechtold JE, Soballe K. In vivo study of the effect of RGD treatment on bone ongrowth on press-fit titanium alloy implants. Biomaterials 2005;26:3521–3526.
91. Elmengaard B, Bechtold JE, Soballe K. In vivo effects of RGD-coated titanium implants inserted in two bone-gap models. J Biomed Mater Res A 2005;75:249–255.

92. Tosatti S, Schwartz Z, Campbell C, Cochran DL, VandeVondele S, Hubbell JA, Denzer A, Simpson J, Wieland M, Lohmann CH, Textor M, Boyan BD. RGD-containing peptide GCRGYGRGDSPG reduces enhancement of osteoblast differentiation by poly(L-lysine)-graft-poly(ethylene glycol)-coated titanium surfaces. J Biomed Mater Res 2004;68A:458–472.
93. Lee MH, Adams CS, Boettiger D, DeGrado WF, Shapiro IM, Composto RJ, Ducheyne P. Adhesion of MC3T3-E1 cells to RGD peptides of different flanking residues: detachment strength and correlation with long-term cellular function. J Biomed Mater Res A 2007;81:150–160.
94. Aota S, Nomizu M, Yamada KM. The short amino acid sequence Pro-His-Ser-Arg-Asn in human fibronectin enhances cell-adhesive function. J Biol Chem 1994;269:24756–24761.
95. Petrie TA, Reyes CD, Raynor JE, Burns KL, Collard DM, Garcia AJ. Biointerfaces promoting tissue healing. J Musculoskelet Neuronal Interact 2007;7:332.
96. Altroff H, Schlinkert R, van der Walle CF, Bernini A, Campbell ID, Werner JM, Mardon HJ. Interdomain tilt angle determines integrin-dependent function of the ninth and tenth FIII domains of human fibronectin. J Biol Chem 2004;279:55995–56003.
97. Krammer A, Craig D, Thomas WE, Schulten K, Vogel V. A structural model for force regulated integrin binding to fibronectin's RGD-synergy site. Matrix Biol 2002;21:139–147.
98. Kokkoli E, Mardilovich A, Wedekind A, Rexeisen EL, Garg A, Craig JA. Self-assembly and applications of biomimetic and bioactive peptide-amphiphiles. Soft Matter 2006;2:1015–1024.
99. Dillow AK, Ochsenhirt SE, McCarthy JB, Fields GB, Tirrell M. Adhesion of alpha5beta1 receptors to biomimetic substrates constructed from peptide amphiphiles. Biomaterials 2001;22:1493–1505.
100. Kokkoli E, Ochsenhirt SE, Tirrell M. Collective and single-molecule interactions of alpha5beta1 integrins. Langmuir 2004;20:2397–2404.
101. Ochsenhirt SE, Kokkoli E, McCarthy JB, Tirrell M. Effect of RGD secondary structure and the synergy site PHSRN on cell adhesion, spreading and specific integrin engagement. Biomaterials 2006;27:3863–3874.
102. Mardilovich A, Kokkoli E. Biomimetic peptide-amphiphiles for functional biomaterials: the role of GRGDSP and PHSRN. Biomacromolecules 2004;5:950–957.
103. Mardilovich A, Craig JA, McCammon MQ, Garg A, Kokkoli E. Design of a novel fibronectin-mimetic peptide-amphiphile for functionalized biomaterials. Langmuir 2006;22:3259–3264.
104. Benoit DS, Anseth KS. The effect on osteoblast function of colocalized RGD and PHSRN epitopes on PEG surfaces. Biomaterials 2005;26:5209–5220.
105. Jensen TW, Hu BH, Delatore SM, Garcia AS, Messersmith PB, Miller WM. Lipopeptides incorporated into supported phospholipid monolayers have high specific activity at low incorporation levels. J Am Chem Soc 2004;126:15223–15230.
106. Petrie TA, Capadona JR, Reyes CD, Garcia AJ. Integrin specificity and enhanced cellular activities associated with surfaces presenting a recombinant fibronectin fragment compared to RGD supports. Biomaterials 2006;27:5459–5470.
107. Richards J, Miller M, Abend J, Koide A, Koide S, Dewhurst S. Engineered fibronectin type III domain with a RGDWXE sequence binds with enhanced affinity and specificity to human alphavbeta3 integrin. J Mol Biol 2003;326:1475–1488.
108. Mould AP, Komoriya A, Yamada KM, Humphries MJ. The CS5 peptide is a second site in the IIICS region of fibronectin recognized by the integrin alpha 4 beta 1. Inhibition of alpha 4 beta 1 function by RGD peptide homologues. J Biol Chem 1991;266:3579–3585.
109. Cutler SM, García AJ. Engineering cell adhesive surfaces that direct integrin alpha5beta1 binding using a recombinant fragment of fibronectin. Biomaterials 2003;24:1759–1770.
110. Raynor JE, Petrie TA, García AJ, Collard DM. Controlling cell adhesion to titanium: functionalization of poly[oligo(ethylene glycol methacrylate] brushes with cell-adhesive peptides. Adv Mater 2007; 19:1724–1728.
111. Petrie TA, Reyes CD, Burns KL, Garcia AJ. Simple application of fibronectin-mimetic coating enhances osseointegration of titanium implants. J Cell Mol Med 2008; Aug. 21.
112. Mizuno M, Kuboki Y. Osteoblast-related gene expression of bone marrow cells during the osteoblastic differentiation induced by type I collagen. J Biochem (Tokyo) 2001;129:133–138.
113. Knight CG, Morton LF, Peachey AR, Tuckwell DS, Farndale RW, Barnes MJ. The collagen-binding A-domains of integrins alpha(1)beta(1) and alpha(2)beta(1) recognize the same specific amino acid sequence, GFOGER, in native (triple-helical) collagens. J Biol Chem 2000;275:35–40.
114. Knight CG, Morton LF, Onley DJ, Peachey AR, Messent AJ, Smethurst PA, Tuckwell DS, Farndale RW, Barnes MJ. Identification in collagen type I of an integrin alpha2 beta1-binding site containing an essential GER sequence. J Biol Chem 1998;273:33287–33294.

115. Morton LF, Peachey AR, Zijenah LS, Goodall AH, Humphries MJ, Barnes MJ. Conformation-dependent platelet adhesion to collagen involving integrin alpha 2 beta 1-mediated and other mechanisms: multiple alpha 2 beta 1-recognition sites in collagen type I. Biochem J 1994;299 (Pt 3):791–797.
116. Emsley J, Knight CG, Farndale RW, Barnes MJ, Liddington RC. Structural basis of collagen recognition by integrin alpha2beta1. Cell 2000;101:47–56.
117. Reyes CD, Garcia AJ. Engineering integrin-specific surfaces with a triple-helical collagen-mimetic peptide. J Biomed Mater Res A 2003;65:511–523.
118. Koide T. Triple helical collagen-like peptides: engineering and applications in matrix biology. Connect Tissue Res 2005;46:131–141.
119. Khew ST, Tong YW. The specific recognition of a cell binding sequence derived from type I collagen by Hep3B and L929 cells. Biomacromolecules 2007;8:3153–3161.
120. Khew ST, Zhu XH, Tong YW. An integrin-specific collagen-mimetic peptide approach for optimizing Hep3B liver cell adhesion, proliferation, and cellular functions. Tissue Eng 2007;13:2451–2463.
121. Khew ST, Tong YW. Template-assembled triple-helical peptide molecules: mimicry of collagen by molecular architecture and integrin-specific cell adhesion. Biochemistry 2008;47:585–596.
122. Baronas-Lowell D, Lauer-Fields JL, Fields GB. Induction of endothelial cell activation by a triple helical alpha2beta integrin ligand, derived from type I collagen alpha1(I)496–507. J Biol Chem 2004;279:952–962.
123. Baronas-Lowell D, Lauer-Fields JL, Borgia JA, Sferrazza GF, Al-Ghoul M, Minond D, Fields GB. Differential modulation of human melanoma cell metalloproteinase expression by alpha2beta1 integrin and CD44 triple-helical ligands derived from type IV collagen. J Biol Chem 2004;279:43503–43513.
124. Malkar NB, Lauer-Fields JL, Juska D, Fields GB. Characterization of peptide-amphiphiles possessing cellular activation sequences. Biomacromolecules 2003;4:518–528.
125. Reyes CD, Petrie TA, Burns KL, Garcia AJ. Mixed ECM ligands synergistically modulate integrin adhesion and signaling. Journal of Cell Physiology 2008;In Press:.
126. Colombi M, Zoppi N, De PG, Marchina E, Gardella R, Tavian D, Ferraboli S, Barlati S. Matrix assembly induction and cell migration and invasion inhibition by a 13-amino acid fibronectin peptide. J Biol Chem 2003;278:14346–14355.
127. Sechler JL, Schwarzbauer JE. Control of cell cycle progression by fibronectin matrix architecture. J Biol Chem 1998;273:25533–25536.
128. Schwarzbauer JE, Sechler JL. Fibronectin fibrillogenesis: a paradigm for extracellular matrix assembly. Curr Opin Cell Biol 1999;11:622–627.
129. Schwarzbauer JE, Sechler JL. Fibronectin fibrillogenesis: a paradigm for extracellular matrix assembly [in process citation]. Curr Opin Cell Biol 1999;11:622–627.
130. Sechler JL, Corbett SA, Schwarzbauer JE. Modulatory roles for integrin activation and the synergy site of fibronectin during matrix assembly. Mol Biol Cell 1997;8:2563–2573.
131. Takahashi S, Leiss M, Moser M, Ohashi T, Kitao T, Heckmann D, Pfeifer A, Kessler H, Takagi J, Erickson HP, Fassler R. The RGD motif in fibronectin is essential for development but dispensable for fibril assembly. J Cell Biol 2007;178:167–178.
132. Sechler JL, Takada Y, Schwarzbauer JE. Altered rate of fibronectin matrix assembly by deletion of the first type III repeats. J Cell Biol 1996;134:573–583.
133. Sechler JL, Cumiskey AM, Gazzola DM, Schwarzbauer JE. A novel RGD-independent fibronectin assembly pathway initiated by alpha4beta1 integrin binding to the alternatively spliced V region. J Cell Sci 2000;113 (Pt 8):1491–1498.
134. Capadona JR, Petrie TA, Fears KP, Latour RA, Collard DM, Garcia AJ. Surface-nucleated assembly of fibrillar extracellular matrices. Adv Mater 2005;17:2604–2608.
135. Gallant ND, Michael KE, Garcia AJ. Cell adhesion strengthening: contributions of adhesive area, integrin binding, and focal adhesion assembly. Mol Biol Cell 2005;16:4329–4340.
136. Koo LY, Irvine DJ, Mayes AM, Lauffenburger DA, Griffith LG. Co-regulation of cell adhesion by nanoscale RGD organization and mechanical stimulus. J Cell Sci 2002;115:1423–1433.
137. Maheshwari G, Brown G, Lauffenburger DA, Wells A, Griffith LG. Cell adhesion and motility depend on nanoscale RGD clustering. J Cell Sci 2000;113:1677–1686.
138. Villard V, Kalyuzhniy O, Riccio O, Potekhin S, Melnik TN, Kajava AV, Ruegg C, Corradin G. Synthetic RGD-containing alpha-helical coiled coil peptides promote integrin-dependent cell adhesion. J Pept Sci 2006;12:206–212.
139. Garanger E, Boturyn D, Dumy P. Tumor targeting with RGD peptide ligands-design of new molecular conjugates for imaging and therapy of cancers. Anticancer Agents Med Chem 2007;7:552–558.
140. Garanger E, Boturyn D, Coll JL, Favrot MC, Dumy P. Multivalent RGD synthetic peptides as potent alphaVbeta3 integrin ligands. Org Biomol Chem 2006;4:1958–1965.
141. Coussen F, Choquet D, Sheetz MP, Erickson HP. Trimers of the fibronectin cell adhesion domain localize to actin filament bundles and undergo rearward translocation. J Cell Sci 2002;115:2581–2590.

8

Ligand-Functionalized Biomaterial Surfaces: Controlled Regulation of Signaling Pathways to Direct Stem Cell Differentiation

Myung Hee Kim and Krishnendu Roy

Appropriate ligand–receptor interactions are critical in triggering signal transduction pathways in cells. This, among other functions, can direct stem cell differentiation into specific lineages. Although many of these ligands act as soluble agents, a significant number are present on surfaces of neighboring cells or as bound components on the extracellular matrix. Recent research has focused on identifying such interactions, quantitatively characterizing them, and incorporating them in synthetic engineered systems to efficiently and intentionally direct differentiation of cells to certain pathways. This chapter presents current research that focuses on directing stem cell differentiation by presenting cell-signaling ligands on biomaterial surfaces. Ligand-directed activation of the Notch signaling pathway is discussed in detail. Additional work in activating the Sonic hedgehog pathway and RhoA signaling are also discussed.

Abbreviations

ALP	alkaline phosphatase
BMP	bone morphogenetic protein
CAM	cell adhesion molecule
CoA	coactivator
CoR	corepressor
CSL	CBF1, Suppressor of Hairless, LAG-1
DL1, DLL1, Delta1	Delta-like ligand 1
DL4, DLL4, Delta4	Delta-like ligand 4
DN	double negative (CD4− CD8−)
DP	double positive (CD4 + CD8+)
DSL	Delta, Serrate/Jagged, Lag-2

M.H. Kim and K. Roy • Department of Biomedical Engineering, The University of Texas at Austin, 1 University Station C0800, Austin TX 78712, USA

D.A. Puleo and R. Bizios (eds.), *Biological Interactions on Materials Surfaces*,
DOI 10.1007/978-0-387-98161-1_8,

ECM	extracellular matrix
ESC	embryonic stem cell
FTOC	fetal thymic organ culture
HES	Hairy Enhancer of Split
hMSC	human mesenchymal stem cell
HSC	hematopoietic stem cell
HSPC	hematopoietic stem cell/progenitor cell
IPN	interpenetrating polymer network
NK	natural killer
Notch IC/ICN	Notch intracellular domain
OP9-DL1	Delta-like ligand 1-transfected OP9 stromal cell
PTCH-1	Patched-1
REEC	rat esophageal epithelial cell
Shh	Sonic hedgehog
SP	single positive (CD4+ CD8− or CD4− CD8+)
VLA	very late antigen

8.1. Introduction

Cell–cell and cell–matrix signaling is essential in directing cell differentiation and thus determining cell fate. Many of these signaling events occur through ligand–receptor interactions between neighboring cells or through ligands that are bound to extracellular matrix (ECM) components. We are only beginning to understand the quantitative implications of these interactions within the stem cell niche, especially in terms of (a) ligand to cell ratios, (b) ligand densities, (c) spatial patterning and ligand gradients, and (d) kinetics of ligand interactions. Designing synthetic cell microenvironments that can mimic such ligand–stem cell interactions *in vitro* is not only critical to understand the fundamental biology of these signaling mechanisms under native-like conditions but also necessary to efficiently direct stem cell differentiation along a desired pathway for cell and tissue engineering applications. One way to achieve such engineered stem cell niches is to present these ligands through two- or three-dimensional (2D or 3D, respectively) biomaterial surfaces in a controllable and quantitative manner. Here we present several examples of current work to present cell-signaling ligands on such surfaces and study directed stem cell differentiation.

8.2. Notch Signaling Pathway

Notch signaling is a cell–cell signaling pathway that is essential during embryonic development and in adult self-renewal systems. The molecules involved in the Notch pathway are highly conserved across species and are vital in developmental cell fate decisions such as neural development, epithelial-to-mesenchymal transitions, vascular development, keratinocyte differentiation, as well as hematopoietic and T-cell differentiation [1–5]. Since it is involved in such numerous cell fate decisions, the Notch signaling pathway is an attractive system that can be mimicked in vitro to study and intentionally control cell fate decisions.

The Notch receptor is a single-pass transmembrane protein that is a heterodimer of two major subunits. The first subunit consists of a large extracellular segment with epidermal growth factor repeats – a common amino acid repeat found in growth factors and ECM proteins. The second subunit consists of a short extracellular domain, a transmembrane domain, and an intracellular domain [6]. Notch is activated by cell surface ligands of the DSL family (Delta, Serrate/Jagged, Lag-2) [7]. Physiologically, ligand binding to Notch initiates proteolytic cleavage of the extracellular and intracellular regions, mediated by action of presenilin-dependent γ-secretase activity [6]. The cleaved extracellular segment is endocytosed by the ligand-presenting cell. The Notch intracellular domain (Notch IC/ICN), containing nuclear localization signals, undergoes nuclear translocation and interacts with a transcriptional coactivator (CoA). CoA, in turn, displaces a corepressor complex (CoR) from its interaction with DNA-binding transcription factor CSL (*C*BF1 in humans, *S*uppressor of Hairless in Drosophila, *L*AG-1 in *Caenorhabditis elegans*) [7, 8]. The Notch effector CSL, once activated, enhances consequent lineage-specific gene expression. A schematic of the Notch signaling pathway in T-cell and B-cell development is provided in Figure 8.1.

In the past few years, studies have focused on mimicking the cell–cell communication involved in Notch signaling for controlling cell fate for biomedical as well as cell and tissue engineering applications. This has been largely done by innovating ways to present various Notch ligands to cells expressing surface Notch receptors, and to initiate the Notch/CSL signaling pathway.

8.2.1. Biological Strategy for Inducing Notch Signaling *In Vitro*: Notch Ligand-Transfected Stromal Cells for T-Cell Differentiation

Hematopoietic stem cells (HSCs) originating from the bone marrow migrate to the thymus and develop into T-cells capable of recognizing nonself antigens. T-cell development and maturation involve complex signaling interactions within the thymic microenvironment that transform CD4− CD8− (double negative [DN]) thymocytes into CD4+ CD8+ (double positive [DP]) cells, and eventually into CD4+ CD8− and CD4− CD8+ (single positive [SP]) mature T-cells [9]. Generation of mature and functional T-cells from HSCs or embryonic stem cells (ESCs) would be a significant breakthrough in cell therapy, especially in adoptive T-cell immunotherapy for cancers and allergic diseases [10, 11]. However, the difficulty of creating synthetic microenvironments for HSCs to efficiently differentiate into T-cells has been a major challenge. Fetal thymic organ cultures (FTOCs) were developed to create an explanted thymic environment for T-cell development, but such systems suffer from low cell yield and seeding efficiency, and are technically challenging for cell or tissue engineering applications [12].

Recent findings demonstrating that Notch signaling in HSCs is required for T-cell lineage commitment led to the development of model systems for T-cell generation from both ESCs and HSCs [5, 13, 14]. Schmitt and Zúñiga-Pflücker developed a coculture system for T-cell differentiation involving stem cells mixed with a mouse bone marrow stromal cell line – OP9 cells – that had been retrovirally transfected to express the Notch ligand, Delta-like ligand 1 (OP9-DL1 cells) [13]. Without Delta-like ligand 1 (DL1, Delta1, DLL1) transfection, OP9 bone marrow stromal cells, in the presence of lymphopoietic cytokines, induce differentiation of HSCs into the B-cell lineage [15, 16]. However, it was found that, when cocultured with the OP9-DL1 cells, ESCs or HSCs can be efficiently driven toward the T-cell lineage. HSCs isolated from mice and cocultured with OP9-DL1 cells were able to generate CD4+ CD8+ double-positive thymocytes after 12 days in culture [13]. Thymocyte development was also demonstrated directly from ESCs [17]. Figure 8.2 shows the flow

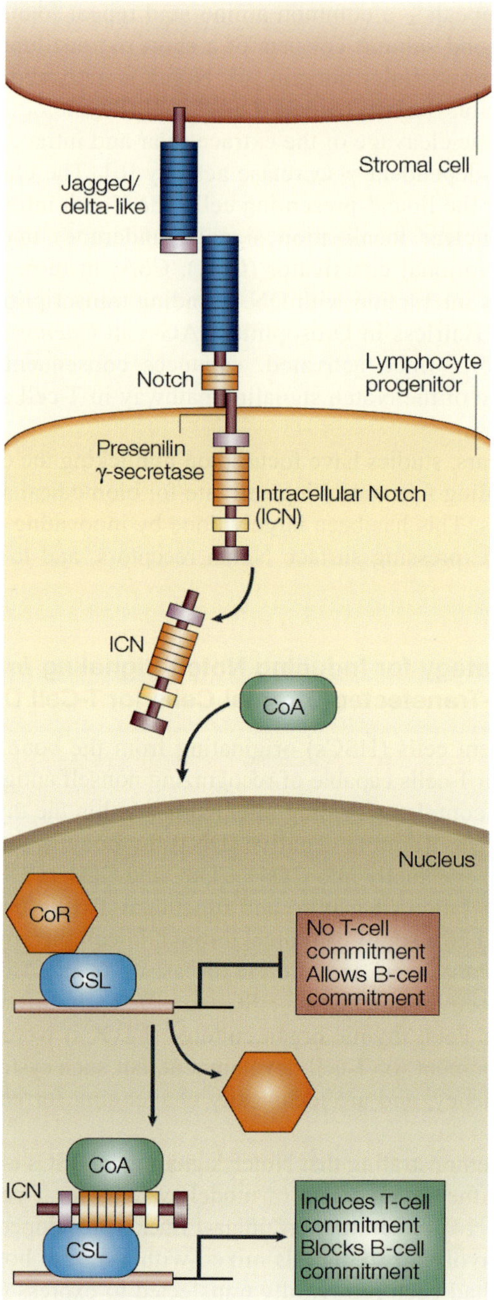

Figure 8.1. The Notch signaling pathway in T-cell development. (Adapted by permission from Macmillan [8]).

cytometry plots of ESCs and HSCs cocultured with either naive OP9 cells or OP9-DL1 cells, respectively. The data show expression of several lymphoid-specific cell surface markers. Early T lineage-committed cells were identified by high CD25 expression. CD4 and CD8 are not present in these early T-cells but are both expressed at later stages of T-cell development.

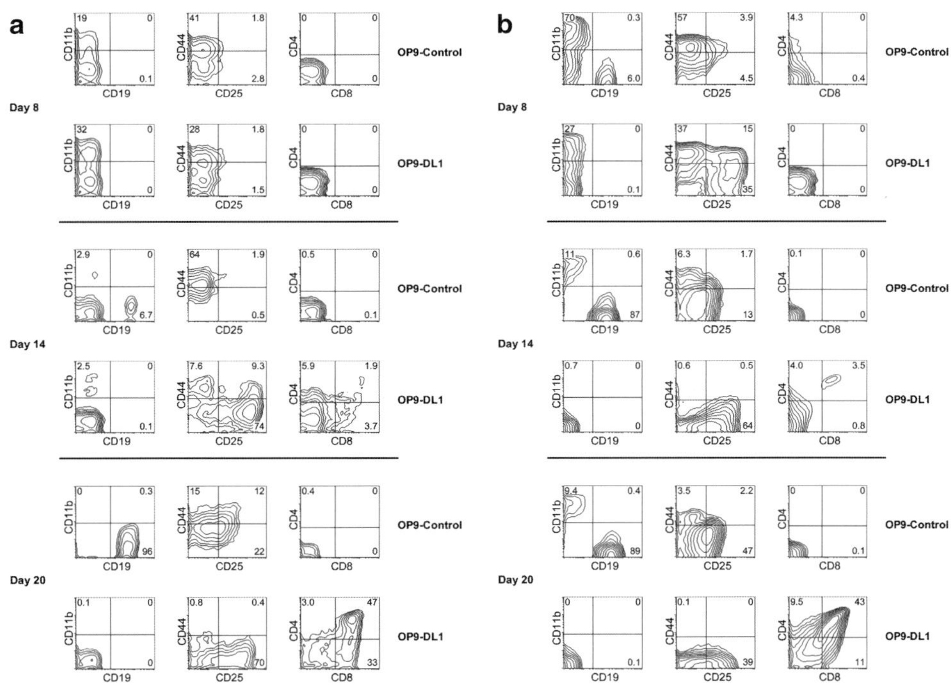

Figure 8.2. T-cell differentiation from (**a**) embryonic stem cells (ESCs) and (**b**) bone marrow-derived HSCs by coculture with OP9-control cells and Delta1-transfected OP9 (OP9-DL1) cells. Nonadherent hematopoietic cells were collected at days 8, 14, and 20, stained for various markers, and analyzed by flow cytometry (Adapted by permission from Macmillan [17]).

Both HSCs and ESCs developed a high percentage of SP cells after 20 days of culture, while the controls did not produce any CD4+ or CD8+ cells. The stains for CD19 (a B-cell marker) showed that these control cells developed instead into the B-cell lineage. The researchers also derived T lineage progenitors from the coculture system and seeded them onto FTOCs. Within the FTOCs, these progenitors were capable of developing into mature T-cells *in vitro*. Surprisingly, these FTOCs were able to function *in vivo* as well. Immunodeficient mice transplanted with the reconstituted FTOCs were able to successfully mount an antigen-specific immune response to a viral infection [17].

More recently, T-cell precursors generated from HSCs on the OP9-DL1 system have been adoptively transferred into lethally irradiated allogeneic mice together with purified HSCs [18]. These transplanted precursors successfully grafted into the host and generated CD4+ CD8− and CD4− CD8+ cells with a normal diverse T-cell receptor repertoire, cytokine response, and proliferative response to foreign antigen. Compared with mice with only HSC transplantation, mice having T-cell precursor transfer showed enhanced resistance to bacterial pathogens. The recipients expressed high graft-versus-tumor activity, and no graft-versus-host disease. These results indicate that the transplanted cells underwent normal positive and negative selection within the thymus to eliminate cells that are alloreactive and maintain cells that respond to tumor-associated antigens.

The Zúñiga-Pflücker group also expanded this work and showed that T-cell development was also possible from human cord blood CD34+ progenitor cells using the same

OP9-DL1 system [19]. Additionally, Plum and coworkers were able to take human bone marrow CD34+ progenitor cells and mature them into T lymphocytes [20].

These studies were able to successfully design and test a simple *in vitro* system to develop T-cells. By identifying a specific ligand required for T-cell commitment – Notch ligand Delta1 - they were able to replace a complex FTOC with a simple coculture system with a stromal cell line expressing that specific ligand. The coculture system was able to develop mature T-cells starting from both HSCs and ESCs, and confirmed the potential of using such stem cells for generating lineage-specific cells for therapeutic use. The *in vivo* experiments started with an FTOC system, and expanded to allogeneic HSC transplantation, both of which demonstrated that cells generated from the coculture system are fully compatible with the host and can potentially generate mature, potentially functional T-cells.

The role of Notch signaling in natural killer (NK) cell development is also being explored by using the Notch ligand-transfected OP9 system. Bevan and coworkers discovered that Jagged1-transfected OP9 cells, as well as Delta1-transfected OP9 cells, have the capability of inhibiting B-cell differentiation [21]. Only Delta1-mediated signaling could induce further proliferation and maturation of T-cell progenitors into mature T-cells, while Jagged1-mediated signaling, instead, directed the differentiation into NK cells expressing the NK marker NK1.1. NK cell development was also possible when Pax5− pro-B cells were provided with transient, not constitutive, Notch signaling through OP9-DL1 stromal cells [22, 23].

8.2.2. Ligand Presentation Through the Cell Surface Is Not Necessary for Notch Signaling: Immobilization of Notch Ligand on Synthetic Surfaces for Notch Signal Activation

The first attempt to design a stromal cell-free system to direct Notch signaling was by Varnum-Finney et al., who studied the effect of surface-immobilized Notch ligand Delta1 on C2C12 myoblast differentiation and Hairy Enhancer of Split (HES)-1 transcription factor activation in U20S cells [24]. Notch activation in C2C12 cells had been previously reported to inhibit myotube formation *in vitro* either through the presence of constitutively active Notch IC/ICN or in coculture with Notch ligand-expressing cells [24, 25]. Notch has also been known to activate the HES-1 gene [26]. HES-related genes are the best-characterized Notch target genes [6]. In their study, Varnum-Finney et al. presented the Notch ligand Delta1 without a cell expressing the ligand. They discovered that immobilization of Delta1 was necessary for Notch signal activation for both C2C12 cell myotube inhibition and HES-1 activation.

C2C12 myoblasts were incubated with either surface-immobilized or soluble Delta1, and stained for myosin heavy chains – a muscle differentiation marker [27]. Surprisingly, 75–80% less differentiated cells (indicating efficient Notch signaling and myotube inhibition) were observed in cultures incubated with immobilized Delta1, in comparison with those with soluble Delta1 (Figure 8.3b and a, respectively). In the case of HES-1 activation, a sixfold increase in HES-1 promoter activity was observed (Figure 8.3c). This outcome is comparable to a normal eightfold to tenfold increase in HES-1 promoter activity in cells cocultured with Notch ligand-expressing cells. It was also found that soluble forms of Delta1 had an inhibitory effect on Notch signaling in C2C12 differentiation in cultures incubated with immobilized Delta1 (data not shown), suggesting that soluble Delta1 still has the capability of binding to Notch receptors but fails to activate the Notch signaling pathway.

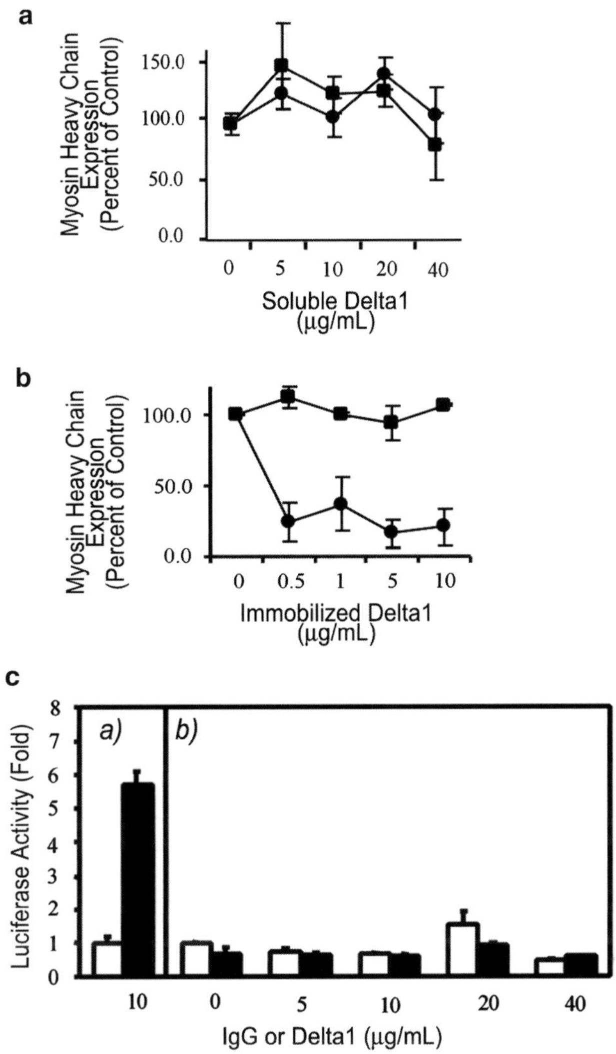

Figure 8.3. Effect of (**a**) soluble and (**b**) immobilized Delta1 at various concentrations (x-axis) on C2 cell differentiation. *Filled circle* denotes experimental group (with Delta1) and *filled square* denotes control group (IgG, instead of Delta1). The y axis indicates myosin heavy chain levels as a percentage of expression in control cells. Effect of (*a*) immobilized (*b*) and nonimmobilized Delta1 on HES-1 activation is shown in (**c**) (*black bars*). *White bars* show control groups with (*a*) immobilized human IgG or (*b*) soluble human IgG. (Reproduced/adapted with permission from The Company of Biologists [24]).

Varnum-Finney's studies concluded that nonimmobilized forms of Delta1 failed to induce Notch signaling, whereas Delta1 immobilized on a plastic surface was able to activate Notch. It was shown that the role of the signaling cell in presenting the Notch ligand can be replaced by the immobilization of the ligand extracellular domain to a plastic surface [24]. Although Varnum-Finney's research was on just two types of cells, it suggests that other cell signaling interactions may also require that the ligand be mechanically immobilized for the signal to be activated. In terms of biomedical research, this information brings up the possibility of replacing cell-cell interactions of ligand–receptor binding with biomaterials expressing the ligands on their surfaces.

Recently, Giachelli and coworkers confirmed the effect of ligand immobilization for Notch signaling using a different ligand, Jagged1, in epithelial differentiation [7]. Jagged1 was immobilized onto polystyrene surfaces. Incubation of rat esophageal epithelial cell (REEC) with either immobilized or soluble Jagged1 resulted in a drastic increase in Notch activation confirmed by gene expression studies. A CBF-1 luciferase assay on REEC incubated with Jagged1 bound to polystyrene surfaces indicated that bound, compared with unbound, Jagged1 yielded more than a ninefold increase in luciferase activity (Figure 8.4a). Notch signaling was also measured by RNA level of genes *caspase 3* and *occludin* – genes upregulated in response to Notch signaling. RNA levels increased by threefold to fourfold with immobilized Jagged1 (Figure 8.4b). Furthermore, differentiation was assayed by expression levels of proteins that are markers for epithelial cell differentiation. Western blots demonstrated that cells incubated with bound Jagged1 had a strong expression of involucrin, CK10, and filaggrin (intermediate and late stage differentiation proteins) compared with soluble Jagged1 (Figure 8.4c). These studies [7] expanded the idea of artificial Notch signaling through a surface to epithelial cells, and suggested its usage in engineering organs that require epithelial barriers.

8.2.3. Method to Scale-Up Notch Signaling and Mimic Cell–Cell Interactions: Microbead-Based Notch Signaling for T-Cell Differentiation

The OP9-DL1 system effectively generates T-cells *in vitro* using a coculture of stem cells with ligand-transfected stromal cells. However, if these cells were to be used therapeutically in the future, a system without mixed coculture of cells that can effectively generate a large, pure population of such cells is necessary, especially since presence of any retrovirally transfected stromal cells is a serious regulatory concern even after separation of the differentiated T-cells from the coculture. In addition, the OP9-DL1/stem cell mixed coculture system is hard to scale-up; it remains difficult to quantify and reproducibly control the amount of ligands expressed on the stromal cells [28].

Although the Varnum-Finney [24] approach described in Sect. 8.2.2 addresses the coculture issue, it is essentially a 2D culture method; scale-up for large cell production is difficult. Further, HSCs and lymphocytes are nonadherent cells and thus more suitable for suspension cultures. In order to address these limitations and better mimic either the HSC or ESC interactions with the Notch ligand, Roy and coworkers developed a microbead-based Notch signaling system [28]. The system replaces the Notch ligand-transfected stromal cells or ligand-coated tissue culture plates with ligand-conjugated magnetic microbeads. These microbeads directly interact with the stem cells and trigger Notch signaling. The magnetic nature of the microbeads allows easy separation of bead-attached cells at various stages for biological assays and better understanding of the kinetic nature of the signaling pathway in T-cell development. This aspect also permits elimination or at least physical separation of the stromal cell line from the developing T-cells thus decreasing chances of contamination with retrovirally transfected cells. Such a system also allows (a) quantitative manipulation of cell-to-bead ratios (i.e., ligand–stem cell ratio), (b) easy control of ligand density on the bead surface, and (c) ready removal of bead (i.e., the ligand) from the culture during various stages, allowing investigation of signaling kinetics on differentiation.

In these studies [28], a different Notch ligand, Delta-like ligand 4 (DL4, Delta4, DLL4), was conjugated onto magnetic microbeads via streptavidin-biotin binding. DLL4 is

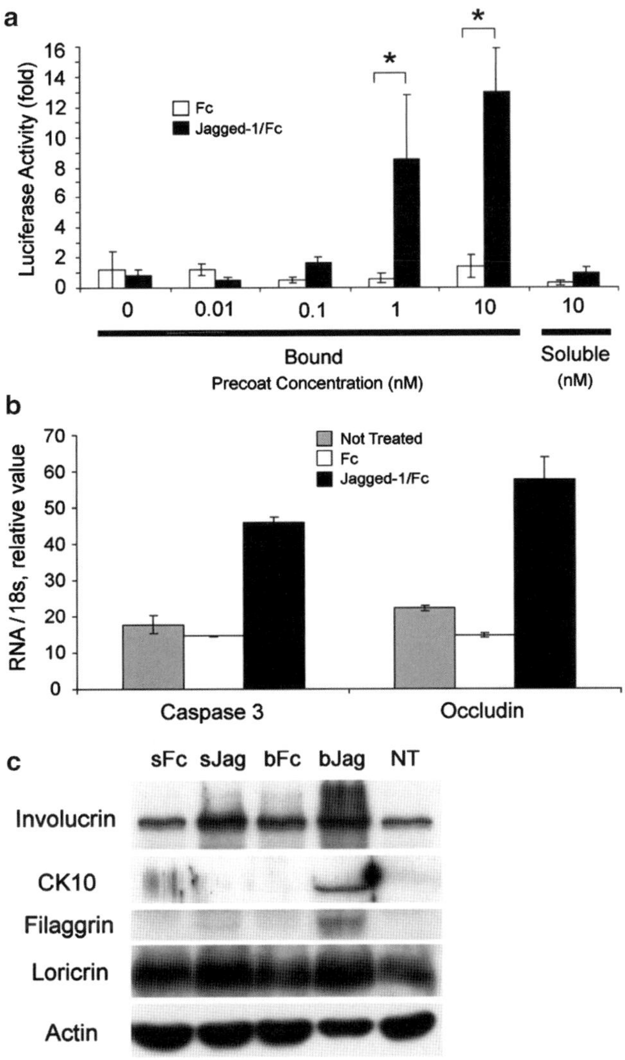

Figure 8.4. Notch signaling in rat esophageal epithelial cells (REECs) assayed by (**a**) luciferase activity, (**b**) quantitative real-time polymerase chain reaction (PCR), and (**c**) Western blot. In (**a**), *white bars* are the negative control groups with either bound or soluble Fc, and *black bars* indicate the experimental groups with bound or soluble jagged-1/Fc. In (**b**), *white bars* are the negative control groups with bound Fc, *gray bars* indicate plates treated with neither Fc nor jagged-1/Fc, and *black bars* indicate experimental groups with bound jagged-1/Fc. In (**c**), Western blots were performed on REECs incubated with soluble Fc (*sFc*), soluble jagged-1/Fc (*sJag*), bound Fc (*bFc*), bound jagged-1/Fc (*bJag*), and not-treated plates (*NT*). *Asterisk* indicates statistically significant ($p < 0.05$) difference from the respective control. (Reprinted with permission of Wiley [7]).

also expressed on thymic stromal cells [29, 30], and is known to serve a function similar to that of Delta-like ligand 1 (DLL1) in Notch signaling [31]. The capability of these DLL4-coated microbeads to trigger Notch signaling was tested with C2C12 cells, as discussed in Sect. 8.2.2. DLL4-conjugated microbeads inhibited myotube formation in C2C12 cells, indicating that effective Notch signaling was present. When these beads were added to an OP9 coculture system of HSCs – a system that would normally induce B-cell differentiation – a significant population of cells positive for T-cell marker was observed, indicating efficient Notch pathway induction by the beads. The strength of this system was that T-cell differentiation was possible even when OP9 cells were physically separated from the stem cells using an insert system. When HSCs were allowed to be in contact with the easily removable microbeads, direct contact with OP9 cells was not necessary for T-cell development, and soluble factors secreted by the stromal cells were enough to support hematopoietic differentiation. HSCs were cultured with DLL4-coated or uncoated beads in either insert culture or in mixed culture. OP9 cells were seeded in Transwell™ inserts for the insert culture or directly on top of HSCs for the mixed culture. Figure 8.5 shows the cell staining results analyzed by flow cytometry for insert culture (Figure 8.5a, b) and for mixed culture (Figure 8.5c, d). Thy1.2, an early T-cell marker, appeared for cells cocultured with DLL4-coated beads in both mixed and insert cultures. The uncoated bead system yielded mostly CD19+ B-cells. Some B-cell differentiation was still present with the DLL4-coated beads, but a sizeable amount of Thy1.2+ cells were observed for both insert and mixed cultures. A drawback of this system is that the microbeads tend to aggregate, providing nonuniform signaling to cells, which might cause some B-cell development. However, this system explored the potential of a scalable method more plausible for *in vitro* generation of therapeutic T-cells [28].

In addition, using microbeads instead of cells, one can easily test the effect of different Notch–HSC ligand ratios and Notch ligand density on cell differentiation and

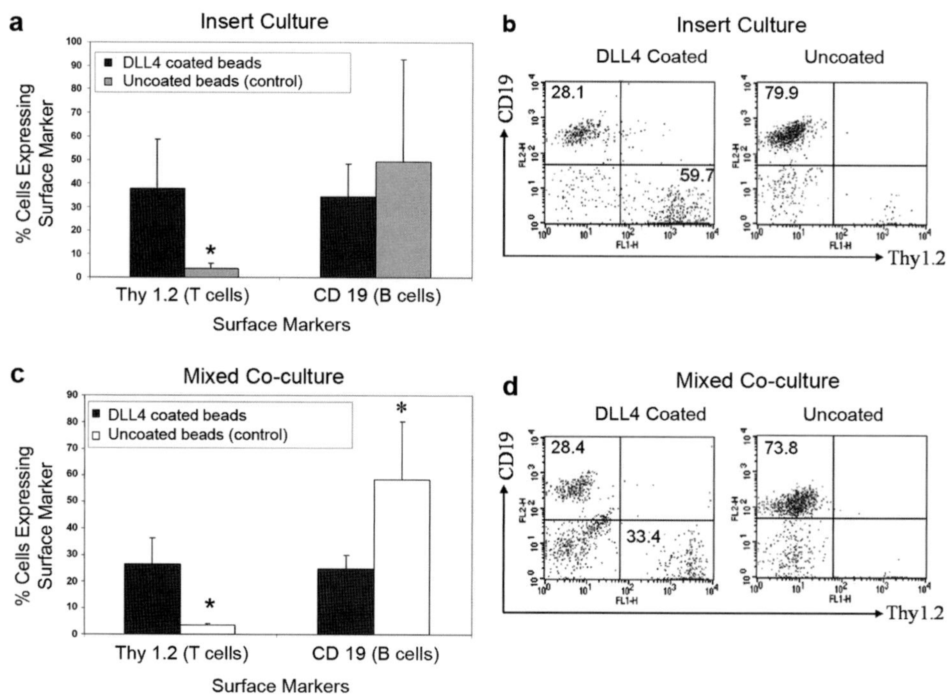

Figure 8.5. T-cell development with delta4 (DLL4)-coated beads from bone marrow-derived HSCs in insert cultures (**a**, **b**) and mixed cultures (**c**, **d**) with OP9 cells. (**a**, **b**) Comparison of the percentage of cells expressing T-cell marker (Thy1.2) and B-cell marker (CD19). (**c**, **d**) Flow cytometry dot plots of cultures stained for Thy 1.2 and CD 19. *Asterisk* indicates statistically significant ($p < 0.05$) difference from DLL4-coated beads. (Reprinted with permission of Wiley [28]).

optimize a method for efficient T-cell generation. The microbead system is yet to be tested either with ESCs or for *in vivo* transplantation. However, the biomaterial-based Notch signaling, instead of a cell-mediated signaling, brings this system closer to therapeutic use since the possibility of implanting retrovirally transfected stromal cells together with therapeutic T-cells is eliminated. The microbead system took a step further into developing a synthetic biomaterial-based system for presenting ligands to direct cell fate. It also proved to be an improvement from the simple 2D immobilization to a 3D ligand presentation that more closely resembles the biological cell–cell contact and allows for a larger-scale cell development system.

8.3. Other Signal Transduction Pathways

8.3.1. Sonic Hedgehog Signaling

Sonic hedgehog (Shh) signaling is a well-known cell-signaling pathway conserved in many mammals. Shh is one of the best-studied ligands that regulates proliferation and differentiation in many cell types [32]. Shh, upon binding to the Patched-1 (PTCH1) receptor, initiates a cascade of molecular signaling involving another molecule, Smoothened . The cascade regulates PTCH1 transcription and also transcriptional effectors of the Gli family [33].

Healy and coworkers [32] sought to apply this signal transduction pathway to a biomaterial-based system for osteoblast differentiation. The signaling domain of Shh was conjugated onto an interpenetrating polymer network (IPN). Culture of rat bone marrow-derived mesenchymal stem cells on Shh-immobilized surfaces directed osteoblast differentiation into the bone lineage when supplemented with bone morphogenetic protein (BMP)-2 and BMP-4. The extent of differentiation, as measured by alkaline phosphatase (ALP) activity, was enhanced with immobilized Shh, in comparison with soluble Shh. RGD conjugation onto the IPN was also necessary for cell attachment. Figure 8.6 shows the ALP activity reported in this study for various experimental conditions. IPN with immobilized RGD and Shh (IPN/RGD/Shh in Figure 8.6a) showed the highest ALP activity among all groups at all

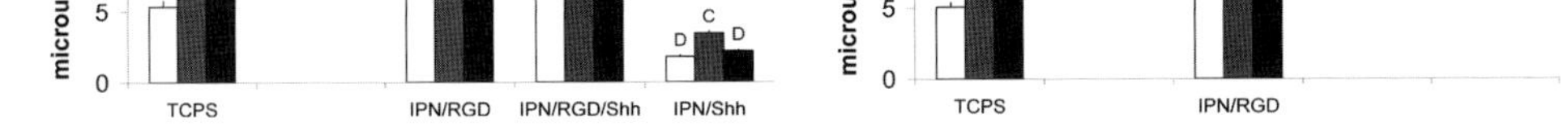

Figure 8.6. Effect of Sonic hedgehog (Shh) signaling on MSC differentiation measured by alkaline phosphatase (ALP) activity. (**a**) ALP activity of MSCs on interpenetrating network (IPN) with immobilized RGD (IPN/RGD), Shh (IPN/Shh), or both (IPN/RGD/Shh). (**b**) ALP activity of MSCs on IPN with immobilized RGD and soluble Shh. *TCPS* denotes the control surface without IPN. Groups not designated by the same letter are significantly different. (Reprinted with permission from Wiley [32]).

time points tested. This ALP activity was higher than either on IPN with immobilized RGD and soluble Shh (IPN/RGD in Figure 8.6b), or on IPN with immobilized RGD alone (IPN/RGD in Figure 8.6a) or immobilized Shh alone (IPN/Shh in Figure 8.6a). This signaling system, an analog of the previously mentioned Varnum-Finney's [24] and Giachelli's [7] system, demonstrated its potential for usage in mesenchymal stem cells in the field of tissue-engineered implants. Edwards and coworkers had developed Shh-expressing cells that enhanced bone regeneration when implanted within a matrix into a rabbit *in vivo* [34]. If such simple immobilization of Shh is enough to enhance bone regeneration, Shh-functionalized biomaterial surfaces could eventually be used *in vivo* for bone regeneration and integration of bone implants. Such an *in vivo* system would be a step further into developing a biomimetic system for peri-implant bone formation and implant fixation [32].

8.3.2. Fibronectin–Immobilized Biomaterial Surface to Induce Mesenchymal Stem Cell Differentiation and HSC Expansion

Fibronectin and the integrin-binding peptide domain of fibronectin – RGD – have been extensively used to promote cell adhesion to material surfaces. However, very few studies have explored in detail their role in triggering specific cell-signaling pathways that could lead to specific cell phenotypes.

Chen and coworkers performed a study on the effect of cell shape on commitment of human mesenchymal stem cells (hMSCs) [35]. The effect of cell shape on cell differentiation had been known, but the molecular basis of this process was unclear [36]. Based on these studies, Chen and coworkers proposed that cell shape directs hMSC differentiation through RhoA signaling. hMSCs were induced to take on either round or spread-out shapes by fibronectin-patterned material surfaces. Fibronectin was microcontact printed onto polydimethylsiloxane surfaces in order to create round-shaped, or spread out "islands" for cell attachment. The cells on the round islands underwent adipogenesis but those on spread-out islands underwent osteogenesis. When the stem cells were assayed for RhoA, it was found that high levels of RhoA were associated with osteogenesis and low levels of RhoA with adipogenesis. When the effect of RhoA alone on differentiation was tested, the results showed that RhoA activity could determine either osteogenic or adipogenic commitment of hMSCs. Furthermore, studies with cell shape revealed that without a specific cell shape, RhoA activity alone could not direct differentiation into a select lineage. In other words, even with full RhoA activity, hMSCs could not differentiate into the osteogenic lineage if the cells were in round shapes. Figure 8.7 shows the effect of cell shape on the kinase activity of ROCK, a RhoA effector. ROCK activity was higher in spread-out cells than in round cells, without treatment with Y-27632, a ROCK inhibitor. In the presence of Y-27632, ROCK activity was minimal. This result demonstrated that cell shape, as a mechanical cue, directed stem cell commitment through RhoA signaling. In terms of biomedical research, this approach is an effective tool to direct cell differentiation into a specific lineage. In the Chen and coworkers study, the fibronectin-immobilized surface was utilized to direct cells to attach and acquire specific shapes. Although direct association of fibronectin binding to RhoA signaling is not known, modification of the ligand-immobilized surface directed cell development in a desired way by controlling cell shape.

Additionally, fibronectin immobilized on material surfaces has been shown to induce not only differentiation but also expansion of stem cells. In an effort to mimic the niche in which hematopoietic stem cell/progenitor cells (HSPCs) grow – bone marrow – Mao and coworkers conjugated fibronectin onto 3D polyethylene terephthalate scaffolds, which were

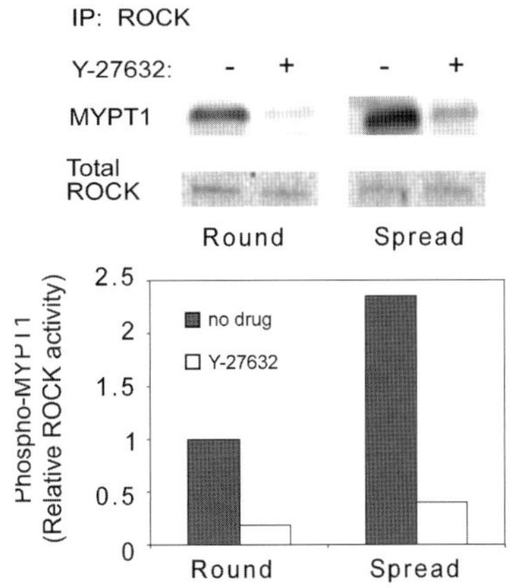

Figure 8.7. Effect of hMSC cell shape on RhoA signaling measured by Western blots. MYPT1 phosphorylation by ROCK kinase was quantified in the presence or absence of ROCK inhibitor Y-27632. (Reprinted with permission from Elsevier [35]).

then used to culture human umbilical cord blood CD34+ cells [37]. After 10 days of culture, the fibronectin-conjugated scaffold yielded a CD34+ expansion of up to 100-fold, compared with less than tenfold expansion in scaffolds with soluble fibronectin. The 3D geometry also enhanced expansion of CD34+ cells. Expansion of CD34+ cells was also possible on scaffolds with immobilized adhesion peptides, such as the RGD and CS-1 binding motifs [38]. Fibronectin and the connecting segments of fibronectin are known to bind to cell adhesion molecules (CAMs) such as very late antigen (VLA)-4 and VLA-5. Although the exact cell signaling mechanism is unknown, binding of fibronectin to either VLA-4 or VLA-5 in HSPCs helps maintain long-term regenerative capacity, support proliferation, and enhance stimulatory growth factor signals in HSPCs [39–41]. The studies with immobilized fibronectin suggest that immobilization of fibronectin onto a biomaterial surface enables it to interact with CAMs to trigger cell signals that control HSPC expansion. The exact pathway by which this occurs is still unknown but these results give engineers a tool to create synthetic, biomimetic environments to control stem cell expansion.

8.4. Conclusions

The work by the Varnum-Finney [24], Giachelli [7], and Roy [28] groups demonstrated that cell-to-cell communication present in Notch signaling can be efficiently mimicked by material surfaces presenting the Notch ligand, thereby inducing ligand-directed stem cell signaling and differentiation. In addition, soluble stromal factors, if necessary for differentiation, can be effectively provided by physically separating stem cells from stromal cells. Not only are these systems cell-free, but they are also highly potent because only small effective amounts of ligands are required on a surface compared with the exogenous amounts needed in solution. These systems allow easy manipulation of ligand–cell ratio and ligand densities providing means, not only to study detailed quantitative effects of ligand-directed cell signaling, but to also create optimal *in vitro* conditions for therapeutic cell generation. Similar studies inducing Shh and RhoA signaling show the versatility of such *in vitro*, surface-directed cell-signaling methods. Taken together, the studies discussed in this chapter provide evidence that surface modification of either a 2D or 3D biomaterial is no longer limited to improving cell attachment or biocompatibility but can also be used to trigger specific cell-signaling pathways and direct cell fate.

References

1. Artavanis-Tsakonas S, Rand MD, Lake RJ. Notch signaling: cell fate control and signal integration in development. Science 1999;284(5415):770.
2. Rangarajan A, Talora C, Okuyama R, Nicolas M, Mammucari C, Oh H, et al. Notch signaling is a direct determinant of keratinocyte growth arrest and entry into differentiation. The EMBO Journal 2001;20: 3427–3436.
3. Lowell S, Jones P, Le Roux I, Dunne J, Watt FM. Stimulation of human epidermal differentiation by delta–notch signalling at the boundaries of stem-cell clusters. Current Biology 2000;10(9):491–500.
4. Noseda M, McLean G, Niessen K, Chang L, Pollet I, Montpetit R, et al. Notch activation results in phenotypic and functional changes consistent with endothelial-to-mesenchymal transformation. American Heart Association 2004:910–917.
5. Radtke F, Wilson A, Mancini SJC, MacDonald HR. Notch regulation of lymphocyte development and function. Nature Immunology 2004;5(3):247–253.
6. Carlson ME, O'Connor MS, Hsu M, Conboy IM. Notch signaling pathway and tissue engineering. Frontiers in Bioscience 2007;12:5143–5156.
7. Beckstead BL, Santosa DM, Giachelli CM. Mimicking cell-cell interactions at the biomaterial-cell interface for control of stem cell differentiation. Journal of Biomedical Materials Research. Part A 2006;79(1):94–103.
8. Zuniga-Pflucker JC. T-cell development made simple. Nature Reviews. Immunology 2004;4(1):67–72.
9. Ladi E, Yin X, Chtanova T, Robey EA. Thymic microenvironments for T-cell differentiation and selection. Nature Immunology 2006;7:338–343.
10. Piersma SJ, Welters MJP, van der Burg SH. Tumor-specific regulatory T-cells in cancer patients. Human Immunology 2008;69(4–5):241–249.
11. Seneviratne SL, Jones L, King AS, Black A, Powell S, McMichael AJ, et al. Allergen-specific CD8+ T-cells and atopic disease. Journal of Clinical Investigation 2002;110(9):1283.
12. Jenkinson EJ, Franchi LL, Kingston R, Owen JJ. Effect of deoxyguanosine on lymphopoiesis in the developing thymus rudiment in vitro: application in the production of chimeric thymus rudiments. European Journal of Immunology 1982;12(7):583–587.
13. Schmitt TM, Zúñiga-Pflücker JC. Induction of T-cell development from hematopoietic progenitor cells by delta-like-1 in vitro. Immunity 2002;17(6):749–756.
14. Radtke F, Wilson A, MacDonald HR. Notch signaling in T-and B-cell development. Current Opinion in Immunology 2004;16(2):174–179.
15. Nakano T, Kodama H, Honjo T. Generation of lymphohematopoietic cells from embryonic stem cells in culture. Science 1994;265(5175):1098–1101.
16. Cho SK, Webber TD, Carlyle JR, Nakano T, Lewis SM, Zuniga-Pflucker JC. Functional characterization of B lymphocytes generated in vitro from embryonic stem cells. Proceedings of the National Academy of Sciences of the United States of America 1999;96(17):9797.
17. Schmitt TM, de Pooter RF, Gronski MA, Cho SK, Ohashi PS, Zuniga-Pfluecker JC. Induction of T-cell development and establishment of T-cell competence from embryonic stem cells differentiated in vitro. Nature Immunology 2004;5(4):410–417.
18. Zakrzewski JL, Kochman AA, Lu SX, Terwey TH, Kim TD, Hubbard VM, et al. Adoptive transfer of T-cell precursors enhances T-cell reconstitution after allogeneic hematopoietic stem cell transplantation. Nature Medicine 2006;12(9):1039–1047.
19. La Motte-Mohs RN, Herer E, Zuniga-Pflucker JC. Induction of T-cell development from human cord blood hematopoietic stem cells by delta-like 1 in vitro. Blood 2005;105(4):1431–1439.
20. De Smedt M, Hoebeke I, Plum J. Human bone marrow CD34+ progenitor cells mature to T-cells on OP9-DL1 stromal cell line without thymus microenvironment. Blood Cells, Molecules and Diseases 2004;33(3): 227–232.

21. Lehar SM, Dooley J, Farr AG, Bevan MJ. Notch ligands delta 1 and jagged1 transmit distinct signals to T-cell precursors. Blood 2005;105(4):1440–1447.
22. Rolink AG, Balciunaite G, Demolière C, Ceredig R. The potential involvement of Notch signaling in NK cell development. Immunology Letters 2006;107(1):50–57.
23. Carotta S, Brady J, Wu L, Nutt SL. Transient Notch signaling induces NK cell potential in Pax5-deficient pro-B cells. European Journal of Immunology 2006;36(12):3294–3304.
24. Varnum-Finney B, Wu L, Yu M, Brashem-Stein C, Staats S, Flowers D, et al. Immobilization of Notch ligand, Delta-1, is required for induction of notch signaling. Journal of Cell Science 2000; 4313–4318.
25. Kopan R, Nye JS, Weintraub H. The intracellular domain of mouse Notch: a constitutively activated repressor of myogenesis directed at the basic helix-loop-helix region of MyoD. Development 1994;120(9):2385–2396.
26. Jarriault S, Brou C, Logeat F, Schroeter EH, Kopan R, Israel A. Signalling downstream of activated mammalian Notch. Nature 1995;377(6547):355–358.
27. Bader D. Immunochemical analysis of myosin heavy chain during avian myogenesis in vivo and in vitro. The Journal of Cell Biology 1982;95(3):763–770.
28. Taqvi S, Dixit L, Roy K. Biomaterial-based notch signaling for the differentiation of hematopoietic stem cells into T-cells. Journal of Biomedical Materials Research. Part A 2006;79(3):689–697.
29. Harman BC, Jenkinson EJ, Anderson G. Microenvironmental regulation of Notch signalling in T-cell development. Seminars in Immunology 2003;15(2):91–97.
30. Parreira L, Neves H, Simoes S. Notch and lymphopoiesis: a view from the microenvironment. Seminars in Immunology 2003;15(2):81–89.
31. Dorsch M, Zheng G, Yowe D, Rao P, Wang Y, Shen Q, et al. Ectopic expression of Delta4 impairs hematopoietic development and leads to lymphoproliferative disease. Blood 2002;100(6):2046.
32. Ho JE, Chung EH, Wall S, Schaffer DV, Healy KE. Immobilized sonic hedgehog N-terminal signaling domain enhances differentiation of bone marrow-derived mesenchymal stem cells. Journal of Biomedical Materials Research. Part A 2007;83(4):1200–1208.
33. Ingham PW. Hedgehog signaling in animal development: paradigms and principles. Genes & Development 2001;15(23):3059–3087.
34. Edwards PC, Ruggiero S, Fantasia J, Burakoff R, Moorji SM, Paric E, et al. Sonic hedgehog gene-enhanced tissue engineering for bone regeneration. Gene Therapy 2005;12:75–86.
35. McBeath R, Pirone DM, Nelson CM, Bhadriraju K, Chen CS. Cell shape, cytoskeletal tension, and RhoA regulate stem cell lineage commitment. Developmental Cell 2004;6(4):483–495.
36. Watt FM, Jordan PW, O'Neill CH. Cell shape controls terminal differentiation of human epidermal keratinocytes. Proceedings of the National Academy of Sciences of the United States of America 1988;85(15):5576.
37. Feng Q, Chai.C, Jiang XS, Leong KW, Mao HQ. Expansion of engrafting human hematopoietic stem/progenitor cells in three-dimensional scaffolds with surface-immobilized fibronectin. Journal of Biomedical Materials Research Part A 2006;78(4):781–791.
38. Jiang XS, Chai C, Zhang Y, Zhuo RX, Mao HQ, Leong KW. Surface-immobilization of adhesion peptides on substrate for ex vivo expansion of cryopreserved umbilical cord blood CD34+ cells. Biomaterials 2006;27(13):2723–2732.
39. Dao MA, Hashino K, Kato I, Nolta JA. Adhesion to fibronectin maintains regenerative capacity during ex vivo culture and transduction of human hematopoietic stem and progenitor cells. Blood 1998;92(12):4612–4621.
40. Yokota T, Oritani K, Mitsui H, Aoyama K, Ishikawa J, Sugahara H, et al. Growth-Supporting activities of fibronectin on hematopoietic stem/progenitor cells in vitro and in vivo: structural requirement for fibronectin activities of CS1 and cell-binding domains. Blood 1998;91(9):3263.
41. Schofield KP, Humphries MJ, de Wynter E, Testa N, Gallagher JT. The effect of alpha4 beta1-integrin binding sequences of fibronectin on growth of cells from human hematopoietic progenitors. Blood 1998;91(9):3230–3238.

9

Growth Factors on Biomaterial Scaffolds

Yoshihiro Ito

Growth factors act on cells via endocrine, paracrine, autocrine, intracrine, juxtacrine, and matricrine mechanisms. Four of these mechanisms are diffusible, and two are nondiffusible. Signal transduction of the diffusible growth factors is significantly different from that of the nondiffusible growth factors. It is important to design growth factors for tissue engineering, regenerative medicine, and cell culture systems by considering the mechanisms of their action at the cellular and molecular level. The action mechanism and the utilization of growth factors are discussed.

Abbreviations

bFGF	Basic fibroblast growth factor
BMP	Bone morphogenic protein
CBD	Collagen-binding domain
CBEGF	Collagen-binding EGF
CB-FGF	Collagen-binding FGF
CHO	Chinese ovary cell
CSF-1	Colony stimulating factor-1
EGF	Epidermal growth factor
ELP	Elastin-like peptide
HB-EGF	Heparin-binding epidermal growth factor
HGF	Hepatocyte growth factor
HUVEC	Human umbilical vein endothelial cell
IL-1ra	Interleukin-1 receptor antagonist
LPS	Lipopolysaccharide
MAPK	Mitogen-activated protein kinase
NGF	Nerve growth factor
PC12	Rat adrenal pheochromocytoma cell line

Y. Ito • Nano Medical Engineering Laboratory, RIKEN Advanced Science Institute, 2-1 Hirosawa Wako-Shi, Saitama 351-0198, Japan

D.A. Puleo and R. Bizios (eds.), *Biological Interactions on Materials Surfaces*,
DOI 10.1007/978-0-387-98161-1_9, © Springer Science+Business Media, LLC 2009

RGD	Arg–Gly–Asp, a cell adhesive peptide sequence
RGDS	Arg–Gly–Asp-Ser, a cell adhesive peptide sequence
TGF-β	Transforming growth factor-β
THP-1	A monocytic cell line
TMD7	A human leukemia cell line established from blast cells
TNF	Tumor necrosis factor
TrkA	Tyrosine kinase A
UT-7/Epo	A cell line established from the bone marrow; growth is supported by erythropoietin
VEGF	Vascular endothelial growth factor

9.1. Introduction

Growth factors play an important role in regenerative medicine. In addition to cell materials (either synthetic substrates or naturally occurring matrices), growth factors are now recognized as a fundamental requirement in the field of tissue engineering (Figure 9.1) [1]. The mechanisms of action and the combination of growth factors with materials are reviewed in this chapter.

9.2. Mechanisms of Action of Growth Factors on Cells

The mechanisms of action of growth factors on cells are shown schematically in Figure 9.2. Diffusible interactions, such as endocrine, paracrine, autocrine, and intracrine, and nondiffusible interactions, such as juxtacrine and matricrine interactions, are known to occur. It is also known that each growth factor does not necessarily have a single mechanism of action at the cellular level.

Figure 9.3 illustrates schematically the diffusible and nondiffusible interaction mechanisms of growth factors with cells. First, the growth factor interacts with the cognate receptor on the cell membrane, and forms a complex. Second, these complexes aggregate on the cell membrane. At the same time, this interaction induces phosphorylation of the intracellular

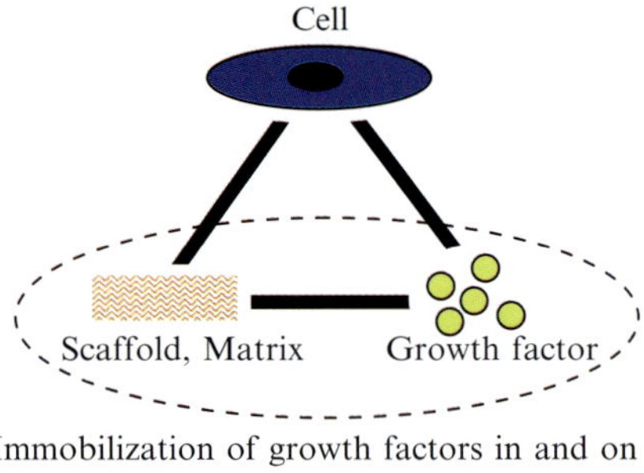

Figure 9.1. The principle aspects of tissue engineering. The present chapter focuses on fusion, i.e., immobilization of growth factors on a scaffold. Reprinted with permission from Soft Matter 2008;4:46-56. Copyright 2008 Royal Society of Chemistry.

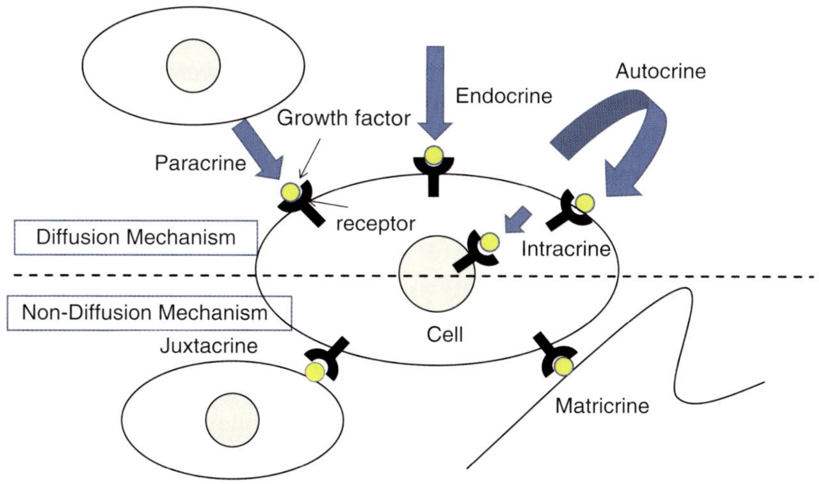

Figure 9.2. Schematic illustration of the mechanisms of action of growth factors on cells.

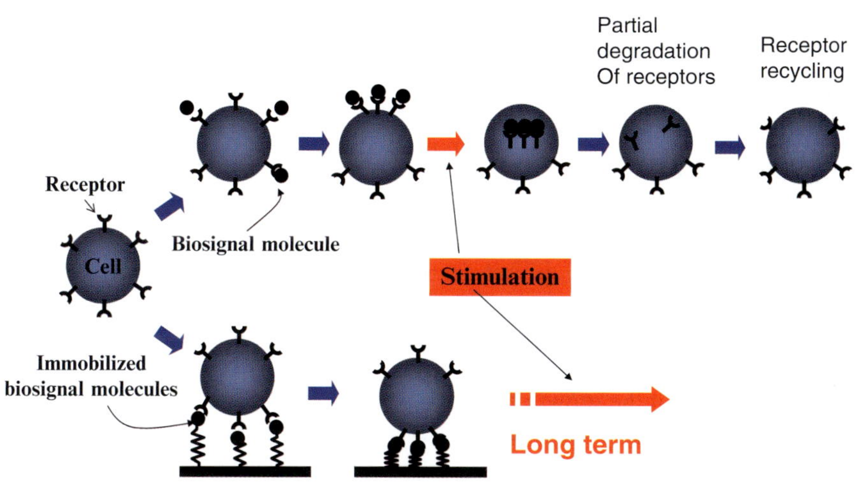

Figure 9.3. Schematic illustration of the interaction of cells with soluble and immobilized growth factors. Reprinted with permission from Soft Matter 2008;4:46–56. Copyright 2008 Royal Society of Chemistry.

domain of the receptor, and triggers signal transduction in the cell. In the case of diffusible interactions, aggregates of the formed complexes are internalized through either a clathrin-dependent or a clathrin-independent mechanism. The internalized complexes are partially decomposed by lysosomes, and are partially recycled back to the cell membrane. Although pertinent mechanisms are not fully understood at the present time, the receptor is considered to be located inside the cell and not on the cell membrane [2–4]. With the exception of an intracrine mechanism, the number of receptors on the cell membrane determines the sensitivity of the cell in diffusible cases. Internalization of the receptors following complex formation reduces the number of receptors on the cell membrane. This is a downregulation mechanism that induces desensitization by reducing excessive response and overstimulation.

The nondiffusible juxtacrine mechanism was elucidated by the discovery of cell membrane-bound growth factor in the 1990s [5, 6]. Heparin-binding epidermal growth factor (HB-EGF), transforming growth factor-β (TGF-β), tumor necrosis factor-α (TNF-α), colony-stimulating factor-1 (CSF-1), and the c-kit ligand are known growth factors that work by a juxtacrine mechanism.

9.3. Immobilized Growth Factors

In the 1990s, chemically immobilized growth factors were devised and investigated by our group and other researchers. The effect of immobilized growth factors has been investigated in the cases of artificial juxtacrine and matricrine mechanisms [1]. Table 9.1 shows the immobilized growth factors reported to date [7–117]. A summary of the evidence of direct interactions between immobilized growth factors and cells follows.

1. In the case of immobilization of biosignal molecules on solid surfaces, achieving sta bility and strength of bonding is very important to prevent release and/or loss of the bioactive molecules. For this reason, the amount of immobilized growth factor must be measured carefully using, for example, radioisotope labeling [7]. Radioisotope labeling is useful to detect small amounts of growth factors on small surface areas. Successful immobilization of growth factors on substrates allowed investigation of their effect on cells in vitro; cells were cultured and the effect was investigated. In the 1960s, the effect of insulin (immobilized on beads) on cells was attempted, but, because of the instability of the immobilized chemical stimulus, the study was stopped. Recently, numerous investigations into immobilized growth factors have been carried out. The efficacy of the respective results, however, has not been established.
2. In order to demonstrate bioactivity of growth factors bound on cell membranes in biology, cells expressing growth factors on their membranes were first chemically fixed using formalin [5, 6], and then other cells were cultured on the chemically fixed cells. Regulation of the function of cells cultured on the fixed cells provided evidence that the cell membrane-binding growth factor was bioactive; in other words, release of the growth factor into the supernatant medium was not necessary for bioactive action. Iwamoto and Mekada [6] discussed the juxtacrine mechanism of HB-EGF. They fixed a stable transfectant of mouse L cells expressing the precursor of HB-EGF (proHB-EGF) with formalin and thus prevented the release of soluble HB-EGF (sHB-EGF); these fixed donor cells were then cultured with EP170.7 cells (an epidermal growth factor [EGF] receptor-ligand-dependent cell line). Under these conditions, EP170.7 cell proliferation was stimulated. Moreover, EP170.7 cell growth depended on the amount of proHB-EGF expressed by the donor cells (6).
3. Specific interactions of immobilized growth factor with a cognate receptor were examined using antibodies against the immobilized growth factor. Inhibition of the growth factor by the antibody tested ascertained a specific interaction. This method has been employed by many researchers. Alternatively, direct interaction with immobilized growth factors can be demonstrated by blocking the cellular receptors using inhibitors. Figure 9.4 shows that adsorption of antibodies against immobilized vascular endothelial growth factor (VEGF) reduced the effect of immobilized VEGF on cell proliferation [79].

Table 9.1. Growth factors that hav3e been chemically immobilized on materials.

Growth factor	Reference
Insulin	[7–38]
Epidermal growth factor (EGF)	[39–57]
Nerve growth factor (NGF)	[40, 58–64]
Bone morphogenic protein (BMP)	[65–74]
Vascular endothelial growth factor (VEGF)	[75–81]
Fibroblast growth factor (FGF)	[51, 82–90]
Insulin growth factor-1 (IGF-1)	[84]
Transforming growth factor-β1 (TGF-β1)	[91–95]
Hepatocyte growth factor (HGF)	[96, 97]
Leukemia inhibitory factor (LIF)	[98, 99]
Stem cell growth factor (SCF)	[100]
Interleukin-2	[101]
Interleukin-1	[102]
Tumor necrosis factor-α (TNF-α)	[103]
Erythropoietin	[104]
Neurotrophin-3	[41]
Transferrin	[105]
E-cadherin	[106]
Osteopontin	[107, 108]
P-selectin	[109]
CXCR3 ligand	[110]
Notch ligand	[111–114]
Sonic hedgehog	[115]
Ephrin-B2	[116]
Osteotropic peptide parathyroid hormone (1–34)	[117]

4. Evidence that growth factors remain bioactive even after repetitive use has been reported in the literature [12, 104]. Figure 9.5 shows the culture of cells on a cytokine immobilized on a micropatterned surface. After removal of the cells, the biomolecule immobilized on the micropatterned surface retained its bioactivity and affected the functions of a fresh batch of cells subsequently cultured on the same surface; the biological activity of the surface was not maintained when the growth factor was released from the surface because of incomplete immobilization. Repeated utilization of immobilized, bioactive molecules (such as enzymes and proteins) is a desirable aspect of bioreactors used to propagate and culture large numbers of cells.
5. Culture of cells far from the immobilized growth factor provides further evidence of the bioactivity of the immobilized biomolecules. Transwells™, which have a semipermeable membrane at one side, have been used for this purpose with media containing nutrients that are permeable but growth factors that are impermeable; the effects of the released growth factor can be investigated in this case. If proliferation of cells cultured on the semipermeable membrane is not enhanced, but the proliferation of cells on a surface containing immobilized growth factors is enhanced, this result provides evidence that the immobilized growth factor is active without being released.

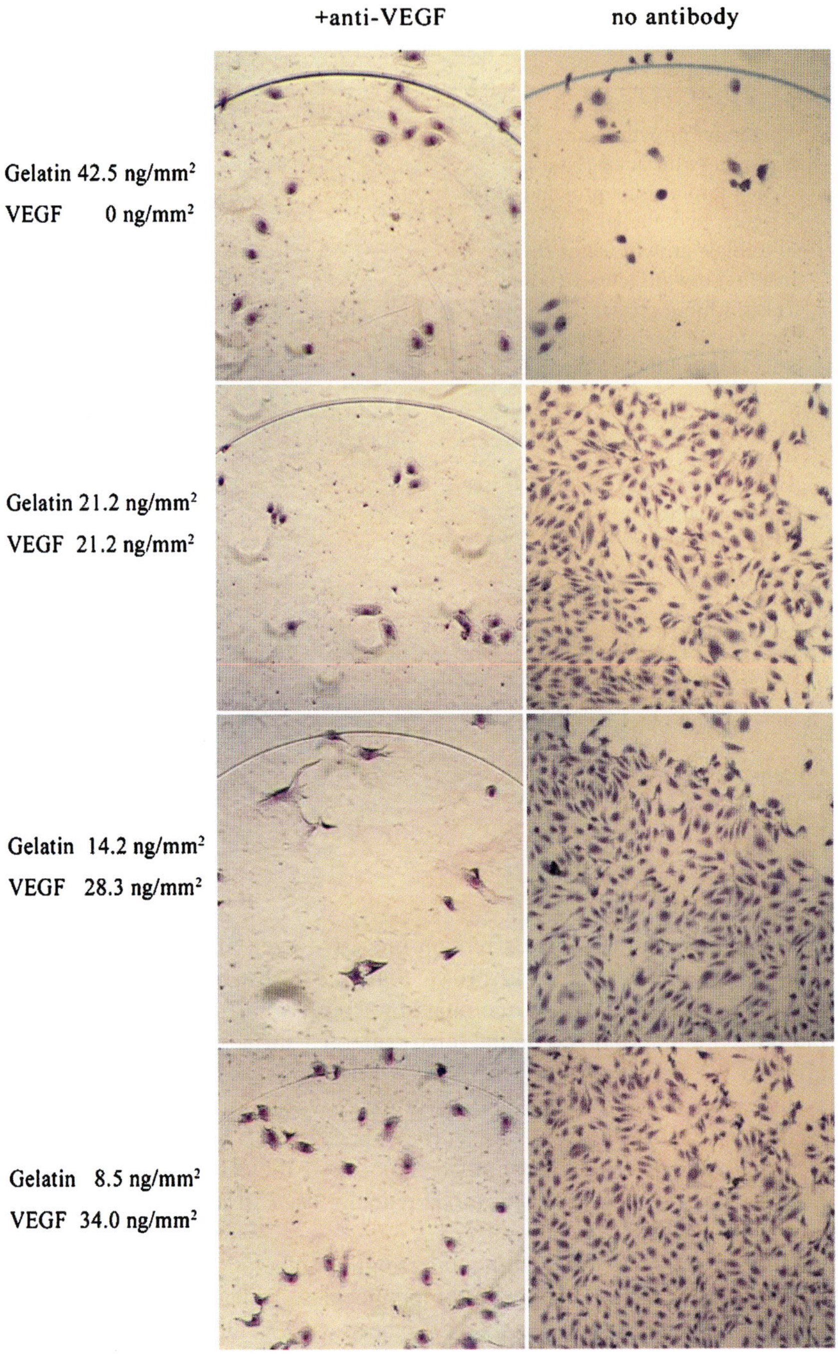

Figure 9.4. HUVEC proliferation on tissue culture polystyrene containing immobilized VEGF in the absence and presence of anti-VEGF antibody. Reprinted with permission from J Biomed Mater Res 2005;74:659–665. Copyright 2005 John Wiley.

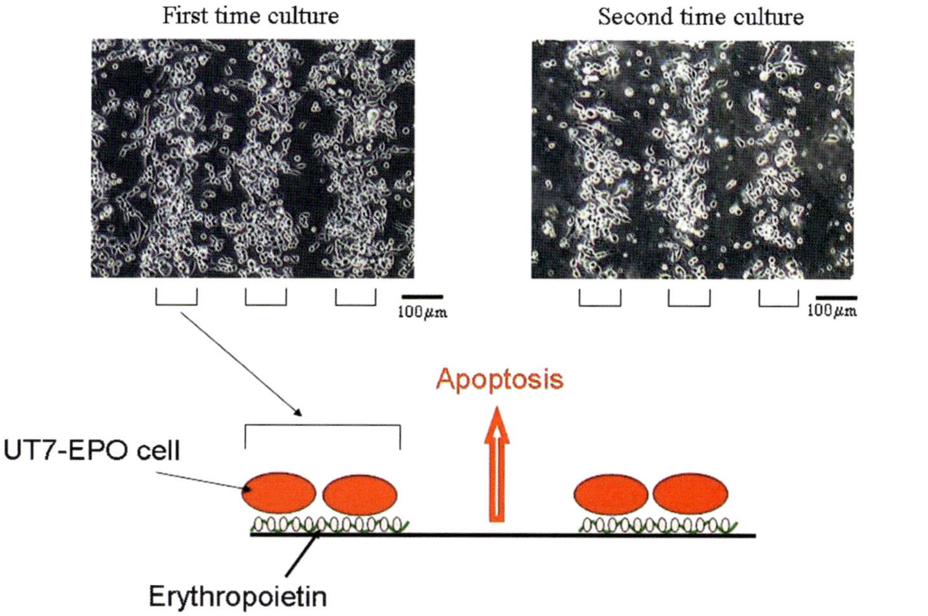

Figure 9.5. UT-7/Epo cultured on immobilized erythropoietin. Reprinted with permission from Biomaterials 2004;25:2293–2298. Copyright 2004 Elsevier.

6. As a modification of the method described in the preceding paragraph (v), visualization of signal transduction and spatial regulation of cell functions has been performed using growth factors immobilized on micropatterned surfaces [34, 45, 46]. In such cases, cell differentiation is enhanced only on areas where the biosignal molecule is immobilized. The immobilized biomolecules influence cell functions, including proliferation and differentiation, without enhancing cell adhesion. Figures 9.6 and 9.7 show that a human leukemia cell line established from blast cells (TMD7) cells and Chinese ovary (CHO) cells overexpressed insulin receptors when cultured on Notch ligand and insulin immobilized on micropatterned surfaces in vitro, respectively. The cells proliferated on the regions that contained immobilized Notch ligand or insulin.
7. To further investigate the interaction of immobilized growth factors, CHO cells overexpressing the cognate receptors were cultured on substrate surfaces modified with narrow (smaller than the cell size) stripes containing immobilized growth factor [46]. The contact area between the cells (stripe width = 2 μm) and the immobilized growth factor were stained using anti-phosphotyrosine antibody (Figure 9.8). Since free lateral diffusion and internalization of the bound growth factor and the receptor complex were prohibited by the immobilization of the growth factor, only the signal proteins in the interaction regions were activated. This finding also indicates that the biological signal was transduced only by cells that interacted with the immobilized growth factor. Utilization of microbeads with immobilized growth factors is an approach that has been used to probe the interaction of cellular receptors with immobilized growth factors.

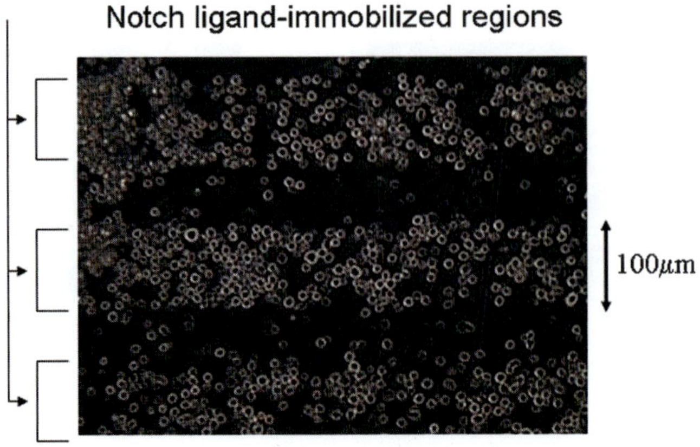

Figure 9.6. TMD7 cultured on a surface containing immobilized notch ligand patterns. Reprinted with permission from Inflamm Regen 2005;25:426–430. Copyright 2005 The Japanese Society of Inflammation and Regeneration.

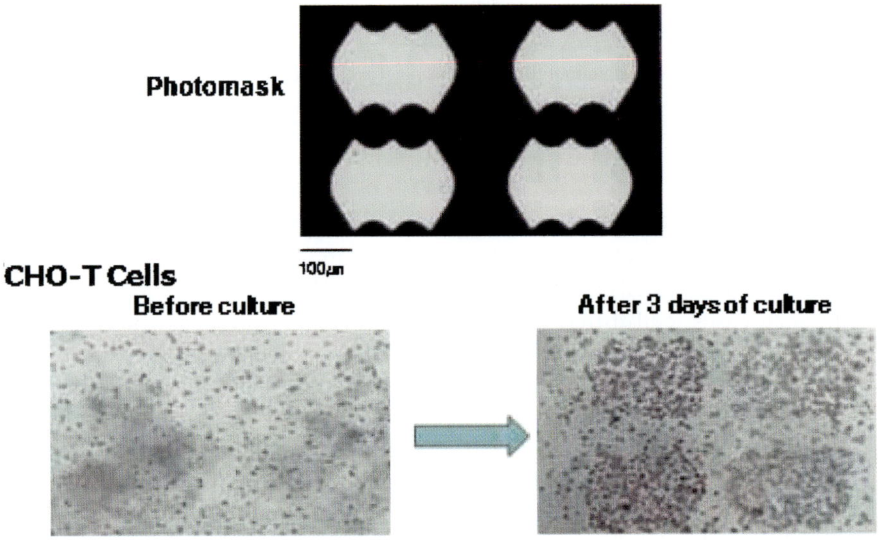

Figure 9.7. CHO cells overexpress insulin receptors when cultured on a surface modified with micropatterns containing insulin. Reprinted with permission from FEBS Lett 1997;403:159–162. Copyright 1997 Elsevier.

8. The effect of immobilized proteins has been either categorized or compared with the effect of extracellular matrix adhesion proteins such as gelatin, collagen, fibronectin, albumin, and immunoglobulins [9]. Recently, microarray systems have been employed by some researchers to investigate the effect of immobilized molecules such as growth factors [41, 84, 118, 119] and provided evidence of the efficacy of these approaches as well as for the stimulation of cell functions under these conditions.
9. Recently, the interaction between immobilized growth factors and cellular receptors has been investigated at the molecular level. For example, the surface density of

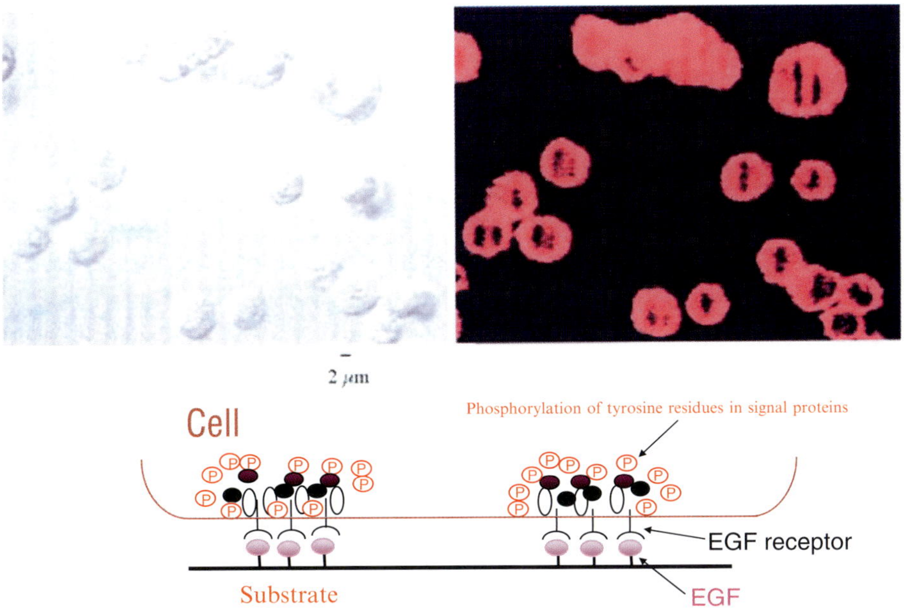

Figure 9.8. CHO cells overexpress EGF receptors when cultured on a surface containing immobilized EGF. Reprinted with permission from Bioconjug Chem 1998;9:277–282. Copyright 1998 American Chemical Society.

immobilized growth factor has been quantitatively investigated [40]. Cellular response decreased markedly when the surface density of the immobilized growth factor was only slightly lower than that of the receptor dimers. The receptor molecules bound to the immobilized growth factor were prevented from lateral diffusion and internalization, and thus remained at their initial positions. In the case of nerve growth factor (NGF)–receptor complexes, it was discovered that they have two distinct diffusive states: one mobile and one immobile [58]. Transition between these two diffusive states occurs reversibly with duration times determined by a single rate-limiting process. Abrupt transition to the immobile phase often occurs simultaneously with clustering of the NGF–receptor complexes. Immobilization depends on phosphorylation of the tyrosine kinase A (TrkA) NGF receptor. Using dual color imaging, it was demonstrated that membrane recruitment of the intercellular signaling protein occurs with NGF–receptor complexes in the immobile phase; this result indicates that signal transduction occurs during this phase. These results provided evidence that NGF signaling is performed through a repetitive random process that induces formation of signaling complexes.

9.4. Effects of Immobilized Growth Factors on Cell Function

The cellular effects of diffusible and nondiffusible delivery of growth factors should be taken into consideration in the design of materials containing growth factors that will be used as scaffolds for cells for tissue engineering applications.

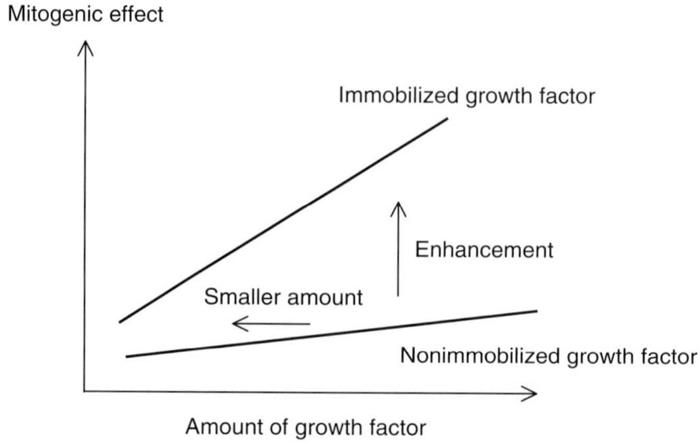

Figure 9.9. Comparison of the mitogenic effect of immobilized growth factor with that of nonimmobilized (soluble) growth factor. The immobilized growth factor exhibits a greater mitogenic effect than does the nonimmobilized growth factor. Even in small amounts, the immobilized growth factor affects cell function because of multivalency and high local concentration (as discussed in Sect. 9.3 (items 1 and 3), respectively). Enhancement of the mitogenic effect is caused by inhibition of downregulation and other aspects (as discussed in Sect. 9.3 (items 2 and 3), respectively). Reprinted with permission from Soft Matter 2008;4:46-56. Copyright 2008 Royal Society of Chemistry.

Stimulation of cells requires smaller amounts of an immobilized growth factor than a soluble growth factor (Figure 9.9). The higher and/or different effect of immobilized (versus soluble) growth factors results from multiple interactions and not only from the formation of a ligand–receptor complex. These effects of an immobilized growth factor are considered to result from the mechanisms presented in Sect. 9.4.1 [1].

9.4.1. High Local Concentration of Growth Factors and Multivalency

The local concentration of immobilized biosignal molecules on a substrate is extremely high (Figure 9.10). In addition, the importance of the multivalency of the immobilized biosignal molecules has been addressed by many researchers. Recently, Kiessling et al. considered synthetic multivalent ligands as probes for signal transduction [120]. Multivalent ligands can bind avidly to multiple receptors on the cell membrane, a process that is facilitated in the fluid lipid bilayer by the two-dimensional diffusion of receptors. The multivalent ligands can activate signaling pathways if they can cluster signaling receptors (Figure 9.11). Kiessling et al. discussed the three major concepts that are critical for the application of multivalent ligands as probes of signal transduction [120]. (1) Since signal transduction cascades are mediated by receptor–receptor interactions, promoting a receptor assembly is critical for signaling. (2) Multivalent ligands can interact with the target receptors through multiple binding modes. (3) The structure of a multivalent ligand determines the favored binding modes. Thus, the structure of growth factor-modified surfaces can be optimized to elicit the desired biological response. Recently, Baker's group reported on the synthesis of a polyamideamine dendrimer–EGF conjugate, which acted as a superagonist by stimulating cell proliferation to a greater degree than free EGF [121].

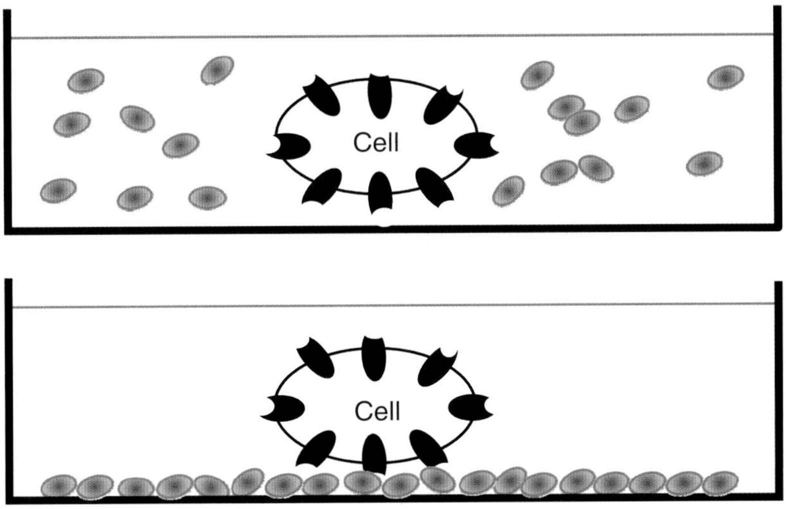

Figure 9.10. Immobilization of a growth factor on the surface of a substrate provides high local concentration foe subsequent cell interactions. Reprinted with permission from Soft Matter 2008;4:46–56. Copyright 2008 Royal Society of Chemistry.

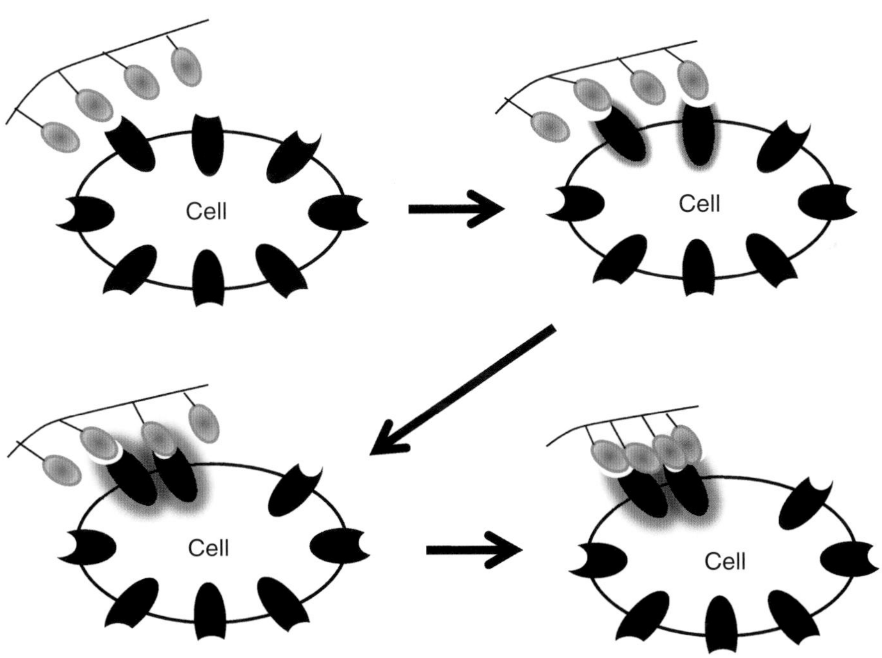

Figure 9.11. Interactions of a multivalent ligand with a cell. The multivalent ligand enhances the formation of a ligand–receptor complex; an interaction of the activated ligand–receptor complexes occurs because of the high local concentration of the ligand. In addition, the complexes formed are stabilized by the multivalent ligands. Reprinted with permission from Soft Matter 2008;4:46–56. Copyright 2008 Royal Society of Chemistry.

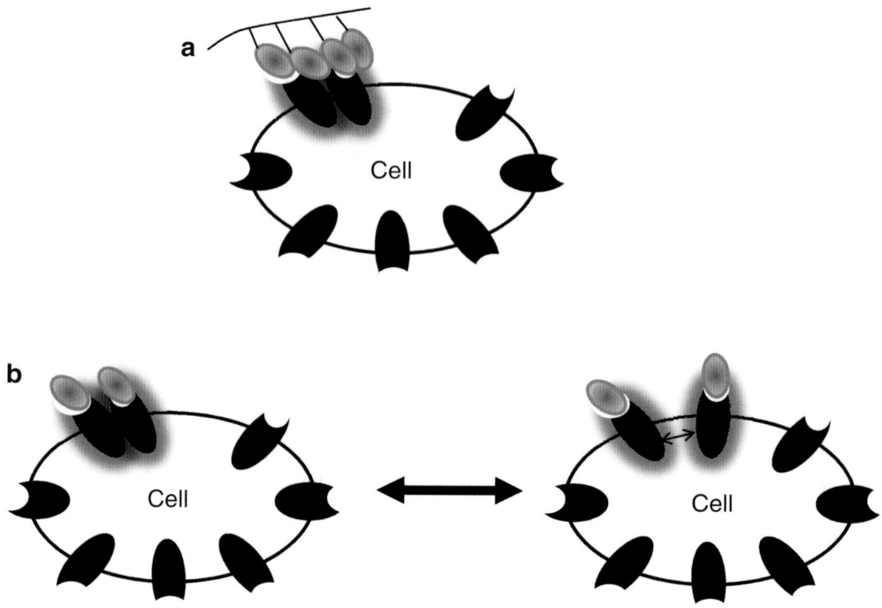

Figure 9.12. (**a**) Multivalent ligands inhibit diffusion of the ligand–receptor complexes. (**b**) Ligand–receptor complexes diffuse in a cell membrane.

In addition to a high local concentration and multivalency, prevention of lateral diffusion of the activated receptor and/or the receptor–ligand complex in the plane of the cell membrane must be taken into consideration (Figure 9.12). Clustering is needed for activation of certain types of receptors, and prevention of lateral diffusion entails stabilization of aggregates of the complexes, and thus leads to a long-lasting effect.

9.4.2. Inhibition of Growth Factor Downregulation

Inhibition of the internalization of growth factors should also be taken into consideration (Figure 9.3). In general, cells decompose biosignal molecules in order to reduce their stimulation, using a process known as downregulation. Immobilization of growth factors inhibits downregulation and, as a result, the stimulation continues for a long period of time without attenuation.

To investigate the aforementioned aspect, activation of cell signaling proteins, insulin receptor β-subunit, or the mitogen-activated protein kinase (MAPK) in the cells was measured in the presence of either immobilized insulin or immobilized EGF [12, 42]. Native insulin and EGF rapidly activated the respective insulin and EGF receptors; this activation, however, was transient. On the other hand, activation by the immobilized insulin and EGF continued to increase up to 12 h; a time lag was needed for the cells to adhere on the surface containing immobilized insulin and EGF. These sustained activations of signaling proteins by the immobilized insulin and EGF explains the high mitogenic effect of the immobilized growth factors.

Sustained activation of the immobilized biosignal molecules had another effect [49]. It is well known that while proliferation of the rat pheochromocytoma cell line PC12 is stimulated by EGF, differentiation of these cells is stimulated by NGF. Immobilized EGF stimulated PC12 differentiation [49]. Immobilized EGF induced long-lasting stimulation of

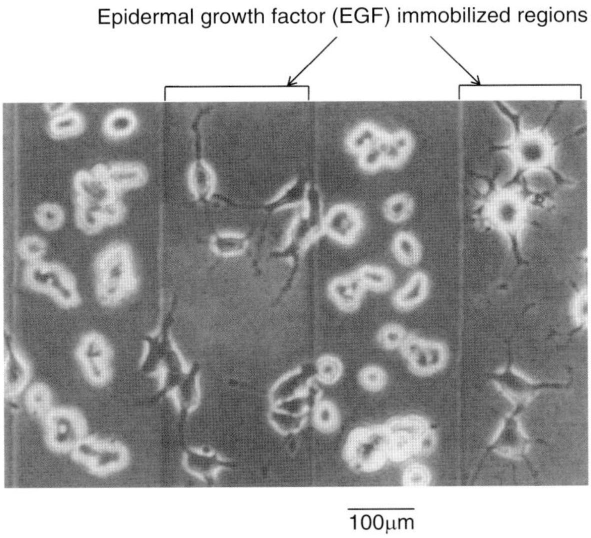

Figure 9.13. PC12 cells cultured on a surface with micropatterns containing immobilized EGF. The immobilized EGF induced differentiation of the PC12 cells. Reprinted with permission from J Biochem 2001;129:733–737. Copyright 2001 Japanese Society of Biochemistry.

MAPK and of a subfamily of the MAPK superfamily in the cells, as did diffusible NGF. This switching between growth stimulation and differentiation is considered to be the result of the duration of exposure of cells to the mitogenic stimulus [49]. Figure 9.13 shows PC12 cells cultured on a micropatterned surface modified with immobilized EGF; in this case, cell differentiation was induced only on the regions containing the immobilized EGF. Figure 9.14 shows the result of this process.

9.4.3. Other Pertinent Mechanisms

The location of the stimulation site, specifically, the cell–medium interface and the cell-matrix interface, should also be taken into consideration (Figure 9.15). Polarity, with receptors presented on either the dorsal or the ventral side, can affect various cell functions.

Immobilization of growth factors may induce a secondary effect on cell function. Kuhl and Griffith-Cima [39] reported on the rounding responses of primary rat hepatocytes on a surface containing immobilized EGF; cells responded differently on surfaces with tethered growth factor compared with cells exposed to nonimmobilized/soluble EGF. Ogiwara et al. [54] reported that A431 cells adhering to surfaces with immobilized EGF-Fc formed filopodia, whereas cells adhering to collagen in the presence of soluble EGF formed lamellipodia. In addition, Ogiwara et al. [55] found that the interaction between photo-immobilized EGF and the receptors on the cell membrane was independent of the Mg^{2+} ion concentration; in contrast, integrin-mediated cell adhesion to natural extracellular matrix depended on the Mg^{2+} ion concentration. Phosphorylation of EGF receptors in A431 cells was induced by immobilized EGF; this process was similar to the process induced by soluble EGF. The DNA uptake decreased when hepatocytes interacted with immobilized EGF but increased in the presence of soluble EGF. Liver-specific functions of hepatocytes on immobilized EGF were maintained for a period of 3 days, but they were not maintained in the presence of soluble

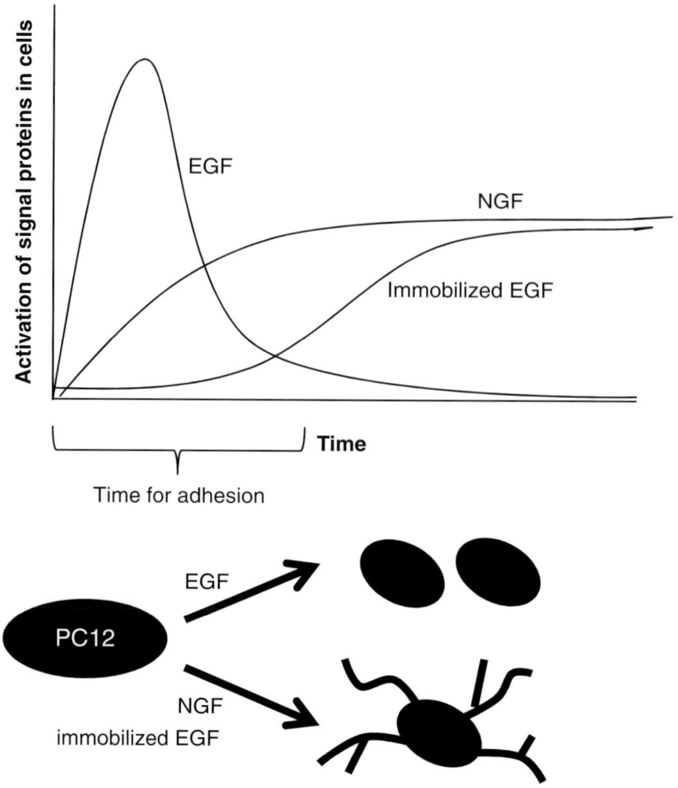

Figure 9.14. Schematic drawing of the time course of activation of signal proteins in cells by immobilized and soluble growth factors. While diffusible EGF and NGF induced PC12 cell proliferation and differentiation, respectively, immobilized EGF induced only cell differentiation.

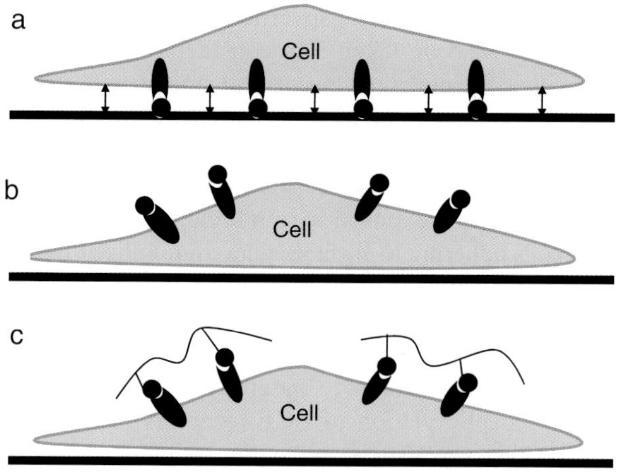

Figure 9.15. An immobilized growth factor is different from a soluble growth factor and provides a different environment for interaction with cells.

EGF; this result indicated that different signal transduction pathways are involved in cells on immobilized EGF than those in the presence of soluble EGF. These differences in the effects of immobilized versus soluble growth factors are due to the complex structure of interfaces surrounding cultured cells.

Reddy et al. [122] developed an approach derived from an understanding of how the attenuation mechanisms, including growth factor depletion and receptor downregulation, arise from ligand–receptor trafficking processes. A recombinant EGF mutant with reduced receptor-binding affinity is a more potent mitogenic stimulus for fibroblasts than natural EGF altered trafficking properties. Optimization of ligand binding parameters requires a systemic integration of processes from the initial binding event to the final cellular response. The experimental data of Reddy et al. [122] showed that consideration of cellular trafficking processes is essential for optimizing the desired cellular response. Immobilized growth factors may have similar effects.

9.5. Biomaterial Design Using Immobilized Growth Factors

Various strategies to incorporate immobilized biosignal molecules in the design of biomaterials have been developed (Figure 9.16). Pertinent aspects are summarized in the sections that follow.

9.5.1. Methods for Growth Factor Immobilization

Immobilization of proteins on material surfaces has been reviewed recently [123, 124]. From the chemistry perspective, these immobilization methods are classified into covalent and noncovalent bonding categories. Noncovalent bonding includes physical adsorption, ionic binding, self-assembly, and biochemical affinity, and does not uphold the immobilization effect for a long period of time. Slow release of proteins from the material surface reduces subsequent biological effects. On the other hand, covalent bonding prolongs the biological effect of proteins for a long period of time.

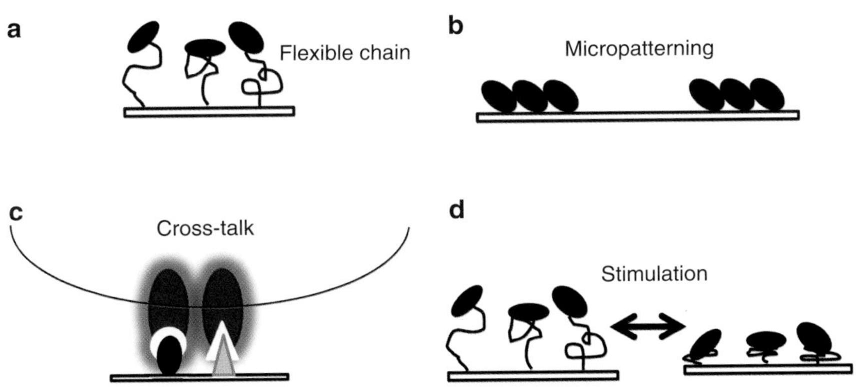

Figure 9.16. Material designs using immobilized growth factors. (**a**) A flexible spacer chain makes the immobilized growth factor mobile. (**b**) Cells interact on the micropatterns containing immobilized growth factors. (**c**) Coimmobilization of different types of ligands (growth factors) induces crosstalk between receptors on the cell membrane. (**d**) Stimuli-responsive polymers add functions pertinent to cells to the immobilized growth factors. Reprinted with permission from Soft Matter 2008;4:46–56. Copyright 2008 Royal Society of Chemistry.

Recently, a novel immobilization method was reported by Mori et al. [88] These researchers immobilized basic fibroblast growth factor (bFGF) onto cubic proteinous microcrystals (*Bombyx mori cypovirus polyhedra*), which are insoluble in a physiological cellular environment.

9.5.2. Spacer Insertion and Surface Stiffness

After cell receptors form complexes with their respective ligands, these complexes are generally considered to aggregate in the cell membrane and to transmit signals to the nuclei (Figure 9.2). In this case, insertion of a spacer between the material surface and the biosignal molecule may enhance diffusion of the complex in the cell membrane (Figure 9.16a). Several studies used spacer chains and reported increased bioactivity [17, 20–23, 39]. Limitations of this approach include steric hindrance in the spacing of the growth factor from the substrate, and the possible impact on the growth factor binding to the cell surface receptor. For example, conjugation with a polyethylene glycol spacer reduces the biological activity of insulin [18].

Insertion of a spacer arm is also related to the surface stiffness. As revealed by recent experiments using soft elastic substrates, cells sense and react to the extracellular matrix stiffness [125, 126]. Therefore, protein immobilization has been performed using polymeric materials with different water contents, and the effect of immobilized growth factor has been examined. Increased flexibility of an immobilized growth factor increased its biological activity.

Iwamoto and Mekada [6] addressed the flexibility of a membrane-anchored growth factor, specifically HB-EGF, in the context of the juxtacrine mechanism. In that study, fixation using formalin perturbed aggregation of the ligand–receptor complexes in the cell membrane; this event led to different biological stimulation of the cells.

9.5.3. Micropatterning

Micropatterning of biomaterials (Figure 9.16b) has been used to regulate cell function such as proliferation and differentiation subsequent to cell adhesion [127–137]. Cells adhering onto micropatterned substrates conform their shape to the underlying adhesion region. This change in morphology is concomitant to changes in the cytoskeleton and affects critical cell functions such as apoptosis and proliferation [138]. Cell shape can also control stem cell differentiation [140]. Gene upregulation and protein synthesis are also altered following changes of the shape of the nucleus [139]. Controlling the cellular microenvironment through micropatterning may be used to modulate cell function pertinent to tissue engineering applications.

Because cells lie in contact and/or in close proximity to other cell types in the tightly controlled structure of tissues inside the body [130], tissue-engineering constructs, which aim at reproducing the structure of biologic tissues, will benefit from incorporating control of cell–cell interactions [141]. Patterned cocultures are a useful tool for tissue-engineering constructs and for studying cell–cell interactions in vitro because they can be used to control the degree of homotypic and heterotypic cell–cell contact. Pioneering work in this area was performed by studying the interaction of hepatocytes and nonparenchymal fibroblasts in cocultures [142]. Recently developed pertinent methods utilize thermally reversible polymers [143], layer-by-layer deposition of ionic polymers [144], microfluidic deposition [145], and molding of hydrogels [146].

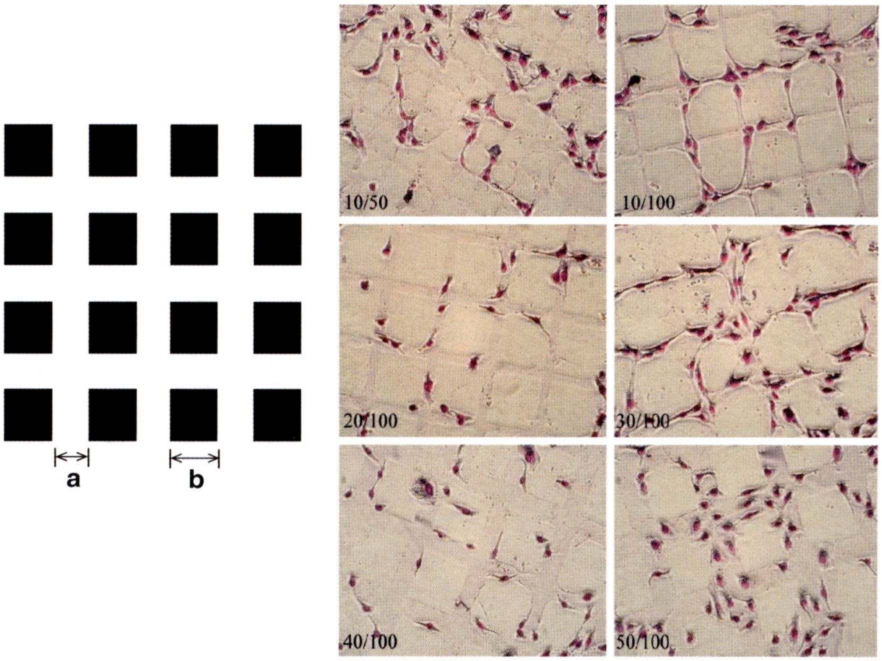

Figure 9.17. HUVECs cultured on glass plates modified with micropatterns containing immobilized VEGF. The numbers (microns) in the photographs indicate **(a)** the width of the immobilized regions and **(b)** the width of the lattice gap. (a/b) is indicated in each micrograph. The surface density of both gelatin and VEGF was 21.2 ng/mm^2. Reprinted with permission from J Biomed Mater Res 2005;74:659–665. Copyright 2008 John Wiley.

As mentioned in Sect. 9.3 (items 6 and 7), micropatterning is also useful for investigating the effect of immobilized growth factors of cell responses. If no signal transduction is observed in the presence of nonimmobilized (soluble) growth factor, then it is possible to conclude that released/soluble growth factor does not affect cells and their function.

In addition, micropatterning has been utilized to regulate cell morphology and tissue formation. Ito et al. [79] used micropatterned substrates and endothelial cells to investigate formation of blood vessels (Figure 9.17). Gomez et al. [64] reported axon extension in neurons on a surface containing covalently immobilized NGF; microtopography was introduced in the form of microchannels. When the two surface stimuli were combined, a synergistic increase in axon length was detected; this response could be the result of faster polarization triggered by the substrate topography and was enhanced by the NGF stimulus.

9.5.4. Coimmobilization

Other macromolecules have been coimmobilized with growth factors to enhance select cell functions (Figure 9.16c). The mitogenic effect of immobilized growth factors on cell function has been reported by various research groups [24–31, 36, 37]. In order to enhance cell adhesion, adhesive extracellular matrix proteins such as collagen, fibronectin, gelatin, and Arg–Gly–Asp–Ser (RGDS) peptide (which is a cell adhesive domain in many extracellular matrix proteins), have been coimmobilized; both cell adhesion and subsequent cell proliferation were remarkably enhanced on the substrate surface containing coimmobilized moieties. Coimmobilization with either other adhesion proteins or different growth factors provides new and unexplored possibilities to regulate cell functions using biomaterials. Cationic polymers, such as poly(allyl amine) and polylysine, have been employed for the physiochemical enhancement of cell adhesion.

In addition, various stimuli-responsive materials have been developed recently and have been used to modulate function of cells interacting with these materials [147–150]. Chen et al. [35] immobilized insulin using a thermoresponsive polymer, poly(*N*-isopropylacrylamide) and reported enhanced cell proliferation in the case of immobilized insulin. Recently, coimmobilization of insulin and a cell adhesive peptide sequence (RGDS) was reported by Hatakeyama et al. [36–38].

9.5.5. Engineering of Proteins for Immobilization

Various gene-engineered proteins have been designed and synthesized and immobilized on substrates (Table 9.2). Nishi et al. [51] constructed fusion proteins consisting of a growth factor moiety and the collagen-binding domain (CBD) of collagenase, which acted as an anchor for the collagen fibrils. They chose EGF and bFGF as parts of the fusion proteins (collagen-binding EGF [CBEGF], collagen-binding bFGF [CB-FGF]). When CBEGF was injected subcutaneously into hairless mice, it remained at the sites of injection for a period of up to 10 days, whereas EGF was not detectable 24 h after injection. Although CBEGF did not exhibit a growth-promoting effect in vivo, CBEGF (but not bFGF), strongly stimulated DNA synthesis in stromal cells at 5 and 7 days after injection. These results indicate that CBD may be used as an anchoring unit to produce fusion proteins that are nondiffusible and long lasting in vivo.

Hayashi et al. [52] developed a recombinant technology to confer a mitogenic activity to type III collagen by fusing it with EGF at the N terminus of collagen. The fusion protein maintained the triple-helical structure of collagen as well as the mitogenic activity of EGF. Furthermore, the chimeric protein could be immobilized in tissue culture dishes in a fibrous form and in collagen fibrils without losing the original mitogenic activity of EGF.

Collagen-binding and fibrin-binding growth factors consisting of EGF, hepatocyte growth factor (HGF), and VEGF and of the binding domains in fibronectin have been reported (Figure 9.18) [53, 96]. The fusion protein bound to gelatin and fibrillar collagen sponges and stimulated substantially cell proliferation after binding to collagen-coated culture plates; in contrast, EGF and HGF alone had no effect. This result indicated that this fusion protein acted as a collagen-associated growth factor. On the other hand, Elloumi et al. [56]

Table 9.2. Gene-engineered growth factors.

Growth factor	Added peptide	Source of peptide motif	References
BMP, FGF, TGF-β	Collagen-binding domain	von Willebrand factor	[71, 82, 91]
EGF, bFGF	Collagen-binding domain	Collagenase	[51, 89]
EGF, HGF	Collagen-binding domain	Fibronectin	[53]
EGF	Collagen	Collagen	[52, 96, 97]
VEGF, NGF, BMP, ephrin-B2	a_2PI_{1-8}	8-Amino acid sequence from a_2 plasmin	[63, 67, 76–78, 116]
EGF	RGD	Adhesion factor	[56]

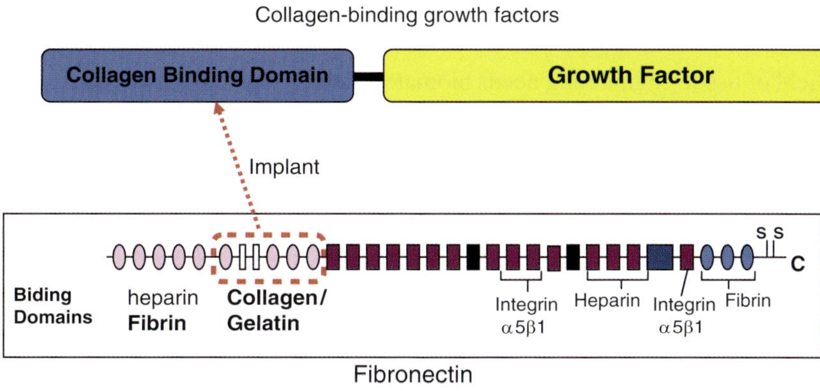

Figure 9.18. Gene-engineered growth factors (e.g., EGF, VEGF, and HGF) containing the collagen-binding domain present in fibronectin. Reprinted with permission from Soft Matter 2008;4:46–56. Copyright 2008 Royal Society of Chemistry.

reported a novel protein consisting of a cell adhesive peptide sequence (Arg–Gly–Asp [RGD]), EGF (a mitogen), and a hydrophobic sequence (an efficient assembler) in a single molecule. This protein was coated on the unmodified hydrophobic surface of a cell culture plate (through the hydrophobic moiety) and retained both cell adhesive activity (through the RGD sequence) and cell growth activity (through the EGF moiety).

Zisch and Hubbell's group engineered fibrin-based hydrogels that were covalently modified with growth factors, including VEGF, NGF, ephrin-B2, and bone morphogenic protein (BMP). They developed a method that allowed the covalent incorporation of exogenous bioactive peptides by the transglutaminase activity of factor XIIIa into fibrin during coagulation. The ability of factor XIIIa to crosslink additional proteins within fibrin was employed to covalently incorporate growth factors.

Recently, Kim et al. [102] formulated a fusion protein of recombinant human interleukin-1 receptor antagonist (IL-1ra) and elastin-like peptide (ELP), and found that immobilized IL-1ra-ELP modulated the inflammatory profile of lipopolysaccharide (LPS)-stimulated human monocytes in vitro. In particular, LPS-stimulated THP-1 monocytes exposed to either soluble or immobilized IL-1ra-ELP did not differentiate; these cells exhibited attenuated expression of pro-inflammatory cytokines but increased expression of anti-inflammatory and wound-healing cytokines. The extent of signaling by either the immobilized or the soluble fusion protein was similar in magnitude, indicating roughly equivalent bioactivity; moreover, signaling in monocytes in vitro is affected by immobilized IL-1ra-ELP.

Other binding methods, such as fusion with the immunoglobulin G Fc region [54], *p*-azido phenylalanine [55], histidine tag [57], cysteine-containing tag [80], and a cellulose-binding domain [100], have been employed. These gene-engineering methods resulted in oriented immobilization of biosignal molecules for efficient interaction with cell receptors to induce signal transduction in cells.

Protein-engineered growth factors can be used to modify the surface of three-dimensional scaffolds. Difficulties encountered with these methods include formation of inclusion bodies and/or denaturation during mass production of the engineered proteins. Further progress in protein engineering is needed before the efficacious use of the pertinent methodologies in tissue engineering is feasible.

9.6. Conclusions

The field of tissue engineering needs biomaterials that provide biofunctional and structural support for living cells outside of the body. Most of the commonly used biomaterials in tissue engineering are designed based on their physicochemical properties, thus providing precise control over mechanical strength, compliance, porosity, and degradation kinetics. Various types of biofunctional signals have been discovered, and new mechanisms of action have been revealed. By tethering, immobilizing, and supplementing biofunctional macromolecules, such as growth factors, directly to a scaffold material, communication between cells on the engineered scaffolds and their extracellular environment can be regulated. For the purpose of surface functionalization, immobilization of the growth factor is a powerful tool for constructing elaborate intelligent biofunctional materials.

References

1. Ito Y. Covalently immobilized biosignal molecule materials for tissue engineering. Soft Matter 2008;4:46–56
2. Wells A, Marti U. Signaling shortcut: cell-surface receptors in the nucleus? Nat Rev Mol Cell Biol 2002;3:697–702
3. Re RN. The intracrine hypothesis and intracellular peptide hormone action. BioEssays 2003;25:401–409
4. Singh AB, Harris RC. Autocrine, paracrine and juxtacrine signaling by EGFR ligands. Cell Signal 2005;17:1183–1193
5. Massague J, Pandiella A. Membrane-anchored growth factors. Annu Rev Biochem 1993;62:515–541
6. Iwamoto R, Mekada E. Heparin-binding EGF-like growth factor: a juxtacrine growth factor. Cytokine Growth Factor Rev 2000;11:335–344
7. Ito Y, Liu SQ, Imanishi Y. Enhancement of cell growth on growth-factor-immobilized polymer film. Biomaterials 1991;12:449–453
8. Liu SQ, Ito Y, Imanishi Y. Cell growth on immobilized cell growth factor. I. acceleration of the growth of fibroblast cells on insulin-immobilized polymer matrix in culture medium without serum. Biomaterials 1992;13:50–58
9. Ito Y, Liu SQ, Nakabayashi M, Imanishi Y. Cell growth on immobilized cell growth factor. II. Fibroblast cell adhesion and growth on the poly(methyl methacrylate) film immobilized with various kinds of proteins. Biomaterials 1992;13:789–794
10. Liu SQ, Ito Y, Imanishi Y. Cell growth on immobilized cell growth factor. 4. Interaction of fibroblast cells with insulin immobilized on poly(methyl methacrylate) membrane. J Biochem Biophys Method 1992;25:139–148
11. Liu SQ, Ito Y, Imanishi Y. Cell growth on immobilized cell growth factor. 7. Protein-free cell culture by using growth factor-immobilized polymer membrane. Enzyme Microb Technol 1993;15:167–172
12. Ito Y, Zheng J, Imanishi Y, Yonezawa K, Kasuga M. Protein-free cell culture on artificial substrate with covalently immobilized insulin. Proc Natl Acad Sci U S A 1996;93:3598–3601
13. Ito Y, Uno T, Liu SQ, Imanishi Y. Cell growth on immobilized cell-growth factor. 8. Protein-free cell culture on insulin-immobilized microcarrie. Biotechnol Bioeng 1992;40:1271–1276
14. Liu SQ, Ito Y, Imanishi Y. Cell growth on immobilized growth factor. 9. Covalent immobilization of cell growth and adhesion factors onto polyurethane surfaces to enhance growth of bovine endothelial cells. J Biomed Mater Res 1993;27:909–915
15. Ito Y, Liu SQ, Orihara T, Imanishi Y. Cell growth on immobilized cell growth factor. Interactions of fibroblast cells with insulin immobilized on 2-hydroxyethyl methacrylate/ethyl methacrylate copolymer membranes. J Bioact Compat Polym 1994;9:170–183
16. Kang IK, Choi SH, Shin DS, Yoon SC. Surface modification of polyhydroxyalkanoate films and their interaction with human fibroblasts. Int J Biol Macromol 2001;28:205–212
17. Ito Y, Inoue M, Liu SQ, Imanishi Y. Cell growth on immobilized cell-growth factor. 6. Enhanced cell growth on cell growth factor and adhesion factor co-immobilized materials. J Biomed Mater Res 1993;27:901–907
18. Chen G, Ito Y, Imanishi Y. Mitogenic activities of water-soluble and water-insoluble insulin conjugates. Bioconjug Chem 1997;8:106–110
19. Bromberg L, Salvati L Jr. Bioactive surfaces via immobilization of self-assembling polymers onto hydrophobic materials. Bioconjug Chem 1999;10:678–686

20. Kim EJ, Kang IK, Jang MK. Park YB. Preparation of insulin-immobilized polyurethanes and their interaction with human fibroblasts. Biomaterials 1998;19:239–249
21. Tessmar J, Mikos A, Gopferich A. The use of poly(ethylene glycol)-block-poly(lactic acid) derived copolymers for the rapid creation of biomimetic surfaces. Biomaterials 2003;24:4475–4486
22. Kellner K, Tessmar J, Milz S, Angele P, Nerlich M, Schulz MB, Blunk T, Gopferich A. PEGylation does not impair insulin efficacy in three-dimensional cartilage culture: an investigation toward biomimetic polymers. Tissue Eng 2004;10:429–440
23. Tessmar J, Kellner K, Schulz MB, Blunk T, Gopferich A. Toward the development of biomimetic polymers by protein immobilization: PEGylation of insulin as a model reaction. Tissue Eng 2004;10:441–453
24. Zheng J, Ito Y, Imanishi Y. Cell growth on immobilized cell-growth factor. 10. Insulin and polyallylamine co-immobilized materials. Biomaterials 1994;15:963–968
25. Ito Y, Zheng J, Imanishi Y. Serum-free cell culture on insulin-immobilized porous collagen beads. Biotechnol Bioeng 1995;45:144–148
26. Zheng J, Ito Y, Imanishi Y. Cell growth on insulin/RGDS-coimmobilized poly(methyl methacrylate) films. J Biomater Sci Polym Ed 1995;7:515–522
27. Zheng J, Ito Y, Imanishi Y. Growth enhancement of anchorage-dependent and anchorage-independent cells by coimmobilization of insulin with poly(allylamine) or gelatin. Biotechnol Prog 1995;11:677–681
28. Li JS, Ito Y, Zheng J, Takahashi T, Imanishi Y. Enhancement of artificial juxtacrine stimulation of insulin by co-immobilization with adhesion factors. J Biomed Mater Res 1997;31:190–197
29. Ito Y, Zheng J, Imanishi Y. Enhancement of cell growth on a porous membrane co-immobilized with cell-growth and cell adhesion factors. Biomaterials 1997;18:197–202
30. Gumusdrelioglu M, Turkoglu H. Biomodification of non-woven polyester fabrics by insulin and RGD for use in serum-free cultivation of tissue cells. Biomaterials 2002;23:3927–3935
31. Gumusdrelioglu M, Karakecili AG. Uses of thermoresponsive and RGD/insulin-modified poly(vinyl ether)-based hydrogels in cell cultures. J Biomater Sci Polym Ed 2003;14:199–211
32. Kim YJ, Kang IK, Huh MW, Yoon SC. Surface characterization and in vitro blood compatibility of poly(ethylene terephthalate) immobilized with insulin and/or heparin using plasma glow discharge. Biomaterials 2000;21:121–130
33. Ito Y, Chen G, Imanishi Y. Photoimmobilization of insulin onto polystyrene dishes for protein-free cell culture. Biotechnol Prog 1996;12:700–703
34. Ito Y, Kondo S, Chen G, Imanishi Y. Patterned artificial juxtacrine stimulation of cells by covalently immobilized insulin. FEBS Lett 1997;403:159–162
35. Chen G, Ito Y, Imanishi Y. Regulation of growth and adhesion of cultured cells by insulin conjugated with thermoresponsive polymers. Biotechnol Bioeng 1997;53:339–344
36. Hatakeyama H, Kikuchi A, Yamato M, Okano T. Bio-functionalized thermoresponsive interfaces facilitating cell adhesion and proliferation. Biomaterials 2006;27:5069–5078
37. Hatakeyama H, Kikuchi A, Yamato M, Okano T. Influence of insulin immobilization to thermoresponsive culture surfaces on cell proliferation and thermally induced cell detachment. Biomaterials 2005;26:5167–5176
38. Hatakeyama H, Kikuchi A, Yamato M, Okano T. Patterned biofunctional designs of thermoresponsive surfaces for spatiotemporally controlled cell adhesion, growth, and thermally induced detachment. Biomaterials 2007;28: 3632–3643
39. Kuhl PR, Griffith-Cima LG. Tethered epidermal growth factor as a paradigm for growth factor-induced stimulation from the solid phase. Nat Med 1996;2:1022–1027
40. Ichinose J, Morimatsu M, Yanagida T, Sako Y. Covalent immobilization of epidermal growth factor molecules for single-molecule imaging analysis of intracellular signaling. Biomaterials 2006;27:3343–3350
41. Nakajima M, Ishimuro T, Kato K, Ko IK, Hirata I, Arima Y, Iwata H. Combinatorial protein display for the cell-based screening of biomaterials that direct neural stem cell differentiation. Biomaterials 2007;28:1048–1060
42. Ito Y, Li JS, Takahashi T, Imanishi Y, Okabayashi Y, Kido Y, Kasuga M. Enhancement of the mitogenic effect by artificial juxtacrine stimulation using immobilized EGF. J Biochem 1997;121:514–520
43. Klenkler BJ, Griffith M, Becerril C, West-Mays JA, Sheardown H. EGF-grafted PDMS surfaces in artificial cornea applications. Biomaterials 2005;26:7286–7296
44. Klenkler BJ, Sheardown H. Characterization of EGF coupling to aminated silicone rubber surfaces. Biotechnol Bioeng 2006;95:1158–1166
45. Chen G, Ito Y, Imanishi Y. Photo-immobilization of epidermal growth factor enhances its mitogenic effect by artificial juxtacrine signaling. Biochim Biophys Acta 1997;1358:200–208
46. Ito Y, Chen G, Imanishi Y. Micropatterned immobilization of epidermal growth factor to regulate cell function. Bioconjug Chem 1998;9:277–282

47. Chen G, Ito Y, Masuda S, Sasaki R. Growth and secretion of erythropoietin of Chinese hamster ovary cells coexpressing epidermal growth factor receptor and erythropoietin genes: design of cells for cell culture matrix. Cytotechnology 2001;35:3–8
48. Chen G, Ito Y. Gradient micropattern immobilization of EGF to investigate the effect of artificial juxtacrine stimulation. Biomaterials 2001;22:2453–2457
49. Ito Y, Chen G, Imanishi Y, Morooka T, Nishida E, Okabayashi Y, Kasuga M. Differential control of cellular gene expression by diffusible and non-diffusible EGF. J Biochem 2001;129:733–737
50. Woo YK, Kwon SY, Lee HS, Park YS. Proliferation of anterior cruciate ligament cells in vitro by photo-immobilized epidermal growth factor. J Orthop Res 2007;25:73–80
51. Nishi N, Matsushita O, Yuube K, Miyanaka H, Okabe A, Wada F. Collagen-binding growth factors: production and characterization of functional fusion proteins having a collagen-binding domain. Proc Natl Acad Sci U S A 1998;95:7018–7023
52. Hayashi M, Tomita M, Yoshizato K. Production of EGF-collagen chimeric protein which shows the mitogenic activity. Biochim Biophys Acta 2001;1528:187–195
53. Ishikawa T, Terai H, Kitajima T. Production of a biologically active epidermal growth factor fusion protein with high collagen affinity. J Biochem 2001;129:627–633
54. Ogiwara K, Nagaoka M, Cho CS, Akaike T. Construction of a novel extracellular matrix using a new genetically engineered epidermal growth factor fused to IgG-Fc. Biotechnol Lett 2005;27:1633–1637
55. Ogiwara K, Nagaoka M, Cho CS, Akaike T. Effect of photo-immobilization of epidermal growth factor on the cellular behaviors. Biochem Biophys Res Commun 2006;345:255–259
56. Elloumi I, Kobayashi R, Funabashi H, Mie M, Kobatake E. Construction of epidermal growth factor fusion protein with cell adhesive activity. Biomaterials 2006;27:3451–3458
57. Nakaji-Hirabayashi T, Kato K, Arima Y, Iwata H. Oriented immobilization of epidermal growth factor onto culture substrates for the selective expansion of neural stem cells. Biomaterials 2007;28:3517–3529
58. Shibata SC, Hibino K, Mashimo T, Yanagida T, Sako Y. Formation of signal transduction complexes during immobile phase of NGFR movements. Biochem Biophys Res Commun 2006;342:316–322
59. Kapur TA, Schoichet MS. Chemically-bound nerve growth factor for neural tissue engineering applications. J Biomater Sci Polym Ed 2003;14:383–394
60. Kapur TA, Schoichet MS. Immobilized concentration gradients of nerve growth factor guide neurite outgrowth. J Biomed Mater Res 2004;68A:235–243
61. Chen PR, Chen MH, Lin FH, Su WY. Release characteristics and bioactivity of gelatin-tricalcium phosphate membranes covalently immobilized with nerve growth factors. Biomaterials 2005;26:6579–6587
62. Ito Y. Regulation of cellular gene expression by artificial biomaterial immobilized with biosignal molecules. Jpn J Artif Organs 1998;27:383–394
63. Sakiyama-Elbert S, Panitch A, Hubbell JA. Development of growth factor fusion proteins for cell-triggered drug delivery. FASEB J 2001;15:1300–1302
64. Gomez N, Lu Y, Chen S, Schmidt CE. Immobilized nerve growth factor and microtopography have distinct effects on polarization versus axon elongation in hippocampal cells in culture. Biomaterials 2007;28:271–284
65. Puleo DA, Kissling RA, Sheu MS. A technique to immobilize bioactive proteins, including bone morphogenetic protein-4 (BMP-4), on titanium alloy. Biomaterials 2002;23:2079–2087
66. Karageorgiou V, Meinel L, Hofmann S, Malhotra A, Volloch V, Kaplan D. Bone morphogenetic protein-2 decorated silk fibroin films induce osteogenic differentiation of human bone marrow stromal cells. J Biomed Mater Res 2004;71A:528–537
67. Schmoekel HG, Weber FE, Schense JC, Graetz KW, Schwalder P, Hubbell JA. Bone repair with a form of BMP-2 engineered for incorporation into fibrin cell ingrowth matrices. Biotechnol Bioeng 2005;89:253–262
68. Schliphake H, Aref A, Scharnweber D, Bierbaum S, Roessler S, Sewing A. Effect of immobilized bone morphogenic protein 2 coating of titanium implants on peri-implant bone formation. Clin Oral Implants Res 2005;16:563–569
69. Tsujigiwa H, Nagatsuka H, Gunduz M, Rodriguez A, Rivera AJ, Legeros RZ, Inoue M, Nagai N. Effects of immobilized recombinant human bone morphogenetic protein-2/succinylated type I atelocollagen on cellular activity of ST2 cells. J Biomed Mater Res 2005;75:210–215
70. Tsujigiwa H, Nagatsuka H, Lee YJ, Han PP, Gunduz M, LeGeros RZ, Inoue M, Yamada M, Nagai N. Immobilized rhBMP-2/succinylated type I atelocollagen gene expression of intracellular signaling molecules on ST2 cells. J Biomed Mater Res 2006;77A:507–511
71. Han B, Perelman N, Tang B, Hall F, Shors EC, Nimni ME. Collagen-targeted BMP3 fusion proteins arrayed on collagen matrices or porous ceramics impregnated with Type I collagen enhance osteogenesis in a rat cranial defect model. J Orthop Res 2006;20:747–755

72. Park YJ, Kim KH, Lee JY, Ku Y, Lee SJ, Min BM, Chung CP. Immobilization of bone morphogenetic protein-2 on a nanofibrous chitosan membrane for enhanced guided bone regeneration. Biotechnol Appl Biochem 2006;43:17–24
73. Liu HW, Chen CH, Tsai CL, Lin IH, Hsiue GH. Heterobifunctional poly(ethylene glycol)-tethered bone morphogenetic protein-2-stimulated bone marrow mesenchymal stromal cell differentiation and osteogenesis. Tissue Eng 2007;13:1113–1124
74. Edlund U, Danmark S, Albertsson AC. A strategy for the covalent functionalization of resorbable polymers with heparin and osteoinductive growth factor. Biomacromolecules 2008;9:901–905
75. Taguchi T, Kishida A, Akashi M, Maruyama I. Immobilization of human vascular endothelial growth factor (VEGF165) onto biomaterials: an evaluation of the biological activity of VEGF165. J Bioact Compat Polym 2000;15:309–320
76. Zisch AH, Schenk U, Schense JC, Sakiyama-Elbert SE, Hubbell JA. Covalently conjugated VEGF--fibrin matrices for endothelialization. J Control Release 2001;72:101–113
77. Ehrbar M, Djonov VG, Schnell C, Tschanz SA, Martiny-Barton G, Schenk U, Wood J, Burri PH, Hubbell JA, Zisch AH. Cell-demanded liberation of VEGF121 from fibrin implants induces local and controlled blood vessel growth. Circ Res 2004;94:1124–1132
78. Ehrbar M, Matters A, Zammaretti P, Hubbell JA, Zisch AH. Endothelial cell proliferation and progenitor maturation by fibrin-bound VEGF variants with differential susceptibilities to local cellular activity. J Control Release 2005;101:93–109
79. Ito Y, Hasuda H, Terai H, Kitajima T. Culture of human umbilical vein endothelial cells on immobilized vascular endothelial growth factor. J Biomed Mater Res 2005;74:659–665
80. Backer MV, Patel V, Jehning BT, Claffey KP, Backer JM. Surface immobilization of active vascular endothelial growth factor via a cysteine-containing tag. Biomaterials 2006;27:5452–5458
81. Ehrbar M, Zeisberger SM, Raeber GP, Hubbell JA, Schnell C, Zisch AH. The role of actively released fibrin-conjugated VEGF for VEGF receptor 2 gene activation and the enhancement of angiogenesis. Biomaterials 2008;29:1720–1729
82. Andrades JA, Santamaria JA, Wu LT, Hall FL, Nimni ME, Becerra J. Production of a recombinant human basic fibroblast growth factor with a collagen binding domain. Protoplasma 2001;218:95–103
83. Ito Y, Hayashi M, Imanishi Y. Gradient micropattern immobilization of heparin and its interaction with cells. J Biomater Sci Polym Ed 2001;12:367–378
84. Watanabe K, Miyazaki T, Matsuda R. Growth factor array fabrication using a color ink jet printer. Zoolog Sci 2003:20;429–434
85. DeLong SA, Moon JJ, West JL. Covalently immobilized gradients of bFGF on hydrogel scaffolds for directed cell migration. Biomaterials 2005;26:3227–3234
86. Inoue S, Tabata Y. Influence of basic fibroblast growth factor in the solution and adsorbed form on the proliferation and differentiation of cells. Inflamm Regen 2006;26:181–184
87. Ohyama T, Nishide T, Iwata H, Sato H, Toda M, Toma N, Taki W. Immobilization of basic fibroblast growth factor on a platinum microcoil to enhance tissue organization in intracranial aneurysms. J Neurosurg 2005;102:109–115
88. Mori H, Shukunami C, Furuyama A, Notsu H, Nishizaki Y, Hiraki Y. Immobilization of bioactive fibroblast growth factor-2 into cubic proteinous microcrystals (*Bombyx mori* cypovirus polyhedra) that are insoluble in a physiological cellular environment. J Biol Chem 2007;282:17289–17296
89. Brewster LP, Washington C, Brey EM, Gassman A, Subramanian A, Calceterra J, Wolf W, Hall CL, Velander WH, Burgess WH, Greisler HP. Construction and characterization of a thrombin-resistant designer FGF-based collagen binding domain angiogen. Biomaterials 2008;29:327–336
90. Shen H, Hu X, Bei J, Wang S. The immobilization of basic fibroblast growth factor on plasma-treated poly(lactide-co-glycolide). Biomaterials 2008;29:2388–2399
91. Andrades JA, Han B, Becerra J, Sorgente N, Hall FL, Nimni ME. A recombinant human TGF-beta1 fusion protein with collagen-binding domain promotes migration, growth, and differentiation of bone marrow mesenchymal cells. Exp Cell Res 1999;250:485–498
92. Fischer U, Hempel U, Becker D, Bierbaum S, Scharnweber D, Worch H, Wenzel KW. Transforming growth factor beta1 immobilized adsorptively on Ti6Al4V and collagen type I coated Ti6Al4V maintains its biological activity. Biomaterials 2003;24:2631–2641
93. Merrett K, Griffith CM, Deslandes Y, Pleizier G, Sheardown H. Interactions of corneal epithelial cells with transforming growth factor b2 modified poly dimethyl siloxane surface. J Biomed Mater Res 2003;67A:981–993
94. Chou CH, Cheng WTK, Lin CC, Chang CH, Tsai CC, Lin FH. TGF-β1 immobilized tri-co-polymer for articular cartilage tissue engineering. J Biomed Mater Res 2006;77B:338–348

95. Mann BK, Schmedlen RH, West JL. Tethered-TGF-beta increases extracellular matrix production of vascular smooth muscle cells. Biomaterials 2001;22:439–444
96. Kitajima T, Terai H, Ito Y. A fusion protein of hepatocyte growth factor for immobilization to collagen. Biomaterials 2007;28:1989–1997
97. Ohkawara N, Ueda H, Shinozaki S, Kitajima T, Ito Y, Asaoka H, Kawakami A, Kaneko E, Shimokado K. Hepatocyte growth factor fusion protein having collagen-binding activity (CBD-HGF) accelerates re-endothelialization and intimal hyperplasia in balloon-injured rat carotid artery. J Atheroscler Thromb 2007;14:185–191
98. Makino H, Hasuda H, Ito Y. Immobilization of leukemia inhibitory factor (LIF) to culture murine embryonic stem cells. J Biosci Bioeng 2004;98:374–379
99. Cetinkaya G, Torkoglu H, Arat S, Odaman H, Onur MA, Gumusderelioglu M, Tumer A. LIF-immobilized nonwoven polyester fabrics for cultivation of murine embryonic stem cells. J Biomed Mater Res 2007;81A:911–919
100. Doheny JG, Jervis EJ, Guarna MM, Humphries RK, Warren RA, Kilburn DG. Cellulose as an inert matrix for presenting cytokines to target cells: production and properties of a stem cell factor-cellulose-binding domain fusion protein. Biochem J 1999;339:429–434
101. Horwitz JI, Toner M, Tompkins RG, Yarmuch ML. Immobilized IL-2 preserves the viability of an IL-2 dependent cell line. Mol Immunol 1993;30:1041–1048
102. Kim DH, Smith JT, Chilkoti A, Reichert WM. The effect of covalently immobilized rhIL-1ra-ELP fusion protein on the inflammatory profile of LPS-stimulated human monocytes. Biomaterials 2007;28:3369–3377
103. Guan Y, Zhong H, Wang X, Zhou T. Synthetic and photo-immobilization of photo-active tumor necrosis factor-alpha. Sheng Wu Yi Xue Gong Cheng Xue Za Zhi 2006;23:346–349
104. Ito Y, Hasuda H, Yamauchi T, Komatsu N, Ikebuchi K. Immobilization of erythropoietin to culture erythropoietin-dependent human leukemia cell line. Biomaterials 2004;25:2293–2298
105. Liu SQ, Ito Y, Imanishi Y. Cell growth on immobilized cell growth factor: 5. Interaction of immobilized transferrin with fibroblast cells. Int J Biol Macromol 1993;15:221–226
106. Nagaoka M, Koshimizu U, Yuasa S, Hattori F, Chen H, Tanaka T, Okabe M, Fukuda K, Akaike T. E-cadherin-coated plates maintain pluripotent ES cells without colony formation. PLoS ONE 2006;e15:1–7
107. Martin SM, Ganapathy R, Kim TK, Leach-Scampavia D, Giachelli CM, Ratner BD. Characterization and analysis of osteopontin-immobilized poly(2-hydroxyethyl methacrylate) surfaces. J Biomed Mater Res 2003;67A:334–343
108. Martin SM, Schwartz JL, Giachelli CM, Ratner BD. Enhancing the biological activity of immobilized osteopontin using a type-1 collagen affinity coating. J Biomed Mater Res 2004;70A:10–19
109. Koike J, Nagata K, Kudo S, Tsuji T, Irimura T. Density-dependent induction of TNF-alpha release from human monocytes by immobilized P-selectin. FEBS Lett 2000;477:84–88
110. Kohrgruber N, Groger M, Meraner P, Kriehuber E, Petzelbauer P, Brandt S, Stingl G, Rot A, Maurer D. Plasmacytoid dendritic cell recruitment by immobilized CXCR3 ligands. J Immunol 2004;173:6592–6602
111. Konno T, Sakano S, Higashida S, Ito Y. Notch ligand-conjugated polymeric matrix. Inflamm Regen 2005;25:426–430
112. Varnum-Finney B, Wu L, Yu M, Brashem-Stain C, Staats S, Flowers D, Griffin JD, Bernstain ID. Immobilization of Notch ligand, Delta-1, is required for induction of notch signaling. J Cell Sci 2000;113:4313–4318
113. Vas V, Szilágyi L, Pálóczi K, Uher F. Soluble Jagged-1 is able to inhibit the function of its multivalent form to induce hematopoietic stem cell self-renewal in a surrogate in vitro assay. J Leukoc Biol 2004;75:714–720
114. Beckstead BL, Santosa DM, Giachelli CM. Mimicking cell-cell interactions at the biomaterial-cell interface for control of stem cell differentiation. J Biomed Mater Res 2006;79A:94–103
115. Ho JE, Chung EH, Wall S, Schaffer DV, Healy KE. Immobilized sonic hedgehog N-terminal signaling domain enhances differentiation of bone marrow-derived mesenchymal stem cells. J Biomed Mater Res 2007;83A:1200–1208
116. Zisch AH, Zeisberger SM, Ehrbar M, Djonov V, Weber CC, Ziemiecki A, Pasquale EB, Hubbell JA. Engineered fibrin matrices for functional display of cell membrane-bound growth factor-like activities: study of angiogenic signaling by ephrin-B2. Biomaterials 2004;25:3245–3257
117. Sharon JL, Puleo DA. The use of N-terminal immobilization of PTH(1–34) on PLGA to enhance bioactivity. Biomaterials 2008;29:3137–3142
118. Ito Y, Nogawa M. Preparation of a protein micro-array using a photo-reactive polymer for a cell-adhesion assay. Biomaterials 2003;24:3021–3026
119. Ito Y. Photoimmobilization for microarrays. Biotechnol Prog 2006;22:924–932
120. Kiessling LL, Gestwicki JE, Strong LE. Synthetic multivalent ligands as probes of signal transduction. Angew Chem Int Ed 2006;45:2348–2368

121. Thomas TP, Shukla R, Kotlyar A, Liang B, Ye JY, Norris TB, Baker JR Jr. Dendrimer-epidermal growth factor conjugate displays superagonist activity. Biomacromolecules 2008;9:603–609
122. Reddy CC, Niyogi SK, Wells A, Wiley HS, Lauffenburger DA. Engineering epidermal growth factor for enhanced mitogenic potency. Nat Biotechnol 1996;14:1696–1699
123. Rusmini F, Zhong Z, Feijen J. Protein immobilization strategies for protein biochips. Biomacromolecules 2007;8:1775–1789
124. Ito Y. Microarray Biochips (Japanese), Ito Y. ed. CMC Press, Tokyo, pp. 91–109, 2007
125. Schwarz U. Soft matters in cell adhesion: rigidity sensing on soft elastic substrates. Soft Matter 2007;3:263–266
126. Levental I, Georges PC, Janmey PA. Fibroblast adaptation and stiffness matching to soft elastic substrates. Soft Matter 2007;3:299–306
127. Ito, Y. Surface micropatterning to regulate cell functions. Biomaterials 1999;20:2333–2342
128. Girard PP, Cavalcanti-Adam EA, Kemkemer R, Spatz JP. Cells and proteins at interfaces. Soft Matter 2007;3:307–326
129. Falconnet D, Csucs G, Grandin HM, Textor M. Surface engineering approaches to micropattern surfaces for cell-based assays. Biomaterials 2006;27:3044–3063
130. Khademhosseini A, Langer R, Borenstein J, Vacanti JP. Microscale technologies for tissue engineering and biology. Proc Natl Acad Sci U S A 2006;103:2480–2487
131. Liu WF, Chen CS. Engineering biomaterials to control cell function. Mater Today 2005;8:28–35
132. Chen CS, Jiang X, Whitesides GM. Microengineering: the environment of mammalian cells in culture. MRS Bull 2005;30:194–201
133. Xu T, Jin J, Gregory C, Hickman JJ, Boland T. Inkjet printing of viable mammalian cells. Biomaterials 2005;26:93–99
134. Curtis ASG. Cell reactions with biomaterials. Eur Cells Mater 2004;8:27–36
135. Park TH, Shuler ML. Integration of cell culture and microfabrication technology. Biotechnol Prog 2003;19:243–253
136. Jung DR, Kapur R, Adams T, Giuliano KA, Mrksich M, Craighead HG. Taylor DL. Topographical and physicochemical modification of material surface to enable patterning of living cells. Crit Rev Biotechnol 2001;21:111–154
137. Folch A, Toner M. Microengineering of cellular interactions. Annu Rev Biomed Eng 2000;2:227–256
138. Chen CS, Mrksich M, Huang S, Whitesides GM, Ingber DE. Geometric control of cell life and death. Science 1997;276:1425–1428
139. Thomas CH, Collier JH, Sfeir CS, Healy KE. Engineering gene expression and protein synthesis by modulation of nuclear shape. Proc Natl Acad Sci U S A 2002;99:1972–1977
140. McBeath R, Pirone DM, Nelson CM, Bhadriraju K, Chen CS. Cell shape, cytoskeletal tension, and RhoA regulate stem cell lineage commitment. Dev Biol 2004;6:483–495
141. Nelson CM, Jean RP, Tan JL, Liu WF, Sniadecki NJ, Spector AA, Chen CS. Emergent patterns of growth controlled by multicellular form and mechanics. Proc Natl Acad Sci U S A 2005;102:11594–11599
142. Bhatia SN, Yarmush UML, Toner M. Controlling cell interactions by micropatterning in co-cultures: hepatocytes and 3T3 fibroblasts. J Biomed Mater Res 1997;34:189–199
143. Yamato M, Konno C, Utsumi M, Kikuchi A, Okano T. Thermally responsive polymer-grafted surfaces facilitate patterned cell seeding and co-culture. Biomaterials 2002;23:561–567
144. Khademhosseini A, Suh KY, Yang JM, Eng G, Yeh J, Levenberg S, Langer R. Layer-by-layer deposition of hyaluronic acid and poly-L-lysine for patterned cell co-cultures. Biomaterials 2004;25:3583–3589
145. Chiu DT, Jeon NL, Huang S, Kane RS, Wargo CJ, Choi IS, Ingber DE, Whitesides GM. Patterned deposition of cells and proteins onto surfaces by using three-dimensional microfluidic systems. Proc Natl Acad Sci U S A 2000;97:2408–2413
146. Tang MD, Golden AP, Tien J. Molding of three-dimensional microstructures of gels. J Am Chem Soc 2003;125:12988–12989
147. Ito Y, Chen G, Guan Y, Imanishi Y. Patterned immobilization. of thermoresponsive polymer. Langmuir 1997;13:2756–2759
148. Nakanishi J, Kikuchi Y, Takarada T, Nakayama H, Yamaguchi K, Maeda M. Photoactivation of a substrate for cell adhesion under standard fluorescence microscopes. J Am Chem Soc 2004;126:16314–16315
149. Nakanishi J, Kikuchi Y, Inoue S, Yamaguchi K, Takarada T, Maeda M. Spatiotemporal control of migration of single cells on a photoactivatable cell microarray. J Am Chem Soc 2007;129:6694–6695
150. Edahiro J, Sumaru K, Tada Y, Ohi K, Takagi T, Kameda M, Shinbo T, Kanamori T, Yoshimi Y. In situ control of cell adhesion using photoresponsive culture surface. Biomacromolecules 2005;6:970–974

10

Cell and Tissue Interactions with Materials: The Role of Growth Factors

Christopher C. Gibson, David A. Puleo, and Rena Bizios

Realization and appreciation that growth factors are essential in tissue development, regeneration, and maintenance as well as crucial mediators/regulators of pertinent cellular- and molecular-level events has motivated new directions in bioengineering research. These endeavors have the potential of seminal contributions in implant biomaterials, tissue engineering, and regenerative medicine.

This chapter summarizes the current knowledge of the role of growth factors in the biology/physiology of two tissues, specifically, vascular tissue and bone. It also provides an overview of research endeavors directed at transferring and applying pertinent growth factor knowledge to biomaterials; delivery of the appropriate type, concentration, and timely sequence of bioactive growth factors, which promote interactions and functions of cells pertinent to neotissue formation, will enhance the long-term performance of implants, including tissue-engineering constructs. Future directions for research and development in these areas are outlined.

Abbreviations

ACS	absorbable collagen sponge
aFGF	acidic fibroblast growth factor (equivalent to FGF-1)
ALK1	activin receptor-like kinase 1
AMD	age-related macular degeneration
AP	alkaline phosphatase
bFGF	basic fibroblast growth factor (equivalent to FGF-2)
BMC	bone marrow cell
BMP	bone morphogenetic protein
ECM	extracellular matrix
FDA	US Food and Drug Administration
FGF	fibroblast growth factor
FGFR1	fibroblast growth factor receptor 1

C.C. Gibson and R. Bizios • Department of Biomedical Engineering, The University of Texas at San Antonio, San Antonio, TX 78249, USA
D.A. Puleo • Center for Biomedical Engineering, University of Kentucky, Lexington, KY 40506, USA

D.A. Puleo and R. Bizios (eds.), *Biological Interactions on Materials Surfaces*,
DOI 10.1007/978-0-387-98161-1_10, © Springer Science+Business Media, LLC 2009

FGFR2	fibroblast growth factor receptor 2
FGF-1	fibroblast growth factor-1 (equivalent to aFGF)
FGF-2	fibroblast growth factor-2 (equivalent to bFGF)
HSC	hematopoietic stem cell(s)
hBMSC	human bone-marrow stromal-cell
HA	hydroxyapatite
HIF-1α	hypoxia-inducible factor-1α
HUVEC	human umbilical vein endothelial cell
IGF	insulin-like growth factor
IGFBP	insulin-like growth factor binding protein
MMP	matrix metalloproteinase
mRNA	messenger ribonucleic acid
Flt-1	fms-related tyrosine kinase-1 (equivalent to VEGFR-1)
Flk-1	kinase insert domain receptor (equivalent to VEGFR-2)
NO	nitric oxide
PDGF	platelet-derived growth factor
PDGFR	platelet-derived growth factor receptor
PDGFR-α	platelet-derived growth factor receptor-α
PDGFR-β	platelet-derived growth factor receptor-β
PDLLA	poly (D,L-lactide)
PLA	poly(lactic acid)
PLGA	poly(lactic-co-glycolic acid)
PEG	polyethylene glycol
rhBMP-2	recombinant human bone morphogenetic protein-2
SCID	severe combined immunodeficiency
SS-PEG-SS	disuccinimidyldiscuccinatepolyethyleneglycol
TGF	transforming growth factor
TGF-β	transforming growth factor-β
TGFβRI	transforming growth factor-β receptor-I
TGFβRII	transforming growth factor-β receptor-II
VEGF	vascular endothelial growth factor
VEGFR	vascular endothelial growth factor receptor

10.1. Introduction

Development and homeostasis of tissues and organs depend on a fascinating, but complex, interplay and function of their constituent cells with their microenvironment. To date, details in the underlying mechanisms of how these cells are directed to specific locations and function in modes characteristic to particular tissues and organs remain unknown but are areas of intensive and promising research.

Advances in developmental, molecular and cell biology, biochemistry, and medicine have identified various chemical compounds such as growth factors, cytokines, and chemokines that are produced by cells; these agents act in autocrine and paracrine fashions in their respective microenvironments, provide essential cues, and regulate crucial events at the cellular molecular levels that are necessary for tissue development, repair, and regeneration. Such contributions are elucidating fundamental aspects of biology and physiology in both health and disease; research in relevant fields is continuing and breakthrough developments are forthcoming.

At the same time, biomedical engineers are seeking to incorporate pertinent discoveries of the aforementioned basic sciences into applications, some of which have focused on mimicking natural processes and aspire to achieve tissue healing, repair, and regeneration. In the case of tissue engineering, for example, these endeavors initially brought together relevant cells and biomaterial matrices. The necessity of growth factors, however, has been recognized and is now considered an integral part of the "triad" of what are unequivocally considered fundamental components/constituents of tissue engineering. Undoubtedly, efficacious use of growth factors in tissue engineering requires complete knowledge of the role of growth factors pertinent to each tissue during development and healing.

Since discussion of all mammalian tissues and all pertinent growth factors is not possible in one single manuscript, in the present chapter we chose to focus on two respective representatives of soft and hard tissues: vascular tissue and bone. In addition to reviewing current knowledge of the role of growth factors in the development, healing, and repair of these two tissues, this chapter includes examples of applications of such information in the design and formulation of the next generations of biomaterials that make available growth factors in order to enhance the long-term performance of implants including tissue-engineering constructs. In addition, cutting-edge research in these fields is highlighted, and future directions for research and development in these areas are outlined.

10.2. Growth Factors in Vascular Network Formation and Repair

Expansion of the vascular system during embryological development, tissue growth, and wound healing in adult mammals is guided in part by the actions of specific growth factors. Growth factors act on endothelial cells and their progenitors, as well as on vascular smooth muscle cells and supportive cells of the stroma, and play the key role in guiding the processes required for new blood vessel formation and growth. Understanding the biological effects of these growth factors as a function of their physiological concentrations, bioactivity, timing of release, and synergism in health and disease is needed in order to advance therapeutic use of growth factors in conjunction with biomaterial implants and engineered constructs for the development of new vasculature in tissues and organs.

10.2.1. Vasculogenesis and Angiogenesis

Formation and expansion of the peripheral vasculature involves two processes: vasculogenesis (the development of new vascular plexus from endothelial progenitors, the angioblast), and angiogenesis (the development of new vessels by budding, sprouting, and anastomoses of existing vessels). Embryological vascular development consists of both vasculogenesis and angiogenesis, while postnatal tissue growth and wound repair primarily consists of angiogenesis.

In humans, "vasculogenesis," the formation of the vascular system, begins during the third week of development. The first step in vasculogenesis is the formation of blood islands in the extraembryonic mesoderm by aggregating angioblasts, which differentiate into endothelial-like cells [1, 2]. Mouse models have demonstrated that proliferation and flattening of the endothelial-like cells (which make up the blood islands) result in the formation of a lumen [3]. As embryo development progresses, the primitive vascular network matures by further differentiation of the endothelial cells. Primitive arteries and veins are formed in the aforementioned fashion, while capillaries are formed by angiogenesis.

Angiogenesis is responsible for the formation of capillary networks that permeate tissues and organs and provide the functional surface for transport phenomena, specifically, supply of oxygen and nutrients and removal of metabolic waste. Angiogenesis is also responsible for the rebuilding or remodeling of capillary networks in adults during wound healing. Angiogenesis is characterized by an endothelial tip cell, which emerges from an established vessel and guides capillary growth [4]. The tip cell migrates without proliferating until a connection is made with another blood vessel [4]. Following the tip cell are one to two supporting stalk cells, which may proliferate as the blood vessel extends [4]. After anastomosis, the blood vessel forms a lumen by cord hollowing and intracellular vacuole formation and fusion [5]. Maturation of the blood vessel requires subsequent processes including recruitment of mural cells and guided apoptosis of redundant vessels [6].

Different combinations of various cell populations (specifically, endothelial cells, smooth muscle cells, fibroblasts, and pericytes) are involved in blood vessel formation and characterize the types (i.e., capillaries, arteries, and veins) of blood vessels formed. Consequently, the composition of the extracellular matrix of the various blood vessels differs. Remodeling of the vascular network occurs in response to pertinent physiological demands, which include biochemical and biomechanical stimuli [7].

10.2.2. Vascular Endothelial Growth Factor (VEGF)

VEGF, the most widely studied stimulator of endothelial cell migration, proliferation, and differentiation, plays a role in nearly every phase of vascular tissue development and maintenance. The VEGF family (a subset of the platelet-derived growth factor [PDGF] superfamily) includes VEGF-A, VEGF-B, VEGF-C, VEGF-D, and placental growth factor. VEGF-A is the primary regulator of vasculogenesis and angiogenesis. There are three primary isoforms of VEGF-A in humans: the 121-, 165-, and 189-amino acid forms. VEGF-A165, the most common subtype, mediates most of the physiological effects of VEGF. VEGF-A121, the shortest subtype, diffuses into serum from its site of release without attachment to nearby extracellular matrix (ECM) proteins. VEGF-A165 has mild affinity for neuropilin-1 and heparin, but VEGF-A189 has significant binding affinity for heparin and is bound to ECM when it is released [8]. Binding of VEGF-A165 to the extracellular matrix results in VEGF gradients in the tissue surrounding the release site of the growth factor [8].

The effect of VEGF on cells is conveyed by its interaction with pertinent receptors on the cell membrane. There are two important VEGF-A receptors (VEGFR) expressed on endothelial cells during angiogenesis: VEGFR-1 (Flt-1) and VEGFR-2 (Flk-1). Compared with VEGFR2, VEGFR1 has an order of magnitude greater affinity for VEGF-A. However, the tyrosine kinase activity of VEGFR-2 is nearly ten times that of VEGFR-1 [8].

Formation of blood islands during development is regulated by VEGF-A and VEGFR-2 [9]. Knockout mice for either of these proteins (even a heterozygous mutation) exhibit a fatal lack of blood island formation and vasculogenesis during early embryogenesis; this result underlines the critical requirement for a minimum VEGF level for normal blood vessel formation [10–12]. Although knockout VEGFR-1 mice also exhibit a lethal phenotype, they form blood vessels that are disorganized and dysfunctional [13]. It has been hypothesized that the high affinity of VEGFR-1 for VEGF-A is a mechanism to regulate the concentration of VEGF-A in order to inhibit overstimulation of VEGFR-2 [14]. Once blood islands have formed, VEGF-A promotes migration and proliferation of newly formed endothelial cells in the island wall, leading to enlargement and coalescence of the primitive blood vessels [15, 16].

Studies in the mouse retina (an established model for angiogenesis) provided evidence that an extracellular VEGF-A gradient guides the extension and migration of endothelial tip cells [4, 17]. VEGFR-2, which is highly expressed on tip cell filopodia, guides migration of endothelial cells in the direction of the highest VEGF-A concentration [4]. Proliferation of the endothelial stalk cells is also mediated by VEGF-A via VEGFR-2 receptors; in this case, however, cell proliferation depends on the local concentration, and not on the concentration gradient, of VEGF-A [4]. In vitro, VEGF-A165 induces bovine microvascular endothelial cells to form capillary-like structures in collagen gels in a dose-dependent manner; in this case, a 100 ng/mL bolus of VEGF-A165 induced maximum such tubule formation [18].

In the human adult, the low levels of VEGF-A expressed in most vascularized tissues play a role in the maintenance of a healthy vasculature [19, 20], while high levels of VEGF-A in adult tissue lead to formation of leaky, hemorrhagic vessels [21–23]. VEGF-A prevents apoptosis in cultured, primary, human umbilical vein endothelial cells (HUVECs) subjected to serum starvation [24, 25]. VEGF-A production, induced by hypoxia in many human cell types such as myocytes, is increased both in surrounding tissues and in blood serum [19, 26, 27]. The mechanism of such VEGF-A upregulation involves increased hypoxia-inducible factor-1α (HIF-1α) [19]. In mice, Flt-1 receptors on endothelial cells are upregulated by hypoxia [28]; this condition promotes angiogenesis and leads to increased perfusion of the hypoxic tissue [28]. The role of VEGF-A in new blood vessel formation in hypoxic tissues is crucial in both health and disease; in the latter case, it is a target of tumor-suppression therapy research [29].

VEGF increases capillary permeability, an important feature of inflammation; the underlying mechanism involves induction of nitric oxide (NO) synthase, which leads to increased NO production and relaxation of vascular smooth muscle cells in the subendothelial layers of vascular tissue [30]. A consequence of increased vascular permeability is deposition of plasma proteins (such as fibrinogen) in the vascular wall tissue; specifically, fibrinogen subsequently facilitates angiogenic sprouting [31]. VEGF also induces expression of extracellular matrix metalloproteinase (MMP); select MMPs degrade the basement membrane, facilitating sprouting angiogenesis [32].

VEGF plays an important role in the recruitment, activation, and differentiation of endothelial progenitor and progenitor cells of the stroma. VEGF receptors are present on hematopoietic stem cells (HSCs); moreover, serum VEGF is chemotactic for HSC from the bone marrow [26, 33]. HSCs, which are recruited by VEGF, have been implicated in angiogenesis [26].

In summary, VEGF is involved in the organization of primitive blood vessel development, guidance of capillary extension, proliferation of endothelial cells, recruitment and commitment of endothelial cell precursors, wound healing, and maintenance of healthy blood vessels. Because VEGF critically affects almost every stage of vascular tissue development, repair, and expansion, it is the most important candidate among growth factors for the development of vascular tissue and of vascularized, engineered constructs of other tissues and organs. VEGF alone, even in high doses, is unlikely to lead to functional blood vessel formation: VEGF must be used in conjunction with other important factors.

10.2.3. Fibroblast Growth Factor (FGF)

FGFs are secreted by many cells, especially macrophages and tumor cells [34]. There are more than 20 FGFs, but most pertinent to the vasculature are the acidic fibroblast growth factor (aFGF or FGF-1) and basic fibroblast growth factor (bFGF or FGF-2) [34].

These growth factors act by activating the tyrosine kinase receptors FGFR1 and FGFR2 on target cell membranes.

Mesodermal aggregation during embryonic development is directed by bFGF [1]. In vitro, physiological concentrations of 1–50 ng/mL bFGF induce endothelial cell proliferation in a dose-dependent manner [18]. Moreover, bFGF promotes migration and proliferation of newly formed endothelial cells in blood islands and causes a primitive vascular network to form as the cavities enlarge and coalesce [15, 16]. bFGF is important in maintaining the newly developed vasculature [16].

FGF affects angiogenesis through both direct and indirect pathways that involve upregulation of receptors specific to other important angiogenic factors such as VEGF. bFGF stimulates expression of VEGF in endothelial cells and stromal cells [35–37]. When combined with VEGF, bFGF acts synergistically to increase endothelial cell migration and proliferation [18, 38].

10.2.4. Platelet-Derived Growth Factor (PDGF)

PDGFs, members of the same superfamily as VEGF, are potent mediators of smooth muscle cell and pericyte proliferation, migration, and recruitment [37, 39–41]. There are five PDGFs: PDGF-AA, PDGF-AB, PDGF-BB, PDGF-C, and PDGF-D. The PDGFs act via two receptors, platelet-derived growth factor receptor-α (PDGFR-α) and platelet-derived growth factor receptor-β (PDGFR-β), located in cell membranes. The most prevalent in vivo interaction of PDGFs and PDGFRs pertinent to the vascular system is that of PDGF-BB and PDGFR-β [41].

PDGF regulation of smooth muscle cells and pericytes is critical to the formation of blood vessels, as these cell types form the supportive mural layer around mature vessels (capillaries, however, lack a layer of vascular smooth muscle cells in their walls). PDGF-BB expression by the tip cells of forming capillaries is hypothesized to recruit progenitors of smooth muscle cells and pericytes that express PDGFR-β; the supportive nature of the mural cells is critical to the stability of new blood vessels [41].

In addition to direct effects on cell function, PDGFs interact with other growth factors. In angiogenesis, the effect of bFGF (which alone causes moderate levels of recruitment and proliferation of smooth muscle cells and pericytes) is enhanced when combined with PDGF-BB [37, 39, 40]. Moreover, the synergistic effect of PDGF-BB and bFGF on smooth muscle cells is thought to lead to maturation and stabilization of the blood vessel stroma [37, 39]. The synergy between these growth factors is due to crosstalk between receptors; bFGF causes FGFR-1 transactivation of the PDGFR of smooth muscle cells and pericytes, inducing their maturation [40].

10.2.5. Transforming Growth Factor-β (TGF-β)

TGF-β is an important regulator of early vascular development and angiogenesis [42]. The TGF-β superfamily consists of nearly 30 proteins including the bone morphogenetic proteins (BMPs). There are two types of TGF-β receptors on cells, both of which are serine-threonine kinases; type-II receptors (like TGFβRII) bind the growth factors and then phosphorylate type-I receptors (such as activin receptor-like kinase 1 [ALK1]), which activate intracellular signaling cascades [31]. Of all the growth factors and receptors in the TGF-β family, TGF-β1 interacting with TGFβRII and ALK1 are the most pertinent ones to vascular tissue [31].

TGF-β1 is expressed by various cell types including endothelial cells and other vascular tissue cells (such as pericytes), and promotes synthesis and deposition of extracellular matrix by these cells [43]. TGF-β1 induces smooth muscle cell differentiation and recruitment from surrounding mesenchymal cells [31, 44–47]. The effects of TGF-β1 on endothelial cells and smooth muscle cells are dose dependent. Upregulation of angiogenic proteins and proteinases by low (0.2–0.5 ng/mL) doses of TGF-β1 leads to increased blood vessel turnover and fosters blood vessel growth. At higher doses, however, TGF-β inhibits endothelial cell proliferation and promotes increased extracellular matrix deposition by endothelial and smooth muscle cells; this contributes to the maturation and strengthening of the blood vessel wall tissue [31, 44–47]. Mutations in either TGF-β or its receptors (TGFβRI and TGFβRII) result in a disorganized vasculature and in frail blood vessels in the mouse embryo [48–50].

Compared with the aforementioned in vivo effects, the in vitro effects of TGF-β are different; proliferation of cultured endothelial cells is inhibited with TGF-β in the range of 10 pg/mL to 10 ng/mL [31, 47, 51]. The differences between in vivo and in vitro effects of TGF-β suggest that the biphasic effects in vivo may be due to TGF-β interaction with other growth factors or growth factor receptors. For example, since TGF-β upregulates the expression of VEGF messenger RNA (mRNA) in mouse fibroblasts as well as in human endothelial cells, VEGF expression (due to TGF-β) may be the driving force for endothelial cell proliferation in vivo [51]. The presence or absence as well as the concentration of growth factors, specifically of TGF-β [52] and VEGF [53], drive cell apoptosis, which is part of the remodeling process of the vascular network.

10.2.6. Growth Factors in the Wound-Healing Process of Vascular Tissue

Vascular infiltration into injured tissue is a requirement for healing, and occurs by the mechanisms of angiogenesis discussed previously in this chapter (Sect. 10.2.1). Infiltrating vessels bud from uninjured capillaries in and around the site of the injury, and this intimately depends on intricate expression of growth factors. VEGF is expressed by many of the cell types involved in the healing response, including endothelial cells, fibroblasts, smooth muscle cells, platelets, macrophages, and neutrophils [54]. VEGF induces vasodilation, degradation of basement membrane proteins, endothelial cell migration, and endothelial cell proliferation [54]. Immediately after injury, during the inflammatory phase of healing, local VEGF levels are increased due to release from activated platelets and inflammatory cells. VEGF secreted by endothelial cells and vascular smooth muscle cells in blood vessels in and around the injured tissue is upregulated within 1 day after injury and is responsible for the growth of capillaries into the site as early as 3 days after injury [19, 54, 55]. VEGF release and diffusion establishes a gradient that guides angiogenesis toward hypoxic tissue [54]. VEGF is critical after the inflammatory phase as well; VEGF expression peaks during the proliferative phase between 2 and 7 days [56]. bFGF is also important in wound healing of vascular tissue. bFGF levels have an initial peak immediately after vascular injury lasting until the proliferative phase at about 3 days. After a period of low expression, bFGF levels increase to their highest point after angiogenesis is complete at about 8 days [54]. It is hypothesized that preformed bFGF in the tissue released after injury accounts for the initial bFGF peak, and is responsible for upregulation of VEGF [54]. The second bFGF peak results from synthesis and release of endothelial cells in the newly formed capillaries; bFGF then regulates lumen formation and basement membrane development [54].

10.2.7. Growth Factors in Select Pathological Conditions Pertinent to Vascularization

Blood vessel formation and degeneration play a critical role in various pathologic conditions. The aberrant function of blood vessels in these conditions relates to the expression of various growth factors. Examples of diseases and syndromes in which growth factors affect the vascular network, and are recognized key aspects of such conditions, include many malignancies, age-related macular degeneration, proliferative retinopathy, rheumatoid arthritis, brain edema, atherosclerosis, and pulmonary hypertension [24, 41].

The most prominent example of the role of vascular growth factors in pathological conditions is in malignancy. Most solid tumor cells secrete abnormally high levels of VEGF; this is the mechanism by which the tumor promotes formation of the vascular network required for its survival and growth [24]. In addition, the cells of the supportive stroma of tumors secrete abnormally high levels of VEGF [24, 57–60]. The crucial role of VEGF in the growth of tumors has been supported by the clinical use of VEGF-neutralizing molecules and VEGFR antagonists, which inhibit formation and maintenance of the vascular network, and arrest tumor growth in some types of cancer [8]. bFGF and TGF-β have been implicated in the angiogenesis of various rodent and human tumors as well [34, 45, 61]. TGF-β is highly excreted in some tumors, and stabilizes newly formed tumor vessels and suppresses the immune response [45]. bFGF specifically increases endothelial cell proliferation, migration, and adhesion in tumor angiogenesis, all critical steps in the formation of new vasculature, making bFGF and its receptors a target of anti-cancer therapy [34]. In fact, administration of anti-bFGF antibodies either slowed or halted the growth of some cancers by inhibiting angiogenesis [62].

10.3. Growth Factors in Bone Development and Repair

Development of new bone and the process of bone growth and repair are guided by an intricate interplay of growth factors and their respective receptors on cell membranes. Growth factors have been implicated in the recruitment of osteoprogenitor cells, differentiation along the osteoblastic pathway, preosteoblast proliferation, and other osteoblast functions (pertinent to new tissue formation), as well as formation of both the organic and inorganic (mineral) phases of bone.

Because the role of growth factors in bone development and repair is complex, an understanding of these processes and events is mandatory in order to succeed in utilizing pertinent growth factors in therapeutic applications in the orthopedic and dental fields. In this respect, the growth factor aspects of critical importance include availability in physiologic and pathologic conditions, localization, concentration, bioactivity, expression and/or activation of their respective receptors on cell membranes, timing of appearance as well as interplay and synergism among various growth factors.

10.3.1. Bone Development

Bone develops in one of two ways: by endochondral formation or intramembranous formation. Endochondral formation is characteristic of long bones (such as the femur), while intramembranous formation is characteristic of flat bones (such as the cranium).

Long bone development, known as "endochondral formation," begins with the interactions of epithelial and mesenchymal cells that result in limb buds [63]. At that point, unknown (to date) signals promote formation of mesenchymal islands in the bud [64]. The cells at the center of these islands differentiate into cartilage-like tissue; they are surrounded by a layer of undifferentiated cells, which later on differentiate into perichondrium and then periosteum [63].

Angiogenesis, and the growth factors regulating it (discussed in Sect. 10.2 of this chapter), is critical to the development of bone; new blood vessels mark a transition from cartilage-like tissue to bone. VEGF gradients guide new blood vessels into the cartilage tissue, thus facilitating bone growth [63]. Simultaneous with invasion by blood vessels, the cartilage is ossified, and a marrow cavity begins to form [63]. Several studies have demonstrated that, by blocking angiogenesis in the developing skeleton, bone formation does not occur [65, 66]. Angiogenesis in the developing skeleton facilitates the breakdown of cartilage; this event allows for subsequent ossification of the newly formed tissue [63]. Osteoprogenitor cells, including osteoclast precursors, travel with newly formed capillaries in developing bone and degrade calcified cartilage, and thus contribute to the initiation of new bone formation at the site of blood vessel invasion and growth [67]. VEGF also regulates the differentiation of osteoblasts via direct cell-cell communication between endothelial and osteoprogenitor cells present in blood flowing through the newly formed capillaries of the developing skeleton [63]. In summary, angiogenesis is a crucial and necessary aspect of new bone formation.

Intramembranous bone formation is characteristic of flat bones such as those found in the skull, and does not have a cartilaginous phase. Initially, mesenchymal stem cells aggregate to form mesenchymal islands in the area of bone development, and subsequently differentiate into osteoprogenitor cells, which secrete collagen type-I and differentiate into osteoblasts. Blood vessels surrounding the islands are incorporated into the newly formed osseous tissue and provide the subsequent vascular supply to bone. As osteoblasts migrate and proliferate, they secrete components of the ECM and form osseous trabeculae, which fuse together to form spongy or cancellous bone. As the trabeculae merge, remodeling of the layer of bone near the periphery leads to formation of the cortical layers that surround a central core of cancellous bone.

The growth factors that are involved in various stages of bone development are reviewed in the sections that follow.

10.3.2. Bone Morphogenetic Protein

The BMPs, consisting of more than 20 isoforms, are members of the TGF superfamily. BMPs are powerful stimulators of bone formation and effectors of mesenchymal cell differentiation to bone cells [68]. BMPs are primarily secreted by osteoblasts [69, 70]. There are two types of BMP receptors in cell membranes, both of which are serine-threonine kinases. BMP binding to the type-II receptor leads to dimerization of that receptor and a type-I receptor, and phosphorylation of the type-I receptor, which then activates downstream signaling [31].

BMP-2, BMP-4, BMP-6, BMP-7, and BMP-9 are the most important BMPs in promoting the differentiation of pluripotent mesenchymal cells to osteoblasts, and lead to increased bone formation [71–73]. Both in vitro and in vivo, BMPs promote the fusion of mononuclear hematopoietic cells of the macrophage-granulocyte lineage; this process results in osteoclast formation and activation, leading to increased bone turnover and remodeling in the late stages of bone repair and development [74]. Details of these processes are still unknown; moreover, some evidence is contradictory. Resolution of such controversies might be provided by a biphasic dose-response of BMPs on various bone cells, as has been recently shown with BMP-5 and BMP-6 in vitro [74, 75]. Specifically, at doses above 300 ng/mL, BMP-5 and BMP-6 inhibited osteoclast formation and function, but at lower doses (0.1–10 ng/mL), osteoclastogenesis was stimulated [75].

10.3.3. Fibroblast Growth Factor (FGF)

In vitro, FGF1 and FGF2 induce proliferation of progenitor cells and osteoblasts as well as osteoblast proliferation, alkaline phosphatase (AP) activity, and collagen type-I production [76–78]. The mitogenic effect of FGF-2 is about ten times that of FGF-1 [79]. FGF-2 inhibits osteoblastic differentiation of murine progenitor cells in a dose-dependent manner [78, 80].

10.3.4. Insulin-Like Growth Factor (IGF)

Two IGFs, namely IGF-1 and IGF-2, play an important role in bone. IGF-2 is present in higher levels in bone, but IGF-1 is more potent [81]. IGF-2 and especially IGF-1 induce increased osteoblast proliferation and enhance other osteoblast functions [82], including chemotaxis [83]. IGFs act in both a paracrine and an autocrine manner when secreted by osteoblasts [84, 85]. Their bioactivity is regulated by a variety of IGF binding proteins (IGFBPs), which sequester excess IGF and, thus, reduce activation of pertinent cell receptors. In addition to being primarily inhibitory regulators, certain IGFBPs can also stimulate and/or enhance IGF signaling [86, 87]. IGF acts synergistically with other growth factors such as TGF-β, PDGF-BB, and BMP-2 to form healed bone [88–90].

10.3.5. Platelet-Derived Growth Factor (PDGF)

Although osteoblasts express only the PDGF-A isoform, PDGF-BB has the greatest biological activity on these cells [91]: it promotes osteoblast proliferation weakly, but it is a potent recruiter of mesenchymal cells [92, 93]. Both osteoblasts and osteoprogenitor cells migrate in response to PDGF [92, 93].

10.3.6. Transforming Growth Factor-β (TGF-β)

There are five isoforms of TGF-β, all part of the TGF superfamily that includes BMPs. TGF-βs are important regulators of cell proliferation, differentiation, synthesis of extracellular matrix proteins, and apoptosis [94]. TGF-β, a ubiquitous growth factor in the body, is present in high levels in bone and cartilage [82]. TGF-β is secreted by osteoblasts and stored in the extracellular matrix of bone [76].

In vitro, TGFs promote increased proliferation of osteoprogenitor cells, which subsequently differentiate into osteoblasts [95, 96]. TGFs also inhibit differentiation of osteoprogenitor cells into osteoblasts during the late stages of bone formation and repair [97]. These contrary effects of TGF on either osteoblasts or their precursors are concentration dependent [82]. High (10 ng/mL bolus or 1 ng/mL delivered four times a week) doses of TGF-β1 induced decreased osteoblast function, but low (1 ng/mL weekly) doses promoted osteoblast function and new bone formation in dogs [98]. TGF-β1 also plays a role in stimulating the migration and organization of cells in the limb buds [94]. TGF-β1 has a protective effect on osteocytes by decreasing their apoptosis [99]. Moreover, TGF-β1 induces increased synthesis of important osteogenic glycosaminoglycans such as chondroitin sulfate and decorin in neonatal rat calvaria and in MC3T3-E1 cell culture [100, 101].

10.3.7. Growth Factors in the Wound-Healing Process of Bone

Growth factors play an important role in the wound-healing process of bone. For example, BMPs act in various combinations and time sequences during healing bone fractures and promote various stages of the bone repair process [102]. BMP-2 is at its highest expression level in mouse bone fracture models 1 day after bone injury [102]. BMP-2 also can induce the differentiation pathway of some myoblasts to the osteoblast lineage [103]. Furthermore, it is the most important factor in the differentiation of osteoprogenitor cells to osteoblasts in the early phases of bone development and repair. BMP-3, BMP-7, and BMP-8 are expressed at high concentrations during resorption of cartilage and osteoblast recruitment (days 14–21 in the murine model) [102]; other studies, however, reported enhanced expression of BMP-7 during the early stages of fracture healing in rats [104, 105]. BMP-5 and BMP-6 are expressed during all phases of bone repair, but not immediately after bone injury [102]. During bone repair in vivo, FGFs are synthesized by several cell types, including macrophages, mesenchymal cells, osteoblasts, and chondroblasts. FGFs have been identified during the early stages of fracture healing in a mouse model [106]. IGFs have been found in human granulation tissue and fracture callus; IGF-2 mRNA is expressed by endothelial and mesenchymal cells and both IGF-1 and IGF-2 are expressed by osteoblasts and chondrocytes [106, 107]. During the hemostatic process following bone fracture, platelets are activated and release PDGFs. PDGFs are also synthesized by macrophages, endothelial cells, and osteoblasts, and exert their effects by both autocrine as well as paracrine mechanisms [108]. During the earliest phase of fracture repair, PDGF was detected in macrophages near the periosteum [106, 109]. In tissue biopsied from human fractures, PDGF was expressed by endothelial and mesenchymal cells in the granulation tissue, and later it was found in osteoblasts, chondrocytes, and osteoclasts [109]. During fracture healing, initial increases in TGF-β levels are accounted for by release from platelets during hematoma formation [110]. This growth factor is subsequently produced by chondrocytes and osteoblasts during endochondral bone formation in mice [106]. TGF-β2 and TGF-β3 are maximally expressed during mouse fracture repair at day 7, an event that coincides with the peak of cartilage formation [102]. TGF-β1 is expressed throughout all phases of bone fracture repair, with peak expression just after the initial bone injury in mice [102].

10.3.8. Growth Factors in Pathological Conditions of Bone

Abnormal expression of growth factors or growth factor receptors has been associated with a variety of pathological conditions of bone, including osteoporosis, skeletal malformations, malignancy, and in the formation of osteophytes (bone spurs). For example, progressive diaphyseal dysplasia, an autosomal dominant condition that leads to muscular dystrophy-like symptoms along with thickening of the cortex in the diaphysis of long bones as well as the skull and pelvis, was traced to a mutation in the TGF-β1 gene [111]. Another example of the importance of growth factors in bone is that mutations in FGFRs lead to abnormal bone formation during development and result in achondroplasia in humans [82, 112]. Osteoporosis has been associated with a variety of growth factor abnormalities; BMP, FGF, and IGFs are involved in the pathogenesis of this disease. BMPs promote and regulate differentiation of osteoprogenitor cells as well as osteoblast proliferation, two functions that are decreased in osteoporosis [113]. Mouse models deficient in BMPs demonstrate an osteoporosis-like condition due to a decrease in the number of osteoblasts and decreased differentiation of osteoprogenitors [113]. In addition, decreased levels of IGF-1 predispose to osteoporosis in mice due to abnormal osteoprogenitor differentiation, and there is a strong correlation between IGF-I levels and bone mineral density in women [114, 115]. FGF also plays a role in osteoporosis; treatment of ovariectomized rats (a model of osteoporosis) with FGF demonstrated a decrease in bone loss due to increases in osteoblast recruitment and ECM component secretion [116–118]. While growth factors are critical in the pathogenesis of a variety of conditions, these specific examples highlight the crucial importance of understanding their roles in both health and disease.

10.4. Future Directions in Growth Factor Research

Development and homeostasis of both the vascular network and of bone depend on an intricate and complex interplay among cells and on interactions of cells with their extracellular milieu including matrices and chemical (such as growth factors) and biophysical (such as mechanical) cues and stimuli. Past research has identified key components and elucidated aspects of important events/steps in fundamental biologic/physiologic processes such as tissue/organ development, function, healing, and repair. Details of the underlying mechanisms (including signaling pathways) as well as complete knowledge of all pertinent aspects are either not known or, at best, partially understood. Ongoing research endeavors, especially regarding the molecular processes involved in tissue formation, have great (but until recently, untapped) potential for seminal contributions to current, dynamic, and cutting-edge fields such as regenerative medicine.

10.5. Applications of Growth Factors to Biomaterials

Despite the fact that growth factors are an integral part of normal (and abnormal) human tissue formation, healing, and regeneration (as discussed briefly in Sects. 10.2 and 10.3), historically they had remained outside the scope and range of research activities pertinent to implant biomaterials and devices. For a long time, research in those areas dealt with biocompatibility issues and focused on producing materials that were not rejected but simply tolerated by the recipient host. Availability and action of growth factors were left at the mercy of the host milieu.

The advent of new biomedical fields, specifically of tissue engineering and regenerative medicine, expanded the requirements for successful outcomes to include promotion of neotissue formation and emphasized the need for directing and controlling availability and function of growth factors. Consequently, considerable research has been devoted to combining various growth factors with biomaterials in order to promote cell functions pertinent to new tissue formation both in vitro and in vivo. These studies improved the outcome in some biomedical applications, but proved limited in their efficacy. In retrospect, such results were inevitable since these studies were based on dated information (and, therefore, limited/incomplete knowledge) about the pertinent physiological systems, specifically, various aspects of the processes of wound healing and neotissue formation.

It should be noted that, in in vitro studies, growth factors (of unknown types and concentrations) were available because they are present in the serum supplements of cell culture media. Later on, select exogenous growth factors were added (usually one at a time either as a bolus or at a sustained-dose level) in the supernatant media in vitro and were either immobilized or incorporated on, and subsequently released from, biomaterials both in vitro and in vivo.

These developments still fell short of simulating the complex tissue healing and regeneration processes in vivo because availability and action of a growth factor was only one aspect of the pertinent processes. Recent advances in the biology and biochemistry of growth factors have revealed that more than one growth factor (as well as their respective receptors on cell membranes) is expressed in preordained sequences at specific time points and concentrations. Following injury, for example, growth factors specific to a tissue are upregulated and delivered at different times during the tissue-healing process; apparently, such timely presence and bioactivity of each growth factor are requisites for the successful outcome of tissue healing and/or regeneration.

Current research strategies and approaches in bioengineering are motivated by advances in biology and physiology (as summarized in Sects. 10.2–10.3 of this chapter) and seek to apply this knowledge in designing and fabricating implant biomaterials capable of delivering specific concentrations of multiple, but select, bioactive molecules in a timely fashion. Such approaches will provide the next generation of "proactive" biomaterials for implant devices and tissue-engineering scaffolds. Undoubtedly, successful outcomes will require continuous research endeavors to achieve delivery of different growth factors at various doses, rates, release profiles, and durations that, at least, simulate (if not enhance) the natural milieu of wound healing and neotissue formation around implants.

10.5.1. Modes of Growth Factor Delivery from Biomaterials

In the physiological milieu, growth factors are excreted from cells and remain soluble in biological fluids, are membrane bound, or bind to various extracellular matrix proteins. For example, certain isoforms of VEGF-A (such as VEGF-A189), as well as the FGFs, attach to extracellular matrix proteins but maintain their bioactivity [8]. A common mechanism for this immobilization is electrostatic interaction between cationic domains of the growth factor and anionic matrix biomolecules, such as heparin sulfate proteoglycans.

Delivery of growth factors from biomaterials has developed based on observations of pertinent events in vivo. In this respect, not only the type and concentration but also the mode of delivery of the growth factor determines its efficacy. Among potential strategies, bolus delivery of growth factors tends to be less successful because of the discrepancy between the duration of such growth factor availability (due to the short plasma half-life of many growth factors) and the time course of, and pertinent requirements for, tissue formation. Alternative strategies have focused on developing methodologies to either immobilize bioactive biomolecules on, or to render them releasable from, implant biomaterials and tissue-engineering scaffolds in patterns of availability and for periods of time necessary to achieve the desirable biological outcome.

Various techniques of immobilization and/or release of bioactive chemical compounds (initially single but, more frequently now, at least two) have been explored, developed, and proven successful in biomedical applications.

Growth Factor Immobilization

Methodologies to immobilize growth factors (typically a single factor per application) on material substrates utilize binding of affinity sites via simple adsorption as well as covalent bonding. An important criterion for the success of immobilization of growth factors on biomaterials is subsequent maintenance of the activity of the biomolecules. This objective is achieved by limiting the potential of conformational changes that affect the bioactive site of the bound growth factor. For example, mouse epidermal growth factor immobilized in one conformation and orientation (namely, by the N terminus) on a substrate surface is bioactive and it promotes enhanced cell proliferation [119]. Retention of the growth factor bioactivity during the immobilization procedures has been a major concern of the various methodologies explored and developed to bind bioactive chemical compounds on various biomaterials (some of which were presented in Chap. 9 of this book).

In addition to utilizing the natural affinity of growth factors for components of the extracellular matrix (for example, the affinity of VEGF and FGF for heparin), other chemical immobilization methods were also explored and developed. For example, recombinant human BMP (rhBMP-2) covalently bound on hydroxyapatite (HA) (via either silanization or mediation by strong calcium-chelating agents [120]) promoted enhanced AP secretion (an index of the osteoblastic differentiation of C2C12 mouse myoblasts in vitro); compared with the results obtained from the same amount of rhBMP-2 adsorbed onto the HA surface, this results was 2.5-times higher when the growth factor was immobilized [120]. In another study, and compared with results obtained in the absence of the growth factor, VEGF covalently immobilized on collagen matrices (using disuccinimidyldiscuccinatepolyethyleneglycol [SS-PEG-SS] as a linker), promoted significant increase in the proliferation of HUVECs; increased (but not significantly) cell proliferation was observed on collagen matrices with nonimmobilized VEGF [121].

In summary, the most important aspect of all successful covalent immobilization techniques is retention of the bioactivity of the immobilized growth factor [120–123]. Other advantages of this mode of growth factor delivery include the requirement for relatively small amounts of these expensive chemical compounds to obtain equal, if not greater, bioactivity than the same growth factor in solution, presumably because of a locally higher effective concentration [120, 121].

Growth Factor Release

Release of single, bioactive, chemical compounds has been an area of both intense research and of major successes in controlled drug delivery. This field has established the physicochemical parameters, such as loading (homogeneous versus heterogeneous, admixed versus entrapped), concentration, and kinetics of release, time of release, mode of material degradation (i.e., bulk versus surface erosion), etc., that govern drug release. Such accomplishments have served as a starting point and "blueprint" for research on growth factor release from materials for other biomedical applications [70, 90, 124–146]. The foremost consideration in these applications was retention of the bioactivity of the released biomolecules. In fact, among the growth factors that affect vascular network formation, VEGF, FGF, PDGF, and TGF are all bioactive in soluble form, as opposed to matrix or membrane bound [50, 54]. In the case of growth factors pertinent to bone, BMP, FGF, IGF, PDGF, and TGF-β retain their bioactivity in solution [147–149].

A number of studies addressed the bioactivity of released growth factors in solution and developed delivery systems to promote pertinent neotissue formation. In the case of new vascular network formation, VEGF was incorporated into various hydrogels, including alginate, chitosan–alginate, and heparin-coated chitosan–alginate [128]. In vitro release of VEGF from alginate and chitosan–alginate hydrogels was diffusion limited during 5 days of monitoring. A heparin coating on the surface of the alginate hydrogel slowed diffusion of

VEGF and resulted in nearly zero-order release kinetics. This study demonstrated that natural materials can be used to deliver growth factors in a controlled fashion. In another in vivo study, FGF-1 and FGF-2 (which are growth factors pertinent to new blood vessel formation) were photocrosslinked into chitosan hydrogels [129] and implanted subcutaneously in the backs of mice. Either FGF-1 or FGF-2 was released from the implants over 7–14 days (the result of chitosan hydrogel degradation by proteinase activity), and promoted significantly ($p < 0.001$) more neovascularization than was observed in response to hydrogels without the growth factors or when FGF alone was injected [129]. In another study, the angiogenic response of VEGF was examined in the chick chorioallantoic membrane and in mice. In this case, cells infiltrating a fibrin matrix in which VEGF-A was covalently immobilized induced activation of matrix metalioproteinases whose proteolytic action in turn released the growth factor; continuous exposure to low amounts of this angiogenic growth factor resulted in numerous new blood vessels that exhibited normal morphology and physiology [150, 151].

Other studies focused on improving new bone formation, and developed various release systems, some of which are already in clinical use. rhBMP-2 released from absorbable collagen sponges (ACSs) was approved by the US Food and Drug Administration (FDA) for use in certain orthopedic (such as spinal fusion and healing complex, open tibia fractures) and dental and oral maxillofacial (sinus augmentation and alveolar ridge augmentation) procedures. BMP-2 treatment improved healing of critical-size bone defects compared with either no treatment or with ACS without rhBMP-2 [152, 153]. A stated advantage of this system is the capability to fine-tune the rhBMP-2 release by varying the amount of growth factor uptake onto the scaffold, buffer pH, collagen crosslinking, and rhBMP-2 isoform chosen for this application [137, 138]. In an effort to improve upon the clinically available growth factor delivery systems, a polymer–ceramic composite delivery system was developed [140]. Fully interconnected porous calcium HA was filled with a 51:49 molar ratio of poly(lactic acid):polyethylene glycol (PEG) and rhBMP-2 dissolved in acetone. Evaporation of the acetone resulted in a homogeneously loaded rhBMP-2 polymer-composite scaffold. Eight weeks after implantation into radial defects in rabbits, the defects repaired with the rhBMP-2-containing constructs exhibited complete healing; in contrast, defects either filled with scaffold material alone or unfilled defects revealed incomplete healing [140].

In summary, the trend established by past successes, and that is the preferred current approach in either vascular network repair or bone healing, is to apply the latest information from advances in biology, physiology, biochemistry, pathology, etc., and imitate the type, relative concentration, sequence of appearance, and action of growth factors that are involved in healing, while avoiding conditions that characterize pertinent pathologies of specific tissues. Harnessing the great potential of these powerful biomolecules to direct and control the outcome of cell interactions with, and leading to neotissue formation around, implant materials may be not only envisioned but also achieved.

10.5.2. Combined Delivery of Growth Factors

The realization that, in vivo, many growth factors act in synergism to either enhance or inhibit their effects on cells and tissues has prompted research on combining at least two growth factors for biomedical applications pertinent to new tissue formation. For this purpose, growth factors may be released simultaneously, in an alternating fashion, or in various other ways that simulate appropriate natural processes, such as wound healing. Although the mechanism(s) underlying the synergism of growth factor action are not well understood, research so far has provided strong evidence that this approach is a development of great promise.

The number of literature reports of both in vitro and in vivo studies examining the effect of multiple growth factors on the formation of various mammalian tissues has been increasing. Because detailed review of this literature is outside the scope of the present chapter, only a few examples will be used to illustrate some pertinent aspects.

In the case of bone-related applications, combined delivery of IGF-1 and PDGF-BB has stimulated bone repair in periodontal defects [154, 155]. The synergism between IGF-1 and TGF-β1 proved osteoinductive in a series of in vitro [156] and in vivo [132–134] studies. In this case, when IGF-1 (5% w/w) and TGF-β1 (1% w/w) were incorporated in poly(D,L-lactide) (PDLLA) films (10-μm thick) on titanium wires (1.0-mm diameter) and then released (approximately 50% of total within the first 48 h, the remaining in a sustained manner), they induced enhanced primary human, osteoblast-like cell proliferation, total protein, and procollagen I synthesis (an index of collagen I synthesis) [156]. Continuous release of these two growth factors, that is, both IGF-1 (5% w/w) and TGF-β1 (1% w/w) (but not of either growth factor tested alone) following incorporation in PDLLA and implantation in rats (midshaft fracture of tibia model) for 28 days resulted in increased remodeling of the fracture callus and more mineralized tissue; most importantly, the new bone formed in the presence of the two growth factors exhibited higher maximum load and torsional stiffness [132–134]. A synergistic effect on bone formation in primates was obtained using implants containing both BMP-2/OP-1 and TGF-β1 [157]. Bone formation was enhanced by combined delivery of BMP-2 and TGF-β3 in the presence of transplanted mesenchymal cells at concentrations an order of magnitude lower than those previously known to be effective [158]. Localized delivery of IGF-I and TGF-β accelerated fracture and osteotomy healing [89, 133]. It should be noted, however, that not all studies using combined growth factor delivery had positive results. For example, a dose-dependent inhibition of bone formation was reported when PDGF-BB was delivered with osteogenic BMP-3 [159]. Undoubtedly, further research is needed not only to resolve discrepancies in the results of various studies but also to elucidate definitively relevant processes and pertinent mechanisms.

In the case of neoangiogenesis, a 3:1 ratio of bFGF and VEGF exhibited angiogenic synergy in a modified murine Matrigel chamber implanted in the neck of mice [160]. Subcutaneous injection of plugs containing two angiogenic growth factors, namely, PDGF-BB and FGF-2 in rat and rabbit ischemic hind limb models (but not either VEGF-A or FGF-2 alone), synergistically induced vascular networks that remained stable for more than a year, even after depletion of the angiogenic factors [125]; these results provided evidence that the two growth factors tested are needed simultaneously for the establishment of the vascular network but not for its long-term maintenance. Furthermore, the role of PDGF-B signaling in the synergistically promoted neoangiogenic effects of VEGF-A and FGF-2 (in vivo using the Matrigel plug assay in mice) was also investigated [124]. This study exemplifies new trends in the research regarding various aspects of the role of growth factors in neotissue formation: investigations of pertinent molecular aspects and signaling mechanisms.

One of the first successful strategies to provide delivery of two growth factors utilized release of VEGF-165 (an angiogenesis-inducing biomolecule) and PDGF-BB (known to induce maturation, and thus stabilization, of the newly formed blood vessels) incorporated in poly(lactide-co-glycolide), a hydrophobic degradable polymer scaffold. VEGF, which was present on the surface of the scaffold, was released quickly (within days). PDGF, which was preincorporated into microparticles before their incorporation into the polymer scaffold, was released more slowly (over weeks). Implantation of the dual-release scaffolds in subcutaneous pockets in rats resulted in formation of numerous, larger, and more mature blood vessels 2 and 4 weeks after implantation [161].

A study pertinent to bone tissue-engineering applications combined delivery of BMP-4 (either directly or using a genetically engineered approach), VEGF (via sustained and localized release), and human bone marrow stromal cells from poly(lactic-co-glycolic acid) (PLGA) scaffolds implanted subcutaneously in mice for a period of 15 weeks [162]. Synergism between the osteogenic and angiogenic growth factors enhanced the function of the bone-forming cells and resulted in the greatest quality and quantity of new bone [162]. This study expanded the scope of previous studies by the same group of researchers [161] to deliver either proteins or plasmid DNA; the novel methodology has potential (untapped to date) for both therapeutic clinical applications and engineering other complex tissues [162].

10.5.3. Combined Angiogenic/Osteogenic Growth Factor Delivery

All tissues (with the exception of a few, such as cartilage) and organs are vascularized; most importantly, they depend on requisite blood supply (that is, supply of oxygen and nutrients and removal of carbon dioxide and metabolic waste) in order to survive as well as to maintain their integrity and homeostasis. Due to diffusion limitations, in vivo, most cells (the constituents of tissue and organs) do not survive more than a few hundred micrometers from the nearest blood capillary [7]. For these reasons, development, and especially optimal growth/regeneration and healing, of vascularized tissues (such as bone) and organogenesis require active blood vessel networks to survive; in the case of tissue-engineered constructs, the vascularized network is also essential for the neotissue to integrate with existing host tissue [163].

Realization and appreciation of the aforementioned requirements has motivated exploration and development of methods for vascularizing bioengineered tissues and organs in vitro and in vivo. These approaches are benefiting from advances in many pertinent scientific areas, including developmental biology, cellular physiology, molecular biology and biochemistry, biomedical engineering, biomaterials, etc., and reflect cutting-edge endeavors in these fields; detailed descriptions can be found in excellent recent reviews of the field [7, 163]. As an illustration of this approach, enhanced formation of both blood vessels and bone, the two tissues that have been the focus of the present chapter, will be briefly discussed in the remainder this section.

The strategies to achieve tissue regeneration in bone, a highly vascularized tissue, utilize matrices/scaffolds, cells, and pertinent growth factors judiciously chosen to promote both angiogenesis and osteogenesis in the new tissue. For example, combining the angiogenic properties of bFGF with osteoinduction by BMP-2 enhanced bone formation in rats [90]. Calcium content (an index of the mineralized extracellular matrix) in fibrous glass membrane implants containing bFGF was three times greater than those without the angiogenic agent; moreover, type II collagen (an index of the cartilaginous tissue), was reduced. These results provided evidence that the presence of bFGF elicited faster invasion of capillaries into the implants; this development promoted accelerated osteoinduction induced by BMP-2.

In another study [164], VEGF (an essential mediator between vascular networks and bone [165]) was incorporated by conventional adsorption onto PLGA scaffolds while BMP-4 was expressed by human bone marrow stromal cells (hBMSCs) following uptake and incorporation of pertinent plasmid DNA. Enhanced BMP-4 expression by the hBMSCs (containing the plasmid), along with the increased vascular network formation induced by VEGF, contributed to the increased new bone formation observed following subcutaneous implantation of these constructs in severe combined immunodeficiency (SCID) mice. Moreover, the elastic modulus of the newly formed bone increased in the presence of the angiogenic and osteogenic growth factors [162].

The time-dependent effects of a combination of VEGF (12 μg) and/or BMP-2 (2 μg) released from gelatin microspheres incorporated into poly(propylene fumarate) scaffolds was investigated in a rat critical-sized calvarial defect model [166]. Compared with results obtained from using the blank and single growth factor scaffolds, the dual growth factor delivery stimulated significantly greater amounts of bone (but similar blood vessel formation) 4 weeks after implantation. At 12 weeks, both the dual growth factor and BMP-2-only animal groups exhibited significantly greater bone content [166].

The aforementioned studies established the importance of multiple growth factor delivery for neotissue formation and they exemplify research activities in a topical biomaterials-related scientific frontier. The insights that research in this area will reveal in the coming years will be of eminent importance for tissue-engineering applications.

10.5.4. Sequential Delivery of Growth Factors

Although simultaneous release of growth factors is technically simpler than sequential delivery, the approach does not appropriately mimic the complex, time-dependent cascades of biomolecule expression that occur during formation of vascular and bone tissues (discussed in Sects. 10.2 and 10.3). In contrast to numerous studies that have explored delivery of growth factor combinations (discussed in Sect. 10.5.2), sequential release of growth factors for biomedical applications has not yet attracted the research attention it deserves.

The first study that investigated the effects of temporal variation of growth factors delivered from biomaterials on bone cell responses used BMP-2 and IGF-1 systematically loaded in two layers of gelatin [167]. The gelatin layers were unloaded, loaded with either BMP-2 or IGF-1, or loaded with BMP-2 and IGF-1. Each gel layer contained between 1.25 and 2.5 μg BMP-2 and/or 50 and 100 ng IGF-1. Growth factors were released in a biphasic fashion; the growth factor loaded in the top gel layer released substantially (maximum 250 ng) during the first day and subsequently decreased to near zero by day 6, while the growth factor in the bottom layer initiated release at day 4, and peaked at day 6 (maximum 175 ng), before trailing off by day 10. Using these materials, profiles were designed for simultaneous, sequential, or sequential release of both growth factors. Osteoblastic responses of either mouse pluripotent C3H10T1/2 (C3H) cells or rat bone marrow cells (BMCs) were monitored in vitro. Results showed that the AP activity of C3H cells differed depending on the sequence of delivery. BMP-2, but not IGF-1, stimulated osteoblast differentiation and proliferation. Importantly, however, IGF-1 modified the effect of BMP-2 on the cells tested; treatment of BMP-2 at day 1 and IGF-1 with BMP-2 release starting on day 4 induced significant increases in DNA content and AP activity. In this case, the presence of BMP may upregulate expression of IGF receptors on CH3 cells. When this regime of delivery of two growth factors was tested on BMCs, treatment with IGF-1 followed by BMP-2 induced increased proliferation and activity of the stromal cells. Treatment with either BMP-2 followed by treatment with IGF-1 or with IGF-1 plus BMP-2 induced the greatest increase in cell proliferation and function. These results clearly demonstrated the importance of sequential delivery of growth factors to induce cell functions pertinent to new tissue formation. Most importantly, this study demonstrated the efficacy of a biomaterials-related strategy that has great potential for mimicking physiologically relevant growth factor delivery profiles.

10.6. State of the Art Summary and Future Directions

Stimulating the body's natural processes for healing using growth factors has great potential for enhancing tissue regeneration in a variety of biomaterials applications. These powerful molecules can be delivered to the interface between cells and tissues with a biomaterial by attachment, either physical or chemical, or by controlled release from the substrate. Encouraging results pertinent to enhanced regeneration of both vascular and bone tissues have been reported from research that explored delivery of two growth factors (for example, one angiogenic and one osteogenic in the case of bone) judiciously selected from the recognized key players in the formation of a specific tissue.

Undoubtedly, these endeavors reflect new directions in the bioengineering field and have the potential for seminal contributions in implant biomaterials, tissue engineering, and regenerative medicine. Elucidation of the underlying mechanisms of action (at the molecular and signal transduction levels) and of the long-term impact of growth factors in health and in disease will provide most valuable insights and facilitate the task of biomaterials scientists and biomedical engineers. A major challenge for the field of growth factor delivery presently, and in the near future, is to "translate" recent and future advancements and contributions in pertinent scientific fields (such as developmental biology, cellular biology, molecular biology and physiology, biochemistry, biomedical engineering, biomaterials, etc.) and, most importantly, to incorporate these findings into functional applications in pertinent biomedical fields. Future successes will, therefore, require innovative, unique, and creative approaches in transferring and incorporating relevant knowledge from physiology (and perhaps even from select pathologies) in the design and fabrication of the next generation of biomaterials and biomaterials-related technologies.

Although accomplishments to date have established the importance of parameters such as the type, concentration, release kinetics, and bioactivity of growth factors in cell functions pertinent to neotissue formation, further research is needed to elucidate the underlying mechanisms. The importance of delivering sequences of more than one biomolecule is increasingly acknowledged as an improved and very promising model for the process of neotissue formation; this area of great excitement and potential requires intensive research and extensive development. In addition to combinations of select biomolecules, their interaction with and/or dependence on other components (such as proteins of the extracellular milieu and cytokines of the relevant microenvironment) that affect the bioactivity of released and/or immobilized growth factors must be examined.

Challenges for future development include choice of proper biomolecule(s) and appropriate modes of delivery of the bioactive agents to cells. To avoid complications arising from the pleiotropic nature of growth factors, judicious choices between endogenous and recombinant formulations must be made. In addition to the release and immobilization modes used thus far, alternative modes of growth factor availability (using gene-related methodologies) must be explored. The complex, time-dependent interplay of multiple biomolecules has not yet been appropriately exploited. Additional research is needed to elucidate key aspects of the local (microenvironmental) and systemic effects of the sequential release of biomolecule cascades in vivo; this knowledge must be subsequently integrated in in vitro studies and systems. In addition to studying differentiated cells of a particular lineage, the effects of growth factors on their progenitors and on pluripotent stem cells for biomaterials-related applications must also be determined. From the biomaterials perspective, development of methods for controlling the orientation/presentation of the immobilized biomolecule so that the receptor-binding domains on the ligand are revealed and/or activated for maximal interaction with cells will be useful.

Undoubtedly, this is the beginning of growth factor applications in biomaterials; the future, however, holds great promise because of the high, but still untapped, potential of this field. The approaches outlined in this section and discussed in this chapter will provide new and important basic science information that will be useful in designing the next generations of biomaterials to deliver growth factors and promote neotissue formation pertinent to various bioengineering applications.

References

1. Risau W, Flamme I. Vasculogenesis. Annu Rev Cell Dev Biol 1995;11:73–91.
2. Gonzalez-Crussi F. Vasculogenesis in the chick embryo. An ultrastructural study. Am J Anat 1971;130: 441–460.
3. Drake CJ, Fleming PA. Vasculogenesis in the day 6.5 to 9.5 mouse embryo. Blood 2000;95:1671–1679.
4. Gerhardt H, Golding M, Fruttiger M, Ruhrberg C, Lundkvist A, Abramsson A, Jeltsch M, Mitchell C, Alitalo K, Shima D, Betsholtz C. VEGF guides angiogenic sprouting utilizing endothelial tip cell filopodia. J Cell Biol 2003;161:1163–1177.
5. Egginton S, Gerritsen M. Lumen formation: in vivo versus in vitro observations. Microcirculation 2003; 10:45–61.
6. Hogan BL, Kolodziej PA. Organogenesis: molecular mechanisms of tubulogenesis. Nat Rev Genet 2002;3: 513–523.
7. Ko HC, Milthorpe BK, McFarland CD. Engineering thick tissues-the vascularisation problem. Eur Cell Mater 2007;14:1–18; discussion 18–9.
8. Shibuya M. Vascular endothelial growth factor-dependent and -independent regulation of angiogenesis. BMB Rep 2008;41:278–286.
9. Gehling UM, Ergun S, Schumacher U, Wagener C, Pantel K, Otte M, Schuch G, Schafhausen P, Mende T, Kilic N, Kluge K, Schafer B, Hossfeld DK, Fiedler W. In vitro differentiation of endothelial cells from AC133-positive progenitor cells. Blood 2000;95:3106–3112.
10. Shalaby F, Rossant J, Yamaguchi TP, Gertsenstein M, Wu XF, Breitman ML, Schuh AC. Failure of blood-island formation and vasculogenesis in Flk-1-deficient mice. Nature 1995;376:62–66.
11. Ferrara N, Carver-Moore K, Chen H, Dowd M, Lu L, O'Shea KS, Powell-Braxton L, Hillan KJ, Moore MW. Heterozygous embryonic lethality induced by targeted inactivation of the VEGF gene. Nature 1996;380:439–442.
12. Carmeliet P, Ferreira V, Breier G, Pollefeyt S, Kieckens L, Gertsenstein M, Fahrig M, Vandenhoeck A, Harpal K, Eberhardt C, Declercq C, Pawling J, Moons L, Collen D, Risau W, Nagy A. Abnormal blood vessel development and lethality in embryos lacking a single VEGF allele. Nature 1996;380:435–439.
13. Fong GH, Rossant J, Gertsenstein M, Breitman ML. Role of the Flt-1 receptor tyrosine kinase in regulating the assembly of vascular endothelium. Nature 1995;376:66–70.
14. Hiratsuka S, Minowa O, Kuno J, Noda T, Shibuya M. Flt-1 lacking the tyrosine kinase domain is sufficient for normal development and angiogenesis in mice. Proc Natl Acad Sci USA 1998;95:9349–9354.
15. Lamalice L, Le Boeuf F, Huot J. Endothelial cell migration during angiogenesis. Circ Res 2007;100: 782–794.
16. Lee SH, Schloss DJ, Swain JL. Maintenance of vascular integrity in the embryo requires signaling through the fibroblast growth factor receptor. J Biol Chem 2000;275:33679–33687.
17. Fruttiger M. Development of the retinal vasculature. Angiogenesis 2007;10:77–88.
18. Pepper MS, Ferrara N, Orci L, Montesano R. Potent synergism between vascular endothelial growth factor and basic fibroblast growth factor in the induction of angiogenesis in vitro. Biochem Biophys Res Commun 1992; 189:824–831.
19. Yla-Herttuala S, Rissanen TT, Vajanto I, Hartikainen J. Vascular endothelial growth factors: biology and current status of clinical applications in cardiovascular medicine. J Am Coll Cardiol 2007;49:1015–1026.
20. Byrne AM, Bouchier-Hayes DJ, Harmey JH. Angiogenic and cell survival functions of vascular endothelial growth factor (VEGF). J Cell Mol Med 2005;9:777–794.

21. Larcher F, Murillas R, Bolontrade M, Conti CJ, Jorcano JL. VEGF/VPF overexpression in skin of transgenic mice induces angiogenesis, vascular hyperpermeability and accelerated tumor development. Oncogene 1998;17:303–311.
22. Pettersson A, Nagy JA, Brown LF, Sundberg C, Morgan E, Jungles S, Carter R, Krieger JE, Manseau EJ, Harvey VS, Eckelhoefer IA, Feng D, Dvorak AM, Mulligan RC, Dvorak HF. Heterogeneity of the angiogenic response induced in different normal adult tissues by vascular permeability factor/vascular endothelial growth factor. Lab Invest 2000;80:99–115.
23. Springer ML, Chen AS, Kraft PE, Bednarski M, Blau HM. VEGF gene delivery to muscle: potential role for vasculogenesis in adults. Mol Cell 1998;2:549–558.
24. Ferrara N. Vascular endothelial growth factor: basic science and clinical progress. Endocr Rev 2004;25: 581–611.
25. Gerber HP, Dixit V, Ferrara N. Vascular endothelial growth factor induces expression of the antiapoptotic proteins Bcl-2 and A1 in vascular endothelial cells. J Biol Chem 1998;273:13313–13316.
26. Miller-Kasprzak E, Jagodzinski PP. Endothelial progenitor cells as a new agent contributing to vascular repair. Arch Immunol Ther Exp (Warsz) 2007;55:247–259.
27. Ohneda O, Nagano M, Fujii-Kuriyama Y. Role of hypoxia-inducible factor-2alpha in endothelial development and hematopoiesis. Methods Enzymol 2007;435:199–218.
28. Brogi E, Schatteman G, Wu T, Kim EA, Varticovski L, Keyt B, Isner JM. Hypoxia-induced paracrine regulation of vascular endothelial growth factor receptor expression. J Clin Invest 1996;97:469–476.
29. Yancopoulos GD, Davis S, Gale NW, Rudge JS, Wiegand SJ, Holash J. Vascular-specific growth factors and blood vessel formation. Nature 2000;407:242–248.
30. Hood JD, Meininger CJ, Ziche M, Granger HJ. VEGF upregulates ecNOS message, protein, and NO production in human endothelial cells. Am J Physiol 1998;274:H1054–8.
31. Holderfield MT, Hughes CC. Crosstalk between vascular endothelial growth factor, notch, and transforming growth factor-beta in vascular morphogenesis. Circ Res 2008;102:637–652.
32. Bergers G, Brekken R, McMahon G, Vu TH, Itoh T, Tamaki K, Tanzawa K, Thorpe P, Itohara S, Werb Z, Hanahan D. Matrix metalloproteinase-9 triggers the angiogenic switch during carcinogenesis. Nat Cell Biol 2000;2:737–744.
33. Ferrara N, Gerber HP, LeCouter J. The biology of VEGF and its receptors. Nat Med 2003;9:669–676.
34. Hillen F, Griffioen AW. Tumour vascularization: sprouting angiogenesis and beyond. Cancer Metastasis Rev 2007;26:489–502.
35. Seghezzi G, Patel S, Ren CJ, Gualandris A, Pintucci G, Robbins ES, Shapiro RL, Galloway AC, Rifkin DB, Mignatti P. Fibroblast growth factor-2 (FGF-2) induces vascular endothelial growth factor (VEGF) expression in the endothelial cells of forming capillaries: an autocrine mechanism contributing to angiogenesis. J Cell Biol 1998;141:1659–1673.
36. Claffey KP, Abrams K, Shih SC, Brown LF, Mullen A, Keough M. Fibroblast growth factor 2 activation of stromal cell vascular endothelial growth factor expression and angiogenesis. Lab Invest 2001;81:61–75.
37. Murakami M, Simons M. Fibroblast growth factor regulation of neovascularization. Curr Opin Hematol 2008;15:215–220.
38. Itoh N, Ornitz DM. Evolution of the Fgf and Fgfr gene families. Trends Genet 2004;20:563–569.
39. Nissen LJ, Cao R, Hedlund EM, Wang Z, Zhao X, Wetterskog D, Funa K, Brakenhielm E, Cao Y. Angiogenic factors FGF2 and PDGF-BB synergistically promote murine tumor neovascularization and metastasis. J Clin Invest 2007;117:2766–2777.
40. Millette E, Rauch BH, Kenagy RD, Daum G, Clowes AW. Platelet-derived growth factor-BB transactivates the fibroblast growth factor receptor to induce proliferation in human smooth muscle cells. Trends Cardiovasc Med 2006;16:25–28.
41. Andrae J, Gallini R, Betsholtz C. Role of platelet-derived growth factors in physiology and medicine. Genes Dev 2008;22:1276–1312.
42. Bohnsack BL, Hirschi KK. Red light, green light: signals that control endothelial cell proliferation during embryonic vascular development. Cell Cycle 2004;3:1506–1511.
43. von Tell D, Armulik A, Betsholtz C. Pericytes and vascular stability. Exp Cell Res 2006;312:623–629.
44. Carmeliet P. Angiogenesis in health and disease. Nat Med 2003;9:653–660.
45. Elliott RL, Blobe GC. Role of transforming growth factor beta in human cancer. J Clin Oncol 2005;23: 2078–2093.
46. Pepper MS. Transforming growth factor-beta: vasculogenesis, angiogenesis, and vessel wall integrity. Cytokine Growth Factor Rev 1997;8:21–43.
47. Pepper MS, Vassalli JD, Orci L, Montesano R. Biphasic effect of transforming growth factor-beta 1 on in vitro angiogenesis. Exp Cell Res 1993;204:356–363.

48. Dickson MC, Martin JS, Cousins FM, Kulkarni AB, Karlsson S, Akhurst RJ. Defective haematopoiesis and vasculogenesis in transforming growth factor-beta 1 knock out mice. Development 1995;121:1845–1854.
49. Goumans MJ, Lebrin F, Valdimarsdottir G. Controlling the angiogenic switch: a balance between two distinct TGF-b receptor signaling pathways. Trends Cardiovasc Med 2003;13:301–307.
50. Otrock ZK, Mahfouz RA, Makarem JA, Shamseddine AI. Understanding the biology of angiogenesis: review of the most important molecular mechanisms. Blood Cells Mol Dis 2007;39:212–220.
51. Pertovaara L, Kaipainen A, Mustonen T, Orpana A, Ferrara N, Saksela O, Alitalo K. Vascular endothelial growth factor is induced in response to transforming growth factor-beta in fibroblastic and epithelial cells. J Biol Chem 1994;269:6271–6274.
52. Molin DG, DeRuiter MC, Wisse LJ, Azhar M, Doetschman T, Poelmann RE, Gittenberger-de Groot AC. Altered apoptosis pattern during pharyngeal arch artery remodelling is associated with aortic arch malformations in Tgfbeta2 knock-out mice. Cardiovasc Res 2002;56:312–322.
53. Stalmans I, Lambrechts D, De Smet F, Jansen S, Wang J, Maity S, Kneer P, von der Ohe M, Swillen A, Maes C, Gewillig M, Molin DG, Hellings P, Boetel T, Haardt M, Compernolle V, Dewerchin M, Plaisance S, Vlietinck R, Emanuel B, Gittenberger-de Groot AC, Scambler P, Morrow B, Driscol DA, Moons L, Esguerra CV, Carmeliet G, Behn-Krappa A, Devriendt K, Collen D, Conway SJ, Carmeliet P. VEGF: a modifier of the del22q11 (DiGeorge) syndrome? Nat Med 2003;9:173–182.
54. Bao P, Kodra A, Tomic-Canic M, Golinko MS, Ehrlich HP, Brem H. The role of vascular endothelial growth factor in wound healing. J Surg Res (in press, 2008).
55. McGrath MH, Emery JM, 3rd. The effect of inhibition of angiogenesis in granulation tissue on wound healing and the fibroblast. Ann Plast Surg 1985;15:105–122.
56. Nissen NN, Polverini PJ, Koch AE, Volin MV, Gamelli RL, DiPietro LA. Vascular endothelial growth factor mediates angiogenic activity during the proliferative phase of wound healing. Am J Pathol 1998;152: 1445–1452.
57. Gerber HP, Kowalski J, Sherman D, Eberhard DA, Ferrara N. Complete inhibition of rhabdomyosarcoma xenograft growth and neovascularization requires blockade of both tumor and host vascular endothelial growth factor. Cancer Res 2000;60:6253–6258.
58. Fukumura D, Xavier R, Sugiura T, Chen Y, Park EC, Lu N, Selig M, Nielsen G, Taksir T, Jain RK, Seed B. Tumor induction of VEGF promoter activity in stromal cells. Cell 1998;94:715–725.
59. Kishimoto J, Ehama R, Ge Y, Kobayashi T, Nishiyama T, Detmar M, Burgeson RE. In vivo detection of human vascular endothelial growth factor promoter activity in transgenic mouse skin. Am J Pathol 2000;157: 103–110.
60. Tsuzuki Y, Fukumura D, Oosthuyse B, Koike C, Carmeliet P, Jain RK. Vascular endothelial growth factor (VEGF) modulation by targeting hypoxia-inducible factor-1alpha–> hypoxia response element–> VEGF cascade differentially regulates vascular response and growth rate in tumors. Cancer Res 2000;60:6248–6252.
61. Presta M, Dell'Era P, Mitola S, Moroni E, Ronca R, Rusnati M. Fibroblast growth factor/fibroblast growth factor receptor system in angiogenesis. Cytokine Growth Factor Rev 2005;16:159–178.
62. Gross JL, Herblin WF, Dusak BA, Czerniak P, Diamond MD, Sun T, Eidsvoog K, Dexter DL, Yayon A. Effects of modulation of basic fibroblast growth factor on tumor growth in vivo. J Natl Cancer Inst 1993;85: 121–131.
63. Colnot C. Cellular and molecular interactions regulating skeletogenesis. J Cell Biochem 2005;95:688–697.
64. Hall BK, Miyake T. All for one and one for all: condensations and the initiation of skeletal development. BioEssays 2000;22:138–147.
65. Colnot C, Lu C, Hu D, Helms JA. Distinguishing the contributions of the perichondrium, cartilage, and vascular endothelium to skeletal development. Dev Biol 2004;269:55–69.
66. Maes C, Carmeliet P, Moermans K, Stockmans I, Smets N, Collen D, Bouillon R, Carmeliet G. Impaired angiogenesis and endochondral bone formation in mice lacking the vascular endothelial growth factor isoforms VEGF164 and VEGF188. Mech Dev 2002;111:61–73.
67. Engsig MT, Chen QJ, Vu TH, Pedersen AC, Therkidsen B, Lund LR, Henriksen K, Lenhard T, Foged NT, Werb Z, Delaisse JM. Matrix metalloproteinase 9 and vascular endothelial growth factor are essential for osteoclast recruitment into developing long bones. J Cell Biol 2000;151:879–889.
68. Urist MR. Bone: formation by autoinduction. 1965. Clin Orthop Relat Res 2002;(395):4–10.
69. Chen D, Zhao M, Mundy GR. Bone morphogenetic proteins. Growth Factors 2004;22:233–241.
70. Kroese-Deutman HC, Ruhe PQ, Spauwen PH, Jansen JA. Bone inductive properties of rhBMP-2 loaded porous calcium phosphate cement implants inserted at an ectopic site in rabbits. Biomaterials 2005;26: 1131–1138.
71. Wang EA, Israel DI, Kelly S, Luxenberg DP. Bone morphogenetic protein-2 causes commitment and differentiation in C3H10T1/2 and 3T3 cells. Growth Factors 1993;9:57–71.

72. Asahina I, Sampath TK, Hauschka PV. Human osteogenic protein-1 induces chondroblastic, osteoblastic, and/or adipocytic differentiation of clonal murine target cells. Exp Cell Res 1996;222:38–47.
73. Cheng H, Jiang W, Phillips FM, Haydon RC, Peng Y, Zhou L, Luu HH, An N, Breyer B, Vanichakarn P, Szatkowski JP, Park JY, He TC. Osteogenic activity of the fourteen types of human bone morphogenetic proteins (BMPs). J Bone Joint Surg Am 2003;85-A:1544–1552.
74. Giannoudis PV, Kanakaris NK, Einhorn TA. Interaction of bone morphogenetic proteins with cells of the osteoclast lineage: review of the existing evidence. Osteoporos Int 2007;18:1565–1581.
75. Wutzl A, Brozek W, Lernbass I, Rauner M, Hofbauer G, Schopper C, Watzinger F, Peterlik M, Pietschmann P. Bone morphogenetic proteins 5 and 6 stimulate osteoclast generation. J Biomed Mater Res A 2006;77:75–83.
76. Canalis E, Raisz LG. Effect of fibroblast growth factor on cultured fetal rat calvaria. Metabolism 1980;29: 108–114.
77. Rodan SB, Wesolowski G, Thomas KA, Yoon K, Rodan GA. Effects of acidic and basic fibroblast growth factors on osteoblastic cells. Connect Tissue Res 1989;20:283–288.
78. Quarto N, Longaker MT. FGF-2 inhibits osteogenesis in mouse adipose tissue-derived stromal cells and sustains their proliferative and osteogenic potential state. Tissue Eng 2006;12:1405–1418.
79. McCarthy TL, Centrella M, Canalis E. Effects of fibroblast growth factors on deoxyribonucleic acid and collagen synthesis in rat parietal bone cells. Endocrinology 1989;125:2118–2126.
80. Kwan MD, Slater BJ, Wan DC, Longaker MT. Cell-based therapies for skeletal regenerative medicine. Hum Mol Genet 2008;17:R93–R98.
81. Lieberman JR, Daluiski A, Einhorn TA. The role of growth factors in the repair of bone. Biology and clinical applications. J Bone Joint Surg Am 2002;84-A:1032–1044.
82. Hughes FJ, Turner W, Belibasakis G, Martuscelli G. Effects of growth factors and cytokines on osteoblast differentiation. Periodontol 2000 2006;41:48–72.
83. Simpson AH, Mills L, Noble B. The role of growth factors and related agents in accelerating fracture healing. J Bone Joint Surg Br 2006;88:701–705.
84. Hadjidakis DJ, Androulakis II. Bone remodeling. Ann N Y Acad Sci 2006;1092:385–396.
85. Hu YY, Zhang C, Lu R, Xu JQ, Li D. Repair of radius defect with bone-morphogenetic-protein loaded hydroxyapatite/collagen-poly(l-lactic acid) composite. Chin J Traumatol 2003;6:67–74.
86. Conover CA. Insulin-like growth factor-binding proteins and bone metabolism. Am J Physiol Endocrinol Metab 2008;294:E10–E14.
87. Hoeflich A, Gotz W, Lichanska AM, Bielohuby M, Tonshoff B, Kiepe D. Effects of insulin-like growth factor binding proteins in bone – a matter of cell and site. Arch Physiol Biochem 2007;113:142–153.
88. Lynch SE, Buser D, Hernandez RA, Weber HP, Stich H, Fox CH, Williams RC. Effects of the platelet-derived growth factor/insulin-like growth factor-I combination on bone regeneration around titanium dental implants. Results of a pilot study in beagle dogs. J Periodontol 1991;62:710–716.
89. Raschke M, Wildemann B, Inden P, Bail H, Flyvbjerg A, Hoffmann J, Haas NP, Schmidmaier G. Insulin-like growth factor-1 and transforming growth factor-beta1 accelerates osteotomy healing using polylactide-coated implants as a delivery system: a biomechanical and histological study in minipigs. Bone 2002;30:144–151.
90. Takita H, Tsuruga E, Ono I, Kuboki Y. Enhancement by bFGF of osteogenesis induced by rhBMP-2 in rats. Eur J Oral Sci 1997;105:588–592.
91. Centrella M, McCarthy TL, Kusmik WF, Canalis E. Relative binding and biochemical effects of heterodimeric and homodimeric isoforms of platelet-derived growth factor in osteoblast-enriched cultures from fetal rat bone. J Cell Physiol 1991;147:420–426.
92. Fiedler J, Etzel N, Brenner RE. To go or not to go: migration of human mesenchymal progenitor cells stimulated by isoforms of PDGF. J Cell Biochem 2004;93:990–998.
93. Mehrotra M, Krane SM, Walters K, Pilbeam C. Differential regulation of platelet-derived growth factor stimulated migration and proliferation in osteoblastic cells. J Cell Biochem 2004;93:741–752.
94. Kanaan RA, Kanaan LA. Transforming growth factor beta1, bone connection. Med Sci Monit 2006;12: RA164–RA169.
95. Centrella M, Horowitz MC, Wozney JM, McCarthy TL. Transforming growth factor-beta gene family members and bone. Endocr Rev 1994;15:27–39.
96. Robey PG, Young MF, Flanders KC, Roche NS, Kondaiah P, Reddi AH, Termine JD, Sporn MB, Roberts AB. Osteoblasts synthesize and respond to transforming growth factor-type beta (TGF-beta) in vitro. J Cell Biol 1987;105:457–463.
97. Maeda S, Hayashi M, Komiya S, Imamura T, Miyazono K. Endogenous TGF-beta signaling suppresses maturation of osteoblastic mesenchymal cells. EMBO J 2004;23:552–563.
98. Lieb E, Vogel T, Milz S, Dauner M, Schulz MB. Effects of transforming growth factor beta1 on bonelike tissue formation in three-dimensional cell culture. II: Osteoblastic differentiation. Tissue Eng 2004;10:1414–1425.

99. Bodine PV, Billiard J, Moran RA, Ponce-de-Leon H, McLarney S, Mangine A, Scrimo MJ, Bhat RA, Stauffer B, Green J, Stein GS, Lian JB, Komm BS. The Wnt antagonist secreted frizzled-related protein-1 controls osteoblast and osteocyte apoptosis. J Cell Biochem 2005;96:1212–1230.
100. Anastassiades TP, Chopra RK, Wood A. Exogenous glycosaminoglycans (GAG) differentially modulate GAG synthesis by anchorage-independent cultures of the outer cells from neonatal rat calvaria in the absence and presence of TGF-beta. Mol Cell Biochem 1996;158:25–32.
101. Takeuchi Y, Matsumoto T, Ogata E, Shishiba Y. Effects of transforming growth factor beta 1 and L-ascorbate on synthesis and distribution of proteoglycans in murine osteoblast-like cells. J Bone Miner Res 1993;8:823–830.
102. Gerstenfeld LC, Cullinane DM, Barnes GL, Graves DT, Einhorn TA. Fracture healing as a post-natal developmental process: molecular, spatial, and temporal aspects of its regulation. J Cell Biochem 2003;88:873–884.
103. Katagiri T, Yamaguchi A, Komaki M, Abe E, Takahashi N, Ikeda T, Rosen V, Wozney JM, Fujisawa-Sehara A, Suda T. Bone morphogenetic protein-2 converts the differentiation pathway of C2C12 myoblasts into the osteoblast lineage. J Cell Biol 1994;127:1755–1766.
104. Bostrom MP. Expression of bone morphogenetic proteins in fracture healing. Clin Orthop Relat Res 1998;(355 Suppl):S116-S123.
105. Ishidou Y, Kitajima I, Obama H, Maruyama I, Murata F, Imamura T, Yamada N, ten Dijke P, Miyazono K, Sakou T. Enhanced expression of type I receptors for bone morphogenetic proteins during bone formation. |J Bone Miner Res 1995;10:1651–1659.
106. Bourque WT, Gross M, Hall BK. Expression of four growth factors during fracture repair. Int J Dev Biol 1993;37:573–579.
107. Andrew JG, Hoyland J, Freemont AJ, Marsh D. Insulinlike growth factor gene expression in human fracture callus. Calcif Tissue Int 1993;53:97–102.
108. Rydziel S, Shaikh S, Canalis E. Platelet-derived growth factor-AA and -BB (PDGF-AA and -BB) enhance the synthesis of PDGF-AA in bone cell cultures. Endocrinology 1994;134:2541–2546.
109. Andrew JG, Hoyland JA, Freemont AJ, Marsh DR. Platelet-derived growth factor expression in normally healing human fractures. Bone 1995;16:455–460.
110. Joyce ME, Jingushi S, Bolander ME. Transforming growth factor-beta in the regulation of fracture repair. Orthop Clin North Am 1990;21:199–209.
111. de Vernejoul MC. Sclerosing bone disorders. Best Pract Res Clin Rheumatol 2008;22:71–83.
112. Ornitz DM. FGF signaling in the developing endochondral skeleton. Cytokine Growth Factor Rev 2005;16: 205–213.
113. Li B. Bone morphogenetic protein-Smad pathway as drug targets for osteoporosis and cancer therapy. Endocr Metab Immune Disord Drug Targets 2008;8:208–219.
114. Niu T, Rosen CJ. The insulin-like growth factor-I gene and osteoporosis: a critical appraisal. Gene 2005; 361:38–56.
115. Langlois JA, Rosen CJ, Visser M, Hannan MT, Harris T, Wilson PW, Kiel DP. Association between insulin-like growth factor I and bone mineral density in older women and men: the Framingham Heart Study. J Clin Endocrinol Metab 1998;83:4257–4262.
116. Fromigue O, Modrowski D, Marie PJ. Growth factors and bone formation in osteoporosis: roles for fibroblast growth factor and transforming growth factor beta. Curr Pharm Des 2004;10:2593–2603.
117. Liang H, Pun S, Wronski TJ. Bone anabolic effects of basic fibroblast growth factor in ovariectomized rats. Endocrinology 1999;140:5780–5788.
118. Dunstan CR, Boyce R, Boyce BF, Garrett IR, Izbicka E, Burgess WH, Mundy GR. Systemic administration of acidic fibroblast growth factor (FGF-1) prevents bone loss and increases new bone formation in ovariectomized rats. J Bone Miner Res 1999;14:953–959.
119. Kuhl PR, Griffith-Cima LG. Tethered epidermal growth factor as a paradigm for growth factor-induced stimulation from the solid phase. Nat Med 1996;2:1022–1027.
120. Schuessele A, Mayr H, Tessmar J, Goepferich A. Enhanced bone morphogenetic protein-2 performance on hydroxyapatite ceramic surfaces. J Biomed Mater Res A 2008.
121. Koch S, Yao C, Grieb G, Prevel P, Noah EM, Steffens GC. Enhancing angiogenesis in collagen matrices by covalent incorporation of VEGF. J Mater Sci Mater Med 2006;17:735–741.
122. Zisch AH, Schenk U, Schense JC, Sakiyama-Elbert SE, Hubbell JA. Covalently conjugated VEGF–fibrin matrices for endothelialization. J Control Release 2001;72:101–113.
123. Zhang G, Suggs LJ. Matrices and scaffolds for drug delivery in vascular tissue engineering. Adv Drug Deliv Rev 2007;59:360–373.
124. Kano MR, Morishita Y, Iwata C, Iwasaka S, Watabe T, Ouchi Y, Miyazono K, Miyazawa K. VEGF-A and FGF-2 synergistically promote neoangiogenesis through enhancement of endogenous PDGF-B-PDGFRbeta signaling. J Cell Sci 2005;118:3759–3768.

125. Cao R, Brakenhielm E, Pawliuk R, Wariaro D, Post MJ, Wahlberg E, Leboulch P, Cao Y. Angiogenic synergism, vascular stability and improvement of hind-limb ischemia by a combination of PDGF-BB and FGF-2. Nat Med 2003;9:604–613.
126. Hao X, Silva EA, Mansson-Broberg A, Grinnemo KH, Siddiqui AJ, Dellgren G, Wardell E, Brodin LA, Mooney DJ, Sylven C. Angiogenic effects of sequential release of VEGF-A165 and PDGF-BB with alginate hydrogels after myocardial infarction. Cardiovasc Res 2007;75:178–185.
127. Rophael JA, Craft RO, Palmer JA, Hussey AJ, Thomas GP, Morrison WA, Penington AJ, Mitchell GM. Angiogenic growth factor synergism in a murine tissue engineering model of angiogenesis and adipogenesis. Am J Pathol 2007;171:2048–2057.
128. Lee KW, Yoon JJ, Lee JH, Kim SY, Jung HJ, Kim SJ, Joh JW, Lee HH, Lee DS, Lee SK. Sustained release of vascular endothelial growth factor from calcium-induced alginate hydrogels reinforced by heparin and chitosan. Transplant Proc 2004;36:2464–2465.
129. Ishihara M, Obara K, Ishizuka T, Fujita M, Sato M, Masuoka K, Saito Y, Yura H, Matsui T, Hattori H, Kikuchi M, Kurita A. Controlled release of fibroblast growth factors and heparin from photocrosslinked chitosan hydrogels and subsequent effect on in vivo vascularization. J Biomed Mater Res A 2003;64:551–559.
130. Zhu XH, Tabata Y, Wang CH, Tong YW. Delivery of basic fibroblast growth factor from gelatin microsphere scaffold for the growth of human umbilical vein endothelial cells. Tissue Eng Part A 2008;14:1939–1947.
131. Yao C, Prevel P, Koch S, Schenck P, Noah EM, Pallua N, Steffens G. Modification of collagen matrices for enhancing angiogenesis. Cells Tissues Organs 2004;178:189–196.
132. Schmidmaier G, Wildemann B, Lubberstedt M, Haas NP, Raschke M. IGF-I and TGF-beta 1 incorporated in a poly(D,L-lactide) implant coating stimulates osteoblast differentiation and collagen-1 production but reduces osteoblast proliferation in cell culture. J Biomed Mater Res B. Appl Biomater 2003;65:157–162.
133. Schmidmaier G, Wildemann B, Bail H, Lucke M, Fuchs T, Stemberger A, Flyvbjerg A, Haas NP, Raschke M. Local application of growth factors (insulin-like growth factor-1 and transforming growth factor-beta1) from a biodegradable poly(D,L-lactide) coating of osteosynthetic implants accelerates fracture healing in rats. Bone 2001;28:341–350.
134. Schmidmaier G, Wildemann B, Gabelein T, Heeger J, Kandziora F, Haas NP, Raschke M. Synergistic effect of IGF-I and TGF-beta1 on fracture healing in rats: single versus combined application of IGF-I and TGF-beta1. Acta Orthop Scand 2003;74:604–610.
135. Kandziora F, Pflugmacher R, Scholz M, Schafer J, Schollmeier G, Schmidmaier G, Duda G, Raschke M, Haas NP. Dose-dependent effects of combined IGF-I and TGF-beta1 application in a sheep cervical spine fusion model. Eur Spine J 2003;12:464–473.
136. Raiche AT, Puleo DA. Cell responses to BMP-2 and IGF-I released with different time-dependent profiles. J Biomed Mater Res A 2004;69:342–350.
137. Friess W, Uludag H, Foskett S, Biron R, Sargeant C. Characterization of absorbable collagen sponges as rhBMP-2 carriers. Int J Pharm 1999;187:91–99.
138. Geiger M, Li RH, Friess W. Collagen sponges for bone regeneration with rhBMP-2. Adv Drug Deliv Rev 2003;55:1613–1629.
139. Murakami N, Saito N, Takahashi J, Ota H, Horiuchi H, Nawata M, Okada T, Nozaki K, Takaoka K. Repair of a proximal femoral bone defect in dogs using a porous surfaced prosthesis in combination with recombinant BMP-2 and a synthetic polymer carrier. Biomaterials 2003;24:2153–2159.
140. Kaito T, Myoui A, Takaoka K, Saito N, Nishikawa M, Tamai N, Ohgushi H, Yoshikawa H. Potentiation of the activity of bone morphogenetic protein-2 in bone regeneration by a PLA-PEG/hydroxyapatite composite. Biomaterials 2005;26:73–79.
141. Wikesjo UM, Lim WH, Thomson RC, Cook AD, Wozney JM, Hardwick WR. Periodontal repair in dogs: evaluation of a bioabsorbable space-providing macroporous membrane with recombinant human bone morphogenetic protein-2. J Periodontol 2003;74:635–647.
142. Mayer M, Hollinger J, Ron E, Wozney J. Maxillary alveolar cleft repair in dogs using recombinant human bone morphogenetic protein-2 and a polymer carrier. Plast Reconstr Surg 1996;98:247–259.
143. PC, Chang BY, Liu CM. Liu 2004: Bone tissue engineering with novel rhBMP-2-PLLA composite scaffolds. Journal of Biomedical Research A 771–780.
144. Mabilleau G, Aguado E, Stancu IC, Cincu C, Basle MF, Chappard D. Effects of FGF-2 release from a hydrogel polymer on bone mass and microarchitecture. Biomaterials 2008;29:1593–1600.
145. Woo BH, Fink BF, Page R, Schrier JA, Jo YW, Jiang G, DeLuca M, Vasconez HC, DeLuca PP. Enhancement of bone growth by sustained delivery of recombinant human bone morphogenetic protein-2 in a polymeric matrix. Pharm Res 2001;18:1747–1753.
146. Ripamonti U, Crooks J, Rueger DC. Induction of bone formation by recombinant human osteogenic protein-1 and sintered porous hydroxyapatite in adult primates. Plast Reconstr Surg 2001;107:977–988.

147. Kim SJ, Kim SY, Kwon CH, Kim YK. Differential effect of FGF and PDGF on cell proliferation and migration in osteoblastic cells. Growth Factors 2007;25:77–86.
148. Ten Dijke P. Bone morphogenetic protein signal transduction in bone. Curr Med Res Opin 2006;22 Suppl 1:S7–S11.
149. Puleo D. Biotherapeutics in orthopaedic medicine: accelerating the healing process? BioDrugs 2003;17: 301–314.
150. Ehrbar M, Djonov VG, Schnell C, Tschanz SA, Martiny-Baron G, Schenk U, Wood J, Burri PH, Hubbell JA, Zisch AH. Cell-demanded liberation of VEGF121 from fibrin implants induces local and controlled blood vessel growth. Circ Res 2004;94:1124–1132.
151. Ehrbar M, Metters A, Zammaretti P, Hubbell JA, Zisch AH. Endothelial cell proliferation and progenitor maturation by fibrin-bound VEGF variants with differential susceptibilities to local cellular activity. J Control Release 2005;101:93–109.
152. Gautschi OP, Frey SP, Zellweger R. Bone morphogenetic proteins in clinical applications. ANZ J Surg 2007;77:626–631.
153. Hollinger JO, Schmitt JM, Buck DC, Shannon R, Joh SP, Zegzula HD, Wozney J. Recombinant human bone morphogenetic protein-2 and collagen for bone regeneration. J Biomed Mater Res 1998;43:356–364.
154. Giannobile WV, Finkelman RD, Lynch SE. Comparison of canine and non-human primate animal models for periodontal regenerative therapy: results following a single administration of PDGF/IGF-I. J Periodontol 1994;65:1158–1168.
155. Howell TH, Fiorellini JP, Paquette DW, Offenbacher S, Giannobile WV, Lynch SE. A phase I/II clinical trial to evaluate a combination of recombinant human platelet-derived growth factor-BB and recombinant human insulin-like growth factor-I in patients with periodontal disease. J Periodontol 1997;68:1186–1193.
156. Wildemann B, Lubberstedt M, Haas NP, Raschke M, Schmidmaier G. IGF-I and TGF-beta 1 incorporated in a poly(D,L-lactide) implant coating maintain their activity over long-term storage-cell culture studies on primary human osteoblast-like cells. Biomaterials 2004;25:3639–3644.
157. Ripamonti U, Duneas N, Van Den Heever B, Bosch C, Crooks J. Recombinant transforming growth factor-beta1 induces endochondral bone in the baboon and synergizes with recombinant osteogenic protein-1 (bone morphogenetic protein-7) to initiate rapid bone formation. J Bone Miner Res 1997;12:1584–1595.
158. Simmons CA, Alsberg E, Hsiong S, Kim WJ, Mooney DJ. Dual growth factor delivery and controlled scaffold degradation enhance in vivo bone formation by transplanted bone marrow stromal cells. Bone 2004;35: 562–569.
159. Marden LJ, Fan RS, Pierce GF, Reddi AH, Hollinger JO. Platelet-derived growth factor inhibits bone regeneration induced by osteogenin, a bone morphogenetic protein, in rat craniotomy defects. J Clin Invest 1993;92: 2897–2905.
160. Ley CD, Olsen MW, Lund EL, Kristjansen PE. Angiogenic synergy of bFGF and VEGF is antagonized by Angiopoietin-2 in a modified in vivo Matrigel assay. Microvasc Res 2004;68:161–168.
161. Richardson TP, Peters MC, Ennett AB, Mooney DJ. Polymeric system for dual growth factor delivery. Nat Biotechnol 2001;19:1029–1034.
162. Huang YC, Kaigler D, Rice KG, Krebsbach PH, Mooney DJ. Combined angiogenic and osteogenic factor delivery enhances bone marrow stromal cell-driven bone regeneration. J Bone Miner Res 2005;20:848–857.
163. Kanczler JM, Oreffo RO. Osteogenesis and angiogenesis: the potential for engineering bone. Eur Cell Mater 2008;15:100–114.
164. Huang W, Carlsen B, Wulur I, Rudkin G, Ishida K, Wu B, Yamaguchi DT, Miller TA. BMP-2 exerts differential effects on differentiation of rabbit bone marrow stromal cells grown in two-dimensional and three-dimensional systems and is required for in vitro bone formation in a PLGA scaffold. Exp Cell Res 2004;299:325–334.
165. Dai J, Rabie AB. VEGF: an essential mediator of both angiogenesis and endochondral ossification. J Dent Res 2007;86:937–950.
166. Patel ZS, Young S, Tabata Y, Jansen JA, Wong ME, Mikos AG. Dual delivery of an angiogenic and an osteogenic growth factor for bone regeneration in a critical size defect model. Bone 2008;43:931–940.
167. Raiche AT, Puleo DA. In vitro effects of combined and sequential delivery of two bone growth factors. Biomaterials 2004;25:677–685.

11

In Vitro and In Vivo Monocyte, Macrophage, Foreign Body Giant Cell, and Lymphocyte Interactions with Biomaterials

James M. Anderson

This chapter focuses on recent developments in our understanding of in vitro and in vivo monocyte, macrophage, foreign body giant cell (FBGC), and lymphocyte interactions with biomaterials in the context of inflammatory and wound-healing responses following implantation of biomaterials, medical devices, and prostheses. Quantitative studies identify the significance of biomaterial surface chemistry in modulating inflammatory cell behaviors such as adhesion, apoptosis, anoikis, fusion, and cytokine secretion. Utilizing genomic and proteomic techniques, cell-cell (juxtacrine) and cytokine-mediated (paracrine) responses between cells have been identified. Consequences of the persistence of the foreign body reaction, i.e., macrophages and FBGCs, at the tissue-material interface are presented. Given the current and future use of new approaches, such as nanotechnology and tissue engineering, the continuing development of a mechanistic understanding of inflammatory cell interactions with biomaterials is necessary.

Abbreviations

DC-STAMP	dendritic cell-specific transmembrane protein
ECM	extracellular matrix
ERK	extracellular signal-regulated kinase
FBGC	foreign body giant cell
FAK	focal adhesion kinase
G-CSF	granulocyte colony-stimulating factor
GM-CSF	granulocyte macrophage colony-stimulating factor
GRO/KC	growth-related oncogene/keratinocyte chemoattractant
ICAM	intercellular adhesion molecule
IFN	interferon
IL	interleukin
IL-1RA	interleukin-1 receptor antagonist

J.M. Anderson • Departments of Pathology and Biomedical Engineering, Case Western Reserve University, Cleveland, OH 44106, USA

D.A. Puleo and R. Bizios (eds.), *Biological Interactions on Materials Surfaces*,
DOI 10.1007/978-0-387-98161-1_11, © Springer Science+Business Media, LLC 2009

LFA	lymphocyte function-associated antigen
LPS	lipopolysaccharide
LT	leukotriene
MCP	monocyte chemoattractant protein
MDC	macrophage-derived chemokine
MIP	macrophage inhibitory protein
MIPs	macrophage inflammatory proteins
MMP	matrix metalloproteinase
mRNA	messenger RNA
NK	natural killer cells
NKT	natural killer T cells
PDGF	platelet-derived growth factor
PEU	polyurethane
PF4	platelet factor 4
RANTES	regulated upon activation, normal T-cell expressed and secreted
RT-PCR	real-time polymerase chain reaction
TCR	T-cell receptor
TGF	transforming growth factor
TIMP	tissue inhibitor of metalloproteinase
TNF	tumor necrosis factor
VLA	very late antigen receptor

11.1. Introduction

Following implantation of biomaterials, a series of biological responses that affect the implanted material and surrounding tissue ensues, including injury, blood-material interactions, provisional matrix formation, acute inflammation, chronic inflammation, granulation tissue development, foreign body reaction, and the development of a fibrous capsule (Figure 11.1) [1–4]. Upon introduction into the physiologic environment, blood-material interactions occur immediately, with serum proteins adsorbing to the surface of the material. Furthermore, vascular injury results in thrombosis on and around the implanted material that coincides with activation of the extrinsic and intrinsic coagulation systems, the complement system, the fibrinolytic system, the kinin-generating system, and platelets. These activation pathways are involved in the protein adsorption and desorption phenomena on material surfaces known as the Vroman Effect [5]. Blood protein deposition on a biomaterial surface may be viewed as the formation of a provisional matrix, which provides structural, biochemical, and cellular components to the processes of wound healing and foreign body reaction. In addition to initiation of thrombus formation, injury that occurs because of the implantation process leads to initiation of inflammatory responses (innate immunity), ultimately resulting in the presence of mitogens, chemoattractants, cytokines, growth factors, and other bioactive agents within the provisional matrix. This provides a rich milieu of activating and inhibiting substances capable of modulating macrophage activity and proliferation and activation of other cell populations in the inflammatory and wound-healing responses. In this regard, the provisional matrix is a biodegradable sustained release system releasing bioactive agents that coordinate the sustained responses to the implanted material.

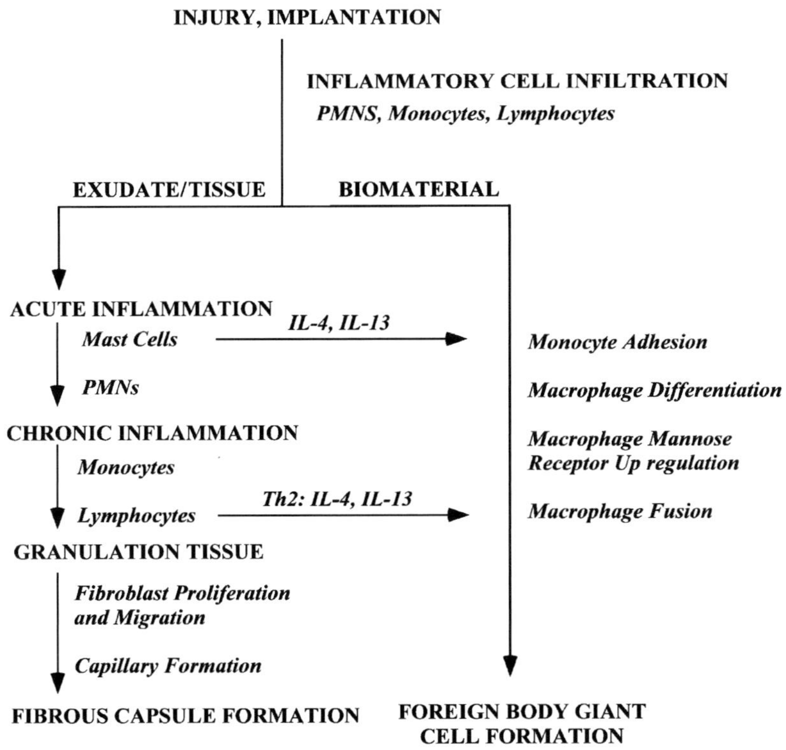

Figure 11.1. Sequence of events involved in inflammatory and wound-healing responses leading to foreign body giant cell formation. This shows the potential importance of mast cells in the acute inflammatory phase and Th2 lymphocytes in the transient chronic inflammatory phase with the production of IL-4 and IL-13, which can induce monocyte/macrophage fusion to form foreign body giant cells.

During, and continuing beyond, the initial blood-material interactions and formation of the provisional matrix, an acute inflammatory response ensues, and eventually gives way to chronic inflammation. Several aspects dictate the extent or degree of acute and chronic inflammation, including the extent of injury due to the implantation procedure, the tissue or organ into which the device is implanted, and the extent of provisional matrix formation. The acute inflammatory response is characterized by infiltration of neutrophils (polymorphonuclear leukocytes) to the implant site as well as mast cell degranulation. Mast cell degranulation releases histamine, which, along with fibrinogen adsorption, mediates the acute inflammatory responses to implanted biomaterials [6, 7]. During degranulation, mast cells also release interleukin (IL)-4 and IL-13, which play roles in determining the extent and degree of the subsequent foreign body reaction [8, 9]. Notably, histamine-mediated phagocyte recruitment and phagocyte adhesion to implant surfaces facilitated by adsorbed fibrinogen and/or complement may modulate inflammatory and wound-healing responses. The acute inflammatory response associated with implanted biomaterials resolves within 1 week after implantation; the extent and timing of resolution is dependent on the extent of injury resulting from implantation.

Chronic inflammation is characterized by the presence of mononuclear cells, such as monocytes and lymphocytes, at the implant site. Chronic inflammation describes a range of cellular responses that is less histologically uniform than acute inflammation, and includes the presence of lymphocytes, plasma cells, monocytes, macrophages, and foreign body giant cells (FBGC). This chronic inflammatory response to implanted biomaterials is usually of short duration (no longer than 2 weeks with biocompatible materials) and is confined to the implant site. Failure of quick resolution of chronic inflammation is likely due to an infection. Following the resolution of chronic inflammation, granulation tissue becomes apparent, being characterized by the presence of macrophages, infiltrating fibroblasts, and neovascularization in the new healing tissue. Granulation tissue is separated from the implanted material by the cellular components of the foreign body reaction: a one- to two-cell layer of monocytes, macrophages, and FBGC. Also, the granulation tissue acts as a precursor to the fibrous capsule, which eventually isolates the implant from the surrounding tissue.

11.2. Monocytes, Macrophages, and FBGCs

11.2.1. Protein Adsorption on Biomaterial Surfaces

Immediately upon introduction of biomaterials and medical devices into the body, a layer of host proteins adsorbs to the surface, mediating further interactions with host cells. It is therefore likely that the types, levels, and surface conformations of adsorbed proteins are critical determinants of the tissue reaction to implanted devices/materials [10]. Furthermore, the physical and chemical surface properties of the implanted material dictate the types, concentrations, and conformations of surface-adsorbed proteins, which consequently control adhesion and survival of cells, especially monocytes, macrophages, and FBGCs, on protein-coated surfaces. Therefore, the interaction of adhesion receptors on inflammatory cell surfaces with adsorbed proteins constitutes the major cellular recognition system for implantable synthetic materials and medical devices and remains a target for potential control of host reactions to implanted materials. Hence, adsorption of proteins, such as albumin, complement, fibronectin, fibrinogen, γ-globulin, vitronectin, and others, may modulate inflammatory cell interactions and adhesion and are linked to the subsequent inflammatory and wound-healing responses [11–14]. It should be noted that adsorbed proteins may also desorb rapidly, leading to time-dependent variations in the types and levels of proteins that cells may encounter at the surface of the material in the in vivo environment [15].

Complement is a group of serum proteins that act to clear pathogens through a series of enzymatic reactions. In terms of implanted biomaterials, complement activation has been associated with adverse side effects on implanted devices, including hemodialyzers, oxygenators, catheters, prostheses, stents, vascular grafts, and others.

Cross-talk between different blood protein cascades and leukocytes and platelets has been well documented. For example, regarding biomaterial surfaces, complement and leukocytes are involved in biomaterial-associated thrombosis, as well as the more commonly identified coagulation factors and platelets. Therefore, the effect of a biomaterial surface on blood components can be identified only when whole blood is used [16, 17].

11.2.2. Monocyte/Macrophage Migration and Adhesion

Monocytes/macrophages are central to the host response to implanted materials, as the progression of events requires their extravasation and migration to the site of the implant. Appropriately, chemokines and other chemoattractants are key factors in guiding the movement of monocytes/macrophages to the site of the implant. Four families of chemokines exist based on the spacing between the first two terminal cysteine residues: CC, CXC, C, and CX_3C [18]. In addition to orchestrating cellular migration and the host response to implanted materials, chemokines also play roles in angiogenesis, hematopoiesis, lymphocyte differentiation, lymphocyte migration, and tumor metastasis [18–20].

The interactions between blood and implanted biomaterials result in chemoattractants being released from platelets and the formed thrombus. These factors attract macrophages to the site of the implant and include CXCL4 (platelet factor 4 [PF4]), IL-1, leukotriene B4 (LTB4), platelet-derived growth factor (PDGF), and transforming growth factor (TGF)-β [21]. Mast cell degranulation and consequent release of histamine play an integral role in recruiting phagocytes, especially macrophages, to the site of the implant [7]. Once at the site of the implanted material, macrophages release additional chemoattractants, including granulocyte colony-stimulating factor (G-CSF), granulocyte macrophage colony-stimulating factor (GM-CSF), IL-6, PDGF, and tumor necrosis factor (TNF)-α [21]. These factors further recruit additional macrophages to the area, acting as a positive feedback mechanism. In fact, CCL2 (monocyte chemoattractant protein [MCP]-1) is expressed by macrophages present in exudates surrounding implanted polyethylene materials [22]. Further studies have shown that CCL2 (MCP-1), CCL4 (macrophage inhibitory protein [MIP]-1β), CCL13 (MCP-4), and CCL22 (macrophage-derived chemokine [MDC]) are released by biomaterial-adherent macrophages [23]. CCL2 (MCP-1) and CCL5 (regulated upon activation, normal T-cell expressed and secreted [RANTES]), CCL3 (MIP-1α), CCL4 (MIP-1β), CCL7 (MCP-3), CCL8 (MCP-2), and CCL13 (MCP-4) have been shown to attract monocytes/macrophages [18, 24]. CCL2 (MCP-1) does not appear to influence the recruitment of monocytes to subcutaneous implant sites [25]. Once at the implant site or biomaterial surface, the macrophages can then adhere and engage in the subsequent events of the foreign body reaction (Figure 11.2). Importantly, as the inflammatory response resolves, the levels of chemoattractants (IL-8 and MIP-1β) being released by material-adherent macrophages decrease, indicating that upon the resolution of inflammation, no further recruitment of macrophages occurs [23].

Once at the site of the implant, the macrophages recognize the adsorbed blood protein-modified material surface as the substrate of adhesion through macrophage surface integrin receptors. Integrins are cell surface receptors that mediate cell-extracellular matrix (ECM) and intercellular interactions, thereby promoting cell migration through the ECM and mediating cellular responses [26, 27]. Integrins are comprised of different α and β subunits through which partnering of different subunit chains conveys specificity and function of the heterodimer [28]. Monocytes/macrophages express integrins with three different β chains,

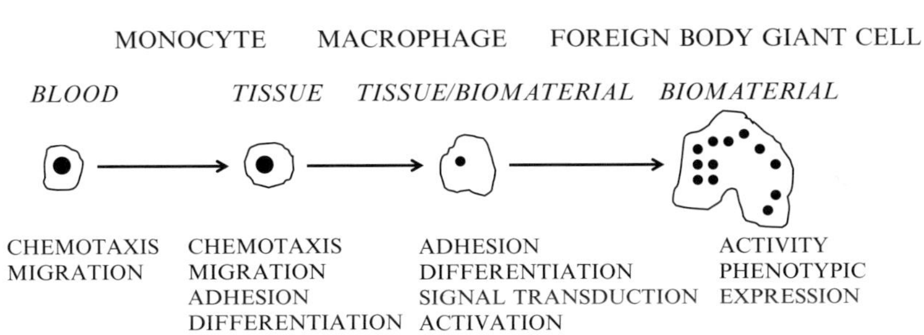

Figure 11.2. In vivo transition from blood-borne monocyte to biomaterial-adherent monocyte/macrophage to body giant cell at the tissue-biomaterial interface.

$\beta1$, $\beta2$, and $\beta3$. Pairing of $\beta1$ integrin includes $\alpha4/\beta1$ and $\alpha5/\beta1$, which bind fibronectin, and $\alpha6/\beta1$, which binds laminin. Of the $\beta2$ integrins, there are $\alpha L/\beta2$, $\alpha M/\beta2$, and $\alpha D/\beta2$, each specific for binding to intercellular adhesion molecules (ICAMs); and $\alpha X/\beta2$, which binds the inactive form of the third complement component fragment (C3bi) and fibrinogen. In addition to binding to ICAMs, $\alpha M/\beta2$ also interacts with fibrinogen, C3bi, and Factor X. Monocytes/macrophages also express $\alpha V/\beta3$ of the $\beta3$ integrins, which binds to vitronectin and other extracellular proteins that contain arginine-glycine-aspartic acid [28].

Initial monocyte adhesion is achieved through $\beta2$ integrins, by binding to various adsorbed protein ligands, including fibrinogen, fibronectin, IgG, and C3bi, while $\beta1$ integrins have been determined to play a role in the subsequent adhesion and fusion to form FBGC [12, 29]. The switch is time dependent since $\beta1$ integrins are not detected on adherent monocytes early but appear later during macrophage development, being strongly expressed on macrophages during fusion and on FBGC [30]. In fact, fused macrophages are characterized by the expression of $\alpha M/\beta2$, $\alpha X/\beta2$, $\alpha5/\beta1$, $\alpha5/\beta1$, $\alpha2/\beta1$, and $\alpha3/\beta1$, indicating the potential interactions with C3b fragments, factor X, fibrin, fibronectin, fibrinogen, and vitronectin at sites of biomaterial implantation [31].

In addition to mediating adhesion, integrin binding to the material-adsorbed protein layer provides intracellular signaling that modulates macrophage behavior through downstream transduction pathways. These pathways affect cytoskeletal rearrangements and formation of adhesion structures that allow spreading of the macrophages across the surface of the material [27, 32]. Macrophage adhesive structures are called "podosomes." Podosomes form early following adhesion, and consist of punctate F-actin on plasma membrane extensions. Podosomes are characterized by a central core of actin with a surrounding ring of vinculin, talin, α-actinin, and paxillin. Other proteins that regulate actin polymerization, such as gelsolin, are also present [33–35]. $\beta2$ integrins are actively recruited to developing podosomes in macrophages, indicating their potential role in podosome formation [35]. Furthermore, podosomes are present in actively fusing macrophages to form FBGC [36], in which αM colocalizes with $\beta2$ [31]. In turn, $\alpha M/\beta2$ integrin podosomes colocalize with paxillin, talin, and vinculin. $\beta2$ integrin ligation results in intracellular signaling via a proline-rich tyrosine kinase 2, which is a member of the focal adhesion kinase (FAK) family [37].

Within podosomes there is extensive interplay between intracellular signaling molecules (such as FAK, src-family kinases, and extracellular signal-regulated kinase [ERK]-1/2) and cytoskeletal proteins (such as vinculin, talin, paxillin, and actin) [38, 39]. FAK regulates cytoskeletal dynamics and assembly/disassembly of focal adhesions following integrin binding and subsequent tyrosine phosphorylation [40]. FAK activation leads to the activation of other protein tyrosine kinases, such as src. Cytoskeletal proteins, such as paxillin, bind FAK and mediate the interactions between the cytoplasmic domains of integrin receptors and other cytoskeletal proteins such as vinculin [41]. ERK signaling is also influenced by integrin signaling, ultimately resulting in the phosphorylation of transcription factors that regulate the cell cycle [42]. Variations in implanted material surface properties may therefore influence the profile of protein adsorption and subsequent integrin interactions and signaling [43–46], including differences in phosphorylation of FAK and ERK [47, 48] and differences in recruitment of talin, α-actinin, and paxillin [49].

In addition to their roles in mediating cell adhesion and cell signaling, integrin interactions are also responsible for influencing the cell cycle and cell death, which are necessary for cell detachment and tissue remodeling [50]. "Anoikis" is a term describing apoptosis induced by cell detachment from a supportive matrix [51]. Proper adhesion to a surface results in FAK-mediated survival signaling, however, disruption of adhesion signals promotes anoikis [52]. Monocytes may strongly adhere to most surfaces initially, but fail to maintain adhesion over extended periods of time [30]. Possibly through influencing variations in protein adsorption profiles, biomaterial surface chemistry dictates apoptosis of adherent macrophages both in vitro and in vivo [12, 53–55]. As a result of shear stress, biomaterial-adherent neutrophils undergo apoptosis (and subsequent detachment) through a caspase-3-mediated mechanism [56]. Accordingly, caspases have been shown to be involved in cleaving gelsolin, disrupting adhesion and leading to anoikis [52]. Therefore, material surface chemistry can promote anoikis by failing to promote adhesion.

11.2.3. Macrophage Fusion/FBGC Formation

The mechanism of cell-cell fusion of macrophages into multinucleated FBGC involves a series of complex, highly orchestrated events [57]. Although the exact molecular mechanisms that lead to macrophage fusion have not been fully elucidated, several potential molecules and pathways have been proposed. For example, IL-4 and IL-13 have been shown to be potent factors for initiating the fusion of macrophages adherent to biomaterial surfaces, as neutralizing IL-4 antibodies ameliorate fusion [58–60]. IL-4 and IL-13 upregulate the expression of mannose receptors on fusing macrophages with localization of the receptors at cell-cell fusion interfaces [59]. Furthermore, inhibition of mannose receptor activity prevents or reduces macrophage fusion [61]. Interestingly, mannose receptors are expressed on macrophages and dendritic cells and mediate endocytosis and phagocytosis, indicating that fusion processes may be related to those observed in phagocytosis [62]. This has been further confirmed since FBGC formation exhibits several features of phagocytosis, such as the presence of the endoplasmic reticulum proteins, calnexin and calregulin, colocalized with actin at fusion interfaces [63]. However, the processes of fusion and phagocytosis do display molecular differences since inhibition of rac1 activation attenuates fusion but not phagocytotic abilities of the macrophages [64]. It should be noted that α-tocopherol (vitamin E) induces the fusion of macrophages through the activation of diacylglycerol kinase [65].

CD44 and CD47 expression are upregulated on macrophage surfaces at the onset of fusion and have been shown to promote fusion and multinucleation, respectively [66, 67]. The presence of dendritic cell-specific transmembrane protein (DC-STAMP) on at least one fusing cell is necessary for fusion in the formation of FBGCs [68]. Although the DC-STAMP ligand is currently unknown, because of its involvement in FBGC formation, CCL2 is a potential candidate [69]. On the other hand, osteopontin, an ECM protein that is upregulated at inflammatory sites, may play an inhibitory role in FBGC formation [70]. Additional molecules that are involved in macrophage fusion are summarized in Table 11.1.

Cell surface molecules induced by IL-4 are needed on both fusing partners in order for fusion to occur [71]. It is believed that in order for fusion to occur, the macrophages must be phenotypically altered by a fusion-inducing stimulus such as IL-4 or IL-13. Due to the necessary monocyte adhesion and macrophage development steps to support fusion, it is not surprising that macrophage fusion on biomaterial surfaces is material dependent [55]. To this end, surfaces that favor vitronectin and fibronectin adsorption favor FBGC formation [72, 73]. FBGC formation is therefore dependent on two main criteria: (1) the presence of fusion-inducing stimuli and (2) material surfaces that promote the adhesion of appropriate proteins in the appropriate conformation.

11.2.4. FBGC Phenotype

FBGC are formed from the fusion of monocyte-derived macrophages, as their phenotype would suggest [74]. FBGC in tissue retrieved from human implant surgeries have been

Table 11.1. Molecular mediators involved in macrophage fusion.[a]

	FBGC	
	In vitro	In vivo
Soluble mediators		
IL-4 [58]	X	X
IL-13 [60]	X	X
MCP-1 [24, 146]	X	X
α-Tocopherol [65]	X	X
Plasma fibronectin [73]	X	
Osteopontin [70]	X	
Receptors		
Mannose receptor [60, 61]	X	X
β1 and β2 Integrins [30]	X	X
SIRPα [147]	X	
CD44 [148]		
CD47 [149]	X	X
DC-STAMP [68]	X	
Tetraspanins [150]	X	X
Signal transductors		
Diacylglycerol kinase [65]	X	X
Rac 1 [64]	X	
V-type ATPase [63, 151]	X	
iPLA2 [63]	X	

[a]Adapted from Themis Kyriakides, private communications

shown to express CD45 (leukocyte common antigen), CD13, CD14, CD15A (Hapten X), CD37, CD39, CD43, and HLA-DR; receptors including CD16 (FcRIII), CD31 (FcRII), CD35 (C3b receptor), and CD71 (transferrin receptor); and adhesion molecules that include CD11a,b,c, CD18 (leukocyte function-associated antigen [LFA] family), CD54 (ICAM-1), and CD44; while CD68 was strongly stained in the cytoplasm of FBGCs [74–77]. Also, the vitronectin receptor (CD51/CD61), very late antigen receptor (VLA-2, CDw49b/CD29), and VLA-4 receptor (CDw49d/CD29) are present on FBGC [78]. FBGC found in tissue surrounding total joint arthroplasties express the osteoclast markers tartrate-resistant acid phosphatase and vitronectin receptor [79].

FBGC derived from human arthroplasties also express cytokine receptors, gp130, IL-1R type 1, IL-2Rα, IL-2Rγ, IL-6R, TNF receptor, M-CSFR, and stem cell factor receptor, with weak expression of receptors for IL-4 and GM-CSF [74, 80]. In a murine model, injection of nitrocellulose particles results in production of IL-1α and TNF-α by FBGC during the first month of the foreign body reaction, with TGF-β being expressed later. However, the production of macrophage inflammatory proteins (MIPs) was not detected [81].

It should be noted that the aforementioned phenotypic characteristics describe those common to FBGC responses to particulate materials. Obviously, materials implanted into varying types of tissues would be expected to evoke varying responses.

11.2.5. Consequences of FBGC Formation on Biomaterial Surfaces

The material-tissue interface between the macrophage/FBGC cell membrane and the surface of the biomaterial is a privileged microenvironment. Within this privileged zone, macrophages and FBGC undergo frustrated phagocytosis, involving the release of potent degradative products, such as reactive oxygen intermediates, degradative enzymes, and acid, in a manner such that these mediators cannot be immediately buffered or inhibited [82, 83]. The contents of phagolysosomes can be introduced into the microenvironment, bringing the pH down to 4, creating an acidic environment [84]. These degradative products will interact with the surface of the implanted material, and the chemistry of the biomaterial surface will dictate its susceptibility to biodegradation. Therefore, medical devices and prostheses may incorporate antioxidants into a surface coating in order to inhibit oxidative processes initiated through the products released by surface-adherent macrophages and FBGC. Other materials are dependent on degradation and hence macrophage/FBGC activity. Enzymatic degradation has been implicated in the degradation of polyester biomaterials, such as resorbable sutures [85]. Polyesters such as polyethylene terephthalate (Dacron®) exhibit degradation at long implant times.

Upon adhesion to a biomaterial surface and subsequent activation, macrophages undergo a respiratory burst that leaves the adherent cell exhausted and unable to perform bactericidal activities. Therefore, cells at the material-tissue interface may not carry sufficient antimicrobial capacity. In addition, apoptosis of material-adherent leukocytes prevents these cells from attacking foreign organisms that may be adherent to the biomaterial, further increasing the possibility of infection [53].

Adhesion of macrophages and fusion into FBGC ultimately results in the degradation of biomaterials, leading to clinical device failure (Figure 11.3). The adverse effects of adherent macrophages and FBGC on implanted devices are well established. For example,

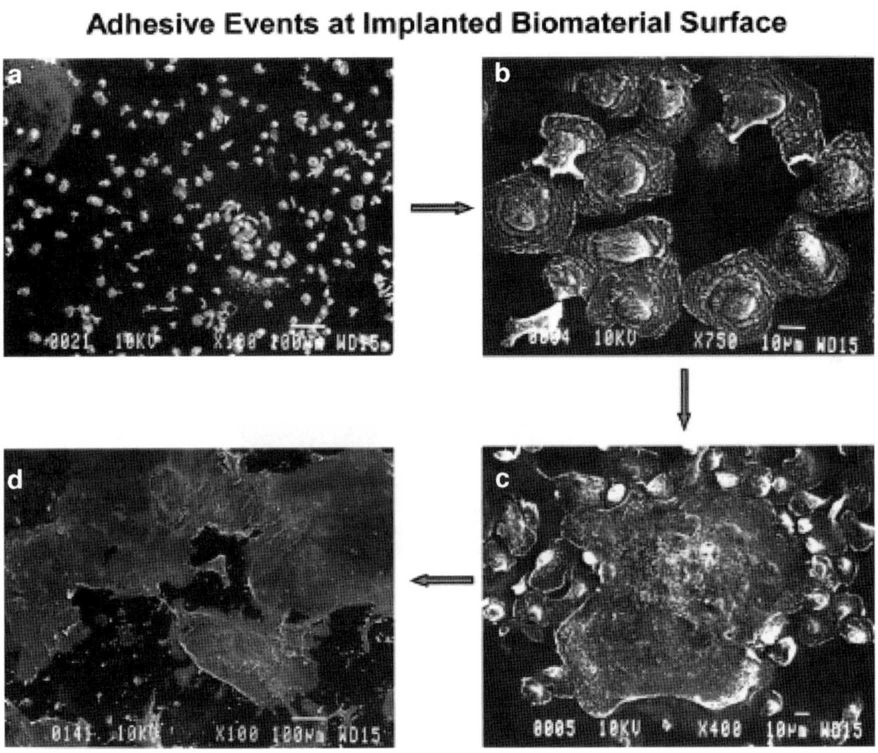

Figure 11.3. Scanning electron microscopy images of an Elasthane 80A PEU surface from an in vivo cage study showing the morphological progression of the foreign body reaction. The sequence of events at the PEU surface includes (**a**) monocyte adhesion (0 days), (**b**) monocyte-to-macrophage development (3 days), (**c**) ongoing macrophage/macrophage fusion (7 days), and (**d**) FBGCs (14 days).

macrophages and FBGC have been shown to be responsible for the in vivo cracking of Pellethane 2363–80A, a material used for pacemaker lead insulation [86, 87] Degradation of Pellethane 2363–80A is a result of oxidative polymer chain cleavage that is facilitated by the adsorption of α2-macroglobulin onto the polyurethane (PEU) surface. α2-Macroglobulin catalyzes the oxidation and chain cleavage of the polyether soft segments by reactive oxygen species and/or hydroxyl radicals [88, 89]. Degradation of PEU by macrophage-derived enzymes has been demonstrated in other studies [90–95].

The oxidation process at the surface of implanted materials is persistent and present for the lifetime of the implant. In general, physical damage may accompany chemical degradation. In the case of pacemaker leads, as chemical degradation proceeds, the polymer surface becomes brittle and more susceptible to physical damage, and as physical damage proceeds, cracks open on the surface and into the bulk of the material, exposing new surfaces to oxidants released by macrophages and FBGC. This suggests that a synergism occurs between chemical degradation and physical stress, leading to device failure.

11.3. Paracrine Interactions Between Macrophages/FBGCs and Inflammatory/Wound-Healing Cells

11.3.1. Macrophage/FBGC and Cytokines

Following activation, macrophages secrete a variety of molecules that drive inflammation and wound healing. Included are cytokines such as IL-1, IL-6, IL-10, IL-12, IL-18, TNF-α, TGF-β, IL-8, MCP-1, and MIP-1α/β [96]. Upon adhesion to biomaterial surfaces, monocytes/macrophages become activated in an attempt to phagocytose the biomaterial and direct the inflammatory and wound-healing response to the biomaterial. Therefore, when describing the foreign body response to implanted materials, cytokines can be classified as being either proinflammatory or prowound healing, depending upon which of these events they promote (Figure 11.4) [97]. For example, IL-2, IL-6, IL-8, and TNF-α may be termed proinflammatory/antiwound-healing cytokines because of their inflammation-promoting properties accomplished through cellular activation and chemotaxis of inflammatory cells.

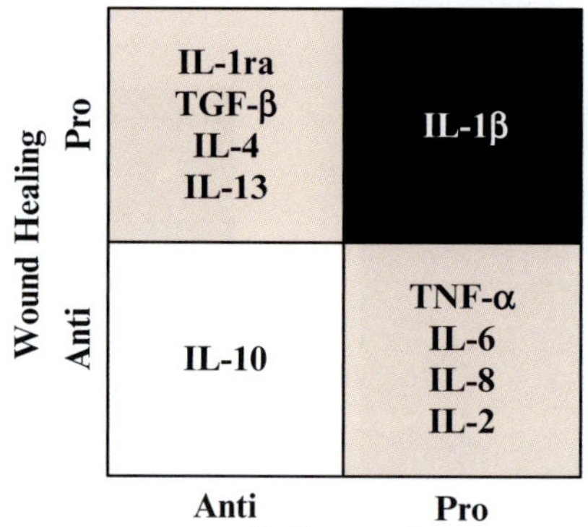

Figure 11.4. Classification of cytokines according to their roles in the foreign body response. Reprinted with permission from [115].

Conversely, cytokines that inhibit the inflammatory response and promote wound healing, such as IL-1 receptor antagonist (IL-1RA), IL-4, IL-13, and TGF-β, are considered antiinflammatory/prowound healing. IL-1β and IL-10 are two unique cytokines in this classification scheme since they represent extremes of the responses. IL-1β is considered a proinflammatory/prowound-healing cytokine because of its ability to activate both inflammatory cells (lymphocytes and monocytes) and wound-healing cells (fibroblasts), while IL-10 acts in the opposite fashion by downregulating the activity of these cell types and suppressing further cytokine production, leading to an antiinflammatory/antiwound-healing effect.

Relatively early studies used IL-1 production from macrophages cultured on biomaterials to assess the reactivity of the polymers as a measure of biocompatibility [98, 99]. Subsequently, IL-1 secretion was found to be dependent on surface characteristics and the proteins adsorbed on the surface [100]. In addition to IL-1, macrophage secretion of IL-6 and TNF-α was found to be dependent on the surface of the biomedical polymers [101]. Secretion of proinflammatory cytokines, such as IL-1β, IL-6, and TNF-α [102–105], from in vitro adherent macrophages on a wide range of materials has been studied extensively [105–113]. Each of these studies has repeatedly shown that surface properties, such as material surface chemistry and surface topography, can dictate macrophage activation [114]. Furthermore, surfaces that inhibit monocyte adhesion and IL-4-mediated macrophage fusion influence different cytokine expression profiles than surfaces that promote fusion [96, 115]. It should be noted that these studies relied mainly on real-time polymerase chain reaction (RT-PCR) analysis, and messenger RNA (mRNA) expression may not reflect the secreted and/or functional level of the cytokines. At the proteomic level, it has been shown that macrophages on a biomaterial surface that does not promote fusion secrete higher levels of proinflammatory cytokines, IL-1β and IL-6, per cell. The same studies demonstrated that biomaterial-adherent macrophages/FBGC undergo a phenotypic switch from a classic activation state to an alternative activation state [23].

Interferon (IFN)-γ and exposure to microbial products, such as lipopolysaccharide (LPS), induce classic macrophage activation [116], which results in killing of intracellular pathogens, upregulation of proinflammatory cytokines, inhibition of antiinflammatory cytokines, and production of nitric oxide. In contrast, exposure to either IL-4 or IL-13 stimulates the development of alternatively activated macrophages, which results in production of glucocorticoids, inhibition of proinflammatory cytokine production, promotion of antiinflammatory cytokine production, and upregulation of mannose receptor expression [117, 118]. Macrophages that have been alternatively activated are also known to play a role in allergic responses, parasite elimination, and matrix remodeling [119]. IL-4-induced alternative activation leads to upregulation of mannose receptors [117], which has been described previously as necessary for the formation of FBGC. Therefore, alternative macrophage activation may be a necessary step prior to fusion [61]. However, the biomaterial-adherent macrophages express classically activated chemokines, such as RANTES and MCP-1. Therefore, biomaterial-adherent macrophages/FBGCs display a different cytokine profile than either classically or alternatively activated macrophages, suggesting that biomaterial activation may be a unique pathway in macrophage activation that leads to the development of a unique phenotype (Table 11.2) [23].

Because of the extensive characterization of cytokine profiles being released from biomaterial-adherent macrophages/FBGC, in vitro proteomic testing of cytokine profiles can now be used as an initial means of assaying biocompatibility [120]. However, due to potential confounding results that may be introduced by contaminating adherent endotoxin (LPS), proper sterilization and processing prior to evaluating adherent macrophage cytokine secretion is necessary.

Table 11.2. A comparison of alternatively activated and classically activated macrophages.

Parameter	Classic	Alternative		Biomaterial effects (Day 3/Day 10)			
	LPS & IFN-γ	IL-4 & IL-13	IL-10	Hydrophobic	Hydrophilic & neutral	Hydrophilic & anionic	Hydrophilic & cationic
Cytokines							
IL-1	↑	↓	↓	↑/↓	↑/↓	↑/↓	↑/↓
IL-6	↑	↓	↓	↑/↓	↑/↓	↑/↓	↑/↓
IL-10	↓	↑		↑/↑	↑/↑	↑/↑	↑/↑
IL-12	↑	↓		↔	↔	↔	↔
TNF	↑	↓	↓	ND	ND	ND	ND
Chemokines							
IL-8	↑		↓	↑/↓	↑/↓	↑/↓	↑/↓
MIP-1β	↑	↓	↓	↓/↓	↓/↓	↑/↓	↓/↓
MDC	↓	↑		↑/↑	↑/↑	↔/↑	↔/↑
TARC	↓	↑		↔	↔	↔	↔
Mig	↑	↓	↓	↔	↔	↔	↔
RANTES	↑	↓	↓	↑/↑	↑/↑	↑/↑	↑/↑
IP-10	↑	↓	↓	↔	↔	↔	↔
ENA-78	↑			↑/↔	↑/↔	↔	↔
MCP-1	↑			↑/↑	↑/↑	↑/↑	↑/↑
MIP-1α	↑	↓	↓	ND	ND	ND	ND
Eotaxin	↑			↔	↔	↔	↔
Eotaxin-2	↓	↑		↑/↑	↑/↑	↔/↑	↔/↑
GRO	↑			↔	↔	↔	↔
MMP/TIMP							
MMP-9		↓		↑/↑	↑/↑	↑/↑	↑/↑
TIMP-1			↑	↑/↑	↑/↑	↑/↑	↑/↑
TIMP-2				↑/↑	↑/↑	↑/↑	↑/↑

Bold notation: measured by enzyme-linked immunosorbent assay (ELISA) and cytokine array; otherwise, measured only by cytokine array
"↑", "↓", and "↔": level of production increased, decreased, or did not change, respectively
ND not determined. Adapted from [116, 118, 119, 144–150]

This potential confounding factor has been demonstrated in studies showing that removal of adherent endotoxin on orthopedic wear particles almost completely eliminated IL-1β, IL-6, and TNF-α production from human monocytes [121], indicating that the presence of LPS may falsely indicate a macrophage classic phenotype and incorrectly predict biomaterial biocompatibility.

Biomaterial-adherent macrophages/FBGCs may greatly influence the behavior of other leukocytes (i.e., neutrophils, monocytes, and lymphocytes) and wound-healing cells (i.e., fibroblasts and keratinocytes) through the secretion of soluble mediators and hence orchestrate the cells' response to implanted materials. In turn, soluble mediators secreted by other inflammatory/wound-healing cells can influence macrophage responses to implanted materials. For example, TNF-α may be secreted by cells surrounding the implant and induce apoptosis of the biomaterial-adherent macrophage [122]. Cytokine expression profiles determined by semiquantitative RT-PCR of exudate leukocytes and biomaterial-adherent macrophages at biomaterial implant sites have demonstrated that the surface chemistry of biomaterials affects cytokine expression profiles of both adherent and surrounding cells [96, 115]. Furthermore, studies at the proteomic level have shown that IL-4, IL-13, IL-10, and growth-related oncogene (GRO)/keratinocyte chemoattractant (KC) secretion are upregulated at the implant site in the presence of biomaterials, while IL-6 and TNF-α are upregulated specifically upon the implantation of PEU and silicone rubber [122].

Proliferation of fibroblasts and angiogenesis are regulated by soluble factors released by biomaterial-adherent macrophages [123]. Fibronectin and other factors are released by alternatively activated macrophages and are believed to be involved in tissue remodeling during wound healing [124]. Alternatively activated macrophages also induce the secretion of profibrogenic factors that enhance fibrogenesis from fibroblasts [125]. Accordingly, the fibrotic response is associated with fibroblast stimulatory potential [126, 127]. Therefore, macrophages adherent to the surface of an implanted biomaterial can coordinate the fibrotic response leading to fibrous capsule formation through release of soluble mediators.

Matrix metalloproteinases (MMPs) are proteolytic proteins that dissolve the ECM through hydrolytic reactions. In this regard, MMPs can control the release of soluble mediators that are "trapped" in the matrix on and surrounding implanted materials. In this regard, MMPs can directly influence the composition of the ECM, as well as cell movement, growth, differentiation, and survival [128]. MMP activity can be inhibited by tissue inhibitors of metalloproteinases (TIMPs) in a 1:1 ratio, dictating the level of MMP activity [129].

Macrophages and FBGC adherent to biomaterial surfaces produce MMP-9, TIMP-1, and TIMP-2 in vitro. Interestingly, MMP-9 concentration increases over time, whereas TIMP concentrations remain constant, indicating a late override by MMP-9 that may result in matrix remodeling. Pharmacological inhibition of MMP-1, -8, -13, and -18 does not affect adhesion but does reduce fusion of macrophages [129]. Therefore, biomaterial-adherent macrophages/FBGCs modulate ECM remodeling and fibrosis, having an impact on implant survival and performance.

11.3.2. Lymphocyte/Macrophage Interactions

IL-4 and IL-13 have been implicated in macrophage fusion to form FBGC on the surface of implanted materials [58, 60]. The presence of lymphocytes surrounding the implant indicates that these cells may be the source driving the fusion and may play a critical role in the foreign body reaction. Furthermore, lymphocytes adhere to biomaterial surfaces in vitro [130, 131] and associate with biomaterial-adherent macrophages and FBGC [132, 133].

Through paracrine-mediated mechanisms, lymphocytes enhance macrophage adhesion and fusion, while conversely, the presence of macrophages may stimulate lymphocyte proliferation [133]. Whereas prolonged direct contact between biomaterial-adherent macrophages and lymphocytes results in enhancement of proinflammatory cytokines, such as TNF-α, IL-6, and IL-8, prevention of such direct interactions results in attenuation of proinflammatory cytokine production through soluble mediators [134]. It is therefore believed that early interactions through soluble mediators may be necessary to initiate the adherent macrophage phenotypic switch, which is later maintained through direct cell-cell (juxtacrine) interactions [134]. Notably, lymphocyte cytokine expression is influenced through surface molecule interactions and can involve T-cell receptor (TCR)-dependent and -independent interactions, such as membrane TNF-α, CD69, CD40, and LFA-1 [135–137]. For example, TCR leads to lymphocyte secretion of TNF-α and IL-10, whereas lymphocytes activated by cytokines such as IL-6, TNF-α, IL-2, or IL-15 induce TNF-α but not IL-10 [135, 136, 138]. Along this line, due to high levels of TNF-α and not IL-10 measured at the sites of biomaterial implants, studies have suggested that the biomaterial interactions are through cytokines [134]. Moreover, the interactions between lymphocytes and biomaterial-adherent macrophages and FBGC and their subsequent responses are differentially influenced by material surface chemistry [139].

However, lymphokines such as IL-2 and IFN-γ are not present during lymphocyte/macrophage interactions, indicating a lack of classic T lymphocyte activation, and macrophage fusion and formation of FBGC still occurs in the absence of lymphocytes, thus indicating that soluble fusion mediators derive from a separate source [140]. Natural killer (NK) cells, natural killer T (NKT) cells, mast cells, eosinophils, and basophils may be the source of IL-4 or IL-13 [141, 142], and lymphocytes may act to promote IL-4/IL-13 secretion by these other cells. It has recently been hypothesized that early activation to the alternatively activated macrophage phenotype may be through the mast cells, with CD4+ T cells promoting long-term phenotype determination at the site of the implant. Regardless of the mechanisms involved, the production of cytokines, chemokines, and ECM proteins remains dependent on biomaterial surface chemistry [143].

11.4. Conclusions and Perspectives

During the past three decades, our knowledge of in vitro and in vivo monocyte, macrophage, FBGC, and lymphocyte interactions with biomaterials has greatly expanded. This area of biomaterials science has moved dramatically from one of description and subjectivity to more quantitative approaches in understanding complex cell-material interactions in the foreign body reaction. Quantitative studies clearly have identified the significance of biomaterial surface chemistry in modulating inflammatory cell behaviors such as adhesion, apoptosis, anoikis, fusion, and cytokine secretion. In addition to juxtacrine or cell–cell interactions, cytokine-mediated paracrine responses between cells have been identified utilizing genomic and proteomic techniques. As biomaterial science advances and new, novel biomaterials, biomedical devices, and tissue-engineered constructs are developed, there will be a continuing need for appreciation and understanding of the inflammatory cell and foreign body reactions with these systems. It must be remembered that inflammatory cell interactions and development of the foreign body reaction is an initial host response to the implantation of any biomaterial, medical device, prosthesis, or tissue-engineered construct. As such, these interactions will continue to define the biocompatibility of new systems, and the continued development of a mechanistic understanding of these interactions is clearly warranted. Simply put, understanding of the early host response to biomaterials, medical devices, prostheses, and tissue-engineered constructs must evolve if we are to adequately engineer these systems such that they perform appropriately in their intended applications.

References

1. Anderson JM (2001) Biological responses to materials. Annu Rev Mater Res 31:81–110
2. Anderson JM (2000) Multinucleated giant cells. Curr Opin Hematol 7(1):40–47
3. Gretzer C, Emanuelsson L, Liljensten E et al. (2006) The inflammatory cell influx and cytokines changes during transition from acute inflammation to fibrous repair around implanted materials. J Biomater Sci Polym Ed 17(6):669–687
4. Anderson JM, Rodriguez A, Chang DT (2008) Foreign body reaction to biomaterials. Semin Immunol 20(2):86–100

5. Horbett T (2004) The role of adsorbed proteins in tissue response to biomaterials. In: Ratner B, et al. (eds) Biomaterials science: An introduction to biomaterials in medicine. Elsevier Academic Press, San Diego, pp 237–246
6. Zdolsek J, Eaton JW, Tang L (2007) Histamine release and fibrinogen adsorption mediate acute inflammatory responses to biomaterial implants in humans. J Transl Med 5:31
7. Tang L, Jennings TA, Eaton J (1998) Mast cells mediate acute inflammatory responses to implanted biomaterials. Proc Natl Acad Sci USA 95(15):8841–8846
8. Keegan AD (2001) IL-4. In: Oppenheim JJ, Feldman M (eds) Cytokine reference. Academic Press, San Diego
9. McKenzie ANJ, Matthews DJ (2001) IL-13. In: Oppenheim JJ, Feldman M (eds) Cytokine reference. Academic Press, San Diego
10. Wilson CJ, Clegg RE, Leavesley DI, et al. (2005) Mediation of biomaterial-cell interactions by adsorbed proteins: a review. Tissue Eng 11(1–2):1–18
11. Jenney CR, Anderson JM (2000) Adsorbed serum proteins responsible for surface dependent human macrophage behavior. J Biomed Mater Res 49(4):435–447
12. Brodbeck WG, Colton E, Anderson JM (2003) Effects of adsorbed heat labile serum proteins and fibrinogen on adhesion and apoptosis of monocytes/macrophages on biomaterials. J Mater Sci Mater Med 14(8):671–675
13. Jenney CR, Anderson JM (2000) Adsorbed IgG: a potent adhesive substrate for human macrophages. J Biomed Mater Res 50(3):281–290
14. Hu WJ, Eaton JW, Ugarova TP, et al. (2001) Molecular basis of biomaterial-mediated foreign body reactions. Blood 98(4):1231–1238
15. Xu LC, Siedlecki CA (2007) Effects of surface wettability and contact time on protein adhesion to biomaterial surfaces. Biomaterials 28(22):3273–3283
16. Nilsson B, Ekdahl KN, Mollnes TE, et al. (2007) The role of complement in biomaterial-induced inflammation. Mol Immunol 44(1–3):82–94
17. Gorbet MB, Sefton MV (2004) Biomaterial-associated thrombosis: roles of coagulation factors, complement, platelets and leukocytes. Biomaterials 25(26):5681–5703
18. Esche C, Stellato C, Beck LA (2005) Chemokines: key players in innate and adaptive immunity. J Invest Dermatol 125(4):615–628
19. Campbell DJ, Kim CH, Butcher EC (2003) Chemokines in the systemic organization of immunity. Immunol Rev 195:58–71
20. Gerard C, Rollins BJ (2001) Chemokines and disease. Nat Immunol 2(2):108–115
21. Broughton G 2nd, Janis JE, Attinger CE (2006) The basic science of wound healing. Plast Reconstr Surg 117 (7 Suppl):12S–34S
22. Rhodes NP, Hunt JA, Williams DF (1997) Macrophage subpopulation differentiation by stimulation with biomaterials. J Biomed Mater Res 37(4):481–488
23. Jones JA, Chang DT, Meyerson H, et al. (2007) Proteomic analysis and quantification of cytokines and chemokines from biomaterial surface-adherent macrophages and foreign body giant cells. J Biomed Mater Res A 83A:585–596
24. Charo IF, Ransohoff RM (2006) The many roles of chemokines and chemokine receptors in inflammation. N Engl J Med 354(6):610–621
25. Kyriakides TR, Foster MJ, Keeney GE, et al. (2004) The CC chemokine ligand, CCL2/MCP1, participates in macrophage fusion and foreign body giant cell formation. Am J Pathol 165(6):2157–2166
26. Delon I, Brown NH (2007) Integrins and the actin cytoskeleton. Curr Opin Cell Biol 19(1):43–50
27. Giancotti FG, Ruoslahti E (1999) Integrin signaling. Science 285(5430):1028–1032
28. Berton G, Lowell CA (1999) Integrin signalling in neutrophils and macrophages. Cell Signal 11(9):621–635
29. McNally AK, Anderson JM (1994) Complement C3 participation in monocyte adhesion to different surfaces. Proc Natl Acad Sci USA 91(21):10119–10123
30. McNally AK, Anderson JM (2002) Beta1 and beta2 integrins mediate adhesion during macrophage fusion and multinucleated foreign body giant cell formation. Am J Pathol 160(2):621–630
31. McNally AK, Macewan SR, Anderson JM (2007) Alpha subunit partners to beta1 and beta2 integrins during IL-4-induced foreign body giant cell formation. J Biomed Mater Res A 82(3):568–574
32. Rose DM, Alon R, Ginsberg MH (2007) Integrin modulation and signaling in leukocyte adhesion and migration. Immunol Rev 218:126–134
33. Marx J (2006) Cell biology. Podosomes and invadopodia help mobile cells step lively. Science 312(5782):1868–1869
34. Buccione R, Orth JD, McNiven MA (2004) Foot and mouth: podosomes, invadopodia and circular dorsal ruffles. Nat Rev Mol Cell Biol 5(8):647–657

35. Calle Y, Burns S, Thrasher AJ, et al. (2006) The leukocyte podosome. Eur J Cell Biol 85(3-4):151–157
36. DeFife KM, Jenney CR, Colton E, et al. (1999) Cytoskeletal and adhesive structural polarizations accompany IL-13-induced human macrophage fusion. J Histochem Cytochem 47(1):65–74
37. Duong LT, Rodan GA (2000) PYK2 is an adhesion kinase in macrophages, localized in podosomes and activated by beta(2)-integrin ligation. Cell Motil Cytoskeleton 47(3):174–188
38. Coppolino MG, Dedhar S (2000) Bi-directional signal transduction by integrin receptors. Int J Biochem Cell Biol 32(2):171–188
39. Miranti CK, Brugge JS (2002) Sensing the environment: a historical perspective on integrin signal transduction. Nat Cell Biol 4(4):E83–90
40. Schober M, Raghavan S, Nikolova M, et al. (2007) Focal adhesion kinase modulates tension signaling to control actin and focal adhesion dynamics. J Cell Biol 176(5):667–680
41. Parsons JT, Martin KH, Slack JK, et al. (2000) Focal adhesion kinase: a regulator of focal adhesion dynamics and cell movement. Oncogene 19(49):5606–5613
42. Juliano RL, Reddig P, Alahari S, et al. (2004) Integrin regulation of cell signalling and motility. Biochem Soc Trans 32(Pt3):443–446
43. Lee MH, Ducheyne P, Lynch L, et al. (2006) Effect of biomaterial surface properties on fibronectin-alpha5beta1 integrin interaction and cellular attachment. Biomaterials 27(9):1907–1916
44. Lan MA, Gersbach CA, Michael KE, et al. (2005) Myoblast proliferation and differentiation on fibronectin-coated self assembled monolayers presenting different surface chemistries. Biomaterials 26(22):4523–4531
45. Keselowsky BG, Collard DM, Garcia AJ (2005) Integrin binding specificity regulates biomaterial surface chemistry effects on cell differentiation. Proc Natl Acad Sci USA 102(17):5953–5957
46. Keselowsky BG, Collard DM, Garcia AJ (2003) Surface chemistry modulates fibronectin conformation and directs integrin binding and specificity to control cell adhesion. J Biomed Mater Res A 66(2):247–259
47. Allen LT, Tosetto M, Miller IS, et al. (2006) Surface-induced changes in protein adsorption and implications for cellular phenotypic responses to surface interaction. Biomaterials 27(16):3096–3108
48. Garcia AJ, Boettiger D (1999) Integrin-fibronectin interactions at the cell-material interface: initial integrin binding and signaling. Biomaterials 20(23–24):2427–2433
49. Keselowsky BG, Collard DM, Garcia AJ (2004) Surface chemistry modulates focal adhesion composition and signaling through changes in integrin binding. Biomaterials 25(28):5947–5954
50. Damsky CH, Ilic D (2002) Integrin signaling: it's where the action is. Curr Opin Cell Biol 14(5):594–602
51. Reddig PJ, Juliano RL (2005) Clinging to life: cell to matrix adhesion and cell survival. Cancer Metastasis Rev 24(3):425–439
52. Frisch SM, Screaton RA (2001) Anoikis mechanisms. Curr Opin Cell Biol 3(5):555–562
53. Brodbeck WG, Shive MS, Colton E, et al. (2001) Influence of biomaterial surface chemistry on the apoptosis of adherent cells. J Biomed Mater Res 55(4):661–668
54. Brodbeck WG, Patel J, Voskerician G, et al. (2002) Biomaterial adherent macrophage apoptosis is increased by hydrophilic and anionic substrates in vivo. Proc Natl Acad Sci USA 99(16):10287–10292
55. Jones JA, Dadsetan M, Collier TO, et al. (2004) Macrophage behavior on surface-modified polyurethanes. J Biomater Sci Polym Ed 15(5):567–584
56. Shive MS, Brodbeck WG, Anderson JM (2002) Activation of caspase 3 during shear stress-induced neutrophil apoptosis on biomaterials. J Biomed Mater Res 62(2):163–168
57. Chen EH, Grote E, Mohler W, et al. (2007) Cell-cell fusion. FEBS Lett 581(11):2181–2193
58. McNally AK, Anderson JM (1995) Interleukin-4 induces foreign body giant cells from human monocytes/macrophages. Differential lymphokine regulation of macrophage fusion leads to morphological variants of multinucleated giant cells. Am J Pathol 147(5):1487–1499
59. DeFife KM, Jenney CR, McNally AK, et al. (1997) Interleukin-13 induces human monocyte/macrophage fusion and macrophage mannose receptor expression. J Immunol 158(7):3385–3390
60. Kao WJ, McNally AK, Hiltner A, et al. (1995) Role for interleukin-4 in foreign-body giant cell formation on a poly(etherurethane urea) in vivo. J Biomed Mater Res 29(10):1267–1275
61. McNally AK, DeFife KM, Anderson JM (1996) Interleukin-4-induced macrophage fusion is prevented by inhibitors of mannose receptor activity. Am J Pathol 149(3):975–985
62. Apostolopoulos V, McKenzie IF (2001) Role of the mannose receptor in the immune response. Curr Mol Med 1(4):469–474
63. McNally AK, Anderson JM (2005) Multinucleated giant cell formation exhibits features of phagocytosis with participation of the endoplasmic reticulum. Exp Mol Pathol 79(2):126–135
64. Jay SM, Skokos E, Laiwalla F, et al. (2007) Foreign body giant cell formation is preceded by lamellipodia formation and can be attenuated by inhibition of Rac1 activation. Am J Pathol 171(2):632–640

65. McNally AK, Anderson JM (2003) Foreign body-type multinucleated giant cell formation is potently induced by alpha-tocopherol and prevented by the diacylglycerol kinase inhibitor R59022. Am J Pathol 163(3):1147–1156
66. Cui W, Ke JZ, Zhang Q, et al. (2006) The intracellular domain of CD44 promotes the fusion of macrophages. Blood 107(2):796–805
67. Han X, Sterling H, Chen Y, et al. (2000) CD47, a ligand for the macrophage fusion receptor, participates in macrophage multinucleation. J Biol Chem 275(48):37984–37992
68. Yagi M, Miyamoto T, Sawatani Y, et al. (2005) DC-STAMP is essential for cell-cell fusion in osteoclasts and foreign body giant cells. J Exp Med 202(3):345–351
69. Vignery A (2005) Macrophage fusion: the making of osteoclasts and giant cells. J Exp Med 202(3):337–340
70. Tsai AT, Rice J, Scatena M, et al. (2005) The role of osteopontin in foreign body giant cell formation. Biomaterials 26(29):5835–5843
71. Helming L, Gordon S (2007) Macrophage fusion induced by IL-4 alternative activation is a multistage process involving multiple target molecules. Eur J Immunol 37(1):33–42
72. McNally AK, Jones JA, Macewan SR, et al. (2008) Vitronectin is a critical protein adhesion substrate for IL-4 induced foreign body giant cell formation. J Biomed Mater Res A 86(2):535–543
73. Keselowsky BG, Bridges AW, Burns KL, et al. (2007) Role of plasma fibronectin in the foreign body response to biomaterials. Biomaterials 28(25):3626–3631
74. Athanasou NA, Quinn J (1990) Immunophenotypic differences between osteoclasts and macrophage polykaryons: immunohistological distinction and implications for osteoclast ontogeny and function. J Clin Pathol 43(12):997–1003
75. Doussis IA, Puddle B, Athanasou NA (1992) Immunophenotype of multinucleated and mononuclear cells in giant cell lesions of bone and soft tissue. J Clin Pathol 45(5):398–404
76. Al-Saffar N, Revell PA, Kobayashi A (1997) Modulation of the phenotypic and functional properties of phagocytic macrophages by wear particles from orthopaedic implants. J Mater Sci Mater Med 8(11):641–648
77. Abbondanzo SL, Young VL, Wei MQ, et al. (1999) Silicone gel-filled breast and testicular implant capsules: a histologic and immunophenotypic study. Mod Pathol 12(7):706–713
78. Quinn JM, Athanasou NA, McGee JO (1991) Extracellular matrix receptor and platelet antigens on osteoclasts and foreign body giant cells. Histochemistry 96(2):169–176
79. Kadoya Y, al-Saffar N, Kobayashi A, et al. (1994) The expression of osteoclast markers on foreign body giant cells. Bone Miner 27(2):85–96
80. Neale SD, Athanasou NA (1999) Cytokine receptor profile of arthroplasty macrophages, foreign body giant cells and mature osteoclasts. Acta Orthop Scand 70(5):452–458
81. Hernandez-Pando R, Bornstein QL, Aguilar Leon D, et al. (2000) Inflammatory cytokine production by immunological and foreign body multinucleated giant cells. Immunology 100(3):352–358
82. Henson PM (1971) The immunologic release of constituents from neutrophil leukocytes II. Mechanisms of release during phagocytosis, and adherence to nonphagocytosable surfaces. J Immunol 107(6):1547–1557
83. Henson PM (1971) The immunologic release of constituents from neutrophil leukocytes I. The role of antibody and complement on nonphagocytosable surfaces or phagocytosable particles. J Immunol 107(6):1535–1546
84. Haas A (2007) The phagosome: compartment with a license to kill. Traffic 8(4):311–330
85. Tokiwa Y, Calabia BP (2006) Biodegradability and biodegradation of poly(lactide). Appl Microbiol Biotechnol 72(2):244–251
86. Zhao QH, McNally AK, Rubin KR, et al. (1993) Human plasma alpha 2-macroglobulin promotes in vitro oxidative stress cracking of Pellethane 2363-80A: in vivo and in vitro correlations. J Biomed Mater Res 27(3):379–388
87. Kao WJ, Zhao QH, Hiltner A, et al. (1994) Theoretical analysis of in vivo macrophage adhesion and foreign body giant cell formation on polydimethylsiloxane, low density polyethylene, and polyetherurethanes. J Biomed Mater Res 28(1):73–79
88. Christenson EM, Anderson JM, Hiltner A (2004) Oxidative mechanisms of poly(carbonate urethane) and poly(ether urethane) biodegradation: in vivo and in vitro correlations. J Biomed Mater Res A 70(2):245–255
89. Wiggins MJ, Wilkoff B, Anderson JM, et al. (2001) Biodegradation of polyether polyurethane inner insulation in bipolar pacemaker leads. J Biomed Mater Res 58(3):302–307
90. Labow RS, Meek E, Santerre JP (2001) Hydrolytic degradation of poly(carbonate)-urethanes by monocyte-derived macrophages. Biomaterials 22(22):3025–3033
91. Labow RS, Meek E, Matheson LA, et al. (2002) Human macrophage-mediated biodegradation of polyurethanes: assessment of candidate enzyme activities. Biomaterials 23(19):3969–3975
92. Labow RS, Tang Y, McCloskey CB, et al. (2002) The effect of oxidation on the enzyme-catalyzed hydrolytic biodegradation of poly(urethane)s. J Biomater Sci Polym Ed 13(6):651–665

93. Matheson LA, Labow RS, Santerre JP (2002) Biodegradation of polycarbonate-based polyurethanes by the human monocytes-derived macrophage and U937 cell systems. J Biomed Mater Res 61(4):505–513
94. Labow RS, Sa D, Matheson LA, et al. (2005) The human macrophage response during differentiation and biodegradation on polycarbonate-based polyurethanes: dependence on hard segment chemistry. Biomaterials 26(35):7357–7366
95. Santerre JP, Woodhouse K, Laroche G, et al. (2005) Understanding the biodegradation of polyurethanes: from classical implants to tissue engineering materials. Biomaterials 26(35):7457–7470
96. Brodbeck WG, Voskerician G, Ziats NP, et al. (2003) In vivo leukocyte cytokine mRNA responses to biomaterials are dependent on surface chemistry. J Biomed Mater Res A 64(2):320–329
97. Fujiwara N, Kobayashi K (2005) Macrophages in inflammation. Curr Drug Targets Inflamm Allergy 4(3):281–286
98. Miller KM, Anderson JM (1988) Human monocyte/macrophage activation and interleukin 1 generation by biomedical polymers. J Biomed Mater Res 22(8):713–731
99. Miller KM, Huskey RA, Bigby LF, et al. (1989) Characterization of biomedical polymer-adherent macrophages: interleukin 1 generation and scanning electron microscopy studies. Biomaterials 10(3):187–196
100. Bonfield TL, Colton E, Anderson JM (1989) Plasma protein adsorbed biomedical polymers: activation of human monocytes and induction of interleukin 1. J Biomed Mater Res 23(6):535–548
101. Bonfield TL, Colton E, Marchant RE, et al. (1992) Cytokine and growth factor production by monocytes/macrophages on protein preadsorbed polymers. J Biomed Mater Res 26(7):837–850
102. Krause TJ, Robertson FM, Liesch JB, et al. (1990) Differential production of interleukin 1 on the surface of biomaterials. Arch Surg 125(9):1158–1160
103. Hwang JJ, Jelacic S, Samuel NT, et al. (2005) Monocyte activation on polyelectrolyte multilayers. J Biomater Sci Polym Ed 16(2):237–251
104. Ma N, Petit A, Yahia L, et al. (2002) Cytotoxic reaction and TNF-alpha response of macrophages to polyurethane particles. J Biomater Sci Polym Ed 13(3):257–272
105. Bonfield TL, Anderson JM (1993) Functional versus quantitative comparison of IL-1 beta from monocytes/macrophages on biomedical polymers. J Biomed Mater Res 27(9):1195–1199
106. Anderson JM, Ziats NP, Azeez A, et al. (1995) Protein adsorption and macrophage activation on polydimethylsiloxane and silicone rubber. J Biomater Sci Polym Ed 7(2):159–169
107. Wagner VE, Bryers JD (2003) Monocyte/macrophage interactions with base and linear- and star-like PEG-modified PEG-poly(acrylic acid) co-polymers. J Biomed Mater Res A 66(1):62–78
108. Yun JK, DeFife K, Colton E, et al. (1995) Human monocyte/macrophage adhesion and cytokine production on surface-modified poly(tetrafluoroethylene/hexafluoropropylene) polymers with and without protein preadsorption. J Biomed Mater Res 29(2):257–268
109. DeFife KM, Yun JK, Azeez A, et al. (1995) Adhesion and cytokine production by monocytes on poly(2-methacryloyloxyethyl phosphorylcholine-co-alkyl methacrylate)-coated polymers. J Biomed Mater Res 29(4):431–439
110. Marques AP, Reis RL, Hunt JA (2004) Cytokine secretion from mononuclear cells cultured in vitro with starch-based polymers and poly-L-lactide. J Biomed Mater Res A 71(3):419–429
111. Xing S, Santerre J, Labow RS, et al. (2002) Differential response to chemically altered polyethylene by activated mature human monocyte-derived macrophages. Biomaterials 23(17):3595–3602
112. Sethi RK, Neavyn MJ, Rubash HE, et al. (2003) Macrophage response to cross-linked and conventional UHMWPE. Biomaterials 24(15):2561–2573
113. Gretzer C, Gisselfalt K, Liljensten E, et al. (2003) Adhesion, apoptosis and cytokine release of human mononuclear cells cultured on degradable poly(urethane urea), polystyrene and titanium in vitro. Biomaterials 24(17):2843–2852.
114. Refai AK, Textor M, Brunette DM, et al. (2004) Effect of titanium surface topography on macrophage activation and secretion of proinflammatory cytokines and chemokines. J Biomed Mater Res A 70(2):194–205.
115. Brodbeck WG, Nakayama Y, Matsuda T, et al. (2002) Biomaterial surface chemistry dictates adherent monocyte/macrophage cytokine expression in vitro. Cytokine 18(6):311–319.
116. Mosser DM (2003) The many faces of macrophage activation. J Leukoc Biol 73(2):209–212
117. Stein M, Keshav S, Harris N, et al. (1992) Interleukin 4 potently enhances murine macrophage mannose receptor activity: a marker of alternative immunologic macrophage activation. J Exp Med 176(1):287–292
118. Gordon S (2003) Alternative activation of macrophages. Nat Rev Immunol 3(1):23–35
119. Mantovani A, Sica A, Sozzani S, et al. (2004) The chemokine system in diverse forms of macrophage activation and polarization. Trends Immunol 25(12):677–686
120. Ainslie KM, Bachelder EM, Borkar S, et al. (2007) Cell adhesion on nanofibrous polytetrafluoroethylene (nPTFE). Langmuir 23(2):747–754

121. Bi Y, Seabold JM, Kaar SG, et al. (2001) Adherent endotoxin on orthopedic wear particles stimulates cytokine production and osteoclast differentiation. J Bone Miner Res 16(11):2082–2091
122. Rodriguez A, Meyerson H, Anderson JM (2008) Quantitative in vivo cytokine analysis at synthetic biomaterial implant sites. J Biomed Mater Res A 89(1):152–159
123. Brodbeck WG, Shive MS, Colton E, et al. (2002) Interleukin-4 inhibits tumor necrosis factor-alpha-induced and spontaneous apoptosis of biomaterial-adherent macrophages. J Lab Clin Med 139(2):90–100
124. Kalltorp M, Oblogina S, Jacobsson S, et al. (1999) In vivo cell recruitment, cytokine release and chemiluminescence response at gold, and thiol functionalized surfaces. Biomaterials 20(22):2123–2137
125. Martin P, Leibovich SJ (2005) Inflammatory cells during wound repair: the good, the bad and the ugly. Trends Cell Biol 15(11):599–607
126. Gratchev A, Guillot P, Hakiy N, et al. (2001) Alternatively activated macrophages differentially express fibronectin and its splice variants and the extracellular matrix protein betaIG-H3. Scand J Immunol 53(4):386–392
127. Song E, Ouyang N, Horbelt M, et al. (2000) Influence of alternatively and classically activated macrophages on fibrogenic activities of human fibroblasts. Cell Immunol 204(1):19–28
128. Miller KM, Rose-Caprara V, Anderson JM (1989) Generation of IL-1-like activity in response to biomedical polymer implants: a comparison of in vitro and in vivo models. J Biomed Mater Res 23(9):1007–1026
129. Miller KM, Anderson JM (1989) In vitro stimulation of fibroblast activity by factors generated from human monocytes activated by biomedical polymers. J Biomed Mater Res 23(8):911–930
130. Jones JA, McNally AK, Chang DT, et al. (2008) Matrix metalloproteinases and their inhibitors in the foreign body reaction on biomaterials. J Biomed Mater Res A 84(1):158–166
131. Sternlicht MD, Werb Z (2001) How matrix metalloproteinases regulate cell behavior. Annu Rev Cell Dev Biol 17:463–516
132. Visse R, Nagase H (2003) Matrix metalloproteinases and tissue inhibitors of metalloproteinases: structure, function, and biochemistry. Circ Res 92(8):827–839
133. Yokoyama M, Nakahashi T, Nishimura T, et al. (1986) Adhesion behavior of rat lymphocytes to poly(ether)-poly(amino acid) block and graft copolymers. J Biomed Mater Res 20(7):867–878
134. Brodbeck WG, Macewan M, Colton E, et al. (2005) Lymphocytes and the foreign body response: lymphocyte enhancement of macrophage adhesion and fusion. J Biomed Mater Res A 74(2):222–229
135. Chang DT, Colton E, Anderson JM (2009) Paracrine and juxtacrine enhancement of adherent macrophage and foreign body giant cell activation. J Biomed Mater Res A 89:490–498
136. Monaco C, Andreakos E, Kiriakidis S, et al. (2004) T-cell-mediated signalling in immune, inflammatory and angiogenic processes: the cascade of events leading to inflammatory diseases. Curr Drug Targets Inflamm Allergy 3:35–42
137. Brennan FM, Foey AD, Feldman M (2006) The importance of cell interactions with macrophages in rheumatoid cytokine production. Curr Top Microbiol Immunol 305:177–194
138. Parry SL, Sebbag M, Feldman M, et al. (1997) Contact with T cells modulates monocyte IL-10 production: role of T cell membrane TNF-α. J Immunol 158:3673–3681
139. Sebbag M, Parry SL, Brennan FM, et al. (1997) Cytokine stimulation of T lymphocytes regulates their capacity to induce monocyte production of tumor necrosis factor-alpha, but not interleukin-10: possible relevance to pathophysiology of rhematoid arthritis. Eur J Immunol 27:624–632
140. Groth T, Altankov G, Klosz K (1994) Adhesion of human peripheral blood lymphocytes is dependent on surface wettability and protein preadsorption. Biomaterials 15(6):423–428
141. Rodriguez A. Macewan SR. Meyerson H. et al. (2008) the foreign body reaction in T-cell deficient mice. J Biomed Mater Res A. Epub PMID 18491378
142. Paul WE (1991) Interleukin-4: a prototypic immunoregulatory lymphokine. Blood 77:1859–1870
143. Gessner A, Mohrs K, Mohrs M (2005) Mast cells, basophils, and eosinophils acquire constitutive IL-4 and IL-13 transcripts during lineage differentiation that are sufficient for rapid cytokine production. J Immunol 174:1063–1072
144. MacEwan MR, Brodbeck WG, Matsuda T, et al.(2005) Student Research Award in the Undergraduate Degree Candidate category, 30th Annual Meeting of the Society for Biomaterials, Memphis, Tennessee, April 27-30, 2005. Monocyte/lymphocyte interactions and the foreign body response: in vitro effects of biomaterial surface chemistry. J Biomed Mater Res A 74(3):285–293
145. Mantovani A, Sozzani S, Locati M, et al. (2002) Macrophage polarization: tumor-associated macrophages as a paradigm for polarized M2 mononuclear phagocytes. Trends Immunol 23(11):549–555
146. Donnelly RP, Fenton MJ, Finbloom D, et al. (1990) Differential regulation of IL-1 production in human monocytes by IFN-gamma and IL-4. J Immunol 145(2):569–575
147. de Waal Malefyt R, Abrams J, Bennett B, et al. (1991) Interleukin 10(IL-10) inhibits cytokine synthesis by human monocytes: an autoregulatory role of IL-10 produced by monocytes. J Exp Med 174(5):1209–1220

148. Fenton MJ, Buras JA, Donnelly RP (1992) IL-4 reciprocally regulates IL-1 and IL-1 receptor antagonist expression in human monocytes. J Immunol 149(4):1283–1288
149. Oswald IP, Wynn TA, Sher A, et al. (1992) Interleukin 10 inhibits macrophage microbicidal activity by blocking the endogenous production of tumor necrosis factor alpha required as a costimulatory factor for interferon gamma-induced activation. Proc Natl Acad Sci USA 89(18):8676–8680
150. Kopydlowski KM, Salkowski CA, Cody MJ, et al. (1999) Regulation of macrophage chemokine expression by lipopolysaccharide in vitro and in vivo. J Immunol 163(3):1537–1544
151. Chizzolini C, Rezzonico R, De Luca C, et al. (2000) Th2 cell membrane factors in association with IL-4 enhance matrix metalloproteinase-1 (MMP-1) while decreasing MMP-9 production by granulocyte-macrophage colony-stimulating factor-differentiated human monocytes. J Immunol 164(11):5952–5960

12

Development and Differentiation of Neural Stem and Progenitor Cells on Synthetic and Biologically Based Surfaces

Erin N. Boote Jones, Donald S. Sakaguchi, and Surya K. Mallapragada

The response of stem and progenitor cells to material surfaces is of particular interest to the fields of biomedical engineering and regenerative medicine for a number of reasons. It is thought that stem and progenitor cells may represent a new treatment option for diseases of large-scale cell loss or malfunction, particularly in the central nervous system, where the environment is not permissive of tissue regeneration. Before treatments with specific outcomes can be reliably planned, however, the interactions between stem and progenitor cells and their environment must be better understood. This chapter discusses the function of neural stem and progenitor cells, as well as their responses to nonbiodegradable and biodegradable polymers, and to biologically derived proteins, which are used both as coatings on material surfaces and as scaffolds.

Abbreviations

AEMA	aminoethyl methacrylate
AHPC	adult hippocampal progenitor cell
AL	alginate lyase
AMF	aligned microfiber
ANF	aligned nanofiber
C17.2	immortalized neural stem cell line from murine postnatal cerebellum
CE3	murine embryonic stem cell line expressing green fluorescent protein
CNPase	oligodendrocyte marker
CNS	central nervous system
Dex-MA	methacrylated dextran

E.N.B. Jones and S.K. Mallapragada • Department of Chemical and Biological Engineering, Iowa State University, Ames, IA 50011, USA

D.S. Sakaguchi • Department of Genetics, Development, and Cell Biology, Iowa State University, Ames, IA 50011, USA

D.A. Puleo and R. Bizios (eds.), *Biological Interactions on Materials Surfaces*,
DOI 10.1007/978-0-387-98161-1_12,

EB	embryoid body
ECM	extracellular matrix
EVAL	poly(ethylene-co-vinyl alcohol)
FMCC	fetal mouse cortical cells
GFAP	glial fibrillary acidic protein
GP	glycerophosphate salt
HiB5	rat hippocampal progenitor cell line
IKVAV	peptide sequence isoleucine–lysine–valine–alanine–valine
K14	keratin-14, a long α-helical peptide
LAS	lysine-alanine sequential polymer
LG3	a globular domain of the laminin α3 chain
LG3K14	a polypeptide composed of keratin-14 fused to the LG3 domain
MC	methacrylamide chitosan
MTT	colorimetric mitochondrial metabolic activity assay
NeuN	mature neuron marker
NGF	nerve growth factor
NPC	neural progenitor cell
NSPC	neural stem and progenitor cell
NT-3	neurotrophin-3
PAA-azido	*N*-4-(azidobenzoyloxy)succinimide conjugated to polyallylamine
PC12	rat cancer cell line derived from a pheochromocytoma of the adrenal medulla
PDL	poly(D-lysine)
PEG	poly(ethylene glycol)
PLA	poly(lactic acid)
PLCL	poly(L-lactide-co-ε-caprolactone)
PLGA	poly(lactic-co-glycolic acid)
PLL	poly(L-lysine)
PLLA	poly(L-lactic acid)
PPy	polypyrrole
PVA	poly(vinyl alcohol)
RA	retinoic acid
RGD	peptide sequence arginine-glycine-aspartate
RMF	random microfiber
RNF	random nanofiber
RW4	murine embryonic stem cell line
SEM	scanning electron microscope
TERA2.cl.SP12	human pluripotent embryonic carcinoma cells
TUJ1	class III β-tubulin
TUNEL	terminal deoxynucleotidyl transferase dUTP nick end labeling
YIGSR	peptide sequence tyrosine–isoleucine–glycine–serine–arginine

12.1. Introduction

The response of stem and progenitor cells to surfaces is of particular interest to the fields of biomedical engineering and regenerative medicine for a number of reasons. Much attention has recently been given to the unique properties of stem and progenitor cells, specifically their capability to self-renew and to differentiate into other types of cells. It is thought that stem and progenitor cells may represent a new treatment option for diseases of large-scale cell loss or malfunction, particularly in the central nervous system (CNS), where the environment is not permissive of tissue regeneration. Before treatments with specific outcomes can be reliably planned, however, the interactions between stem and progenitor cells and their environment must be better understood. Both soluble and contact-mediated extracellular signals that regulate the cell differentiation fate have been identified. This chapter will examine the interactions between neural stem and progenitor cells (NSPCs) and material surfaces.

12.2. Neural Stem and Progenitor Cells

NSPCs, which are multipotential progenitors of neurons and glia, are capable of self-renewal and have been isolated from the developing and adult CNS [1–4] (Figure 12.1). In addition, they have been produced in vitro from embryonic stem cells and from induced pluripotent stem cells by differentiation toward a neuroepithelial fate [5–8]. Under appropriate conditions, neural stem cells can give rise to a complement of more specialized cells found in the CNS. NSPCs possess several characteristics that make them ideal vectors for CNS repair. Novel therapeutic strategies are being developed to take advantage of the ability of NSPCs to proliferate in culture and to survive following transplantation into the brain, spinal cord, and retina, where they may integrate and stably express foreign genes, or repopulate the damaged or diseased nervous system. NSPCs can be clonally expanded in culture in order to provide a renewable supply of cellular material for transplantation. In addition, they may be engineered to express exogenous genes for neurotrophic growth factors that may facilitate neuronal survival and regeneration. Thus, NSPCs may be capable of functionally integrating into the host neural circuitry where they may serve as cellular sources supplying trophic, or other, factors to facilitate cell survival and neuroprotection, as well as to stimulate regenerative events.

NSPCs have been isolated from various regions of the CNS, such as the hippocampus, subventricular zone, spinal cord ependyma, and retina [4, 9–12]; in vitro studies show that NSPCs can adopt a variety of cellular fates. The discovery of NSPCs in the adult brain and retina has encouraged research into their role during neurogenesis in the healthy mature brain and after injury.

Multiple studies suggest that adult bone marrow-derived stem cells possess the ability, under some circumstances, to transdifferentiate into neurons following transplantation into rodents [13–15]. This potentially extraordinary ability to generate other lineages is not without controversy. Other studies have suggested that some of these findings may perhaps result from cell fusion events [16]. Nevertheless, the multipotent nature of marrow stem cells makes them attractive candidates for cell-based therapies using autologous transplantation. In addition, stem-like cells have been isolated from other nonneural tissue sources including but not limited to skin, muscle, fat, and dental pulp; these sources may have the potential to produce neural cell types as well [17–20]. While it may seem most appropriate to use NSPCs derived from the same region of the brain that will be targeted for transplantation, it is imperative to support exploration of multiple stem/progenitor cell types as potential therapeutic vectors for repair strategies.

Recent studies have shown that a variety of signals, including the molecular microenvironment (substrate, humoral, and contact-mediated signals), influence the proliferation,

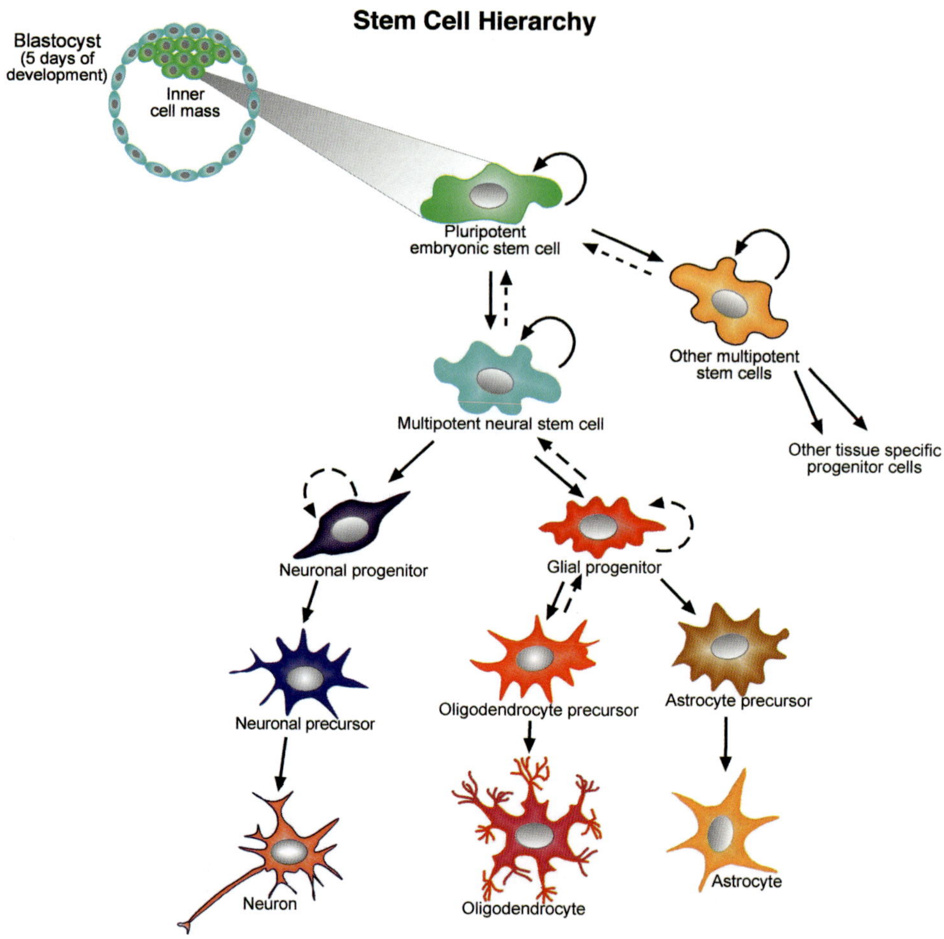

Figure 12.1. Stem cell hierarchy and multipotency of neural stem cells. A blastocyst, at approximately 5 days of development, contains a small cluster of cells referred to as the "inner cell mass" that are pluripotent stem cells. These pluripotent stem cells produce other tissue-specific stem cells and progenitors. This schematic illustration depicts the generation of differentiated cell types from multipotent neural stem cells. Different specialized multipotent stem cells are produced from pluripotent embryonic stem cells. Multipotent neural stem cells produce the three lineages found in the CNS: neurons, oligodendrocytes, and astrocytes. In this schematic illustration, the *curved arrows* represent self-renewing ability and the *dashed lines/arrows* represent putative cellular reprogramming events. As additional research is conducted and understanding of the true potential of stem cells and progenitor cells is improved, modifications of this illustration will be necessary. Illustration by D.S. Sakaguchi.

differentiation, and survival of NSPCs and their progeny. Neural progenitor cells (NPCs) have been cultured and transplanted into various CNS regions. When grafted into regions where neurogenesis is known to occur, these cells can differentiate into neurons. A provocative scenario is that, when exposed to optimal microenvironmental conditions, NSPCs will differentiate in a fashion appropriate to that brain region. However, there is limited understanding of the regulation of stem cell behavior in the healthy or diseased/damaged brain. Thus, by gaining a better understanding of environmental cues, including substrate composition and topography, we will be in a much better position to direct NSPC differentiation and integration. Understanding how microenvironmental factors influence NSPCs will be essential for the development of safe and effective stem cell-based therapies to repair and regenerate the damaged and diseased CNS.

12.3. Synthetic Material Surfaces

Much of what we know about the proliferation and differentiation potential of NSPCs comes from in vitro experiments where cells are cultured on synthetic surfaces such as polystyrene. It is imperative to understand how cells react to these synthetic surfaces and what differences arise because cells are cultured in vitro instead of existing in the in vivo environment.

12.3.1. Nonbiodegradable Substrates

For longer than 30 years, tissue culture-treated polystyrene has been used to culture cells. Polystyrene surfaces are exposed to high-energy oxygen ions, making the hydrophobic polystyrene surface hydrophilic and negatively charged [21]. Polystyrene can then be coated with cell adhesive proteins such as poly(L-lysine) (PLL), laminin, or fibronectin to further enhance cell adhesion and subsequent functions. This substrate is convenient and inexpensive, but the homogeneity of the surface restricts exploration of its effects on stem and progenitor cell proliferation and differentiation. Although very different from the complex in vivo milieu, polystyrene is nevertheless an important tool in the in vitro study of all types of cells. By culturing cells on surfaces that possess some of the same cell adhesive properties as tissue culture-treated polystyrene, the effects of specific surface chemistry on stem cell proliferation and differentiation can be explored.

Young and Hung [22] cultured neural stem cells on poly(vinyl alcohol) (PVA) and poly(ethylene-co-vinyl alcohol) (EVAL) to compare cell function on different synthetic surfaces. PVA is a very hydrophilic polymer, while polyethylene is more hydrophobic. The EVAL substrate contains regions of both PVA and ethylene. Neural stem cells in the form of neurospheres were cultured at varying cell densities on either PLL-coated glass, PVA, or EVAL substrates under either growth or differentiation conditions. Neurospheres seeded on glass attached, spread, and differentiated in a manner consistent with previously published results [23]. On PVA substrates, cells did not adhere, but aggregated and died whether they were seeded as whole neurospheres or had been dissociated before seeding. Primary neurons had similar growth on EVAL substrates as reported in a previous study [24]. Dissociated nestin-positive progenitors at a broad range of cell densities, on the other hand, adhered and survived on the EVAL substrate. Cell function was consistent across the entire substrate and was not due to localization on either the philic or phobic domains, most likely due to the fact that the scale of the philic and phobic domains was smaller than the cell body. After 4 days, the cells had not extended neurites; cell count analysis revealed that the number of cells on the substrate had not changed from any of the original seeding densities, whether basic fibroblast growth factor (FGF)-2 was present or not. Neurospheres behaved differently when plated at low and high densities; at low seeding density, the neurospheres attached, dissociated, and extended neurites. At high seeding density, the neurospheres did not attach, but instead the number of neurospheres increased, tripling in 4 days of culture. These differences were further supported by immunocytochemical evidence, with the dissociated cells expressing astrocyte-specific glial fibrillary acidic protein (GFAP) and neuron-specific enolase, while the high-density neurosphere cells predominantly expressed nestin. Thus, in this study, dissociated neural stem cells entered a stasis state, low-density neurospheres differentiated, and high-density neurospheres multiplied [22]. Young and Hung concluded that the EVAL substrate could control cell-substrate and cell-cell interactions [22].

In addition to chemical modification of the substrate surface, the effects of topography changes have been explored. Substrate topography and the interaction between contact guidance and chemical cues [25] as well as the effect of micropattern scale on cell outgrowth [26, 27] have been investigated. Recknor showed that topographical changes combined with astrocyte coculture could also change growth and differentiation of NPCs [28]. In experiments by Recknor et al., rat adult hippocampal progenitor cells (AHPCs) were cultured on micropatterned polystyrene grooves with dimensions on the same order of magnitude as the cell body: 13-μm wide and 4-μm deep, spaced 16-μm apart. The substrates used in that study were gravity-cast from a toluene solution on etched silicon wafers and made with an internal smooth control zone. More than 75% of AHPCs and astrocytes grew along, and aligned to within 10° of the three-dimensional (3D) pattern [29]. When AHPCs and astrocytes were both present, neuronal differentiation (as assessed by immunoreactivity for class III β-tubulin [TUJ1 antibody labeling] expression) was significantly increased on the patterned side but not on the smooth side of the substrates tested [28]. Recknor et al. concluded that the synergistic effects of the astrocytes and patterning on the substrate surface contributed to the increased neuronal differentiation [28].

Cells on patterned substrates (which have some topographical changes) still grow in a monolayer culture. True 3D culture can be achieved through the use of porous scaffolds. One method for making a 3D scaffold is a high internal phase emulsion. In this technique, an aqueous phase is added drop-wise to an organic phase, while constantly stirring, and the resulting emulsion is molded and baked at 60°C for 5 h, resulting in a styrene foam block with large numbers of interconnecting pores [30]. This technique is well known, but generally results in pores too small for cell culture. By adding a small amount of water-miscible tetrahydrofuran to the emulsion, Hayman et al. [31] were able to increase the mean void diameter from 5-20 to 50-100 μm (Figure 12.2) [31]. In order to increase cell adhesion and

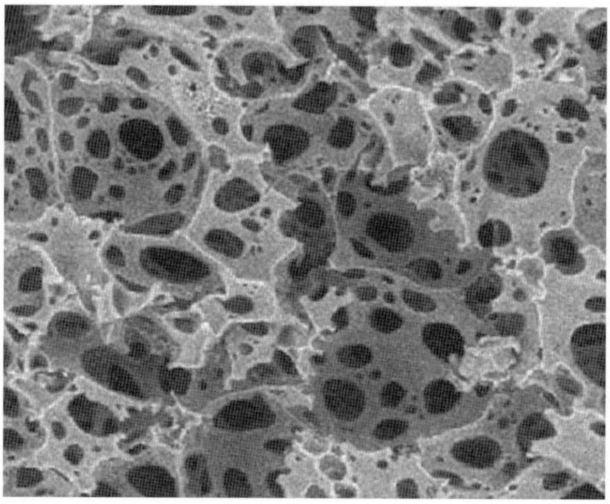

Figure 12.2. Scanning electron micrograph of a polyHIPE foam. Large numbers of interconnecting pores are formed during the manufacturing process, providing increased surface area and greater opportunity for cell function in a 3D environment. Scale bar: 50 μm. Reproduced with permission from Elsevier [31].

neurite outgrowth after sterilization and protein coating, neurally induced human pluripotent embryonic carcinoma cells (TERA2.cl.SP12) were seeded on these scaffolds. PLL and laminin were used both separately and together as scaffold coatings to enhance cell adhesion and neurite outgrowth. Small cell aggregates entered the pores of the scaffolds, but large aggregates (300–500 μm) remained on the surface where they were easily visualized. Scaffolds coated with PLL had up to 40% more adherent cells than scaffolds coated with laminin only; best adhesion was achieved with both PLL and laminin present. Average neurite extension increased by a factor of four from 25 to 100 μm, with a maximum observed length of 500 μm on substrates with both coatings [31]. Hayman et al. concluded that these durable and easily manufactured scaffolds could prove useful for in vitro study of 3D cultures of NPCs [31].

In the nervous system specifically, another factor that may influence cell proliferation and differentiation is electrical activity. Polypyrrole (PPy) is a biocompatible, electrically conductive polymer that has been used in several biological applications including neural prosthetics [32] and substrates for nerve growth [33]. While bioactive molecules have been incorporated into PPy as dopants, low loading and decreased conductivity limit the effectiveness of these methods [34]. An alternative is to tether molecules to the surface of a polymer. Gomez et al. used an arylazido-containing compound, which reacts with UV light to create singlet nitrenes that insert into C–H, N–H, and other bonds to decorate PPy with nerve growth factor (NGF) [35]. PPy films were deposited on a working electrode from an aqueous solution of pyrrole monomer and sodium salt of poly(styrene sulfonate). NGF was conjugated with fluoroisothiocyanate for fluorescent visualization, then immobilized on the PPy substrate by preparing *N*-4-(azidobenzoyloxy)succinimide conjugated to polyallylamine (PAA-azido), casting PAA-azido and NGF on a glass microscope slide, and exposing to UV light. Approximately 5% of the cast protein was tethered to the surface of the slide through this procedure. A surface concentration of approximately 0.98 ng protein/mm^2 induced neurite extension from PC12 cells comparable to 50 ng/mL soluble NGF. The conductivity of PPy was reduced by 30% after the surface modification, a nonsignificant difference comparable to other reported conductivity changes. Cell adhesion and neurite outgrowth were increased on NGF-tethered surfaces compared with PAA-azido only controls. Neurite outgrowth from PC12 cells was maintained at the same level on PPy with tethered NGF and PPy plus soluble NGF for 2 days; after 10 days, however, fewer cells had neurites on the tethered substrates. Gomez [35] attributed this result to the restriction of endocytosis-dependent pathways. When current was passed through PPy-NGF slides at 100 mV for 2 h, the neurite length measured 24 h later was significantly increased on PPy-NGF that had been exposed to current compared with PPy-NGF with no stimulation. Gomez suggested that this result shows a synergistic effect between chemical and electrical stimuli [35].

12.3.2. Biodegradable Polymer Substrates

As mentioned previously in Sect. 12.3.1, polystyrene is commonly used for in vitro cell culture experiments. However, in most cases it is unsuitable for in vivo implantation, where degradable polymers are more desirable. For these cases, biodegradable polymers offer several advantages, including tailored material substrates, degradation rates, and the opportunity for synchronized tissue remodeling to occur. Several biodegradable polymers have been approved for specific clinical applications and are attractive candidates for stem cell applications. Bhang tested the interaction of a hippocampal progenitor cell line with several common biodegradable polymers; these studies used various materials as culture surfaces and used their degradation products as media additives [36]. The polymers chosen were poly(lactic-co-glycolic acid) (PLGA), poly(L-lactide-co-ε-caprolactone) (PLCL), and poly (L-lactic acid) (PLLA) due to their extensive use as biological scaffolds. HiB5 cells, a hippocampal progenitor line, were assayed for cell viability, mitochondrial metabolic activity, apoptotic activity, proliferation, and differentiation. In all assays, PLGA performed best. PLCL and PLLA were significantly different from PLGA at the same time point, with PLLA supporting the lowest cell viability and mitochondrial metabolic activity, and the highest apoptotic activity. In terms of the cell proliferation assay, PLGA was not significantly different from the positive control glass substrates. PLCL and PLLA, however, supported significantly lower cell proliferation than either glass or PLGA, but were not significantly different from each other. Cell differentiation was assessed on all three substrates at 7 days by immunoreactivity for neurofilament and class III β-tubulin. Again, PLGA performed best both in percent of cells expressing neuronal proteins and number of neurites extending more than two cell bodies per cell. From these results, Bhang concluded that PLGA is the most suitable scaffold material for NPCs [36].

Solvent casting with salt leaching is a well-known method for creating biodegradable scaffolds [37, 38]. Levenberg prepared biodegradable scaffolds using a salt-leaching method and examined the growth of embryonic stem cells under neuronal differentiation conditions [39]. Scaffolds prepared from a 50% PLA/poly(glycolic acid) mixture with pore sizes from 200 to 250 μm degraded in a time scale commensurate with the growth rate of the H9 clone cells tested and supported ingrowth of dissociated either 4- or 9-day-old H9 clone embryoid bodies (EBs) [40]. Neural rosettes were formed during 2 weeks of culture in all media tested, but significantly more and better defined rosettes were formed when neurotrophins were present in the media. Four-day-old EB had the largest increase in rosette numbers from NGF and neurotrophin (NT)-3, but NT-3 alone and NT-3 plus retinoic acid (RA) induced the most rosettes in 9-day-old EB. Nine-day-old EB cultures that received NT-3 and RA in combination with either NGF or NT-3 had significantly enhanced neuronal differentiation over control medium conditions (as measured by class III β-tubulin immunoreactivity), but no increased differentiation after 4 days of EB culture [39]. Levenberg concluded that the observed differential response emphasized the importance of cell age, timing in signaling, and cell differentiation [39].

Salt leaching creates a porous sponge-like architecture, which may be desirable in some situations. In other cases, a more directed architecture may be preferred. In the latter cases, electrospun scaffolds may provide a better alternative substrate. Briefly, to create electrospun scaffolds, a PLLA solution was ejected from a syringe fitted with an 18-gauge needle. A 12-kV voltage was applied to create the polymer jet. A rotating disc was used to collect aligned nanofibers and a stationary disc was used to collect random fibers [41]. Solution concentration had the greatest effect on fiber morphology. The average fiber diameter was 300 nm for a 2% and 1.5 μm for a 5% solution. These two average fiber diameters were denoted as "aligned nanofiber" (ANF) and "aligned microfiber" (AMF) and used for cell growth studies. Aligned fibers had a smaller diameter than random fibers from the same solution concentration; Yang [42] attributed this result to stretching caused by the tangential forces exerted by the high- speed rotation of the collection plate. Since the purpose of the study was to examine cell response to nanoscale surface topography, the C17.2 neuronal precursor cell line was selected. C17.2 cells can differentiate without surface protein molecules, which would affect fiber topography. Cells seeded on ANF and AMF substrates exhibited bipolar morphology with approximately 90% of neurites extending parallel to the aligned fibers (Figure 12.3) [42]. This result was significantly different from results obtained on random nanofibers (RNF) and random microfibers (RMF), where cells exhibited multipolar

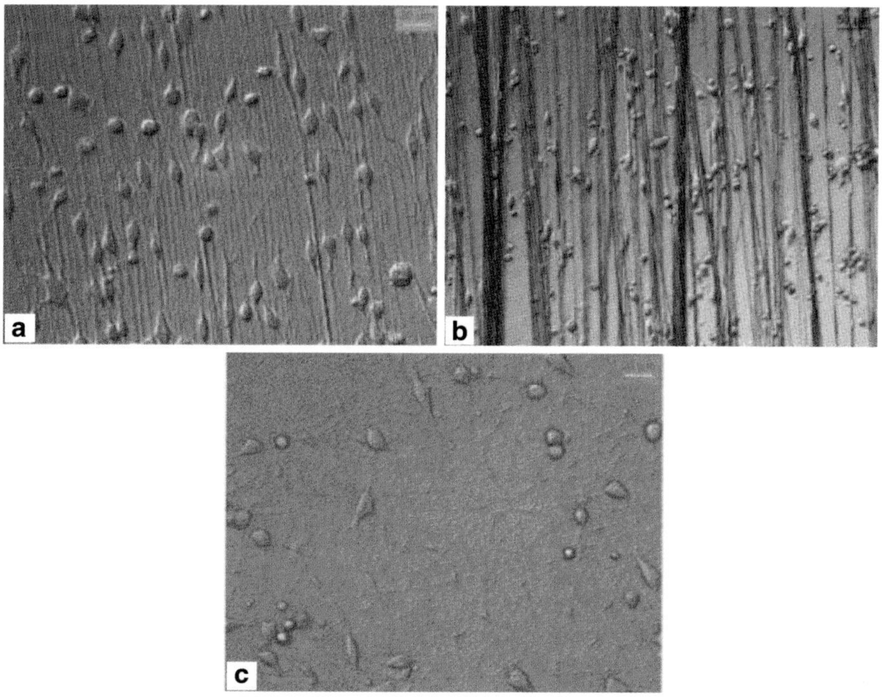

Figure 12.3. Phase contrast light micrographs showing neural stem cell attachment on electrospun fibrous PLLA (**a**) aligned nanofibers, (**b**) aligned microfibers, and (**c**) random nanofibers, after 1 day of culture. Reproduced with permission from Elsevier [42].

processes and randomly oriented neurites. Average neurite length was significantly longer on ANF than on AMF, RMF, or RNF, which were all similar to each other. Neuronal differentiation assessment by neurofilament immunocytochemistry and morphology showed that 80% of cells differentiated on ANF and RNF, while only 40% differentiated on AMF and RMF. The interpretation of Yang of this result was that the scale of the fibers, rather than their alignment, stimulated cell differentiation [42].

12.4. Biologically Derived Surfaces

Up to this point, the discussion in this chapter has revolved around synthetic surfaces that may or may not have been modified with proteins, but several biologically derived molecules can be prepared into scaffolding surfaces as well. Amino acid polymers, blood clotting proteins, extracellular matrix (ECM) molecules, and other biologically derived molecules can form 3D hydrogels that are able to support adhesion and various other functions of cells.

12.4.1. Poly(L-lysine)-Based Substrates

PLL is often used to increase cell adhesion. Where PLL is a homopolymer, lysine-alanine sequential polymer (LAS) is a polymer with alternating lysine and alanine residues. Under serum-free conditions, LAS promoted increased hippocampal neuron survival and axon formation more than PLL [43]. Wang and coworkers [44] tested the response of neural stem cell neurospheres to this polymer substrate. When neurospheres in serum-free media were plated at low density, the cells did not migrate but instead sent out long, straight neurites, forming a network within 4 days of culture [44]. Neurospheres seeded on LAS substrates had significantly longer neurites than those on PLL substrates, confirming LAS as the superior substrate for neurite outgrowth. LAS was used for subsequent experiments. At higher neurosphere density in serum-free media, rather than elaborating neurites, the cells migrated out and the neurosphere dissociated. When serum was added to the media, low-density neurospheres changed their response: cells migrated away from the neurosphere and did not form a neurite network. A mitochondrial metabolic activity assay (MTT colorimetric assay) showed increased viability and also enhanced differentiation outcomes in the presence of serum. Wang [44] concluded that the neurospheres were changing their communication strategy based on the density and presence of serum proteins. A minimum threshold concentration of soluble growth factors such as hormones and cytokines appeared to be necessary before cells began migrating away from the neurospheres [44].

Through incorporation of cell adhesive molecules such as PLL, the range of materials used in scaffolds can be expanded. Royce Hynes [45] developed a photopolymerized hydrogel system that included poly(ethylene glycol) (PEG) and PLL. The main advantage of this hydrogel system over prefabricated biodegradable scaffolds is the ability to polymerize in situ. To form the hydrogel, macromers composed of monacrylated PEG chains covalently bound to a PLL backbone by 1,1′-carbonyldiimidazole chemistry were photopolymerized using Irgacure 2959 photoinitiator and 4–10 min of UV exposure. The brief UV exposure had no significant effect on neural stem cell survival. While the photoinitiator in solution caused a significant decrease in cell survival, cells encapsulated in PEG/PLL hydrogels that employed the photoinitiator showed excellent cell survival. Royce Hynes et al. [45] attributed this result to the rapid utilization of the photoinitiator-free radicals during the polymerization process. Cell differentiation was assessed by immunoreactivity for nestin, neurofilament, and GFAP. Nestin indicates undifferentiated neural progenitors, while neurofilaments and GFAP are expressed by mature neurons and astrocytes, respectively. The cells initially expressed nestin. After 7 and 17 days, approximately 50% of the cells expressed neurofilaments while the rest continued to express nestin [45]. Since the PEG/PLL scaffold tested is nontoxic, can be polymerized in situ, and allows cells to survive and differentiate to neural cell fates, it is an attractive substrate for stem cell delivery.

12.4.2. Laminin-Based Substrates

Along similar lines as the PEG/PLL hydrogel, Nakaji-Hirabayashi designed a self-assembling keratin-based hydrogel with the intent of protecting stem cells, transplanted into the brain, from inflammatory responses [46]. The protein designed contained keratin-14 (K14), a long α-helical peptide, fused to a globular domain of the laminin $\alpha 3$ chain (LG3). The resulting polypeptide could integrate with native keratin chains through coiled-coil interactions and coordinate with integrins through the laminin domain. Proteins with, and without, the laminin domain were expressed in *Escherichia coli*, purified, and combined with keratins extracted from human hair to form hydrogels. Synthetic proteins with the laminin domain were designated LG3K14, and those without the laminin domain were designated K14. Neural stem cell (NSC) adhesion was increased by a factor of three on LG3K14, the hydrogel incorporating the laminin domain, over both native keratin and keratin–K14 hydrogel. After 10 days in culture, NSC proliferation was also significantly increased on the LG3K14 hydrogel compared with results obtained on the K14 hydrogel and native keratin. The majority of cells expressed nestin after 10 days on all three types of hydrogels tested. Nakaji-Hirabayashi concluded that incorporation of the laminin LG3 domain significantly increased NSC adhesion and proliferation, and that similar molecules designed from the ground up could provide tuned biomaterials for the promotion of NSC survival after transplantation [46]. This method shows potential for 3D scaffolds, but current fabrication techniques do not allow for encapsulation of living cells during scaffold formation. Further modifications to design polypeptide scaffolds and fabrication protocols could potentially overcome this limitation.

The peptide sequence isoleucine–lysine–valine–alanine–valine (IKVAV) from the laminin molecule is known to be a bioactive site that promotes neurite sprouting and directs neurite outgrowth [47]. Silva [48] designed an amphiphilic molecule that self-assembles into a nanofibrous scaffold when added to cell suspensions. The chemical structure of the peptide amphiphile includes the IKVAV sequence, a Glu residue (which gives the molecule a net-negative charge at pH 7.4), four Ala residues, three Gly residues, and an alkyl tail of 16 carbons (Figure 12.4.). A control amphiphile that contained a nonphysiological sequence instead of IKVAV was also synthesized. When electrolytes in cell media screened the electrostatic repulsions caused by the Glu residue, the molecules self-assembled into nanofibers. The IKVAV sequences were presented along the outside of the nanofibers. When NPCs were added to a 1 wt% solution of the amphiphile, a clear gel formed within seconds; encapsulated cells survived for the entire 22 days of the study [48]. The scaffolds contain 99.5% water, allowing diffusion of nutrients and metabolic waste for large numbers of cells. Cell migration

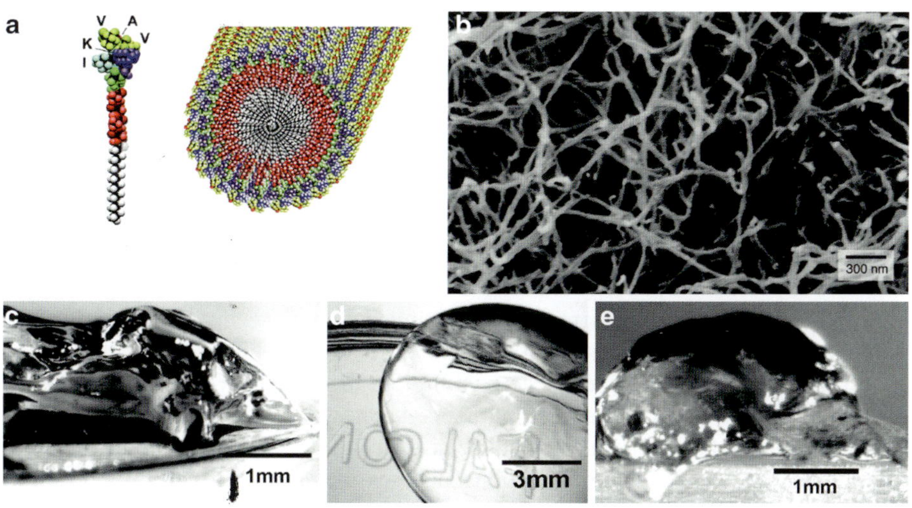

Figure 12.4. (**a**) Molecular graphics illustration of an IKVAV-containing peptide amphiphile molecule and its self-assembly into nanofibers. (**b**) Scanning electron micrograph of an IKVAV nanofiber network formed by adding cell medium to a peptide amphiphile aqueous solution. The sample in the image was obtained by network dehydration and critical-point drying of samples caged in a metal grid (to prevent network collapse); this sample was sputtered with gold-palladium films and imaged at 10 kV. (**c** and **d**) Micrographs of the gel formed by adding to IKVAV peptide amphiphile solutions (**c**) cell culture media and (**d**) cerebral spinal fluid. (**e**) Micrograph of an IKVAV nanofiber gel surgically extracted from an enucleated rat eye after intraocular injection of the peptide amphiphile solution. Reproduced with permission from [48].

out of the neurospheres and length of neurite outgrowth were significantly higher on the IKVAV scaffolds than on the control scaffolds. Cells on IKVAV scaffolds rapidly differentiated into neurons, with 35% of the cells expressing class III β-tubulin after only 1 day in culture. Compared with NPCs cultured on laminin-coated substrates, neuronal differentiation was higher and astrocytic differentiation was lower after 1 and 7 days of culture in self-assembled hydrogels. Self-assembled IKVAV molecular sheets on two-dimensional (2D) surfaces had the same effect on cell differentiation, while soluble nonassembled IKVAV had no effect. Silva interpreted this result as indicating that the density of the epitope, rather than its presentation, led to the high cell differentiation rates [48]. The ability of the amphiphile solution to self-assemble upon contact with tissue makes the amphiphilic hydrogel an excellent candidate for NPC delivery in vivo.

12.4.3. Fibrin and Collagen

Fibrin, a protein involved in the clotting of blood that forms a mesh structure when fibrinogen and thrombin are mixed, provides a permissive environment for cell growth, but is most effective when the scaffold properties are tailored for specific cell types. Optimization has been achieved for chick dorsal root ganglia [49], fibroblasts [50], and mesenchymal stem cells [51]. Willerth optimized the concentration of fibrinogen, thrombin, and aprotinin for the culture of EB for neural differentiation [52]. Two murine embryonic cell lines, RW4 and CE3 (which are similar except that CE3 expresses green fluorescent protein) were induced to form EB by RA treatment. Whole or dissociated EBs were then seeded in or on the fibrin scaffolds. Saturated fibrinogen solutions made the longest lasting fibrin scaffolds. Varied thrombin concentration in the starting solutions did not affect the degradation rate of the fibrin scaffold. High levels of aprotinin preserved the scaffold structure for longer than twice the time required for complete degradation without aprotinin. The more cells seeded on the scaffold, the more quickly the scaffold degraded. Dissociated cells formed networks on the surface of the scaffold but did not penetrate and colonize the bulk of the structure; this result allowed for easy visualization of cells but did not produce the desired 3D structure for tissue-engineering applications. In contrast, whole EBs exhibited much more differentiation and migration throughout the bulk of the scaffold when embedded into the scaffolds than when placed on the scaffold surface. Immunocytochemical analysis revealed differentiation into neurons (as measured by class III β-tubulin) and astrocytes (as measured by GFAP). Stained cells exhibited the expected morphology for their respective cell types. Willerth concluded that the optimized fibrin scaffolds represented a promising potential vector for tissue-engineering applications utilizing embryonic stem cells directed to a neural lineage [52].

Collagen is another biologically derived protein that has been tested for use as a scaffolding material. Watanabe determined optimal cell adhesion and proliferation conditions for NSPCs in collagen gels over a range of collagen concentrations and NSPC seeding densities [53]. Parameters examined for optimization purposes included cell viability and migration distance. Seeding cells at densities that were too low or high caused reduced cell viability. Watanabe attributed this result to lack of cell-to-cell interactions at low density and diffusional limitations at high densities [53]. Cell and neurite outgrowth decreased as collagen concentration increased; this observation was consistent with previously published reports [54]. Immunocytochemical differentiation assay revealed that, shortly after seeding, nearly all the cells expressed nestin, but after seven divisions these cells had differentiated to neurons (class III β-tubulin), astrocytes (GFAP), and oligodendrocytes (CNPase) in proportions that were similar to those observed in 2D culture. In addition, after 14 divisions, NeuN-expressing cells were observed, indicating the presence of mature neurons. Watanabe concluded that collagen gels at the appropriate concentration and NSPC seeding density provided a stable environment in which cells could remain viable, migrate, and differentiate [53].

Brannvall et al. produced gels from a mixture of collagen and hyaluronate [55]. Hyaluronan is an ECM component of the developing brain, and has been studied for its effects on neurite outgrowth [56]. A 50% volume mixture of collagen-1 and hyaluronate solution formed a gel at 37°C. NSPCs from the subventricular zone of embryonic, postnatal, and adult mice were dissociated and seeded in the collagen-hyaluronate gel or on a glass coverslip, for comparison with 2D culture. While embryonic cells displayed the most proliferation (as assayed by bromodeoxyuridine incorporation), they also had the highest rates of apoptosis (as measured by a TUNEL assay, which detects nicked DNA). When growth factors were removed from the media, cell morphology changed in all three donor ages tested. Initially all cells were immunopositive for nestin. After 6 days without growth factors, differentiation was assessed. In 3D culture, postnatal NSPCs displayed the most differentiation, with only 10% of cells continuing to express nestin. Embryonic NSPCs had 47% nestin expression, and adult NSPCS had 80% nestin expression. Neuronal differentiation, as measured by class III β-tubulin immunostaining, was 26% for embryonic but only 9% for adult cells. In postnatal cells, however, the proportion of cells differentiating to neurons was 70%. This result is much higher than the result obtained with cells from either embryonic or adult source cells and much higher than the same postnatal cells on 2D culture; in the latter case, only 14% of cells expressed class III β-tubulin. Follow-up studies indicated that, after 11 days, postnatal neurons expressed both gamma-aminobutyric acid and synapsin I, as well as the mature neuronal marker neurofilament-H and the neurotransmitter glutamate. Brannvall highlighted the differential regulation observed at different cell donor ages and concluded that, despite the low growth rate of postnatal cells, the neurogenesis observed made postnatal cells the best choice for generating neurons in the collagen-hyaluronan matrix [55].

12.4.4. Chitosan

Chitosan, formed from the deacetylation of chitin (which is extracted from the shells of crustaceans and exoskeletons of arthropods), is biocompatible, cell adhesive, and enzyme degradable, but only soluble in dilute acid. Yu synthesized a novel water-soluble methacrylamide chitosan (MC) by reacting approximately 23% of the chitosan amine groups with methacrylic anhydride [57]. The MC formed a transparent gel by crosslinking the reactive double bonds of the methacryl groups (Figure 12.5). Scaffolds were thiolated to add peptides containing the bioactive cell-adhesion and neurite-outgrowth sequences tyrosine-isoleucine-glycine-serine-arginine (YIGSR) and IKVAV through the thiol-maleimide coupling reaction. Addition of these sequences to the hydrogel improved primary neuron adhesion and neurite outgrowth. Neurites were able to penetrate the gels formed with 3 wt% crosslinking agent, but not the gels formed with 7 wt% crosslinking agent. Yu attributed this result to the smaller pore size found in the more tightly crosslinked 7% scaffolds [57].

By adding a polyol salt such as glycerophosphate salt (GP) to chitosan solution at low temperatures, a pH-neutral solution may be formed [58]. A macroporous hydrogel is then formed upon temperature increase to 37°C. To improve neuronal compatibility, Crompton covalently bonded poly(D-lysine) (PDL) to the chitosan surface through a photoreactive species [59]. *N*-(3-dimethylaminopropyl)-*N'*-ethylcarbodiimide hydrochloride and 4-azidoaniline were reacted with PDL to form a photoreactive molecule that could be conjugated to chitosan through UV exposure [59]. Fetal mouse cortical cells (FMCC) were suspended within the

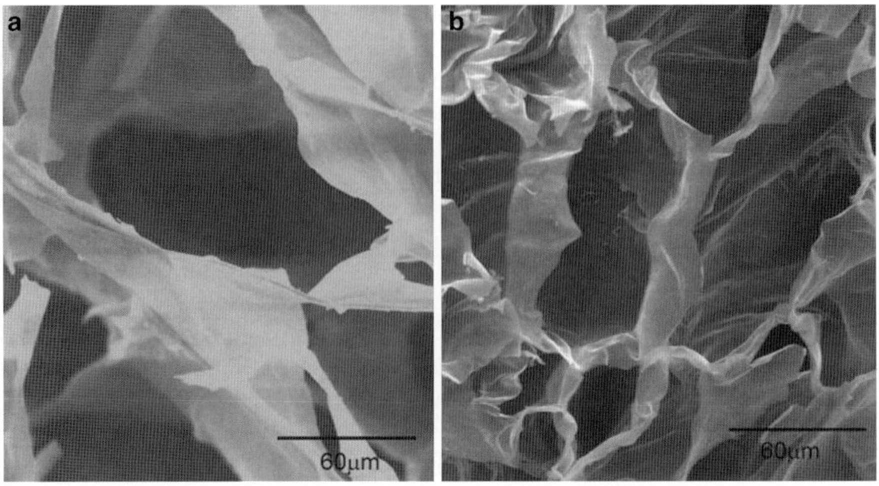

Figure 12.5. Scanning electron microscopy (SEM) revealed large (**a**) and smaller (**b**) pores when MC was crosslinked with 3 and 7 wt% APS/SMBS, respectively. Reproduced with permission from Wiley [57].

chitosan/GP solution prior to gelation of the scaffold. Two-dimensional chitosan/GP films with biological levels of osmolarity promoted high adhesion of FMCC. Cells cultured on PDL-modified chitosan did not exhibit changes in attachment and survival, but did show significantly decreased neurite extensions. Polylysine is known to be cytotoxic at high concentrations due to surface charge effects, but in this case Crompton et al. [59] did not attribute the observed decrease in neurite extension to cell damage since there was no observed effect on cell survival. Instead, these researchers proposed that polylysine has a mechanism for inhibiting neurite outgrowth that is unrelated to surface charge-based cytotoxicity [59]. Chitosan/GP hydrogels with 0-100% PDL-modified chitosan were manufactured. In 3D gel culture, cell survival and the number of cells with neurites were greatest on 0.1% PDL; in contrast, gels with concentrations greater than 10% supported lower cell numbers and neurites per cell than control chitosan/GP without PDL. Compared with results obtained with 2D cultures, cell survival and neurite outgrowth decreased in 3D cultures. Crompton et al. attributed this result to the greater number of PDL interactions when cells were completely surrounded by chitosan/GP in the 3D cultures [59].

12.4.5. Polysaccharides from Other Sources

Ashton et al. [60] utilized alginate hydrogels for hippocampal NPC encapsulation. Alginate, a polysaccharide found in brown seaweed that can be produced by bacteria [61], is compatible with CNS tissue [62], can encapsulate cells under mild conditions [63], and can be crosslinked to form a hydrogel by the addition of divalent cations [64]. Since alginate cannot be enzymatically degraded in mammals, high molecular weight alginate may persist for months in vivo [65]. The strategy of Ashton et al. was to include biodegradable poly(lactide-co-glycolide) (PLGA) microspheres loaded with alginate lyase (AL) in the scaffolds, giving tunable control over substrate degradation [60]. By varying the AL loading and concentration of microspheres, degradation rates could be tuned from 0% to 70% in 7 days. When NPCs were encapsulated, a 30-fold increase in cell number was observed in 1% alginate hydrogels containing 10 μg microspheres with 1:10 w/w ratio of AL to PLGA. This result is similar to the rate of cell proliferation observed when cells were cultured without any restriction on spreading, as in 2D cultures. As the rate of scaffold degradation decreased, so did the rate of cell proliferation, down to only a twofold increase in hydrogels containing only control PLGA microspheres without AL. Ashton et al. [60] attributed this increase in proliferation to the less-constrained growth possible in partially degraded hydrogels. No negative impact on NPC proliferation rates was observed at the presence of even 500 μg/mg microspheres and 50 mU/mL AL; these concentrations were much higher than those used to determine cell growth to scaffold degradation relationships. This result led Ashton to conclude that the microspheres tested are compatible with the NPC culture, and that the alginate hydrogel system shows promise for in vivo stem and progenitor cell localization [60].

Dextran is a bacterial polysaccharide generally produced by certain strains of *Leuconostoc* or *Streptococcus*. Due to its resistance to protein adsorption and cell adhesion, dextran has been used in chromatography applications (Sephadex®) [66, 67] and for microcarrier cell culture (Cytodex®) [68]. Lévesque synthesized dextran scaffolds by exploiting the immiscibility of methacrylated dextran (Dex-MA) and PEG solutions during polymerization [69]. The scaffolds thus obtained appeared promising as a 3D cell culture substrate due to their morphology, but did not deliver due to poor cell adhesion. By copolymerizing Dex-MA with aminoethyl methacrylate (AEMA), primary amine groups were introduced to immobilize cell-adhesion moieties while preserving the bioactivity of the peptide sequence [70]. Either porous scaffolds or hydrogels could be made by polymerization with or without PEG, respectively. By adding increasing amounts of AEMA, the number of amine sites was increased up to approximately 36 μmol/cm^3 in a 0.5 wt% AEMA gel or scaffold. Peptides incorporating either the bioactive arginine–glycine–aspartate (RGD) sequence of fibronectin or the YIGSR and IKVAV sequences of laminin were added to these hydrogels and scaffolds. Dex-MA-co-AEMA scaffolds greatly increased cell adhesion compared with the Dex-MA-only scaffolds. When dorsal root ganglia were seeded on the copolymer scaffolds, cell bodies were found up to 600 μm below the surface of the construct. The average number of cells was significantly increased on both RGD and YIGSR/IKVAV over the results obtained on Dex-MA-co-AEMA, but neurite outgrowth was only significantly increased on the YIGSR/IKVAV peptide. When cells were preincubated with soluble RGD and YIGSR/IKVAV peptides (to test the specificity of the interaction), cell adhesion and neurite outgrowth were both markedly decreased [70]. Lévesque and Shoichet concluded that this novel cell-adhesive dextran showed promise for neural tissue-engineering applications [70].

12.5. Conclusion

From 2D to 3D substrates, and from synthetic to naturally derived biomaterials, the response of NSPCs is complex and, in many cases, depends on cell age and seeding density as well as material surface chemistry and topography. By elucidating stem and progenitor cell responses and function on various natural and synthetic surfaces, the current knowledge base is expanded and the goal of directed growth and differentiation of stem and progenitor cells comes closer to fruition.

Acknowledgments

The authors gratefully acknowledge the NIH (RO1GM072005) for financial support.

References

1. McKay R. Stem cells in the central nervous system. Science. 1997;276(5309):66–71.
2. Alvarez-Buylla A, Temple S. Stem cells in the developing and adult nervous system. J Neurobiol. 1998;36(2):105–110.
3. Gage FH, Coates PW, Palmer TD, Kuhn HG, Fisher LJ, Suhonen JO et al. Survival and differentiation of adult neuronal progenitor cells transplanted to the adult brain. Proc Natl Acad Sci USA. 1995;92(25):11879–11883.
4. Weiss S, Dunne C, Hewson J, Wohl C, Wheatley M, Peterson AC et al. Multipotent CNS stem cells are present in the adult mammalian spinal cord and ventricular neuroaxis. J Neurosci. 1996;16(23):7599–7609.
5. Guillaume DJ, Johnson MA, Li XJ, Zhang SC. Human embryonic stem cell-derived neural precursors develop into neurons and integrate into the host brain. J Neurosci Res. 2006;84(6):1165–1176.
6. Johnson MA, Weick JP, Pearce RA, Zhang SC. Functional neural development from human embryonic stem cells: accelerated synaptic activity via astrocyte coculture. J Neurosci. 2007;27(12):3069–3077.
7. Takahashi K, Tanabe K, Ohnuki M, Narita M, Ichisaka T, Tomoda K et al. Induction of pluripotent stem cells from adult human fibroblasts by defined factors. Cell. 2007;131(5):861–872.
8. Yu J, Vodyanik MA, Smuga-Otto K, Antosiewicz-Bourget J, Frane JL, Tian S et al. Induced pluripotent stem cell lines derived from human somatic cells. Science. 2007;318(5858):1917–1920.
9. Coles BL, Angenieux B, Inoue T, Del Rio-Tsonis K, Spence JR, McInnes RR et al. Facile isolation and the characterization of human retinal stem cells. Proc Natl Acad Sci USA. 2004;101(44):15772–15777.
10. Klassen HJ, Ng TF, Kurimoto Y, Kirov I, Shatos M, Coffey P et al. Multipotent retinal progenitors express developmental markers, differentiate into retinal neurons, and preserve light-mediated behavior. Invest Ophthalmol Vis Sci. 2004;45(11):4167–4173.
11. Palmer TD, Takahashi J, Gage FH. The adult rat hippocampus contains primordial neural stem cells. Mol Cell Neurosci. 1997;8(6):389–404.
12. Shihabuddin LS, Ray J, Gage FH. FGF-2 is sufficient to isolate progenitors found in the adult mammalian spinal cord. Exp Neurol. 1997;148(2):577–586.
13. Keene CD, Ortiz-Gonzalez XR, Jiang Y, Largaespada DA, Verfaillie CM, Low WC. Neural differentiation and incorporation of bone marrow-derived multipotent adult progenitor cells after single cell transplantation into blastocyst stage mouse embryos. Cell Transplant. 2003;12(3):201–213.
14. Mezey E, Key S, Vogelsang G, Szalayova I, Lange GD, Crain B. Transplanted bone marrow generates new neurons in human brains. Proc Natl Acad Sci USA. 2003;100(3):1364–1369.
15. Woodbury D, Schwarz EJ, Prockop DJ, Black IB. Adult rat and human bone marrow stromal cells differentiate into neurons. J Neurosci Res. 2000;61(4):364–370.
16. Terada N, Hamazaki T, Oka M, Hoki M, Mastalerz DM, Nakano Y, et al. Bone marrow cells adopt the phenotype of other cells by spontaneous cell fusion. Nature. 2002;416(6880):542–545.
17. Jiang Y, Jahagirdar BN, Reinhardt RL, Schwartz RE, Keene CD, Ortiz-Gonzalez XR et al. Pluripotency of mesenchymal stem cells derived from adult marrow. Nature. 2002;418(6893):41–49.
18. Miura M, Gronthos S, Zhao M, Lu B, Fisher LW, Robey PG et al. SHED: stem cells from human exfoliated deciduous teeth. Proc Natl Acad Sci USA. 2003;100(10):5807–5812.
19. Prockop DJ, Azizi SA, Phinney DG, Kopen GC, Schwarz EJ. Potential use of marrow stromal cells as therapeutic vectors for diseases of the central nervous system. Prog Brain Res. 2000;128:293–297.
20. Toma JG, Akhavan M, Fernandes KJ, Barnabe-Heider F, Sadikot A, Kaplan DR et al. Isolation of multipotent adult stem cells from the dermis of mammalian skin. Nat Cell Biol. 2001;3(9):778–784.
21. Hudis M. Techniques and Applications of Plasma Chemistry. New York: John Wiley, 1974.
22. Young TH, Hung CH. Behavior of embryonic rat cerebral cortical stem cells on the PVA and EVAL substrates. Biomaterials. 2005;26(20):4291–4299.
23. Tsai RY, McKay RD. Cell contact regulates fate choice by cortical stem cells. J Neurosci. 2000;20(10):3725–3735.
24. Young TH, Hu WW. Covalent bonding of lysine to EVAL membrane surface to improve survival of cultured cerebellar granule neurons. Biomaterials. 2003;24(8):1477–1486.
25. Gomez N, Chen S, Schmidt CE. Polarization of hippocampal neurons with competitive surface stimuli: contact guidance cues are preferred over chemical ligands. J R Soc Interface. 2007;4(13):223–233.

26. Song M, Uhrich KE. Optimal micropattern dimensions enhance neurite outgrowth rates, lengths, and orientations. Ann Biomed Eng. 2007;35(10):1812–1820.
27. Tai HC, Buettner HM. Neurite outgrowth and growth cone morphology on micropatterned surfaces. Biotechnol Prog. 1998;14(3):364–370.
28. Recknor JB, Sakaguchi DS, Mallapragada SK. Directed growth and selective differentiation of neural progenitor cells on micropatterned polymer substrates. Biomaterials. 2006;27(22):4098–4108.
29. Recknor JB, Sakaguchi DS, Mallapragada SK. Growth and differentiation of astrocytes and neural progenitor cells on micropatterned polymer films. Ann N Y Acad Sci. 2005;1049:24–27.
30. Busby W, Cameron NR, Jahoda CA.Emulsion-derived foams (PolyHIPEs) containing poly(epsilon-caprolactone) as matrixes for tissue engineering. Biomacromolecules. 2001;2(1):154–164.
31. Hayman MW, Smith KH, Cameron NR, Przyborski SA. Growth of human stem cell-derived neurons on solid three-dimensional polymers. J Biochem Biophys Methods. 2005;62(3):231–240.
32. George PM, Lyckman AW, LaVan DA, Hegde A, Leung Y, Avasare R et al. Fabrication and biocompatibility of polypyrrole implants suitable for neural prosthetics. Biomaterials. 2005;26(17):3511–3519.
33. Kotwal A, Schmidt CE. Electrical stimulation alters protein adsorption and nerve cell interactions with electrically conducting biomaterials. Biomaterials. 2001;22(10):1055–1064.
34. Hodgson, AJJ. M. J.; Campbell, T.; Georgevich, A.; Woodhouse, S.; Aoki, T.; Ogata, N.; Wallace, G. G.; Integration of biocomponents with synthetic structures: use of conducting polymer polyelectrolyte composites. Proc SPIE Int Soc Opt Eng. 1996;2716:164–176.
35. Gomez N, Lu Y, Chen S, Schmidt CE. Immobilized nerve growth factor and microtopography have distinct effects on polarization versus axon elongation in hippocampal cells in culture. Biomaterials. 2007;28(2):271–284.
36. Bhang SH, Lim JS, Choi CY, Kwon YK, Kim BS. The behavior of neural stem cells on biodegradable synthetic polymers. J Biomater Sci Polym Ed. 2007;18(2):223–239.
37. Wake MC, Gupta PK, Mikos AG. Fabrication of pliable biodegradable polymer foams to engineer soft tissues. Cell Transplant. 1996;5(4):465–473.
38. Liao CJ, Chen CF, Chen JH, Chiang SF, Lin YJ, Chang KY. Fabrication of porous biodegradable polymer scaffolds using a solvent merging/particulate leaching method. J Biomed Mater Res. 2002;59(4):676–681.
39. Levenberg S, Burdick JA, Kraehenbuehl T, Langer R. Neurotrophin-induced differentiation of human embryonic stem cells on three-dimensional polymeric scaffolds. Tissue Eng. 2005;11(3–4):506–512.
40. Levenberg S, Huang NF, Lavik E, Rogers AB, Itskovitz-Eldor J, Langer R. Differentiation of human embryonic stem cells on three-dimensional polymer scaffolds. Proc Natl Acad Sci USA. 2003;100(22):12741–12746.
41. Yang F, Xu CY, Kotaki M, Wang S, Ramakrishna S. Characterization of neural stem cells on electrospun poly(l-lactic acid) nanofibrous scaffold. J Biomater Sci Polym Ed. 2004;15(12):1483–1497.
42. Yang F, Murugan R, Wang S, Ramakrishna S. Electrospinning of nano/micro scale poly(l-lactic acid) aligned fibers and their potential in neural tissue engineering. Biomaterials. 2005;26(15):2603–2610.
43. Brewer GJ, Deshmane S, Ponnusamy E. Precocious axons and improved survival of rat hippocampal neurons on lysine-alanine polymer substrates. J Neurosci Methods. 1998;85(1):13–20.
44. Wang JH, Hung CH, Young TH. Proliferation and differentiation of neural stem cells on lysine-alanine sequential polymer substrates. Biomaterials. 2006;27(18):3441–3450.
45. Royce Hynes S, McGregor LM, Ford Rauch M, Lavik EB. Photopolymerized poly(glycol)/poly(l-lysine) hydrogels for the delivery of neural progenitor cells. J Biomater Sci Polym Ed. 2007;18(8):1017–1030.
46. Nakaji-Hirabayashi T, Kato K, Iwata H. Self-assembling chimeric protein for the construction of biodegradable hydrogels capable of interaction with integrins expressed on neural stem/progenitor cells. Biomacromolecules. 2008;9(5):1411–1416.
47. Schense JC, Bloch J, Aebischer P, Hubbell JA. Enzymatic incorporation of bioactive peptides into fibrin matrices enhances neurite extension. Nat Biotechnol. 2000;18(4):415–419.
48. Silva GA, Czeisler C, Niece KL, Beniash E, Harrington DA, Kessler JA, Selective differentiation of neural progenitor cells by high-epitope density nanofibers. Science. 2004;303(5662):1352–1355.
49. Schense JC, Hubbell JA. Three-dimensional migration of neurites is mediated by adhesion site density and affinity. J Biol Chem. 2000;275(10):6813–6818.
50. Cox S, Cole M, Tawil B. Behavior of human dermal fibroblasts in three-dimensional fibrin clots: dependence on fibrinogen and thrombin concentration. Tissue Eng. 2004;10(5–6):942–954.
51. Bensaid W, Triffitt JT, Blanchat C, Oudina K, Sedel L, Petite H. A biodegradable fibrin scaffold for mesenchymal stem cell transplantation. Biomaterials. 2003;24(14):2497–2502.
52. Willerth SM, Arendas KJ, Gottlieb DI, Sakiyama-Elbert SE. Optimization of fibrin scaffolds for differentiation of murine embryonic stem cells into neural lineage cells. Biomaterials. 2006;27(36):5990–6003.

53. Watanabe K, Nakamura M, Okano H, Toyama Y. Establishment of three-dimensional culture of neural stem/progenitor cells in collagen Type-1 Gel. Restor Neurol Neurosci. 2007;25(2):109–117.
54. Balgude AP, Yu X, Szymanski A, Bellamkonda RV. Agarose gel stiffness determines rate of DRG neurite extension in 3D cultures. Biomaterials. 2001;22(10):1077–1084.
55. Brannvall K, Bergman K, Wallenquist U, Svahn S, Bowden T, Hilborn J et al. Enhanced neuronal differentiation in a three-dimensional collagen-hyaluronan matrix. J Neurosci Res. 2007;85(10):2138–2146.
56. Deister C, Aljabari S, Schmidt CE. Effects of collagen 1, fibronectin, laminin and hyaluronic acid concentration in multi-component gels on neurite extension. J Biomater Sci Polym Ed. 2007;18(8):983–997.
57. Yu LM, Kazazian K, Shoichet MS. Peptide surface modification of methacrylamide chitosan for neural tissue engineering applications. J Biomed Mater Res A. 2007;82(1):243–255.
58. Chenite A, Chaput C, Wang D, Combes C, Buschmann MD, Hoemann CD et al. Novel injectable neutral solutions of chitosan form biodegradable gels in situ. Biomaterials. 2000;21(21):2155–2161.
59. Crompton KE, Goud JD, Bellamkonda RV, Gengenbach TR, Finkelstein DI, Horne MK, Polylysine-functionalised thermoresponsive chitosan hydrogel for neural tissue engineering. Biomaterials. 2007;28(3):441–449.
60. Ashton RS, Banerjee A, Punyani S, Schaffer DV, Kane RS. Scaffolds based on degradable alginate hydrogels and poly(lactide-co-glycolide) microspheres for stem cell culture. Biomaterials. 2007;28(36):5518–5525.
61. Remminghorst U, Rehm BH. Bacterial alginates: from biosynthesis to applications. Biotechnol Lett. 2006;28(21):1701–1712.
62. Kataoka K, Suzuki Y, Kitada M, Hashimoto T, Chou H, Bai H et al. Alginate enhances elongation of early regenerating axons in spinal cord of young rats. Tissue Eng. 2004;10(3–4):493–504.
63. Li X, Liu T, Song K, Yao L, Ge D, Bao C, Culture of neural stem cells in calcium alginate beads. Biotechnol Prog. 2006;22(6):1683–1689.
64. Prang P, Muller R, Eljaouhari A, Heckmann K, Kunz W, Weber T, The promotion of oriented axonal regrowth in the injured spinal cord by alginate-based anisotropic capillary hydrogels. Biomaterials. 2006;27(19):3560–3569.
65. Bouhadir KH, Lee KY, Alsberg E, Damm KL, Anderson KW, Mooney DJ. Degradation of partially oxidized alginate and its potential application for tissue engineering. Biotechnol Prog. 2001;17(5):945–950.
66. Bollag DM. Ion-exchange chromatography. Methods Mol Biol. 1994;36:11–22.
67. Bollag DM. Gel-filtration chromatography. Methods Mol Biol. 1994;36:1–9.
68. Hirtenstein M, Clark J, Lindgren G, Vretblad P. Microcarriers for animal cell culture: a brief review of theory and practice. Dev Biol Stand. 1980;46:109–116.
69. Lévesque SG, Lim RM, Shoichet MS. Macroporous interconnected dextran scaffolds of controlled porosity for tissue-engineering applications. Biomaterials. 2005;26(35):7436–7446.
70. Lévesque SG, Shoichet MS. Synthesis of cell-adhesive dextran hydrogels and macroporous scaffolds. Biomaterials. 2006;27(30):5277–5285.

13

Toward Osteogenic Differentiation of Marrow Stromal Cells and In Vitro Production of Mineralized Extracellular Matrix onto Natural Scaffolds

Ana M. Martins, Catarina M. Alves, Rui L. Reis, Antonios G. Mikos, and F. Kurtis Kasper

Tissue engineering has emerged as a new interdisciplinary field for the repair of various tissues, restoring their functions by using scaffolds, cells, and/or bioactive factors. A temporary scaffold acts as an extracellular matrix analog to culture cells and guide the development of new tissue. In this chapter, we discuss the preparation of naturally derived scaffolds of polysaccharide origin, the osteogenic differentiation of mesenchymal stem cells cultured on biomimetic calcium phosphate coatings, and the delivery of biomolecules associated with extracellular matrix mineralization.

Abbreviations

BMP	bone morphogenetic protein
BMP-2	bone morphogenetic protein-2
BMSC	bone marrow stromal cell
BMSSC	bone marrow stromal stem cell
CaP	calcium phosphate
ECM	extracellular matrix
FGF-1	fibroblast growth factor-1
FGF-2	fibroblast growth factor-2
IGF-2	insulin-like growth factor-2
MSC	mesenchymal stem cell
RGD	arginine–glycine–aspartic acid

A.M. Martins, C.M. Alves, and R.L. Reis • 3B's Research Group – Biomaterials, Biodegradables and Biomimetics, University of Minho, Headquarters of the European Institute of Excellence on Tissue Engineering and Regenerative Medicine, AvePark, 4806-909, Taipas, Guimarães, Portugal
IBB – Institute for Biotechnology and Bioengineering, PT Government Associated Laboratory, Guimarães, Portugal
A.M. Martins, A.G. Mikos, and F.K. Kasper • Department of Bioengineering, Rice University, 6100 Main Street, Houston, TX 77005-1892, USA

D.A. Puleo and R. Bizios (eds.), *Biological Interactions on Materials Surfaces*,
DOI 10.1007/978-0-387-98161-1_13,

SBF	simulated body fluid
SPCL	blend of starch and poly(ε-caprolactone)
TGF-β	transforming growth factor-β
TGF-β1	transforming growth factor-β1

13.1. Introduction

Bone is a dynamic, highly vascularized tissue with a unique capacity to heal and remodel without leaving a scar. It is the structural framework of the body and is composed of an inorganic mineral phase of hydroxyapatite and an organic phase of mainly type I collagen. Bone continuously resorbs and reforms in a remodeling process that is carried out by two types of bone cells: the bone-building osteoblasts and the bone-resorbing osteoclasts. Slowly and insidiously, bone deteriorates, losing minerals and structure. Bone injuries produced as a result of disease and/or trauma present a major health concern. A fracture, usually of the hip, wrist, or a vertebra, is often the first indication that osteoporosis has been weakening the bones of a patient for years [1]. Treatment options include transplantation, surgical repair, prostheses, mechanical devices, and drug therapy [2]. However, major damage to a tissue or organ can neither be repaired nor long-term recovery effected in a truly satisfactory way using these methods.

In this context, an emerging field of science termed "tissue engineering," defined as an "interdisciplinary field that applies the principles of engineering and life sciences toward the development of biological substitutes that restore, maintain, or improve tissue function" [3] has been gaining significant recognition. Tissue engineering often uses stem cells, progenitor cells and/or organ-specific cells for seeding a scaffold ex vivo, however it may also involve the implantation of an acellular construct for guided tissue regeneration [4]. Indeed, a wide range of strategies exists for tissue engineering in general, and bone tissue engineering specifically.

Bone tissue engineering is a rapidly expanding field, full of innovative ideas for treating bone trauma and pathologies. Selection of the most appropriate material to produce a scaffold in bone-related applications is a very important step toward the construction of a tissue engineered construct. There is an increasing interest in the production of novel scaffolds from renewable resources. Natural polymers are an attractive alternative to synthetic polymers for various clinical applications partly due to their biocompatibility and also because they are typically biodegraded by "normal" and/or enzymatic hydrolysis (carried out, in the majority of cases, by specific enzymes also present in the human body). Some of the advantages associated with naturally derived biomaterials are their cost effectiveness as well as the wide range of properties and structures attainable with these materials. A large number of different naturally derived biomaterials have been studied and proposed for bone tissue-engineering applications, namely polysaccharides (chitosan, starch, alginate, hyaluronic acid, and cellulose, among others) and proteins (soy, collagen, and fibrin). Polysaccharides, in particular, have some attractive properties, such as nontoxicity (pertinent monomer residues are not hazardous to health), high swelling ability, and stability over a range of pH values.

For successful bone replacement, the ideal scaffold should be biocompatible [5–8] with the surrounding biological fluids and tissues to avoid any detrimental tissue response. The scaffolding material should degrade into nontoxic residues that can be easily removed from the body through normal excretion processes [5, 6, 8, 9]. The scaffolds serve as temporary substrates for living cells as well as physical supports for tissue regeneration [10]. Adequate surface area and appropriate surface energy are also needed to permit cell adhesion, promote cell proliferation, and allow retention of differentiated cell functions [5–8, 10]. In addition, sufficient mechanical stability of the scaffold material is necessary to maintain the desired shape and structure during cell culture in vitro and transplantation in vivo. Control of scaffold pore morphology is critical for controlling cell colonization rates and maintaining transport of oxygen, nutrients, and metabolic waste, as well as for supporting organization of the engineered tissue. Furthermore, angiogenesis, a requirement for the survival and success of vascularized tissues, can be affected by the porosity of the scaffold. Pore morphology can also be expected to significantly affect scaffold degradation kinetics and the mechanical properties of the developing tissue [6, 11].

The scaffolds used for tissue engineering purposes mimic the extracellular matrix (ECM) of the regenerating bone environment. Thus, in addition to serving as a mechanical support, a tissue engineering scaffold may also be "informative" to the cells. An ideal three-dimensional (3D) construct for bone tissue engineering, above all other pertinent characteristics, should be simultaneously osteoinductive (capable of recruiting osteoprogenitor cells and stimulating their differentiation along the bone-forming cell lineage), osteoconductive (capable of supporting the formation of bone at the surface of the scaffold), and also resorbable and amenable to gradual replacement by newly formed bone [12]. In the medical field, consideration of biodegradation is a priority on the list of safety standards when choosing polymers as potential biomaterials for tissue engineering applications. Naturally derived materials have recently gained interest, as they are structurally similar to the native ECM of many tissues; exhibit excellent biocompatibility; and induce minimal inflammatory response and tissue damage. Natural polymers may present a biologically active environment to the cells, since they usually contain domains that provide cues and can send important signals to guide cells at various stages of development [10].

A method to potentially increase the biological activity of a bone tissue engineering scaffold is to coat the surface of the scaffold with calcium phosphate (CaP). One of the main goals of using CaP coatings on bone tissue engineering scaffolds is to promote osteoconduction by enhancing adhesion of osteogenic cells and ingrowth of bone into porous biomaterials [13]. New technologies have been developed to promote osteogenic activity of bone tissue-engineering scaffolds. These approaches tend to integrate into the coatings osteoinductive or bioactive agents (e.g., enzymes and antibiotics), to immobilize constitutional elements of bone (e.g., growth factors, including bone morphogenetic proteins (BMPs) and other members of the transforming growth factor (TGF)-β superfamily), adhesion proteins (e.g., collagen, fibronectin, laminin, and vitronectin) and peptides (e.g., the arginine-glycine-aspartic acid (RGD) sequence) on the surface of biomaterials. Immobilization and/or delivery of bioactive molecules at specific sites have been exploited to enhance cell adhesion, differentiation, and other cell functions as well as to promote mineralization of the ECM of the tissue-engineered bone constructs.

13.2. Scaffolds of Natural Origin – Polysaccharides

A large number of natural polymers, including polysaccharides, have been suggested as candidates for the production of scaffolds for bone tissue engineering purposes. Polysaccharides are relatively complex carbohydrates. They are high molecular weight polymers having one or more monosaccharide repeating-units joined together by glycosidic bonds. Polysaccharides tend to be amorphous and insoluble in water. Some of the main advantages associated with this class of polymers are wide availability, cost effectiveness, good hemocompatibility (probably because of their similarities with heparin), nontoxicity, and a wide range of properties and structures suitable for biomedical applications. These polymers have been proposed as scaffolds for bone tissue engineering applications as well as carriers for cells and bioactive molecules (e.g., proteins, enzymes, and growth factors) for controlled-release systems.

Chitosan, starch, and alginate, three examples of polysaccharide materials, will be described in detail in the sections that follow.

13.2.1. Chitosan

Chitosan, a naturally derived polymer, is a partially deacetylated derivative of chitin found in crustacea exoskeletons (e.g., shrimp, crab, and lobster), cell walls of fungi, and cuticles of insects [14, 15]. Depending on the source and preparation procedure, the molecular weight of chitosan may range from 300 to more than 1,000 kDa [11]. Chitosan is a suitable functional biomaterial because it is biocompatible, biodegradable, minimally immunogenic, nontoxic, and hydrophilic. Moreover, it has adsorption properties with remarkable affinity for proteins, and is inexpensive [16–21]. Some studies report that chitosan enhanced osteogenesis [22–24] and improved wound healing [25, 26]. In addition, chitosan is a hemostatic agent [11, 16] with antithrombogenic properties [27]. It has proved to be a useful excipient in various drug delivery systems due to its nontoxicity, high cohesive and hydrophilic properties, and polycationic character resulting from primary amine groups, which provide a high charge density in acidic solutions (pH < 6.5) [18, 28]. It is soluble in dilute or weak acids (such as acetic and formic acid), but it is normally insoluble in aqueous solutions above pH 6.5.

Chitosan is a binary polyheterosaccharide of *N*-acetylglucosamine and glucosamine with a $\beta 1 \rightarrow 4$ linkage. The superior tissue compatibility of chitosan can be partially attributed to its structural similarity to glycosaminoglycans, which are major components of the ECM of bone and cartilage [15, 29]. Chitosan is easily hydrolyzed by various chitosanases [30], which are completely absent in mammals, and is biodegraded in the presence of lysozyme in aqueous media in vitro [17, 31–35]; this degradation process depends on the degree of deacetylation [31], which represents the proportion of *N*-acetyl-D-glucosamine units with respect to the total number of units [30]. Chitosan degradation kinetics are inversely related to the degree of deacetylation [31, 32]. In vitro and in vivo, chitosan is degraded by enzymatic hydrolysis; the primary agent of this process is lysozyme, which targets acetylated residues [36]. Chitosan and glucosamine, its biodegradation product, are not toxic in vivo [37]. Lysozyme, or muramidase, is an enzyme that catalyzes the hydrolysis of the peptidoglycan layer of bacterial cell walls [38]. This enzyme is active over a broad pH range (from 3 to 8) and hydrolyzes its substrates both inside and outside cells. Lysozyme is widely distributed in the human body [39]. It is found in the nose, bronchus, bronchiole, middle ear, lacrimal gland, bone marrow, and digestive tract [16], and in lymphocytes; lysozyme is also secreted by monocytes, macrophages, and granulocytes, which are the largest source of the enzyme [40, 41]. Monocytes and macrophages are the primary contributors to the lysozyme content in human serum [41]; the concentration in serum is in the range of 7–13 mg/L [39]. The susceptibility of chitosan to degradation induced by lysozyme make the protein an attractive target for incorporation into this biodegradable material [29, 42–44].

Incorporation of active biomolecules, such as growth factors, has been used as a highly beneficial strategy for improving bone regeneration in tissue engineering applications.

The biological activity of chitosan on bone regeneration has been confirmed in many studies [18, 45]. Chitosan can be easily fabricated into bulk porous scaffolds, films, microparticles, sponges, and beads. The feasibility of forming porous scaffolds permits wide application of this polymer in tissue engineering. This is mainly true for bone tissue engineering applications because chitosan supports osteoblast proliferation and phenotypic expression [15]. Chitosan fiber meshes with appropriate mechanical properties, developed by Tuzlakoglu et al. [46], exhibited bioactivity; this is a very important aspect for biomaterials used as bone tissue engineering scaffolds. Martins et al. [35] proposed the development of chitosan-based scaffolds with the capability of forming porous structures in situ following attack by specific enzymes (namely, α-amylase and lysozyme) present in the human body. In addition to the capability of forming pores in situ, other advantages these scaffolds have when compared with other conventional materials are their suitable mechanical properties and lack of toxicity. Coutinho et al. [47] studied the function of an osteoblastic-like cell line (SaOs-2) on chitosan blends with synthetic biodegradable polymers, and reported enhanced osteoblastic activity. Costa-Pinto et al. [48] formulated scaffolds based on blends of chitosan and synthetic polyesters, and provided evidence that these scaffolds are cytocompatible. Furthermore, chitosan-based scaffolds promoted the attachment and proliferation of mouse mesenchymal stem cells (MSCs) [48], which exhibited high levels of alkaline phosphatase activity and produced a mineralized ECM [48].

13.2.2. Starch

Starch is one of the most abundant naturally occurring polymers with properties that make it attractive for several biomedical applications. Starch is found as insoluble granules of α-amylose (20–30%) and amylopectin (70–80%) [49]. Amylopectin polymers are highly branched structures containing (1→4)-α-D-glucose and (1→6)-α-D-glucose linkages, whereas amylose is much more linear with long stretches of (1→4)-α-D-glucose-linked monomer units. Starch is extremely difficult to process and is brittle when used without the addition of a plasticizer [49]. Over the years, several other materials have been blended with starch to improve its processability, including several synthetic [50–54] and natural polymers, such as polysaccharides [35, 55] and proteins [56]. Reis and coworkers [35, 57–70] have proposed use of starch-based scaffolds for biomedical applications. Starch exhibits low toxicity [35, 64], biodegradability [35, 70–72], and biocompatibility [73–75], which are excellent characteristics for bone tissue engineering applications. Compared with other biodegradable polymers available, starch is inexpensive, and above all, reusable. Specific enzymes present in the human body, namely α-amylase in the blood plasma, can easily degrade starch. The main enzymes involved in starch degradation are α-amylases, β-amylases, α-glucosidases, and other debranching enzymes.

An important consideration of biodegradable materials of natural origin being considered for use in the biomedical field is the host response to the degradation products. Starch degradation products are oligosaccharides that can be metabolized to produce energy. Due to their degradation by α-amylases, this constitutes another strategy to control and tailor the degradation of starch-based scaffolds. Martins et al. [35] developed a novel biodegradable matrix based on chitosan and starch, with the capability of forming a porous structure in situ following attack by specific enzymes (namely α-amylase and lysozyme) present in the human body. These researchers showed that pore size and distribution in the chitosan matrix is controlled by the location of the "sacrificial" phase (i.e., native starch) that is enzymatically degraded [35]. This same study reported an interesting approach for the control of matrix degradation in situ and consequent pore formation, which could result in scaffolds with mechanical properties appropriate for the initial stage of implantation [35]. Martins et al. [76] also studied the influence of α-amylase on the degradation of fiber-mesh scaffolds based on a blend of starch and poly(ε-caprolactone) (SPCL) and demonstrated enhanced scaffold porosity and pore size and decreased average fiber diameter with time. Furthermore, culture of rat marrow stromal cells on SPCL fiber meshes (in medium supplemented with α-amylase) resulted in enhanced cell proliferation [76].

13.2.3. Alginate

Alginate (alginic acid or algin) is a linear polyuronate containing D-mannuronic acid and L-guluronic acid that is abundant in the cell walls of brown algae. Due to the biocompatibility and gelation of alginate with certain divalent cations, it is widely used for cell immobilization and encapsulation. Alginate is soluble in aqueous solutions at room temperature and forms stable gels in the presence of calcium, barium, and strontium without chemical crosslinking agents [77]; for this reason, the viability and biological activity of entrapped cells and biochemical agents are maintained in alginate gels. As a biomaterial, alginate has a number of advantages including biocompatibility and nonimmunogenicity, which are related to its hydrophilicity [78, 79].

Several studies examined alginate sponges as scaffolds for tissue engineering applications [78] and reported that their structural and morphological properties are appropriate for cell culture and proliferation as well as for neovascularization [78]. Other studies reported that alginate supports synthesis of pertinent ECM components by various cell types, and provides an amenable environment for cell encapsulation, drug delivery, and gene delivery [80]. Alginate also permits cotransplantation of multiple cell types and appropriate growth stimuli to promote, for example, the osteogenic phenotype [81]. Encapsulated bone marrow stromal cells (BMSCs) were studied for the purpose of healing bone defects in orthopedics [82]. Studies with gels containing MSCs and alginate beads loaded with vancomycin (a treatment for bone infections), reported that bone marrow-derived MSCs proliferated and expressed alkaline phosphatase, osteopontin, and collagen 1A1 genes [83]. Cai et al. [84] reported expression of bone-specific ECM markers when they examined the ectopic bone-forming ability of BMSCs in combination with scaffolds made from alginate gel and implanted subcutaneously in nude mice for 8 weeks. Moreover, hydrogels such as alginate are effective substrates for both two-dimensional [85] and 3D [78, 85] cell cultures, indicating the suitability of alginate for tissue engineering applications.

13.3. CaP Biomimetic Coatings

Ideally, tissue engineering scaffolds should mimic, to the greatest degree possible, the properties of the native target tissue in an effort to promote, direct, and control regeneration of a specific, desired type of tissue. The term "biomimetics" is used to describe a branch of science that seeks to produce such "bioinspired" materials for a variety of applications.

Compared with other biomaterials, CaPs have a unique characteristic for bone mimicry and substitution. Their composition resembles that of bone mineral; most importantly, they can induce a biological response similar to that generated during bone remodeling, which involves resorption and formation of new bone tissue [86]. Osteoclasts are responsible for bone mineral degradation, resulting in bone resorption [86]. During bone resorption, the degradation products of CaP (calcium and phosphate ions) are naturally metabolized but do not cause abnormally increased calcium and phosphate levels in urine, serum, or organs [87]. It should be noted that osteoclasts degrade CaP in a similar fashion as they degrade natural bone [88–90].

In 1972 Hench et al. [91] showed that "Bioglass" (that is, glass in the Na_2O–CaO–SiO_2–P_2O_5 system), spontaneously bonded to living bone without formation of surrounding fibrous tissue. In the early 1990s, Kokubo and coworkers [92, 93] proposed that the essential requirement for a biomaterial to bond to living bone is the formation of bone-like apatite on the surface of the biomaterial when implanted in vivo. This in vivo apatite formation can be reproduced in vitro using simulated body fluid (SBF), which is a solution containing inorganic ion concentrations similar to those of human extracellular fluids but without any cells or proteins [94]. Under such in vitro conditions, the formed layer consists of carbonate apatite with small crystallites and low crystallinity [94]. This apatite is referred to as "bone-like apatite" due to its similarity to apatite present in natural bone.

Biomimetic methodology for coating biomaterials with a bone-like apatite layer has been described in several publications [92, 95–98]. This technique mimics the natural biomineralization processes, which involve controlled crystal phase nucleation and growth. The main advantage of the biomimetic methodology is the use of physiological conditions (pH 7.4 at 37°C) simulating the conditions under which apatite is formed in bone. Moreover, this technique allows incorporation of proteins and bioactive agents into CaP coatings without compromising bioactivity of the organic compounds [96, 98–101]. In 1997, Reis et al. [95] adapted the methodology developed by Kokubo and used bioactive glass as a precursor to nucleation and growth of CaP films on starch-based polymers. Briefly, for the preparation of biomimetic CaP coatings based on the methodology previously developed by Abe et al. [92] and Kokubo [93] and adapted by Reis et al. [95], the materials under consideration were first impregnated with bioactive glass, and were then immersed in SBF solution for several days at 37°C; this phase is known as the "nucleation stage" and allows formation of CaP nuclei. In order to accelerate apatite formation, the biomaterials were subsequently immersed at 37°C in simulated body fluid solution (1.5 SBF) with an ionic concentration 1.5-fold greater than physiological levels; this condition enhances CaP nuclei growth. The CaP biomimetic coatings, which are thus formed, exhibit osteoconductive properties that will be discussed later in this chapter.

13.3.1. Osteoconductivity

Scaffolds for bone tissue engineering should be osteoconductive; that is, able to support formation of bone within and/or upon the scaffold. Osteoconductivity has been observed when porous structures were implanted into or adjacent to bone. In such cases, osteoprogenitor cells migrated into pores and filled the porous structure with newly formed bone. This process is characterized by an initial ingrowth of fibrovascular tissue that invades the porous structure followed by later development of new bone directly within it [102]. Hydroxyapatite-based materials are osteoconductive, provided that fully differentiated osteogenic cells are available at the site of implantation [12]. Adsorption of growth factors from the local milieu and from the blood circulation contributes to the osteoconductivity of hydroxyapatite by creating suitable conditions for bone formation when implanted in an osseous environment in vivo. Many relatively insoluble CaP materials are osteoconductive, and, in some cases, may induce extraskeletal new bone formation (i.e., they are osteoinductive).

13.3.2. Osteoinductivity

Osteoinduction is the process by which stem and osteoprogenitor cells are recruited to the bone-healing site and stimulated to undergo osteogenic differentiation [103]. Osteoinductivity implies the ability of chemical compounds to induce osteogenic differentiation of uncommitted progenitor cells [12]. It has been proposed that biomaterials do not have an osteoinductive character in the absence of appropriate osteoinductive agents, such as certain BMPs and other bioactive molecules [104]. However, several studies have reported that some CaP biomaterials [105–107], namely CaP coatings [107, 108], may be osteoinductive. These CaP biomaterials may induce bone formation at extraskeletal sites without addition of osteogenic cells or bioactive agents. Hydroxyapatite is not osteoinductive because it cannot induce osteogenic differentiation of progenitor cells when implanted in a nonosseous environment, such as skin and muscle [12].

13.3.3. Incorporation of Biomolecules into CaP Biomimetic Coatings

Numerous attempts have been made to improve the osteoconductivity of biomaterials. Coatings of CaP expedite osteoconduction and bone ingrowth at the surface of bone substitutes and, therefore, are useful strategies in tissue engineering endeavors for the regeneration of bone tissue. However, a methodology that enables regeneration of bone tissue should not only expedite osteoconduction, but also osteoinduction through biochemical pathways [109–112]. It is known that BMPs can be incorporated into CaP implants (with adequate 3D geometry) to promote osteogenesis [112, 113]; the surface of such implants, however, will be rapidly conditioned by several highly concentrated molecules [114]. For this reason, other types of delivery-specific approaches have been investigated as alternatives that further functionalize and enhance the potential of CaP coatings. Specifically, the CaP biomimetic coatings have been used as a carrier of various molecules, including osteoinductive agents such as BMPs [115–117], other proteins [101, 118–120], enzymes [96, 98, 101], and antibiotics [13, 121].

Biomimetic CaP coatings, produced as described in earlier parts of this chapter, are deposited onto surfaces under physiological temperature and pH [110], enabling coprecipitation and consequent incorporation of biologically active molecules [99]. This approach circumvents difficulties common to plasma spraying techniques. By using low temperatures, biomimetic processes can be applied not only to highly resistant materials (e.g., metallic alloys) but also to polymeric and naturally derived materials (e.g., chitosan, starch, and collagen) for implantation [122].

The major objective of CaP coatings is to provide appropriate biological composition and to improve the quality of the surfaces of various materials used for orthopedic applications. The conditions under which such a coating is prepared affect conformational stability of incorporated biomolecules, and thus the bioactivity and shelf-life of the final product. Such coatings, which are structurally and chemically comparable to the mineral component of bone, can possesses favorable bioactive properties that may facilitate outcomes in cases of critical clinical need [13, 123].

This alternative coating technique may be used to produce systems with several advantages, such as reduction of burst release of incorporated molecules into the biological milieu. In this case, biomolecules incorporated in the inorganic phase are gradually released as the latticework undergoes degradation. The advent of the slow degradation of the coating modulates delivery of bioactive agents. Slow release of these chemical compounds may improve the osteoinductive capacity of the implant material [100, 124].

One of the potential applications of CaP coatings pertains to the incorporation of bioactive agents and proteins. Azevedo et al. [101] used a biomimetic technique and successfully incorporated bovine serum albumin and α-amylase into a CaP coating on the surface of a starch-based polymer. In that study, the properties of the resultant biomaterial were tailored by judicious choice of specific enzymes and their incorporation at different compositions and combinations into CaP coatings that retained their bioactivity [101]. Efficient incorporation of active α-amylase into biomimetic coatings controlled the degradation rate of starch-based biomaterials. Similar results and applications were achieved with chitosan scaffolds after incorporation of lysozyme [96, 98]. Martins et al. [96, 98] incorporated lysozyme into CaP coatings on the surface of chitosan scaffolds in order to control the degradation rate of chitosan and subsequent formation of pores. Furthermore, since lysozyme has antibacterial properties, these coatings may be used as a carrier for its sustained release, potentially mitigating infection at the implantation site. Several studies reported in the literature addressed incorporation of BMPs into biomimetic CaP layers [110, 116, 123, 125]. These studies indicated that CaP coatings have the potential for sustained delivery of many other bioactive agents. Liu and coworkers [99] demonstrated that BMP-2 retained its osteoinductivity when delivered from biomimetic systems and that the osteoconductivity of implant material surfaces was affected by BMP-2 and its delivery mode [123].

In summary, the results discussed in this section support the strategy of adding osteoinductive signaling molecules into CaP biomimetic coatings for the purpose of inducing bone growth.

13.4. Osteogenic Differentiation of Marrow Stromal Cells and Mineralized ECM Production In Vitro

Biomaterials and scaffolds considered for bone tissue engineering are often evaluated in vitro for their ability to support adhesion, proliferation, and differentiation of progenitor cells along the osteogenic pathway prior to being evaluated in vivo. In vitro cell–scaffold interactions are determined using osteoblasts, osteosarcoma cell lines, and osteoprogenitor cells. The scaffolds used for this purpose mimic the ECM of bone and play a crucial role in supporting cell functions and differentiation, but may also be used to deliver biomolecules.

Osteoblastic differentiation of MSCs comprises cell proliferation, cell maturation, and matrix mineralization. During these phases, cells synthesize and secrete alkaline phosphatase, type I collagen, and other noncollagenous ECM proteins, such as osteocalcin, osteopontin, osteonectin, and bone sialoprotein. Mineralization occurs through accumulation of calcium and phosphorous in the ECM.

13.4.1. BMSCs Versus MSCs

The osteoprogenitor cells used for bone tissue engineering purposes are derived from various tissue sources. Bone marrow stroma consists of a heterogeneous cell population that provides structural and physiological support for hematopoietic cells [126]. Bone marrow contains three main cell types: endothelial cells, hematopoietic cells, and stromal cells. Friedenstein [127, 128] was the first to identify in bone marrow cell populations with strong osteogenic potential. When marrow cells are plated at low cell densities, BMSCs form colonies known as "colony-forming unit fibroblasts"; this term indicates that each colony derives from a single proliferating progenitor [129]. The term “BMSCs” is applied to isolated bone marrow cells with potential to form connective tissues [129].

Due to their high proliferation potential, BMSCs can be expanded in culture to obtain large numbers of cells starting from a small sample of bone marrow aspirate. The BMSC population contains precursor cells capable of extensive proliferation and differentiation into several phenotypes. Furthermore, BMSCs maintain their multipotential capacity during prolonged culture and multiple passages in vitro. Among these BMSCs there is a subpopulation of undifferentiated multipotent cells able to generate “mesenchyme,” the mass of tissue that develops from the mesoderm of an embryo. This cell population is present in all postnatal tissues and is referred to as “MSCs” [130, 131]. In the past, researchers working with cells from the bone marrow used different names to refer to the same cells. This practice lead to nomenclature confusion; for example, BMSCs have been referred to as multipotent adult progenitor cells, MSCs, bone marrow stromal stem cells (BMSSCs), and mesodermal progenitor cells [132]. What is presently known is that, if appropriately induced, these cells can also differentiate along pathways different from those associated with the cells’ tissues of origin [133].

Stem cells are able to provide replacements for various differentiated cell types. The use of MSCs has several advantages, as they have unique biological properties, are capable of extensive replication in culture in an undifferentiated state, and can differentiate along multiple pathways to form various cells from a number of tissues, including bone, cartilage, and fat [4]. Identification of stem cells using surface markers has not been definitive either, because similar markers are also present on nonstem cells, or because a particular marker may only be temporarily expressed on a stem cell at a certain stage or under specific conditions.

13.4.2. Osteogenic Differentiation

In addition to being osteoconductive and osteoinductive, an ideal scaffold for bone tissue engineering applications should also be osteogenic that is, containing living cells capable of differentiation into osteoblasts. Differentiation of MSCs along the osteoblastic lineage in vitro starts with a period of cell proliferation followed by synthesis and deposition of ECM components by the cells; accumulation of calcium finally leads to mineralization of the ECM. To induce osteogenic differentiation in MSCs, the culture medium is usually supplemented with osteogenic agents such as dexamethasone, β-glycerophosphate, and ascorbic acid.

Dexamethasone, a synthetic glucocorticoid, stimulates MSC proliferation and supports osteogenic lineage differentiation [134–136]. Organic phosphates, such as β-glycerophosphate, also support osteogenesis by contributing to mineralization of the ECM and modulating osteoblast function [136–138]. Free phosphates can also induce expression of osteogenic protein markers, such as osteopontin [136, 139]. Other supplements, such as ascorbic acid, enhance collagen synthesis and upregulate alkaline phosphatase expression in bone cells. Ascorbic acid stimulates marrow stromal cells to differentiate along the osteoblast lineage [139–141]. Furthermore, ascorbic acid promotes osteogenic induction evidenced by increased alkaline phosphatase activity and production of osteocalcin in osteogenic cultures [142].

Martins et al. [76] used marrow stromal cells cultured on starch-poly(ε-caprolactone) blend scaffolds in static cultures and reported that the enzyme lipase enhanced osteogenic differentiation and promoted deposition of a mineralized ECM. The BMP family of growth factors is frequently used for osteoinduction. BMP-2 increases calcium-containing nodule formation and the calcium content of osteogenic cultures in vitro [136]. The TGF-β superfamily contains a large number of growth factors with different functions, many of which regulate cell proliferation and ECM production. Fibroblast growth factors (FGFs), namely

FGF-1 and FGF-2, are produced by oseoblasts and are constituents of the bone matrix. Insulin-like growth factors (IGF) stimulate osteogenesis; IGF-2 is the most abundant growth factor found in bone matrix. Gomes et al. [143] demonstrated that an in vitro generated bone-like ECM produced by marrow stromal cells contains bioactive growth factors including TGF-β1, FGF-2, vascular endothelial growth factor, and BMP-2. Pham et al. [144] reported that the gene expression profiles of various bone-related growth factors and ECM proteins in MSCs cultured in osteogenic media were upregulated; these chemical compounds are present in native bone tissue. Costa-Pinto et al. [48] studied the osteogenic differentiation of a mouse MSC line (BMC9) cultured on novel melt-based chitosan/polyester scaffolds and reported high levels of alkaline phosphatase activity and formation of a calcified ECM; these results are evidence of differentiation of the cells along the osteogenic pathway.

Expression of osteoblast phenotype markers in culture defines three different phases of bone-related activities: cell proliferation, ECM maturation, and ECM mineralization. During active cell proliferation, growth-related genes are expressed, and minimal levels of type I collagen are observed [145]. Following this phase, a period of matrix maturation occurs when alkaline phosphatase is maximally expressed. Finally, the ECM becomes mineralized, the third period of the bone developmental sequence [145]. There are two transition periods between the aforementioned developmental periods: the first occurs at the end of the proliferative period and the second when expression of osteoblastic phenotype markers (such as osteocalcin and osteopontin) becomes significantly elevated with the onset of mineralization [145].

Alkaline phosphatase activity, an early marker of the osteoblastic phenotype, is upregulated at the onset of cell differentiation but subsequently decreases as cell differentiation progresses. Another marker of bone formation is calcium-containing mineral deposits in the ECM. To detect mineral deposition, tetracycline-HCl, a fluorochrome-labeling agent for bone tissues [146], is added to the osteogenic culture media [147]. Tetracycline accumulates at sites of bone formation and fluoresces brightly when activated with appropriate fluorescent light. Qualitative (or semiquantitative) analysis of calcium-containing mineral deposits in bone cell cultures uses the von Kossa, alizarin red, and methylene blue/basic fuchsin staining methods [147, 148]. An important artifact, which should be kept in mind when using these analyses, is that the ECM uptakes calcium independently from cell-mediated mineral deposition. For this reason, confirmation of the results obtained using the aforementioned staining methods should be complemented with data from either diffraction or spectroscopy methods such as thin-film X-ray diffraction and Fourier-transformed infrared spectroscopy [76, 148, 149].

Expression of osteopontin occurs during the mid- to late-stages of osteogenic differentiation of MSCs [150]. Osteopontin is an extracellular protein secreted by differentiating osteoblasts that is upregulated both during cell proliferation and at the onset of ECM mineralization. Osteocalcin, another late-stage marker of osteoblastic differentiation, can be assessed using commercially available immunoassays. Immunohistochemistry using specific antibodies to detect the presence of growth factors, bone- and ECM-related proteins, and enzymes is well established and widely used. Real-time reverse transcriptase polymerase chain reaction is used to determine expression of bone-related genes, such as osteoblast marker genes, growth factors, and ECM biomolecules, in MSCs [144].

13.4.3. Bone-Specific Matrix Proteins

The bone matrix is not only composed of a mineralized phase, but also of an organic phase containing collagenous and noncollagenous proteins, matrix metalloproteinases, proteoglycans, and glycoproteins. Bone formation involves regulated secretion, deposition, and removal of a complex array of these matrix proteins, which appear in a defined temporal and spatial sequence [12]. Mineralization also dictates the spatial orientation of matrix deposition [12]. Most proteins originally thought to be unique to the bone ECM were subsequently proven to be expressed in many other tissues of the body. Osteocalcin is the only protein still considered to be bone-specific in bone mineralization [12].

As discussed previously, alkaline phosphatase is considered an early-stage marker of osteoblastic differentiation [145] and is expressed during the postcell proliferative period of ECM deposition. Type I collagen, the major ECM protein of bone, provides a template for subsequent mineralization [151]. Alkaline phosphatase, collagen, and osteonectin are expressed at high levels near the end of cell proliferation and during the period of ECM deposition and maturation [139].

Osteopontin and bone sialoprotein, *N*-linked glycoproteins containing integrin-binding RGD motifs, are involved in cell-matrix interactions. Osteopontin is widely distributed in different tissues, whereas bone sialoprotein is highly enriched in bone and skeletal cartilage [152]. Osteopontin, a phosphorylated glycoprotein associated with the early stages of osteogenesis that precede mineralization, is secreted by osteoblasts into the mineralizing ECM during bone development [139, 153]. In bone, bone sialoprotein is expressed by fully mature osteogenic cells capable of depositing mineralized matrix [152]. Extracellular bone sialoprotein localizes to newly formed, mineralized bone matrix; its distribution coincides with that of mineral deposits [154]. Bone sialoprotein, a protein expressed during the early phases of bone deposition, controls both mineral formation and cell-matrix interactions [155]. This protein is used as a marker of initial bone formation [155]. The function of bone sialoprotein in bone, which has not been completely elucidated yet, may be related to the regulation of physiological mineralization of skeletal ECMs [154, 156]. Osteocalcin is another marker of late-term osteogenic differentiation associated with osteoblast-mediated matrix deposition and mineralization [157, 158]. Expression of osteopontin, osteocalcin, and bone sialoprotein occurs later during the third period of ECM mineralization.

13.5. Summary

Surface modification of biomaterials uses methods that mimic biomineralization and enable incorporation of bioactive molecules and agents; such treatments can improve both in vitro and in vivo osteogenic differentiation. The main objective of CaP coatings is osteoconduction and enhanced adhesion of osteogenic cells onto biomaterial surfaces. Because CaP coatings have structures and chemical properties similar to those of native bone, they have great potential and promise to increase bone ingrowth in areas of clinical need.

Because they lack essential properties, such as bioactivity and osteoinductivity, most currently available polymers present limitations for bone-related biomedical applications. In this respect then, the biomimetic coating technique discussed in the present chapter has the potential to impart these essential properties to biomaterials. Since CaP layers can be applied on 3D scaffolds, the biomimetic-coating approach has been receiving increased attention in the bone tissue engineering field.

Moreover, CaP coatings have been considered as a potential carrier for the delivery of various biomolecules, chosen for their physicochemical and biological properties as well as for their osteoconductivity. Complementing the CaP biomimetic coating approach, incorporation of biomolecules provides osteoinductive properties to biomaterials. Since this method is carried out under physiological conditions, proteins, enzymes, and other bioactive agents can be incorporated into CaP layers without loss of their bioactivity. A major advantage is the fact that the biomaterial CaP coating biomolecule can simultaneously exhibit osteoinductive and osteoconductive properties, because it can act as a carrier system for the controlled release of multiple biologically active proteins. Incorporation of enzymes into CaP layers coated on the surface of scaffolds (using the biomimetic-coating technique) can be also used to control the degradation rate of the material substrate in vivo. An integrated approach combining a material scaffold, CaP coatings, bioactive molecules and/or enzymes, and in vitro cell cultures may provide an optimal environment for cell adhesion and osteogenic differentiation as well as to generate a mineralized ECM containing select bioactive molecules.

Incorporation of bioactive molecules into CaP coatings on scaffolds for tissue engineering applications has the potential to provide advanced, tissue-specific constructs to promote improved alternative treatment of bone pathologies and trauma. The present chapter summarized the results of studies that used biomolecules important to bone tissue engineering. Further research is needed to elucidate important aspects such as details of the release profiles of entrapped bioactive molecules, retention of their bioactivity, etc. Establishment and further development of nature-inspired techniques to design and formulate novel biomaterials could provide the next generation of effective scaffolds for bone tissue engineering.

Acknowledgments

The authors would like to acknowledge European NoE EXPERTISSUES (NMP3-CT-2004-500283) (R.L.R), Project HIPPOCRATES (NMP3-CT-2003-505758) (R.L.R), and grants from the US National Institutes of Health to A.G.M. (R01 AR42639, R01 DE15164 and R01 DE17441).

References

1. Marx J. Coming to grips with bone loss. Science. 2004;305(5689):1420–1422.
2. Persidis A. Tissue engineering. Nature Biotechnology. 1999;17(5):508–510.
3. Langer R, Vacanti JP. Tissue engineering. Science. 1993;260(5110):920–926.
4. Bianco P, Robey PG. Stem cells in tissue engineering. Nature. 2001;414(6859):118–121.
5. Hutmacher DW. Scaffolds in tissue engineering bone and cartilage. Biomaterials. 2000;21(24):2529–2543.
6. Salgado AJ, Coutinho OP, Reis RL. Bone tissue engineering: state of the art and future trends. Macromolecular Bioscience. 2004;4(8):743–765.
7. Atala A. Engineering tissues, organs and cells. Journal of Tissue Engineering and Regenerative Medicine. 2007;1(2):83–96.
8. Hutmacher DW, Schantz JT, Lam CX, Tan KC, Lim TC. State of the art and future directions of scaffold-based bone engineering from a biomaterials perspective. Journal of Tissue Engineering and Regenerative Medicine. 2007;1(4):245–260.
9. Langer R, Peppas NA. Advances in biomaterials, drug delivery, and bionanotechnology. AIChE Journal. 2003;49(12):2990–3006.
10. Langer R. Selected advances in drug delivery and tissue engineering. Journal of Controlled Release. 1999;62 (1–2):7–11.
11. Madihally SV, Matthew HW. Porous chitosan scaffolds for tissue engineering. Biomaterials. 1999;20(12):1133–1142.
12. Riminucci M, Bianco P. Building bone tissue: matrices and scaffolds in physiology and biotechnology. Brazilian Journal of Medical and Biological Research = Revista brasileira de pesquisas medicas e biologicas/Sociedade Brasileira de Biofisica et al. 2003;36(8):1027–1036.
13. Daculsi G, Laboux O, Le Geros R. Outcome and perspectives in bioactive coatings: what's new, what's coming. ITBM-RBM. 2002;23:317–325.

14. Suh JK, Matthew HW. Application of chitosan-based polysaccharide biomaterials in cartilage tissue engineering: a review. Biomaterials. 2000;21(24):2589–2598.
15. Lahiji A, Sohrabi A, Hungerford DS, Frondoza CG. Chitosan supports the expression of extracellular matrix proteins in human osteoblasts and chondrocytes. Journal of Biomedical Materials Research. 2000;51(4):586–595.
16. Muzzarelli RAA. Biochemical significance of exogenous chitins and chitosans in animals and patients. Carbohydrate Polymers. 1993;20(1):7–16.
17. Tomihata K, Ikada Y. In vitro and in vivo degradation of films of chitin and its deacetylated derivatives. Biomaterials. 1997;18(7):567–575.
18. Kumar MNVR. A review of chitin and chitosan applications. Reactive & Functional Polymers. 2000;46(1):1–27.
19. Benesch J, Tengvall P. Blood protein adsorption onto chitosan. Biomaterials. 2002;23(12):2561–2568.
20. Krajewska B. Application of chitin- and chitosan-based materials for enzyme immobilizations: a review. Enzyme and Microbial Technology. 2004;35(2–3):126–139.
21. Kim IY, Seo SJ, Moon HS, Yoo MK, Park IY, Kim BC, et al. Chitosan and its derivatives for tissue engineering applications. Biotechnology Advances. 2008;26(1):1–21.
22. Klokkevold PR, Vandemark L, Kenney EB, Bernard GW. Osteogenesis enhanced by chitosan (poly-N-acetyl glucosaminoglycan) in vitro. Journal of Periodontology. 1996;67(11):1170–1175.
23. Pound JC, Green DW, Chaudhuri JB, Mann S, Roach HI, Oreffo RO. Strategies to promote chondrogenesis and osteogenesis from human bone marrow cells and articular chondrocytes encapsulated in polysaccharide templates. Tissue Engineering. 2006;12(10):2789–2799.
24. Pound JC, Green DW, Roach HI, Mann S, Oreffo ROC. An ex vivo model for chondrogenesis and osteogenesis. Biomaterials. 2007;28(18):2839–2849.
25. Ueno H, Yamada H, Tanaka I, Kaba N, Matsuura M, Okumura M, et al. Accelerating effects of chitosan for healing at early phase of experimental open wound in dogs. Biomaterials. 1999;20(15):1407–1414.
26. Ueno H, Mori T, Fujinaga T. Topical formulations and wound healing applications of chitosan. Advanced Drug Delivery Reviews. 2001;52(2):105–115.
27. Hoekstra A, Struszczyk H, Kivekas O. Percutaneous microcrystalline chitosan application for sealing arterial puncture sites. Biomaterials. 1998;19(16):1467–1471.
28. Khor E. Chitin: a biomaterial in waiting. Current Opinion in Solid State & Materials Science. 2002;6(4):313–317.
29. Lee JY, Nam SH, Im SY, Park YJ, Lee YM, Seol YJ, et al. Enhanced bone formation by controlled growth factor delivery from chitosan-based biomaterials. Journal of Controlled Release. 2002;78(1–3):187–197.
30. Chatelet C, Damour O, Domard A. Influence of the degree of acetylation on some biological properties of chitosan films. Biomaterials. 2001;22(3):261–268.
31. Pangburn SH, Trescony PV, Heller J. Lysozyme degradation of partially deacetylated chitin, its films and hydrogels. Biomaterials. 1982;3(2):105–108.
32. Sashiwa H, Saimoto H, Shigemasa Y, Ogawa R, Tokura S. Lysozyme susceptibility of partially deacetylated chitin. International Journal of Biological Macromolecules. 1990;12(5):295–296.
33. Varum KM, Myhr MM, Hjerde RJ, Smidsrod O. In vitro degradation rates of partially N-acetylated chitosans in human serum. Carbohydrate Research. 1997;299(1–2):99–101.
34. Martins AM, Pham QP, Malafaya PB, Raphael RM, Kasper FK, Reis RL, et al. "Smart" and stimulus responsive chitosan-based scaffolds/cells for bone tissue engineering: influence of lysozyme upon scaffold degradation and osteogenic differentiation of cultured marrow stromal cells induced by CaP coatings. Tissue Engineering Part A. 2008;14(5):795.
35. Martins AM, Santos MI, Azevedo HS, Malafaya PB, Reis RL. Natural origin scaffolds with in situ pore forming capability for bone tissue engineering applications. Acta Biomaterialia. 2008;4(6):1637–1645.
36. Hirano S, Tsuchida H, Nagao N. N-acetylation in chitosan and the rate of its enzymic hydrolysis. Biomaterials. 1989;10(8):574–576.
37. Ma J, Wang H, He B, Chen J. A preliminary in vitro study on the fabrication and tissue engineering applications of a novel chitosan bilayer material as a scaffold of human neofetal dermal fibroblasts. Biomaterials. 2001;22(4):331–336.
38. Brouwer J, van Leeuwen-Herberts T, Otting-van de Ruit M. Determination of lysozyme in serum, urine, cerebrospinal fluid and feces by enzyme immunoassay. Clinica Chimica Acta; International Journal of Clinical Chemistry. 1984;142(1):21–30.
39. Hankiewicz J, Swierczek E. Lysozyme in human body fluids. Clinica Chimica Acta; International Journal of Clinical Chemistry. 1974;57(3):205–209.

40. Teijón C, Olmo R, Dolores Blanco M, Romero A, María Teijón J. Effects of lead administration at low doses by different routes on rat spleens. Study of response of splenic lymphocytes and tissue lysozyme. Toxicology. 2003;191(2–3):245–258.
41. Torsteinsdottir I, Hakansson L, Hallgren R, Gudbjornsson B, Arvidson NG, Venge P. Serum lysozyme: a potential marker of monocyte/macrophage activity in rheumatoid arthritis. Rheumatology (Oxford, England). 1999;38(12):1249–1254.
42. Garcia Cruz DM, Escobar Ivirico JL, Gomes MM, Gomez Ribelles JL, Sanchez MS, Reis RL, et al. Chitosan microparticles as injectable scaffolds for tissue engineering. Journal of Tissue Engineering and Regenerative Medicine. 2008;2(6):378–380.
43. Abarrategi A, Civantos A, Ramos V, Sanz Casado JV, Lopez-Lacomba JL. Chitosan film as rhBMP2 carrier: delivery properties for bone tissue application. Biomacromolecules. 2008;9(2):711–718.
44. Park YJ, Lee YM, Park SN, Sheen SY, Chung CP, Lee SJ. Platelet derived growth factor releasing chitosan sponge for periodontal bone regeneration. Biomaterials. 2000;21(2):153–159.
45. Di Martino A, Sittinger M, Risbud MV. Chitosan: a versatile biopolymer for orthopaedic tissue-engineering. Biomaterials. 2005;26(30):5983–5990.
46. Tuzlakoglu K, Alves CM, Mano JF, Reis RL. Production and characterization of chitosan fibers and 3-D fiber mesh scaffolds for tissue engineering applications. Macromolecular Bioscience. 2004;4(8):811–819.
47. Coutinho DF, Pashkuleva IH, Alves CM, Marques AP, Neves NM, Reis RL. The effect of chitosan on the in vitro biological performance of chitosan-poly(butylene succinate) blends. Biomacromolecules. 2008;9(4): 1139–1145.
48. Costa-Pinto AR, Salgado AJ, Correlo VM, Sol P, Bhattacharya M, Charbord P, et al. Adhesion, proliferation, and osteogenic differentiation of a mouse mesenchymal stem cell line (BMC9) seeded on novel melt-based chitosan/polyester 3D porous scaffolds. Tissue Engineering Part A. 2008;14(6):1049–1057.
49. Malafaya PB, Silva GA, Reis RL. Natural-origin polymers as carriers and scaffolds for biomolecules and cell delivery in tissue engineering applications. Advanced Drug Delivery Reviews. 2007;59(4–5):207–233.
50. Trimnell D, Fanta GF. Formulations prepared from polyacrylamide and starch. Journal of Polymer Materials. 1994;11(4):271–277.
51. Bastioli C, Cerutti A, Guanella I, Romano GC, Tosin M. Physical state and biodegradation behavior of starch-polycaprolactone systems. Journal of Environmental Polymer Degradation. 1995;3(2):81–95.
52. Villar MA, Thomas EL, Armstrong RC. Rheological properties of thermoplastic starch and starch poly(ethylene-co-vinyl alcohol) blends. Polymer. 1995;36(9):1869–1876.
53. Kotnis MA, Obrien GS, Willett JL. Processing and mechanical-properties of biodegradable poly(hydroxybutyrate-co-valerate)-starch compositions. Journal of Environmental Polymer Degradation. 1995;3(2):97–105.
54. Mayer JM, Elion GR, Buchanan CM, Sullivan BK, Pratt SD, Kaplan DL. Biodegradable blends of cellulose-acetate and starch – Production and properties. Journal of Macromolecular Science. Pure and Applied Chemistry. 1995;A32(4):775–785.
55. Nakamatsu J, Torres FG, Troncoso OP, Min-Lin Y, Boccaccini AR. Processing and characterization of porous structures from chitosan and starch for tissue engineering scaffolds. Biomacromolecules. 2006;7(12): 3345–3355.
56. Bajpai AK, Shrivastava J. alpha-Amylase induced enhanced enzymatic degradation of binary polymeric blends of crosslinked starch and gelatin. Journal of Macromolecular Science. Pure and Applied Chemistry. 2004; A41(8):949–969.
57. Gomes ME, Ribeiro AS, Malafaya PB, Reis RL, Cunha AM. A new approach based on injection moulding to produce biodegradable starch-based polymeric scaffolds: morphology, mechanical and degradation behaviour. Biomaterials. 2001;22(9):883–889.
58. Malafaya PB, Elvira C, Gallardo A, San Roman J, Reis RL. Porous starch-based drug delivery systems processed by a microwave route. Journal of Biomaterials Science. 2001;12(11):1227–1241.
59. Elvira C, Mano JF, San Roman J, Reis RL. Starch-based biodegradable hydrogels with potential biomedical applications as drug delivery systems. Biomaterials. 2002;23(9):1955–1966.
60. Espigares I, Elvira C, Mano JF, Vazquez B, San RJ, Reis RL. New partially degradable and bioactive acrylic bone cements based on starch blends and ceramic fillers. Biomaterials. 2002;23(8):1883–1895.
61. Leonor IB, Sousa RA, Cunha AM, Reis RL, Zhong ZP, Greenspan D. Novel starch thermoplastic/bioglass composites: mechanical properties, degradation behavior and in-vitro bioactivity. Journal of Materials Science. 2002;13(10):939–945.
62. Baran ET, Mano JF, Reis RL. Starch-chitosan hydrogels prepared by reductive alkylation cross-linking. Journal of Materials Science. 2004;15(7):759–765.

63. Boesel LF, Fernandes MH, Reis RL. The behavior of novel hydrophilic composite bone cements in simulated body fluids. Journal of Biomedical Materials Research. Part B Applied Biomaterials. 2004;70(2):368–377.
64. Salgado AJ, Coutinho OP, Reis RL. Novel starch-based scaffolds for bone tissue engineering: cytotoxicity, cell culture, and protein expression. Tissue Engineering. 2004;10(3–4):465–474.
65. Oliveira AL, Reis RL. Pre-mineralisation of starch/polycrapolactone bone tissue engineering scaffolds by a calcium-silicate-based process. Journal of Materials Science. 2004;15(4):533–540.
66. Pavlov MP, Mano JF, Neves NM, Reis RL. Fibers and 3D mesh scaffolds from biodegradable starch-based blends: production and characterization. Macromolecular Bioscience. 2004;4(8):776–784.
67. Tuzlakoglu K, Bolgen N, Salgado AJ, Gomes ME, Piskin E, Reis RL. Nano- and micro-fiber combined scaffolds: a new architecture for bone tissue engineering. Journal of Materials Science. 2005;16(12):1099–1104.
68. Malafaya PB, Stappers F, Reis RL. Starch-based microspheres produced by emulsion crosslinking with a potential media dependent responsive behavior to be used as drug delivery carriers. Journal of Materials Science. 2006;17(4):371–377.
69. Silva GA, Coutinho OP, Ducheyne P, Shapiro IM, Reis RL. Starch-based microparticles as vehicles for the delivery of active platelet-derived growth factor. Tissue Engineering. 2007;13(6):1259–1268.
70. Gomes ME, Azevedo HS, Moreira AR, Ella V, Kellomaki M, Reis RL. Starch-poly(epsilon-caprolactone) and starch-poly(lactic acid) fibre-mesh scaffolds for bone tissue engineering applications: structure, mechanical properties and degradation behaviour. Journal of Tissue Engineering and Regenerative Medicine. 2008;2(5):243–252.
71. Azevedo HS, Gama FM, Reis RL. In vitro assessment of the enzymatic degradation of several starch based biomaterials. Biomacromolecules. 2003;4(6):1703–1712.
72. Balmayor ER, Tuzlakoglu K, Marques AP, Azevedo HS, Reis RL. A novel enzymatically-mediated drug delivery carrier for bone tissue engineering applications: combining biodegradable starch-based microparticles and differentiation agents. Journal of Materials Science. 2008;19(4):1617–1623.
73. Mendes SC, Reis RL, Bovell YP, Cunha AM, van Blitterswijk CA, de Bruijn JD. Biocompatibility testing of novel starch-based materials with potential application in orthopaedic surgery: a preliminary study. Biomaterials. 2001;22(14):2057–2064.
74. Marques AP, Reis RL, Hunt JA. The biocompatibility of novel starch-based polymers and composites: in vitro studies. Biomaterials. 2002;23(6):1471–1478.
75. Salgado AJ, Coutinho OP, Reis RL, Davies JE. In vivo response to starch-based scaffolds designed for bone tissue engineering applications. Journal of Biomedical Materials Research. 2007;80(4):983–989.
76. Martins AM, Pham QP, Malafaya PB, Sousa RA, Gomes ME, Raphael RM, et al. The role of lipase and α-amylase in both the degradation of starch/poly(ε-caprolactone) fiber meshes and the osteogenic differentiation of cultured marrow stromal cells. Tissue Engineering Part A. 2009;15(2):295–305.
77. Gu F, Amsden B, Neufeld R. Sustained delivery of vascular endothelial growth factor with alginate beads. Journal of Controlled Release. 2004;96(3):463–472.
78. Shapiro L, Cohen S. Novel alginate sponges for cell culture and transplantation. Biomaterials. 1997;18(8): 583–590.
79. Augst AD, Kong HJ, Mooney DJ. Alginate hydrogels as biomaterials. Macromolecular Bioscience. 2006;6(8): 623–633.
80. Orive G, Hernandez RM, Gascon AR, Calafiore R, Chang TM, de Vos P, et al. Cell encapsulation: promise and progress. Nature Medicine. 2003;9(1):104–107.
81. Alsberg E, Anderson KW, Albeiruti A, Rowley JA, Mooney DJ. Engineering growing tissues. Proceedings of the National Academy of Sciences of the United States of America. 2002;99(19):12025–12030.
82. Wong M. Alginates in tissue engineering. Methods in Molecular Biology (Clifton, NJ). 2004;238:77–86.
83. Hou T, Xu J, Li Q, Feng J, Zen L. In vitro evaluation of a fibrin gel antibiotic delivery system containing mesenchymal stem cells and vancomycin alginate beads for treating bone infections and facilitating bone formation. Tissue Engineering Part A. 2008;14(7):1173–1182.
84. Cai X, Lin Y, Ou G, Luo E, Man Y, Yuan Q, et al. Ectopic osteogenesis and chondrogenesis of bone marrow stromal stem cells in alginate system. Cell Biology International. 2007;31(8):776–783.
85. Barralet JE, Wang L, Lawson M, Triffitt JT, Cooper PR, Shelton RM. Comparison of bone marrow cell growth on 2D and 3D alginate hydrogels. Journal of Materials Science. 2005;16(6):515–519.
86. Barrere F, van Blitterswijk CA, de Groot K. Bone regeneration: molecular and cellular interactions with calcium phosphate ceramics. International Journal Of Nanomedicine. 2006;1(3):317–332.
87. den Hollander W, Patka P, Klein CP, Heidendal GA. Macroporous calcium phosphate ceramics for bone substitution: a tracer study on biodegradation with 45Ca tracer. Biomaterials. 1991;12(6):569–573.
88. Lu J, Descamps M, Dejou J, Koubi G, Hardouin P, Lemaitre J, et al. The biodegradation mechanism of calcium phosphate biomaterials in bone. Journal of Biomedical Materials Research. 2002;63(4):408–412.

89. Wenisch S, Stahl JP, Horas U, Heiss C, Kilian O, Trinkaus K, et al. In vivo mechanisms of hydroxyapatite ceramic degradation by osteoclasts: fine structural microscopy. Journal of Biomedical Materials Research. 2003;67(3):713–718.
90. Zerbo IR, Bronckers AL, de Lange G, Burger EH. Localisation of osteogenic and osteoclastic cells in porous beta-tricalcium phosphate particles used for human maxillary sinus floor elevation. Biomaterials. 2005;26(12):1445–1451.
91. Hench LL, Splinter RJ, Allen WC, Greenee TK. Bonding mechanisms at the inetrface of ceramics prosthetic materials. Journal of Biomedical Materials Research. 1972;2:117–141.
92. Abe Y, Kokubo T, Yamamuro T. Apatite coating on ceramics, metals and polymers utilizing a biological process. Journal of Materials Science. Materials in Medicine. 1990;1(4):233–238.
93. Kokubo T. Bioactive glass ceramics: properties and applications. Biomaterials. 1991;12(2):155–163.
94. Ohtsuki C, Kamitakahara M, Miyazaki T. Coating bone-like apatite onto organic substrates using solutions mimicking body fluid. Journal of Tissue Engineering And Regenerative Medicine. 2007;1(1):33–38.
95. Reis RL, Cunha AM, Fernandes MH, Correia RN. Treatments to induce the nucleation and growth of apatite-like layers on polymeric surfaces and foams. Journal of Materials Science. 1997;8(12):897–905.
96. Martins AM, Salgado AJ, Azevedo HS, Leonor IB, Reis RL. Lysozyme incorporation in biomimetic coated chitosan scaffolds: development and behaviour in contact with osteoblastic-like cells. Tissue Engineering. 2006;12(4):1018–1019.
97. Tuzlakoglu K, Reis RL. Formation of bone-like apatite layer on chitosan fiber mesh scaffolds by a biomimetic spraying process. Journal of Materials Science. 2007;18(7):1279–1286.
98. Martins AM, Pham QP, Malafaya PB, Raphael RM, Kasper FK, Reis RL, et al. Natural stimulus responsive scaffolds/cells for bone tissue engineering: influence of lysozyme upon scaffold degradation and osteogenic differentiation of cultured marrow stromal cells induced by CaP coatings. Tissue Engineering Part A, in press, (DOI: 10/1089/ten.tea.2008.0023).
99. Liu Y, Hunziker EB, Layrolle P, de Bruijn JD, de Groot K. Bone morphogenetic protein 2 incorporated into biomimetic coatings retains its biological activity. Tissue Engineering. 2004;10(1–2):101–108.
100. Liu Y, Hunziker EB, Randall NX, de Groot K, Layrolle P. Proteins incorporated into biomimetically prepared calcium phosphate coatings modulate their mechanical strength and dissolution rate. Biomaterials. 2003;24(1):65–70.
101. Azevedo HS, Leonor IB, Alves CM, Reis RL. Incorporation of proteins and enzymes at different stages of the preparation of calcium phosphate coatings on a degradable substrate by a biomimetic methodology. Materials Science & Engineering C. Biomimetic and Supramolecular Systems. 2005;25(2):169–179.
102. Cornell CN. Osteoconductive materials and their role as substitutes for autogenous bone grafts. Orthopedic Clinics of North America. 1999;30(4):591–598.
103. Albrektsson T, Johansson C. Osteoinduction, osteoconduction and osseointegration. European Spine Journal. 2001;10(Suppl 2):S96–S101.
104. Fujibayashi S, Neo M, Kim HM, Kokubo T, Nakamura T. Osteoinduction of porous bioactive titanium metal. Biomaterials. 2004;25(3):443–450.
105. Ripamonti U. Osteoinduction in porous hydroxyapatite implanted in heterotopic sites of different animal models. Biomaterials. 1996;17(1):31–35.
106. Yuan H, Li Y, de Bruijn JD, de Groot K, Zhang X. Tissue responses of calcium phosphate cement: a study in dogs. Biomaterials. 2000;21(12):1283–1290.
107. Habibovic P, de Groot K. Osteoinductive biomaterials - Properties and relevance in bone repair. Journal of Tissue Engineering and Regenerative Medicine. 2007;1(1):25–32.
108. Barrere F, van der Valk CM, Dalmeijer RA, Meijer G, van Blitterswijk CA, de Groot K, et al. Osteogenecity of octacalcium phosphate coatings applied on porous metal implants. Journal of Biomedical Materials Research. 2003;66(4):779–788.
109. Liu Y, de Groot K, Hunziker EB. Osteoinductive implants: the mise-en-scene for drug-bearing biomimetic coatings. Annals of Biomedical Engineering. 2004;32(3):398–406.
110. Liu Y, Li JP, Hunziker EB, de Groot K. Incorporation of growth factors into medical devices via biomimetic coatings. Philosophical Transactions. Series A. Mathematical, Physical, and Engineering Sciences. 2006;364(1838):233–248.
111. El-Ghannam A. Bone reconstruction: from bioceramics to tissue engineering. Expert Review of Medical Devices. 2005;2(1 doi:10.1586/17434440.2.1.87):87–101.
112. Paul W, Sharma CP. Ceramic drug delivery: a perspective. Journal of Biomaterials Applications. 2003;17(4):253–264.
113. LeGeros RZ. Properties of osteoconductive biomaterials: calcium phosphates. Clinical Orthopaedics and Related Research. 2002 (395):81–98.

114. Vroman L, Adams AL. Rapid identification of proteins on flat surfaces, using antibody-coated metal oxide suspensions. Journal of Immunological Methods. 1986;93(2):213–216.
115. Maus U, Andereya S, Gravius S, Ohnsorge JA, Niedhart C, Siebert CH. BMP-2 incorporated in a tricalcium phosphate bone substitute enhances bone remodeling in sheep. Journal of Biomaterials Applications. 2008;22(6):559–576.
116. Liu Y, de Groot K, Hunziker EB. BMP-2 liberated from biomimetic implant coatings induces and sustains direct ossification in an ectopic rat model. Bone. 2005;36(5):745–757.
117. Ripamonti U, Yeates L, van den Heever B. Initiation of heterotopic osteogenesis in primates after chromatographic adsorption of osteogenin, a bone morphogenetic protein, onto porous hydroxyapatite. Biochemical and Biophysical Research Communications. 1993;193(2):509–517.
118. Onuma K, Kanzaki N, Kobayashi N. Association of calcium phosphate and fibroblast growth factor-2: a dynamic light scattering study. Macromolecular Bioscience. 2004;4(1):39–46.
119. Liu Y, Layrolle P, de Bruijn J, van Blitterswijk C, de Groot K. Biomimetic coprecipitation of calcium phosphate and bovine serum albumin on titanium alloy. Journal of Biomedical Materials Research. 2001;57(3):327–335.
120. Wen HB, de Wijn JR, van Blitterswijk CA, de Groot K. Incorporation of bovine serum albumin in calcium phosphate coating on titanium. Journal of Biomedical Materials Research. 1999;46(2):245–252.
121. Radin S, Campbell JT, Ducheyne P, Cuckler JM. Calcium phosphate ceramic coatings as carriers of vancomycin. Biomaterials. 1997;18(11):777–782.
122. de Groot K, Geesink R, Klein CP, Serekian P. Plasma sprayed coatings of hydroxylapatite. Journal of Biomedical Materials Research. 1987;21(12):1375–1381.
123. Liu Y, Enggist L, Kuffer AF, Buser D, Hunziker EB. The influence of BMP-2 and its mode of delivery on the osteoconductivity of implant surfaces during the early phase of osseointegration. Biomaterials. 2007;28(16): 2677–2686.
124. Klein CP, Patka P, Wolke JG, de Blieck-Hogervorst JM, de Groot K. Long-term in vivo study of plasma-sprayed coatings on titanium alloys of tetracalcium phosphate, hydroxyapatite and alpha-tricalcium phosphate. Biomaterials. 1994;15(2):146–150.
125. Liu Y, Huse RO, de Groot K, Buser D, Hunziker EB. Delivery mode and efficacy of BMP-2 in association with implants. Journal of Dental Research. 2007;86(1):84–89.
126. Krebsbach PH, Kuznetsov SA, Bianco P, Robey PG. Bone marrow stromal cells: characterization and clinical application. Critical Reviews in Oral Biology and Medicine. 1999;10(2):165–181.
127. Friedenstein AJ. Precursor cells of mechanocytes. International Review of Cytology. 1976;47:327–359.
128. Friedenstein AJ. Marrow stromal fibroblasts. Calcified Tissue International. 1995;56(Suppl 1):S17.
129. Owen M, Friedenstein AJ. Stromal stem cells: marrow-derived osteogenic precursors. Ciba Foundation Symposium. 1988;136:42–60.
130. Caplan AI. Mesenchymal stem cells. Journal of Orthopaedic Research. 1991;9(5):641–650.
131. Caplan AI. Review: mesenchymal stem cells: cell-based reconstructive therapy in orthopedics. Tissue Engineering. 2005;11(7–8):1198–1211.
132. Derubeis AR, Cancedda R. Bone marrow stromal cells (BMSCs) in bone engineering: limitations and recent advances. Annals of Biomedical Engineering. 2004;32(1):160–165.
133. Bianco P, Riminucci M, Gronthos S, Robey PG. Bone marrow stromal stem cells: nature, biology, and potential applications. Stem Cells. 2001;19(3):180–192.
134. Bellows CG, Heersche JN, Aubin JE. Determination of the capacity for proliferation and differentiation of osteoprogenitor cells in the presence and absence of dexamethasone. Developmental Biology. 1990;140(1): 132–138.
135. Liu F, Aubin JE, Malaval L. Expression of leukemia inhibitory factor (LIF)/interleukin-6 family cytokines and receptors during in vitro osteogenesis: differential regulation by dexamethasone and LIF. Bone. 2002;31(1):212–219.
136. Tuan RS, Boland G, Tuli R. Adult mesenchymal stem cells and cell-based tissue engineering. Arthritis Research & Therapy. 2003;5(1):32–45.
137. Chung CH, Golub EE, Forbes E, Tokuoka T, Shapiro IM. Mechanism of action of beta-glycerophosphate on bone cell mineralization. Calcified Tissue International. 1992;51(4):305–311.
138. Tenenbaum HC, Limeback H, McCulloch CA, Mamujee H, Sukhu B, Torontali M. Osteogenic phase-specific co-regulation of collagen synthesis and mineralization by beta-glycerophosphate in chick periosteal cultures. Bone. 1992;13(2):129–138.
139. Beck GR, Jr., Zerler B, Moran E. Phosphate is a specific signal for induction of osteopontin gene expression. Proceedings of the National Academy of Sciences of the United States of America. 2000;97(15):8352–8357.

140. Quarles LD, Yohay DA, Lever LW, Caton R, Wenstrup RJ. Distinct proliferative and differentiated stages of murine Mc3t3-E1 cells in culture - An invitro model of osteoblast development. Journal of Bone and Mineral Research. 1992;7(6):683–692.
141. Franceschi RT, Iyer BS, Cui Y. Effects of ascorbic acid on collagen matrix formation and osteoblast differentiation in murine MC3T3-E1 cells. Journal of Bone and Mineral Research. 1994;9(6):843–854.
142. Liu P, Oyajobi BO, Russell RG, Scutt A. Regulation of osteogenic differentiation of human bone marrow stromal cells: interaction between transforming growth factor-beta and 1,25(OH)(2) vitamin D(3) in vitro. Calcified Tissue International. 1999;65(2):173–180.
143. Gomes ME, Bossano CM, Johnston CM, Reis RL, Mikos AG. In vitro localization of bone growth factors in constructs of biodegradable scaffolds seeded with marrow stromal cells and cultured in a flow perfusion bioreactor. Tissue Engineering. 2006;12(1):177–188.
144. Pham QP, Kurtis Kasper F, Scott Baggett L, Raphael RM, Jansen JA, Mikos AG. The influence of an in vitro generated bone-like extracellular matrix on osteoblastic gene expression of marrow stromal cells. Biomaterials. 2008;29(18):2729–2739.
145. Owen TA, Aronow M, Shalhoub V, Barone LM, Wilming L, Tassinari MS, et al. Progressive development of the rat osteoblast phenotype in vitro: reciprocal relationships in expression of genes associated with osteoblast proliferation and differentiation during formation of the bone extracellular matrix. Journal of Cellular Physiology. 1990;143(3):420–430.
146. Frost HM. Tetracycline-based histological analysis of bone remodeling. Calcified Tissue Research. 1969;3(3):211–237.
147. Gomes ME, Sikavitsas VI, Behravesh E, Reis RL, Mikos AG. Effect of flow perfusion on the osteogenic differentiation of bone marrow stromal cells cultured on starch-based three-dimensional scaffolds. Journal of Biomedical Materials Research. 2003;67(1):87–95.
148. Boskey AL. Biomineralization: conflicts, challenges, and opportunities. Journal of Cellular Biochemistry. Supplementary. 1998;30–31:83–91.
149. Gomes ME, Holtorf HL, Reis RL, Mikos AG. Influence of the porosity of starch-based fiber mesh scaffolds on the proliferation and osteogenic differentiation of bone marrow stromal cells cultured in a flow perfusion bioreactor. Tissue Engineering. 2006;12(4):801–809.
150. Holtorf HL, Jansen JA, Mikos AG. Flow perfusion culture induces the osteoblastic differentiation of marrow stroma cell-scaffold constructs in the absence of dexamethasone. Journal of Biomedical Materials Research. 2005;72(3):326–334.
151. Hoshi K, Kemmotsu S, Takeuchi Y, Amizuka N, Ozawa H. The primary calcification in bones follows removal of decorin and fusion of collagen fibrils. Journal of Bone and Mineral Research. 1999;14(2):273–280.
152. Bianco P, Fisher LW, Young MF, Termine JD, Robey PG. Expression of bone sialoprotein (BSP) in developing human tissues. Calcified Tissue International. 1991;49(6):421–426.
153. Mark MP, Butler WT, Prince CW, Finkelman RD, Ruch JV. Developmental expression of 44-kDa bone phosphoprotein (osteopontin) and bone gamma-carboxyglutamic acid (Gla)-containing protein (osteocalcin) in calcifying tissues of rat. Differentiation. 1988;37(2):123–136.
154. Bianco P, Riminucci M, Silvestrini G, Bonucci E, Termine JD, Fisher LW, et al. Localization of bone sialoprotein (BSP) to Golgi and post-Golgi secretory structures in osteoblasts and to discrete sites in early bone matrix. The Journal of Histochemistry and Cytochemistry. 1993;41(2):193–203.
155. Cancedda R, Castagnola P, Cancedda FD, Dozin B, Quarto R. Developmental control of chondrogenesis and osteogenesis. The International Journal of Developmental Biology. 2000;44(6):707–714.
156. Hunter GK, Goldberg HA. Nucleation of hydroxyapatite by bone sialoprotein. Proceedings of the National Academy of Sciences of the United States of America. 1993;90(18):8562–8565.
157. Lian JB, Stein GS. Concepts of osteoblast growth and differentiation: basis for modulation of bone cell development and tissue formation. Critical Reviews in Oral Biology and Medicine. 1992;3(3):269–305.
158. Lian JB, Stein GS. The developmental stages of osteoblast growth and differentiation exhibit selective responses of genes to growth factors (TGF beta 1) and hormones (vitamin D and glucocorticoids). The Journal of Oral Implantology. 1993;19(2):95–105; discussion 36–37.

14

Biomimetic Nanophase Materials to Promote New Tissue Formation for Tissue-Engineering Applications

Xiaohua Liu, Ian O. Smith, and Peter X. Ma

One important aspect of tissue engineering is the development of new biomaterials to facilitate cell–material interactions, which can be achieved by mimicking certain advantageous features of natural extracellular matrix (ECM). Biomimetic nanoscale materials mimic the natural ECM and can be used as scaffolds to provide a better environment for new tissue formation. This chapter focuses on biomimetic nanoscale materials for tissue engineering. First, several techniques are introduced for the fabrication of nanophase scaffolds, which mimic the physical architecture of natural ECM. Second, the role of surface modification in mimicking the chemical composition of natural ECM is discussed. Two novel surface-modification techniques are presented to illustrate the process of fabricating biomimetic scaffolds that mimic both the physical architecture and chemical composition of natural collagen type I. Finally, the biological effects of the nanoarchitecture of scaffolds, such as protein interaction and cell function, including cell attachment, proliferation, and differentiation, are discussed.

Abbreviations

2D	two-dimensional
3D	three-dimensional
BSP	bone sialoprotein
CaP	calcium phosphates
ECM	extracellular matrix
HA	hydroxyapatite
NF-PLLA	nanofibrous poly(L-lactic acid)
PDAC	poly(diallyldimethylammonium chloride)
PHB	poly(3-hydroxybutyrate)
PLA	poly(lactic acid)
PLGA	poly(lactic-co-glycolic acid)
PLLA	poly(L-lactic acid)
RGD	Arg–Gly–Asp

X. Liu, I.O. Smith, and P.X. Ma • Department of Biologic and Materials Sciences, The University of Michigan, Ann Arbor, MI 48109-1078, USA

D.A. Puleo and R. Bizios (eds.), *Biological Interactions on Materials Surfaces*,
DOI 10.1007/978-0-387-98161-1_14,

SBF	simulated body fluid
THF	tetrahydrofuran
TIPS	thermally induced phase separation
TRITC	tetramethyl rhodamine iso-thiocyanate

14.1. Introduction

Tissue engineering is an interdisciplinary field that applies the principles of engineering and life sciences toward the development of biological substitutes that restore, maintain, or improve organ or tissue function [1]. Since the concept of tissue engineering emerged two decades ago, scientific advances in biomaterials, biology, and medicine have created unique opportunities to fabricate tissues in the laboratory by combining scaffolds (artificial extracellular matrices), cells, and biologically active molecules. In an ideal situation, scaffolds would incorporate the functions of natural extracellular matrix (ECM), specifically, they would provide support for cell attachment and proliferation, deliver and retain biochemical factors, enable diffusion of nutrients for cells, and exert suitable mechanical and other stimuli for cell function. For these reasons, scaffolds should mimic the natural ECM in order to provide the optimal physiological environment for cells [2, 3].

Two approaches are generally employed to fabricate biomimetic scaffolds: mimicking the physical architecture of ECM and mimicking the chemical composition of ECM. For example, collagen (type I) is the most abundant ECM component of many tissues and is composed of a nanoscale fibrillar structure in vivo. This structure has been determined to be important for cell attachment, proliferation, and differentiation [4, 5]. With the development of nanobiotechnology, the ability to mimic collagen and form a similar nanofibrous matrix has emerged. A number of synthetic and natural polymers have been employed to mimic the fibrillar structure of collagen [6]. On the other hand, many natural ECM molecules possess biological recognition cues within their molecular chains, which facilitate the interaction between cells and ECM. While synthetic materials possess many advantages such as batch-to-batch consistency, controlled mechanical properties, and no risk of pathogen transmission, they do not have the same biological cues within their molecular chains. Therefore, considerable efforts are ongoing for synthetic materials to mimic the chemical composition of ECM.

14.2. Fabrication of Biomimetic Scaffolds with Nanoscale Architecture

The design of a biomimetic scaffold leads to a need for nanoscale components. Biomimetic scaffolds should integrate a nanoarchitecture similar to that of the natural ECM into the highly porous scaffold structure aiming to better regulate cell function [7, 8]. The scaffold must provide a surface that promotes cell attachment, proliferation, and differentiation, as well as a highly porous network to promote tissue regeneration in three dimensions. High interconnectivity between pores is desirable for uniform cell seeding, distribution, as well as for oxygen and nutrient diffusion [3, 9]. The scaffold must also have the mechanical properties necessary to provide structural support until new tissue formation is complete. Additionally, the scaffold must be biocompatible, at implantation as well as during its degradation. The scaffold must promote a suitable biological response, and must degrade at a rate matching that of the new tissue formation. Scaffolds for tissue engineering tend to fall into three broad material categories: polymer scaffolds, ceramic scaffolds, and composite scaffolds.

14.2.1. Nanofibrous Polymeric Scaffolds

The nanofibrous structure of the ECM is mimicked in scaffold design, because this physical feature may play an important role in regulating cell function [7, 8]. For example, during bone tissue formation, collagen fibers on the scale of between 50 and 500 nm in diameter are deposited and form the basis of the ECM, which then acts as the substrate where apatite crystals are deposited to form woven bone [10]. Nanofibrous polymer scaffolds are commonly fabricated using electrospinning, thermally induced phase separation (TIPS), or molecular self-assembly.

Molecular self-assembly is a technique where noncovalent interactions between biomolecules are utilized to arrange them into well-defined structures, such as nanofibers [11, 12]. However, these nanofiber networks are not easily arranged into macroporous structures.

The oldest, and currently the most commonly used, technique for forming nanofibers is electrospinning. Electrospinning is a technique that originated in the early 1900s; in this case, a polymer is drawn to a target by using an electrical field [13]. This method results in fibers with diameters from micron to nanometer range, which can be spun into fiber "mats" of varying density. While it is an effective method for creating individual nanofibers on the scale needed to mimic the ECM (a few hundred nanometers in diameter), the resulting fiber network lacks significant thickness and is roughly two-dimensional (2D) in structure. To address this aspect, new techniques that introduce three-dimensional (3D) aspects to electrospun nanofiber networks, including the stacking of individual nanofiber sheets into 3D constructs [14], as well as spinning fibers directly onto a 3D printed scaffold template have been employed [15]. However, these 3D constructs lack a suitable cohesive interconnected macroporous structure and further refinement is necessary.

To better incorporate a 3D interconnected macroporous network into nanofibrous scaffold design, our group has developed a novel technique for incorporating nanoscale fiber dimensions, similar to those of collagen fibers, into a highly porous, interconnected scaffold network, using a TIPS technique [7]. TIPS creates polymer-rich and solvent-rich phases out of a polymer solution and then removes the solvent through sublimation under low pressure. This method yields a continuous network of nanoscale polymer fibers. Macropore size and degree of interconnectivity are further tailored using a series of molding techniques (Figure 14.1) [16–18].

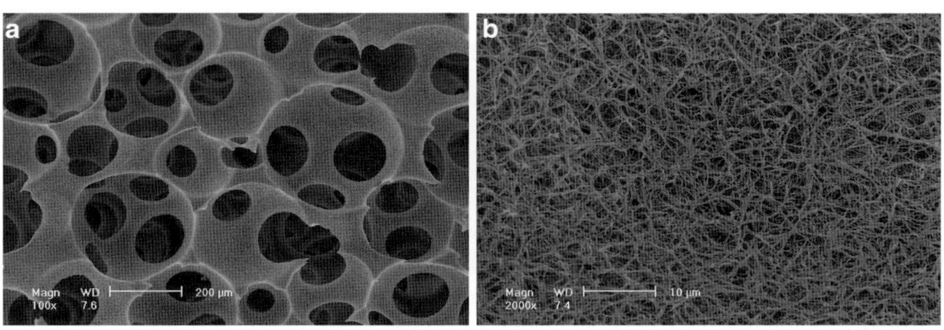

Figure 14.1. Scanning electron micrographs of a 3D nanofibrous PLLA scaffold fabricated using the TIPS process: **(a)** overall pore structure at low magnification; **(b)** the pore-wall structure in **(a)** at a higher magnification. From Liu et al. [19], reprinted with permission of the American Scientific Publishers.

14.2.2. Nanophase Ceramic Scaffolds

Ceramic materials including calcium phosphates (CaP), alumina (Al_2O_3), and Zirconia (ZrO_2) have been investigated for use in hard tissue-engineering applications. CaP materials, especially hydroxyapatite (HA; $Ca_{10}[PO_4]_6[OH]_2$), are obvious candidates due to their chemical similarity to the naturally occurring bone mineral, their high level of biocompatibility, and good interaction with bone-forming cells in vitro and in vivo. However, ceramics as candidates for tissue-engineering scaffolds present a set of inherent limitations that cannot be ignored. Although they are often strong in compression, ceramics are not very tough, are very brittle, and are prone to fracture under complex loading conditions [20]. This brittleness is made worse in the case of the highly porous preparations of ceramic materials, needed and used as tissue-engineering scaffolds.

Compared with the performance of polymer scaffolds in the biological environment, ceramic materials are good candidates for bone tissue engineering because of their favorable biological response in vitro and in vivo. Nanoscale crystalline HA is a biocompatible material with a minimal inflammatory response [21, 22], but not without potential cytotoxicity [21]. Recent research in ceramic materials for tissue engineering has focused on developing a scaffold structure similar to that of natural cancellous bone, in order to promote bone regeneration in vitro and in vivo [23].

Several studies have been conducted using dense ceramic substrates with easily controlled grain sizes, designed specifically for use in cell culture [24–26]. This work has yielded extensive in vitro cell response data as a function of ceramic grain size down to the nanometer scale. These data are useful to study the link between the two, but they do not take into account the need for a porous architecture in tissue-engineering scaffolds.

The data supplied for cells on a bioinert ceramic like Al_2O_3 corroborate the hypothesis that improved cell attachment can be attributed to increased physical surface irregularities that further promote cell fixation [27]. Fixation also likely plays a role in improved cell attachment on nanograined HA, but this cell function may also occur due to other factors, including protein adsorption. This aspect is discussed in Sect. 14.4. of this chapter.

Techniques for fabricating highly porous ceramic scaffolds include foaming and sintering [28], porogen casting [29], and 3D printing [30]. Examples of scaffolds fabricated using these techniques are shown in Figure 14.2. While these techniques provide scaffolds with interconnected networks of pores, potential porosity is limited to ~70%, compared with >90% of polymer and composite scaffolds. In addition, the level of control over the crystalline grain size during sintering is limited, and special techniques, including microwave heating, are required to reduce the scale of the sintered grains [31].

14.2.3. Nanocomposite Scaffolds

One method used to avoid the pitfalls associated with fabricating porous ceramic scaffolds while maintaining some of their inherent advantages, is to use ceramic/polymer composite scaffolds. HA has been combined with several natural and synthetic polymers, most commonly poly(lactic acid) (PLA) [32–37], poly(lactic-co-glycolic acid) (PLGA) [38, 39], gelatin [40], and collagen [32, 41, 42] to produce such scaffolds. These combinations retain many of the benefits of HA, while offering improved control of porosity (as high as 90%, vs. 70% for pure ceramic), degradability, and mechanical properties when compared with pure ceramic scaffolds [28, 34]. Inclusion of nanocrystalline HA in natural polymer

Figure 14.2. HA scaffolds, fabricated using **(a)** foaming and sintering [29] (reprinted with permission by the Dove Press), **(b)** porogen casting [30] (reprinted with permission by Trans Tech Publications), and **(c)** 3D printing [31] (reprinted with permission by Wiley).

scaffolds led to improved mechanical properties versus those of pure polymers. In addition, the presence of HA in vivo may potentially reduce some negative effects arising from the degradation of some polymers. The presence of HA may reduce cell death by enhancing protein adsorption [43] and under certain circumstances by alleviating the acidity of the surrounding milieu [44, 45].

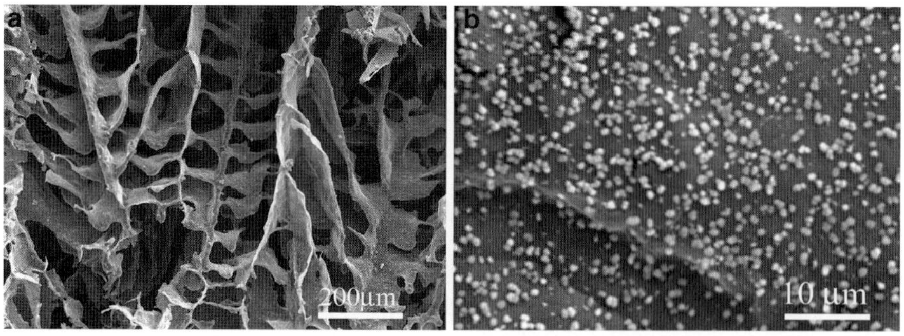

Figure 14.3. PLLA/nano HA composite scaffold fabricated using the TIPS process, followed by incubation in a SBF. **(a)** Low magnification, **(b)** high magnification. [47] (reprinted with permission by John Wiley and Sons).

In one successful example, nanocrystalline HA was combined with poly(L-lactic acid) (PLLA) and TIPS was then used to fabricate composite scaffolds with porosities between 89% and 95% (Figure 14.3) [34]. Inclusion of nanoscale crystals resulted in a more regular pore network of the scaffolds compared with that of scaffolds containing micron-scale crystals, since nanoscale crystals do not significantly alter the pore formation process of PLLA solution during phase separation.

Currently, the most commonly used method of incorporating a ceramic component into a fibrous polymer scaffold system is by direct growth onto an already existing scaffold [37, 39]. By soaking the porous scaffold, or nanofibrous network, in simulated body fluid (SBF), apatite crystals are formed on the scaffold surface. This method is different from the incorporation of crystals during scaffold processing, which, while somewhat effective, often leads to a breakdown of the scaffold architecture with increased volume fraction of HA. Since the scaffolds constructed by this method contain much of the HA away from the pore surfaces, HA is not available for subsequent biological interactions. When the apatite content is concentrated on the pore surface, the biologic interactions may be enhanced.

The SBF technique used to grow apatite crystals on scaffold surfaces is straightforward and requires no special equipment. Such polymer scaffolds are fabricated in similar fashion to those used for pure polymer scaffolds. TIPS has been developed to fabricate 3D nanofibrous scaffolds [37, 39]. By varying the solvent and polymer concentration during processing, porosity and pore morphology are optimized to allow apatite deposition throughout the 3D scaffold.

The structure and material properties, such as inclusion of functional groups (such as COOH and OH) [37], variations in polymer crystallinity, hydrophobicity/hydrophilicity of the surface [39], and architectural features (such as pore size, shape, and interconnectivity), affect apatite formation and distribution.

14.3. Surface Modification of Nanofibrous Scaffolds

In tissue engineering, the interactions between cells and scaffolds take place on the scaffold surface. Therefore, the nature of the scaffold surface can directly affect cellular response, ultimately influencing the rate and quality of new tissue formation. As mentioned earlier (Sect. 14.1.), natural ECM components possess biological recognition moieties within their molecule chains. For example, the Arg–Gly–Asp (RGD) sequence, which is the best-known peptide sequence for promoting cell adhesion, is a cell recognition motif found in many ECM proteins. In contrast, synthetic biodegradable materials do not contain biological recognition cites in their molecular chains. Therefore, one important strategy in preparing biomimetic scaffolds is to modify the scaffold surface and incorporate bioactive molecules to promote cell–scaffold interaction.

14.3.1. Surface-Modification Methods for Scaffolds

Several methods have been developed to modify the scaffold surface. Surface hydrolysis of poly(glycolic acid) scaffolds under strong alkaline conditions has been used to increase cell-seeding density and improve cell–biomaterial interactions [47]. This method is limited because it is technique sensitive and is difficult to preserve the surface topography and mechanical properties of the original scaffold.

Plasma exposure (i.e., a gas discharge process) is an effective procedure for specific surface etching. Low-pressure ammonia plasma treatment has been utilized for the modification of thin films of poly(3-hydroxybutyrate) (PHB) [48]. Amine functions were introduced in order to permit subsequent protein immobilization reactions needed for adjustment of the biointerfacial properties of PHB. Such plasma treatment of PHB induced a durable conversion of a hydrophobic into a hydrophilic material but did not cause significant changes in the topography of the analyzed thin films. Besides PHB, other materials have also been treated with this technique in order to enhance cell–material interactions. For example, PLA films and porous scaffolds have been modified with NH_3 plasma treatment, followed by attachment of poly(L-lysine) and RGD peptides [49]. Because of the limitation of plasma penetration, this method can only be used for 2D films and for very thin 3D scaffolds.

Another strategy for surface modification is to directly incorporate functional groups or peptides into the polymer chains by copolymerization or graft polymerization. For example, poly(L-lactic acid-co-L-lysine) with the peptide sequence RGD chemically attached to the lysine residue was synthesized [50, 51]. This approach combines advantages of both natural and synthetic materials. Since these poly(alpha-hydroxy acid)-associated copolymers can have a variety of functional groups, they can be further modified by chemical attachment of various biologically active molecules. However, bulk chemical modification usually results in changes of the degradation behavior and of the mechanical properties of polymers; such changes are often undesirable.

Surface modification is advantageous for biomaterials, but, to date, it is focused on 2D constructs and very thin 3D constructs. True 3D scaffolding surface modification, especially the surface modification of nanofibrous scaffolds, is still a challenge. In our laboratory, several effective techniques, including self-assembly and molecular entrapment, have recently been developed for 3D surface modification of nanofibrous scaffolds [19, 52].

14.3.2. Surface Engineering of Nanofibrous Scaffolds Using Self-Assembly Techniques

Electrostatic layer-by-layer self-assembly, which is based on alternating adsorption of cationic and anionic species, is a novel and promising technique to prepare well-defined surfaces [53, 54]. This new layer-by-layer deposition process provides the means to create polycation–polyanion complexes one molecular layer at a time, thereby allowing an unprecedented level of control over the composition and surface functionality of materials.

Nanofibrous PLLA (NF-PLLA) scaffolds have been fabricated to mimic the physical architecture of natural collagen matrix at the nanoscale level [7]. To further mimic the chemical composition of collagen matrix, gelatin (derived from collagen by hydrolysis), was incorporated onto the surface of NF-PLLA scaffolds by the electrostatic layer-by-layer self-assembly technique. The pretreated NF-PLLA scaffolds were first activated in an aqueous poly(diallyldimethylammonium chloride) (PDAC) solution (to obtain stable positively charged surfaces). After washing the scaffolds with water, the scaffolds were dipped into gelatin solution and then washed again with water. Further growth of the PDAC/gelatin bilayers was accomplished by repeating these procedures.

In order to visualize the attachment of gelatin on the scaffold surface, tetramethyl rhodamine iso-thiocyanate (TRITC)-labeled gelatin was added in the self-assembly solution. Development of color on the scaffold after the self-assembly process indicated that the gelatin was successfully incorporated on the scaffold surface (Figure 14.4). Confocal images provided further evidence that gelatin was distributed evenly throughout the entire (both outer and inner) scaffold surfaces (Figure 14.5).

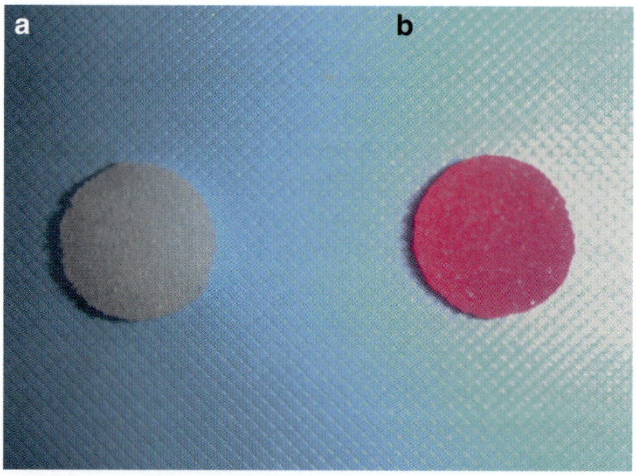

Figure 14.4. Photographs of NF-PLLA scaffolds. **(a)** Before surface modification; **(b)** after surface modified with TRITC-labeled gelatin.

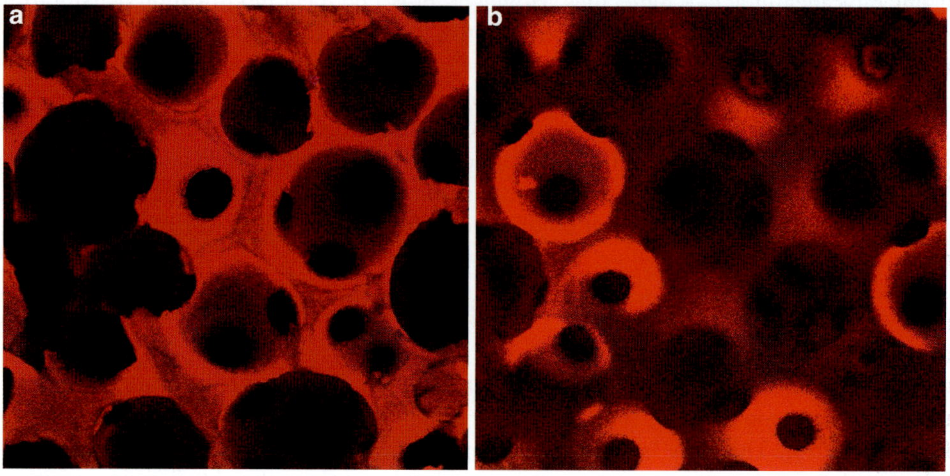

Figure 14.5. Confocal images of NF-PLLA scaffolds after surface modification with TRITC-labeled gelatin. **(a)** Exterior surface; **(b)** center of the scaffold.

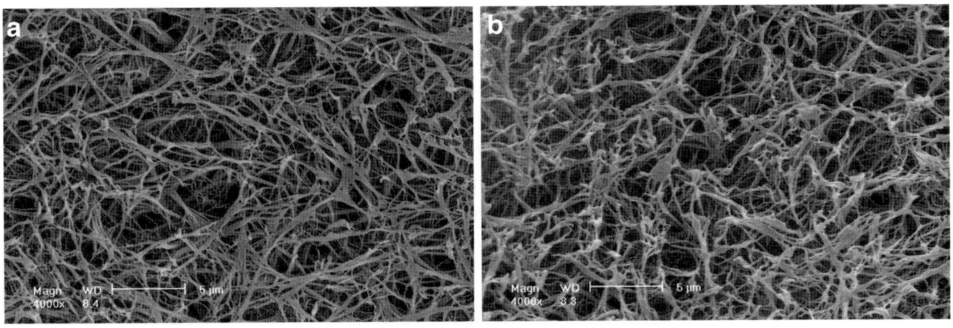

Figure 14.6. Scanning electron micrographs of NF-PLLA matrices (**a**) before and (**b**) after surface modification with gelatin using the electrostatic layer-by-layer self-assembly technique.

Quantitative analysis by X-ray photoelectron spectroscopy showed that the amount of gelatin adsorbed on the NF-PLLA surface was controlled by the number of self-assembled bilayers. After the first two bilayers, the adsorbed amount of elemental nitrogen on the NF-PLLA surface increased roughly linearly with the bilayer numbers [18]. A nitrogen elemental content as high as 9.4% was reached as the surface of NF-PLLA was deposited with eight bilayers of PDAC/gelatin polyelectrolytes. This self-assembly method does not affect the material topography and its mechanical properties; such effects are often inevitable when using other surface-modification methods such as plasma exposure or alkaline/acidic etching. As shown in Figure 14.6, the surface-modified NF-PLLA scaffolds maintained similar nanofibrous pore-wall structure as that of unmodified controls. After gelatin surface modification, the diameters of the nanofibers still ranged from 50 to 500 nm, which are dimensions similar to natural collagen fibers. The surface-modified NF-PLLA scaffold combines the advantages of both natural ECM and synthetic polymer materials. Since gelatin is denatured collagen, its use also circumvents concerns of pathogen transmission and immunorejection associated with collagen from animal and cadaver sources.

This self-assembly technique is effective for biomaterial surface modification. Advantages include the high degree of molecular control over surface chemistry, coating thickness, and maintenance of the 3D scaffold architecture. Such a technique also allows incorporation of proteins and other biomolecules into the layers without loss of bioactivity. Furthermore, this process uses an aqueous solution, is easy to carry out, and is universally applicable to any 3D geometry as long as the scaffold pores are interconnected.

14.3.3. Porogen-Induced Surface Modification for Nanofibrous Scaffolds

In this strategy, a surface entrapment method was integrated into a scaffold fabrication process (the TIPS technique) to prepare surface-modified NF-PLLA scaffolds in a single step.

Gelatin spheres (which acted both as a porogen and a surface-modification agent) were fabricated using a surfactant-free emulsification method. Gelatin particles were first dissolved in distilled water. The gelatin solution was then added to mineral oil. The mixture was vigorously stirred and was quickly transferred into an ice–salt bath. Cold acetone was poured into the stirred suspension. After stopping the stirring, the gelatin spheres settled on the bottom of the mixture and were washed with anhydrous acetone (to remove residual mineral oil), soaked in 1,4-dioxane (to facilitate solvent exchange), freeze-dried, and sifted (to separate them into different size ranges).

To fabricate porogen-induced surface-modified NF-PLLA scaffolds, the gelatin spheres were added to Teflon molds, which were then heat-treated in a saturated water vapor atmosphere for designated times. PLLA solution (in H_2O/tetrahydrofuran [THF] solvent mixture) was cast onto each newly pretreated gelatin sphere assembly. After the solution evenly penetrated the entire gelatin sphere assembly, the mold was quickly transferred into a freezer to allow phase separation and entrapment of the gelatin. The scaffolds were then immersed in distilled water (to leach out the gelatin), frozen at −18°C, and freeze-dried.

The use of a solvent mixture was crucial to the surface entrapment of gelatin. Introduction of the solvent mixture (H_2O/THF) in the PLLA solution ensured that gelatin is entrapped onto the surface of the PLLA scaffolds. Water is a good solvent for gelatin, but THF is a poor solvent for gelatin. When gelatin spheres were in contact with the PLLA solution, the surface of the gelatin spheres swelled. PLLA and gelatin molecules could diffuse into each other and become entangled. After leaching the gelatin spheres, the entangled gelatin molecules were entrapped on the pore surface of the polymer scaffolds, and remained stable when soaked in water at 40°C (a temperature above the gelation temperature of gelatin).

In this one-step process, gelatin spheres acted as both porogen and surface-modification agent. As porogen, the gelatin spheres controlled the pore size and shape of the scaffold. As surface-modification agents, some gelatin molecules were entrapped onto the surface of the scaffold during the fabrication process.

There are several advantages for using this porogen-induced surface-modification method. First, fabrication and surface modification of the scaffold proceed at the same time, so that only one, instead of two steps (scaffold fabrication and subsequent surface modification of the scaffold), is needed. Second, the surface-entrapment procedure is simple and effective; the entrapped agents are stable on the modified surface for longer than 2 months. No functional groups in the polymer chains are needed for this surface-modification method. Any scaffold can be modified using this entrapment method as long as a suitable solvent mixture is employed. Meanwhile, in contrast to the method of incorporating functional groups by copolymerization, the gelatin entrapment method maintains the bulk properties of the modified materials.

14.4. Effects of the Nanoarchitecture of Scaffolds on Cell Function and New Tissue Formation

14.4.1. Protein Adsorption

The natural ECM contains several proteins, including collagen, elastin, fibronectin, vitronectin, and laminin, which aid in cell adhesion through integrin–ligand interactions. For this reason, adsorption of ECM proteins onto the surfaces of synthetic scaffolds plays a crucial role in tissue-engineering applications. Scaffolds that mimic the natural ECM promote adsorption of ECM proteins and subsequent cell functions.

One way to engineer better biomimetic polymer, ceramic, and composite scaffolds that exhibit improved protein adsorption is to incorporate nanoscale components into biomaterials. For example, our group has shown that adsorption of key proteins on nanofibrous PLLA scaffolds (fabricated using TIPS in combination with porogen leaching techniques) increased more than 300% over that on solid-walled scaffolds [35]. In this study, scaffolds were wetted and immersed into solutions of fibronectin, vitronectin, albumin, and laminin for a period of 4 h and then washed; the adsorbed proteins were collected and assayed. The nanofibrous scaffolds adsorbed significantly higher levels of each protein tested [35].

Additional improvement in protein adsorption is possible by decreasing the diameter of the nanofibers. Ko et al. reported in their study of PLGA electrospun nanofibrous mats that albumin adsorption increased exponentially with decreased nanofiber diameter; fibers <100 nm in diameter exhibited the greatest adsorption levels [55]. It was theorized that this trend was due to the increased surface area of the mats consisting of thinner fibers.

Protein adsorption is also enhanced on nanocrystalline HA [56]. Recent research has focused on better understanding the underlying mechanism [57], as well as on ways, including functionalization of the nano-HA substrate, to improve protein adsorption [58]. Inclusion of nanoscale HA particles in solid-walled PLLA scaffolds led to increased adsorption of fetal bovine serum proteins over PLLA scaffolds containing no HA, as well as over scaffolds containing micron-scale HA [34].

14.4.2. Cell Attachment

The biological effect of the nanofibrous architecture was investigated by comparing nanofibrous scaffolds with solid-walled scaffolds (those with smooth pore walls). After 24 h, cell (MC3T3-E1) attachment on the nanofibrous scaffolds was 70% higher than that obtained on the solid-walled scaffolds [35]. Meanwhile, the morphology of cells adhering onto nanofibrous scaffolds was different from that on the solid-walled scaffolds. Mouse calvarial cells cultured on solid-walled scaffolds were flat and exhibited relatively smooth cell surfaces with few cell processes. In contrast, cells on the nanofibrous scaffolds formed many cell processes [59]. These results suggest that the nanofibrous architecture provides better substrates for cells to attach.

14.4.3. Cell Differentiation and Tissue Formation

The effect of the synthetic nanofibrous matrix on the osteoblastic gene expression was investigated using reverse transcription polymerase chain reaction. In this case, the level of runx2, a major regulator of the osteoblast phenotype, was consistently higher in the cells cultured on the nanofibrous than on solid-walled scaffolds [59]. At day 7, expression of the osteocalcin gene was much higher in cells cultured on nanofibrous scaffolds than in cells cultured on solid-walled scaffolds. The most dramatic difference between nanofibrous and solid-walled scaffolds was observed with bone sialoprotein (BSP) expression; cells on nanofibrous scaffolds expressed much higher levels of BSP than cells cultured on solid-walled scaffolds. Histology with von Kossa staining after 2 weeks of culture revealed that there was mineral deposition throughout the nanofibrous scaffolds, while there was little observable mineral content in the solid-walled scaffolds. Calcium content analysis revealed 13-fold greater amount of calcium in the nanofibrous scaffolds than that in the solid-walled scaffolds. Transmission electron microscopy observation also showed no mineral deposition on osteoblast-seeded solid-walled scaffolds, while large amounts of small globular mineral deposits attached to the nanofibers. These results suggest that, compared with the outcome obtained on solid-walled scaffolds, neonatal murine calvarial cells cultured on nanofibrous scaffolds exhibited early and enhanced expression of the osteoblast phenotype as well as enhanced calcium-containing mineral deposits in the ECM.

14.5. Conclusion

The design, fabrication, and implementation of biomimetic scaffolds for tissue engineering are at the cutting edge of the field. Inclusion of nanophase design aspects in these scaffolds is one way to better mimic the ECM. These approaches lead to more effective polymer, ceramic, and composite scaffold design and new and useful methods of surface modification, which result in improved protein adsorption and cell responses in vitro and in vivo. Nanoscale biomimetic scaffold design continues to be an area of great interest; ongoing research activities aim at improving the current state of this scientific field, whose potential and promise are great but yet is untapped for biotechnology and tissue-engineering applications.

References

1. Langer R, Vacanti JP. Tissue Engineering. Science 1993;260(5110):920–926.
2. Ma PX. Biomimetic materials for tissue engineering. Advanced Drug Delivery Reviews 2008;60(2):184–198.
3. Liu XH, Ma PX. Polymeric scaffolds for bone tissue engineering. Annals of Biomedical Engineering 2004;32(3):477–486.
4. Elsdale T, Bard J. Collagen substrata for studies on cell behavior. The Journal of Cell Biology 1972;54(3):626–637.
5. Grinnell F. Cell-collagen interactions – Overview. Methods in Enzymology 1982;82:499–503.
6. Smith LA, Ma PX. Nano-fibrous scaffolds for tissue engineering. Colloids and Surfaces B-Biointerfaces 2004;39(3):125–131.
7. Ma PX, Zhang RY. Synthetic nano-scale fibrous extracellular matrix. Journal of Biomedical Materials Research 1999;46(1):60–72.
8. Zhang RY, Ma PX. Synthetic nano-fibrillar extracellular matrices with predesigned macroporous architectures. Journal of Biomedical Materials Research 2000;52(2):430–438.
9. Chen VJ, Ma PX. Nano-fibrous poly(L-lactic acid) scaffolds with interconnected spherical macropores. Biomaterials 2004;25(11):2065–2073.
10. Fratzl P, Gupta HS, Paschalis EP, Roschger P. Structure and mechanical quality of the collagen-mineral nanocomposite in bone. Journal of Materials Chemistry 2004;14:2115–2123.
11. Zhang S. Fabrication of novel biomaterials through molecular self-assembly. Nature Biotechnology 2003;21(10):1171–1178.
12. Niece K, Hartgerink JD, Donners JJJM, Stupp SI. Self-assembly combining two bioactive peptide-amphiphile molecules into nanofibers by electrostatic attraction. Journal of the American Chemical Society 2003;125(24): 7146–7147.
13. Morton WJ, inventor. Method of dispersing fluids. United States Patent No. 705,691, 1902.
14. Srouji S, Kizhner T, Suss-Tobi E, Livne E, Zussman E. 3-D nanofibrous electrospun multilayered construct is an alternative ECM mimicking scaffold. Journal of Materials Science. Materials in Medicine 2008;19:1249–1255.
15. Moroni L, Schotel R, Hamaan D, de Wijn JR, van Bitterswijk CA. 3D fiber-deposited electrospun integrated scaffolds enhance cartilage tissue formation. Advanced Functional Materials 2008;18:53–60.
16. Chen VJ, Smith LA, Ma PX. Bone regeneration on computer-designed nano-fibrous scaffolds. Biomaterials 2006;27(21):3973–3979.
17. Ma PX, Choi JW. Biodegradable polymer scaffolds with well-defined interconnected spherical pore network. Tissue Engineering 2001;7(1):23–33.
18. Wei GB, Ma PX. Macroporous and nanofibrous polymer scaffolds and polymer/bone-like apatite composite scaffolds generated by sugar spheres. Journal of Biomedical Materials Research Part A 2006;78A(2):306–315.
19. Liu XH, Smith LA, Wei G, Won YJ, Ma PX. Surface engineering of nano-fibrous poly(L-lactic acid) scaffolds via self-assembly technique for bone tissue engineering. Journal of Biomedical Nanotechnology 2005;1(1):54–60.
20. Barsoum M. Fundamentals of Ceramics. McGraw-Hill, New York; 1996.
21. Huang J, Best SM, Bonfield W, Brooks RA, Ruchton N, Jayasinghe SN, et al. In vitro assessment of the biological response to nano-sized hydroxyapatite. Journal of Materials Science. Materials in Medicine 2004;15:441–445.
22. Silva C, Pinheiro AG, de Oliveira RS, Goes JC, Aranha N, de Oliveira LR, et al. Properties and in vivo investigation of nanocrystalline hydroxyapatite obtained by mechanical alloying. Materials Science and Engineering C 2004;24:549–554.

23. Longsworth J, Eppell SJ. Design and assembly of a sterile nanoscale orthopedic biomaterial. Biomaterials, Proceedings of Northeast Conference; 2002; 2002. pp. 173–174.
24. Webster SR, Bizios R. Design and evaluation of nanophase alumina for orthopaedic/dental applications. Nanostructured Materials 1999;12:983–986.
25. Webster SR, Bizios R. Osteoblast adhesion on nanophase ceramics. Biomaterials 1999;20:1221–1227.
26. Webster T, Ergun, C, Doremus RH, Siegel RW, Bizios R. Nanocrystalline hydroxyapatite enhances osteoblast function. First Joint BMES/EMBS Conference, 1999, Atlanta, GA; 1999. p. 744.
27. Hench L. Bioceramics. Journal of the American Ceramic Society 1998;81:1705–1728.
28. Smith IO, McCabe LR, Baumann MJ. MC3T3-E1 osteoblast attachment and proliferation on porous hydroxyapatite scaffolds fabricated with nanophases powder. International Journal of Nanomedicine 2006;1:189–194.
29. Catagne J, Chamousset J, Minois LM, Poniset A, Bouler JM, Tancret F. Mechanical properties of macroporous biphasic calcium phosphate bioceramics fabricated using a porogen. Key Engineering Materials 2005;280–283:1549–1553.
30. Seitz H, Rieder W, Irsen S, Leukers B, Tille C. Three-dimensional printing of porous ceramic scaffolds for bone tissue engineering. Journal of Biomedical Materials Research 2005;74B:782–788.
31. Seo D, Hwang KH, Lee JK. Nanostructured hydroxyapatite by microwave sintering. Journal of Nanoscience and Nanotechnology 2008;8(2):944–948.
32. Liao S, Cui, FZ, Zhang W, Feng QL. Hierarchically biomimetic bone scaffold materials: Nano-HA/collagen/PLA composite. Journal of Biomedical Materials Research 2004;69B:158–165.
33. Ma PX, Zhang R, Xiao G, Franceschi R. Engineering new bone tissue in vitro on highly porous poly(α-hydroxyl acids)/hydroxyapatite composite scaffolds. Journal of Biomedical Materials Research 2001;54:284–293.
34. Wei G, Ma PX. Structure and properties of nano-hydoxyapatite/polymer composite scaffolds for bone tissue engineering. Biomaterials 2004;25:4749–4657.
35. Woo KM, Chen VJ, Ma PX. Nano-fibrous scaffolding architecture selectively enhances protein adsorption contributing to cell attachment. Journal of Biomedical Materials Research Part A 2003;67A(2):531–537.
36. Zhang R, Ma PX. Poly(α-hydroxyl acids)/hydroxyapatite porous composites for bone-tissue engineering. I. Preparation and morphology. Journal of Biomedical Materials Research 1999;44:446–455.
37. Zhang R, Ma PX. Porous poly(L-lactic acid)/apatite composites created by biomimetic process. Journal of Biomedical Materials Research 1999;45:285–293.
38. Kim SS, Park MS, Gwak SJ, Choi CY, Kim BS. Accelerated bonelike apatite growth on porous polymer/ceramic composite scaffolds in vitro. Tissue Engineering 2006;12(10):2997–3006.
39. Zhang R, Ma PX. Biomimetic polymer/apatite composite scaffolds for mineralized tissue engineering. Macromolecular Bioscience 2004;4:100–111.
40. Kim H-W, Kim H-E, Salih V. Stimulation of osteoblast responses to biomimetic nanocomposites of gelatin-hydroxyapatite for tissue engineering scaffolds. Biomaterials 2005;26:5221–5230.
41. Du C, Cui FZ, Feng QL, Zhu XD, deGroot K. Tissue response to nano-hydroxyapatite/collagen composite implants in marrow cavity. Journal of Biomedical Materials Research 1998;42:540–548.
42. Du C, Cui FZ, Zhu XD, deGroot K. Three-dimensional nano-HAp/collagen matrix loading with osteogenic cells in organ culture. Journal of Biomedical Materials Research 1999;44:407–415.
43. Woo KM, Seo J, Zhang RY, Ma PX. Suppression of apoptosis by enhanced protein adsorption on polymer/hydroxyapatite composite scaffolds. Biomaterials 2007;28(16):2622–2630.
44. Yoneda S, Guthrie WF, Bright DS, Khatri CA, Wang FW. In vitro biocompatibility of hydrolytically degraded poly(d,l-lactic acid). Seventh World Biomaterials Congress, 2004, Sydney, NSW; 2004. p. 1324.
45. Zhang R, Ma PX. Degradation behavior of porous poly (alpha-hydroxy acids)/hydroxyapatite composite scaffolds. Polymer Preprints 2000;41(2):1618–1619.
46. Zhang RY, Ma PX. Porous poly(L-lactic acid)/apatite composites created by biomimetic process. Journal of Biomedical Materials Research 1999;45(4):285–293.
47. Gao JM, Niklason L, Langer R. Surface hydrolysis of poly(glycolic acid) meshes increases the seeding density of vascular smooth muscle cells. Journal of Biomedical Materials Research 1998;42(3):417–424.
48. Neff JA, Caldwell KD, Tresco PA. A novel method for surface modification to promote cell attachment to hydrophobic substrates. Journal of Biomedical Materials Research 1998;40(4):511–519.
49. Hu YH, Winn SR, Krajbich I, Hollinger JO. Porous polymer scaffolds surface-modified with arginine-glycine-aspartic acid enhance bone cell attachment and differentiation in vitro. Journal of Biomedical Materials Research Part A 2003;64A(3):583–590.
50. Barrera DA, Zylstra E, Lansbury PT, Langer R. Synthesis and Rgd peptide modification of a new biodegradable copolymer – Poly(lactic acid-co-lysine). Journal of the American Chemical Society 1993;115(23):11010–11011.
51. Cook AD, Hrkach JS, Gao NN, Johnson IM, Pajvani UB, Cannizzaro SM, et al. Characterization and development of RGD-peptide-modified poly(lactic acid-co-lysine) as an interactive, resorbable biomaterial. Journal of Biomedical Materials Research 1997;35(4):513–523.

52. Liu XH, Won YJ, Ma PX. Porogen-induced surface modification of nano-fibrous poly(L-lactic acid) scaffolds for tissue engineering. Biomaterials 2006;27(21):3980–3987.
53. Decher G. Fuzzy nanoassemblies: Toward layered polymeric multicomposites. Science 1997;277(5330): 1232–1237.
54. Hammond PT. Recent explorations in electrostatic multilayer thin film assembly. Current Opinion in Colloid & Interface Science 1999;4(6):430–442.
55. Ko H, Kim JH, Chung DJ. Synthesis of biodegradable scaffold from nano fibers and evaluation of protein adsorption behaviors on its surface. Key Engineering Materials 2007;342–343:341–344.
56. Webster T, Ergun C, Doremus RH, Siegel RW, Bizios R. Specific proteins mediate enhanced osteoblast adhesion on nanophase ceramics. Journal of Biomedical Materials Research 2000;51:475–483.
57. dos Santos E, Farina, M, Soares GA, Anselme K. Surface energy of hydroxyapatite and (beta)-tricalcium phosphate ceramics driving serum protein adsorption and osteoblast adhesion. Journal of Materials Science. Materials in Medicine 2008;19:2307–2316.
58. Nelson M, Balasundaram G, Webster TJ. Increased osteoblast adhesion on nanoparticulate crystalline hydroxyapatite functionalized with KRSR. International Journal of Nanomedicine 2006;1(3):339–349.
59. Woo KM, Jun JH, Chen VJ, Seo JY, Baek JH, Ryoo HM, et al. Nano-fibrous scaffolding promotes osteoblast differentiation and biomineralization. Biomaterials 2007;28(2):335–343.

15

Photofunctionalization of Materials to Promote Protein and Cell Interactions for Tissue-Engineering Applications

Shalu Suri, Ankur Singh, and Christine E. Schmidt

Photofunctionalization has been utilized extensively for cell and tissue-engineering research, most commonly in the form of photopolymerization and photografting. Photopolymerization can be performed in vivo, in a minimally invasive manner, and with spatial and temporal control permitting the fabrication of complex scaffolds. A number of natural as well as synthetic polymers have been photofunctionalized to engineer tissues such as bone, cartilage, and skin. In this chapter, we describe the basic mechanism of photofunctionalization and different photoinitiators utilized in the biomedical field. The chapter also focuses on the different photofunctionalization strategies including photopolymerization, photografting, and some advanced techniques, and how these techniques have been explored to study cell, protein, and scaffold interactions. Some of the applications of photofunctionalization in the field of neural, bone, and cartilage tissue engineering are also discussed.

Abbreviations

BDNF	brain-derived neurotrophic factor
CaM	calmodulin
CQ	camphorquinone
CS	chondroitin sulphate
CTNF	ciliary neurotrophic factor
DMPA	2,2-dimethoxy-2-phenylacetophenone
DTC	dithiocarbamate
ECM	extracellular matrix
GAG	glycosaminoglycans
HA	hyaluronic acid
HAp	hydroxyapatite
I2959	irgacure 2959
IPN	interpenetrating polymeric network
ITX	isopropyl-thioxanthone
LRP	living radical polymerization

S. Suri, A. Singh, and C.E. Schmidt • Department of Biomedical Engineering, The University of Texas at Austin, Austin, TX 78712, USA

D.A. Puleo and R. Bizios (eds.), *Biological Interactions on Materials Surfaces*,
DOI 10.1007/978-0-387-98161-1_15,

MSC	mesenchymal stem cells
NGF	nerve growth factor
NT-3	neurotrophin-3
PAA	polyallylamine
PEG	poly(ethylene glycol)
PGA	poly(glycolic acid)
PI	photoinitiator
PLA	poly(lactic acid)
PM(PMA-ala)	poly(methacrylated pyromellitylimidoalanine)
RGD	arginine-glycine-aspartic acid tripeptide
TED	tetra-ethylthiuram disulfide
TGF	transforming growth factor

15.1. Introduction

Photofunctionalized materials have been explored extensively in tissue engineering in an attempt to restore, replace, and enhance either tissue or organ function. Photofunctionalization offers a unique way to fabricate complex three-dimensional (3D) scaffolds with precisely defined geometry and patterns by controlling the crosslinking mechanism and radiation exposure. Photofunctionalization offers rapid conversion (within a few minutes) of a liquid hydrogel precursor solution into a crosslinked polymeric network with spatial and temporal control, thus decreasing the potential invasiveness of implantation procedures.

The fabrication parameters and conditions of photofunctionalized materials make them attractive for tissue-engineering applications. Photoinitiated materials have been researched in many tissue-engineering fields including ocular [1, 2], neural [3–5], dermal [6], cardiovascular [7, 8], dental [9, 10], and cartilage- and bone-related applications [11–14]. Further, photoinitiated systems have also been employed extensively to modulate surface and bulk properties of biomaterials to control cell and protein interactions. More recently, photofunctionalization has been utilized to encapsulate cells for the purpose of immunoisolation [15].

The present chapter details the mechanisms, different strategies, and some of the applications of photofunctionalization in the field of biomedical engineering.

15.2. Mechanisms of Photofunctionalization

Photochemical processes utilize electromagnetic radiation to initiate a chemical reaction [16, 17]. Either ultraviolet (UV) light (~200 to 400 nm) or visible light (~400 to 800 nm) is employed to excite a photoinitiator (PI), thus creating initiation species, such as free radicals or cations, which in turn initiate the desired photochemical reaction. These reactions include polymerization of monomers or oligomers (photopolymerization), crosslinking of functional unsaturated moieties on macromolecules to create a 3D network (photocrosslinking), covalent attachment of functional additives to either the surface or bulk of an already existing polymer matrix (photografting), and degradation of a compound (photodegradation) [17] (see Table 15.1).

Photodegradation is not a commonly used technique in the biomedical field and is, thus, beyond the scope of the present chapter. The present chapter primarily focuses on photopolymerization and photografting, since these are the major photochemical techniques employed in the field of tissue engineering to specifically control scaffold shape and surface properties for guiding protein and cell interactions.

Table 15.1. Techniques of photofunctionalization pertinent to biomedical applications.

Technique	Definition (electromagnetic radiation used for)	Examples in biomedical applications	References
Photopolymerization[a]	Polymerization of monomers or oligomers	Fabrication of 3D hydrogel scaffolds, cell encapsulation, drug delivery carriers, stimuli-responsive materials	[11, 32, 33, 36]
Photocrosslinking[a]	Crosslinking of functional unsaturated moieties on macromolecules to create a 3D network	Fabrication of 3D hydrogels networks, drug delivery carriers, stimuli responsive hydrogels	[13, 22, 75]
Photografting	Covalent attachment of functional additives to either the surface or bulk of an already existing polymer matrix	Attachment of cell adhesive moieties, nonfouling surfaces, photografted antibodies for antigen detection	[42–45]

[a] Photopolymerization is often used in a broad sense that includes photocrosslinking.

15.2.1. Photoinitiators

A PI is the small chromophore-containing compound that absorbs light and undergoes excitation to create either reactive radicals or ionic species, which then initiate or propagate a reaction. Radiation excites the PI chromophore to a higher energy level (PI*) followed by cleavage and formation of active species (radical $PI^{\cdot}$, ion pair PI^+/PI^-) (Scheme 15.1a). The absorption maximum (λ_{max}) of the PI chromophore should be in the same range as the emission spectrum of the incident light to achieve significant absorption. Absorption can further be increased by selecting a PI with a higher molar extinction coefficient (ε), which dictates how strongly the PI will absorb radiation at a particular wavelength. The ε is constant for a fixed wavelength. The absorbed light intensity (I_a) can be calculated using the following equation:

$$I_a = \varepsilon Cl, \tag{1}$$

where C and l represent the concentration of PI and path length, respectively [16, 18]. Thus, the rate of a photoinitiation reaction is dependent on parameters such as the molar extinction coefficient of PI, concentration of PI, presence of quenching agents such as oxygen, and initiation efficiency of either free radical or ionic species.

Oxygen inhibits the polymerization reaction by quenching the excited PI and by scavenging free radicals. The effect of oxygen scavenging is more pronounced in thin layers of materials because oxygen can easily diffuse in the hydrogels or hydrogel films; however, in thick samples (e.g., 3D hydrogels) the effect of scavenging is negligible. To minimize the scavenging effect, photopolymerization can be performed either in a vacuum or in the presence of inert gases such as argon or nitrogen [18].

Possible basic mechanisms of photoinitiation include free radical-induced photoinitiation, photoinitiation by hydrogen abstraction, and cationic photoinitiation [16]. A number of PIs such as Irgacure 2959 (I2959), Irgacure 651 (I651), camphorquinone (CQ), and fluorenone have been investigated for their potential application in the biomedical field during the past decade [9, 19–22]. The criteria for selecting a PI for tissue-engineering applications are not limited to cytocompatibility but also include water solubility, hydrophilicity, shelf life, and light absorption characteristics.

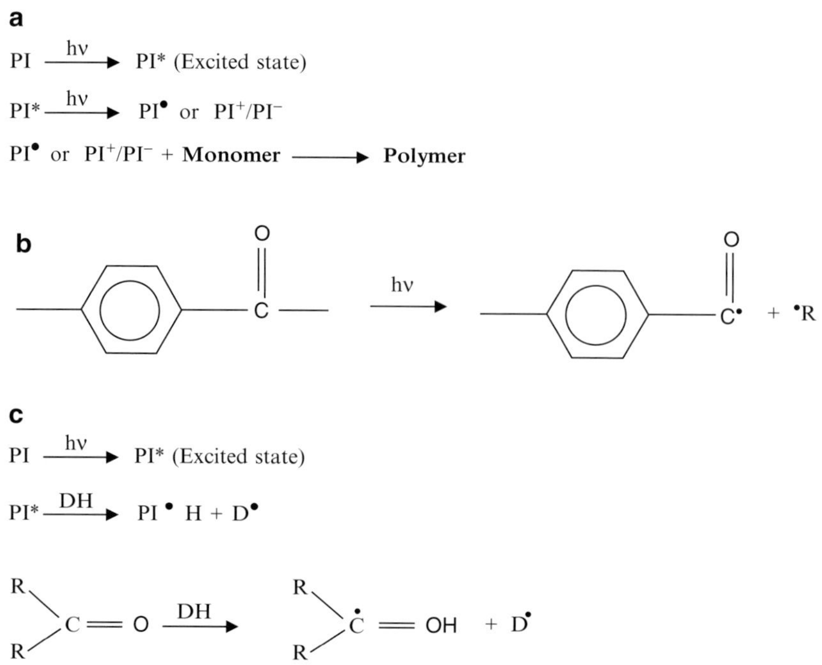

Scheme 15.1. (**a**) Schematic showing the excitation of photoinitiators (PI) and generation of photoinitiating species (PI or PI+/PI_). Photoinitiating species then react with the monomer or macromolecule (M) and grow the polymer chain via photopolymerization, (**b**) example of free radical photopolymerization. The PI, upon exposure to radiation, undergoes photolysis and generates free radicals, which initiate the polymerization process, (**c**) radical photopolymerization by hydrogen abstraction. The PI abstracts a hydrogen atom from the donor molecule (DH) and generates free radicals.

Radical-induced photoinitiation involves cleavage between C–C, C–X (where X is a halogen), C–O, and C–S bonds to generate free radicals. These PIs function primarily by absorbing radiation selectively in the UV wavelength range (Scheme 15.1b). Some examples of this category of PIs are benzoin ethers, hydroxyl alkyl phenyl ketones (I2959) [20, 21, 23], benzyl dimethyl ketal (I651) [9, 23], benzoyl cyclohexanol (Irgacure 184) [23], methyl thio phenyl morpholino (Irgacure 907), dialkoxy acetophenones, and sulfonyl ketones [16].

Examples of PIs that function by hydrogen abstraction are benzophenones, thioxanthones, and quinones. These PIs react by abstracting a hydrogen from a donor (DH) to generate a ketyl radical and a donor radical (D˙) (Scheme 15.1c) [16, 18]. Some of the members of this category are CQ and 9-fluorenone, both of which have been investigated for biomedical applications [22].

Cationic photoinitiation is different from radical photopolymerization, as it requires radiation in the initiation step to generate cationic species, rather than for propagation [24]. The most commonly used cationic PIs are diazonium salts, onium salts, and organometallic complexes. PIs in this category react by ring opening of an oxirane group. Cationic PIs have been used for dental applications to cure resins with minimal shrinkage. For example, Tilbrook et al. photocured epoxy-polyoyl matrix using onium/CQ initiators [10]. However, cationic PIs are not as commonly used in other biomedical engineering applications because of the harsh reaction conditions, such as acid formation, toxicity, and high temperature.

15.2.2. Photosensitizers

Photosensitizers are added sometimes in conjunction to PIs since photosensitizers assist and/or accelerate the chemical reaction by absorbing radiation and transferring the radiation to the PI [18]. Some examples of photosensitizers are isopropyl thioxanthone (ITX) and 1-phenyl-1,2-propanedione [25]. For example, ITX has been employed as a photosensitizer with the PI CQ [20].

15.3. Photoinitiators in Biomaterials and Tissue Engineering

Several critical issues must be taken into consideration prior to selecting a PI for the photofunctionalization of biomaterials. The reactivity of the PI with other active species in the system, toxicity of initiator intermediates, effect of radiation on acellular and cellular components, and the solubility of the PI are all important factors that determine the applicability of a PI. Among different types of PIs, free radical PIs are most widely used in the biomedical and tissue-engineering fields. A detailed list of UV and visible light-activated PIs commonly utilized to polymerize biomaterials is provided in Table 15.2 (adapted from [20]).

Several studies have exclusively explored the adverse effects of different PIs in conjunction with UV and visible radiation on cell compatibility [22, 26]. Such studies have shown that, by altering the irradiation time and intensity, one can achieve improved cell viability and function. Exploration of cytocompatible PIs initially began for dentistry applications and laid the foundation for soft tissue-engineering applications [22, 27]. One of the preliminary studies conducted by Atsumi et al. evaluated the sensitivity of two visible-light PIs, CQ and 9-fluorenone on human submandibular gland cells and reported that increasing concentrations of PI resulted in elevated levels of reactive oxygen species and decreased cell survival [22]. Williams et al. compared the combined effect of different UV irradiation times and PI concentrations on cell viability [23]. Three different UV light PIs – I2959 (1-[4-(2-hydroxyethoxy)-phenyl]-2-hydroxy-2-methyl-1-propane-1-one), I651 (2,2-dimethoxy-1,2-diphenylethan-1-one), and Irgacure 184 (I184; 1-hydroxy-cyclohexyl-phenyl-ketone) (see Table 15.2 for additional information) [20] – and six cell lines were compared; this study reported that I2959 was the least cytotoxic PI [23]. In addition, depending on the cell type, the concentration tolerated by the cells varied from 0.03 to 0.1% [23], although cell proliferation was affected. Another study by Bryant et al. compared different UV and visible PIs, and found that I2959 at ≤0.05% (concentration, w/w) with ~6 mW/cm^2 of ~365 nm UV light and CQ at ≤0.01% (w/w) with ~60 mW/cm^2 of ~470 to 490 nm visible light were most cytocompatible to NIH/3T3 fibroblasts [20]. To further support their findings, the authors demonstrated excellent viability of chondrocytes when these cells were photoencapsulated in acrylated poly(vinyl alcohol) hydrogels with 0.05% I2959 and 10 min UV (~365 nm) exposure.

Photoinitiated systems are intriguing for in situ polymerization since the prepolymer solution can be injected in vivo at the tissue defect site and can be polymerized by either direct or indirect exposure to light. Long-wave UV light (>365 nm) and visible light used in most of the photopolymerizable systems are not damaging to cells and tissues and can penetrate human skin. Langer, Elisseeff, and colleagues polymerized a monomer solution of poly(ethylene oxide)

Table 15.2. Some photoinitiators (with their molar extinction coefficients and wavelengths) employed in biomedical applications (adapted and modified from [21]).

Photoinitiator	Molar extinction coefficient (ε) ($l\ mol^{-1}\ cm^{-1}$)	Wavelength (λ) (nm)	Chemical structure	References
Darocur 2959 (D2959)	6.7	365	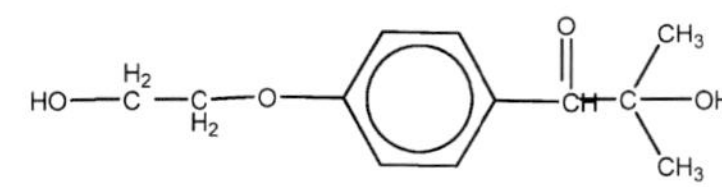	[13, 14, 16]
Irgacure 651 (I651)	94.6	365	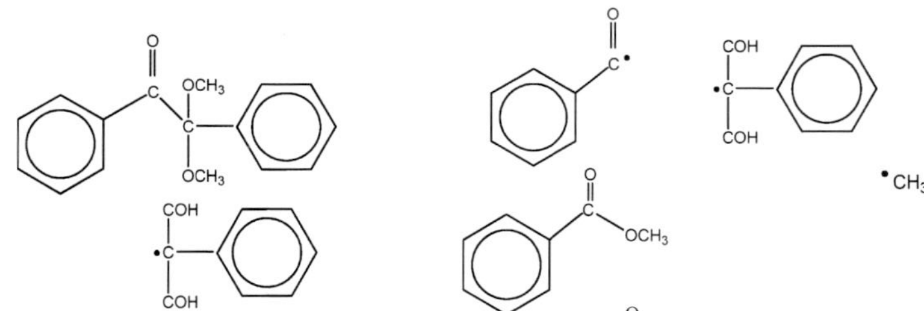	[11, 16]
Irgacure 184 (I184)	7.7	365	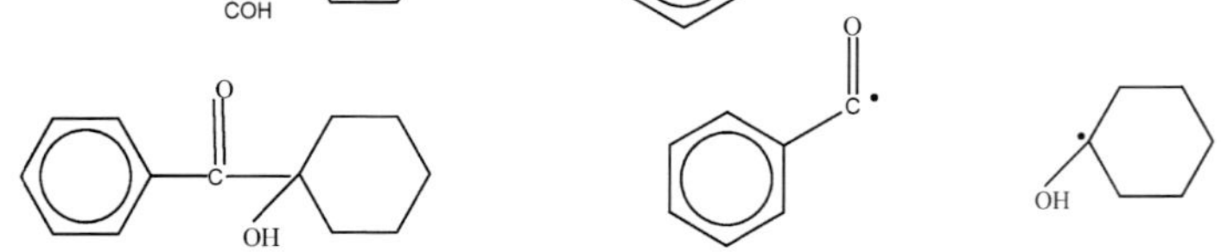	[13, 16]
Irgacure 907 (I907)	120.8	365	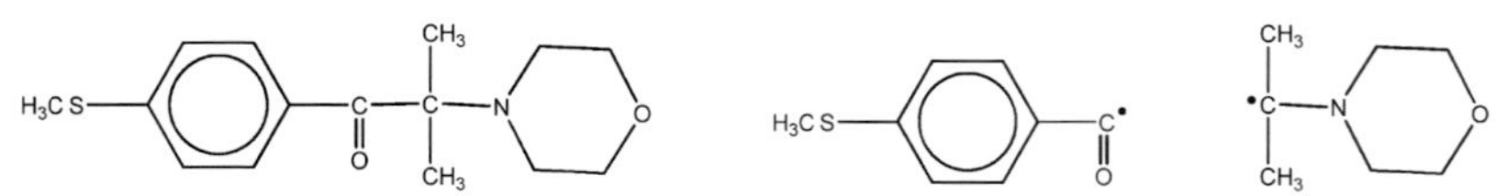	[9]
Camphorquinone (CQ)	24.8	470	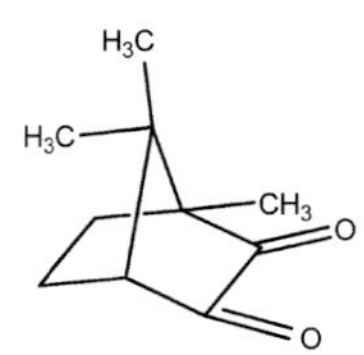	[13, 15]
ITX		365	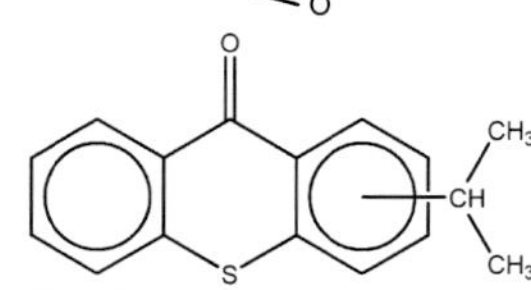	[13]

dimethacrylate in a minimally invasive manner by transdermal illumination following subcutaneous injection in mice [28]. This approach not only has the advantage of creating complex scaffolds that conform to the defect site, but also allows the scaffolds to establish close contacts with surrounding cells and tissues, and thus facilitates cell-cell and cell-material interactions.

In addition to commonly used PIs, thermal radical initiators such as peroxides and azides have also been exploited as PIs since they can be decomposed photochemically. However, these PIs are associated with some disadvantages such as hydrophobicity, low photo-absorption, and low polymerization efficiency and thus are not very commonly employed for biomedical applications. Examples of thermal radical initiators successfully exploited for biomaterials applications include hydroperoxide [29] and *N*-4-(azidobenzoyloxy)succinimide (contains a phenylazido group) [4, 30].

In summary, since, compared with conventional polymerization techniques, photopolymerization can be performed at physiological temperature and pH, is fast, uses low levels of organic crosslinkers and solvents, does not harm cells and tissues, and can be performed with minimally invasive techniques, photopolymerization is an attractive methodology for in vivo applications.

15.4. Strategies to Fabricate Photofunctionalized Materials for Biomedical Applications

Photofunctionalized materials have been used in the development of commercial applications such as electronic and optical materials, membranes, battery electrodes, adhesives and sealants, coatings, and for material surface modifications [16]. In the biomedical field, photofunctionalization of materials is achieved by: (1) photopolymerizing hydrogels and scaffolds; (2) photografting bioactive molecules on the surface and/or bulk of polymer matrices; and (3) using advanced techniques to obtain "smart scaffolds."

15.4.1. Photopolymerized Biomaterial Scaffolds

Hydrogels represent an important class of photopolymerized biomaterial scaffolds, which are not only used for clinical tissue-engineering applications, but are also used in vitro to study different phenomena such as cell function in 3D environments, drug release from hydrogels, and the effects of stimuli-responsive hydrogels. Hydrogels are attractive materials because of their tissue-like hydration properties, which are characterized by a high swelling ratio and high water content present in their crosslinked network without undergoing dissolution. The hydrophilic nature and high swelling ratio also make hydrogels permeable to oxygen, nutrients, metabolites, and waste products. A number of materials and crosslinking techniques have been explored to design hydrogels with varying properties such as pore size, mechanical strength, and degradation rate.

Photopolymerization involves the use of either hydrophilic monomers that possess unsaturated moieties, or monomers that have been functionalized with unsaturated groups (e.g., vinyl, [meth]acrylate) and water-soluble PIs. Photopolymerized hydrogels are advantageous for tissue-engineering applications because liquid monomers can be converted into solid hydrogels by first injecting the solution in vivo and then polymerizing it at the defect site under physiological conditions. Furthermore, cells have been incorporated into liquid monomer solutions and photopolymerization performed under cytocompatible conditions so as to maintain excellent cell survival and cell function.

Considerable effort has been devoted to modifying synthetic materials as well as natural polymers such as polysaccharides to render them photocrosslinkable. Among synthetic materials, a great deal of research has been conducted on poly(ethylene glycol) (PEG) hydrogels. PEG is inert, nonimmunogenic, non-cell adhesive, FDA approved, and can be modified easily by incorporation of different groups to target specific applications [31]. For example, various cell-adhesive peptide sequences have been incorporated into PEG gels to promote cell adhesion, proliferation, and differentiation. Photocrosslinking of PEG can be achieved by attaching different unsaturated groups (the most common being the acrylates and methacrylates) to create PEG mono-, PEG di-, and PEG multi-meth(acrylate). These unsaturated moieties can be activated by PIs to initiate crosslinking. Different molecular weight PEG-*n*-meth(acrylates) (where *n* represents the number of meth[acrylates] per PEG molecule) have been crosslinked at different concentrations resulting in hydrogels of varying pore sizes and mechanical strengths. The factors that contribute to crosslinking density and mechanical strength of PEG hydrogels are the concentration of monomer, number of unsaturated groups per PEG chain, and molecular weight of monomer in the hydrogel precursor solution. Other than methacrylates and acrylates, PEG hydrogels have also been modified with photolabile groups that do not require the use of PIs such as the cinnamylidene acetyl group [32] and the thiol-ene group [33].

A number of studies have provided evidence supporting the advantages of PEG hydrogels for various applications; however, native PEG hydrogel is nondegradable. For this reason, different modification strategies have been devised to render PEG biodegradable by adding degradable functional groups such as poly(lactic acid) (PLA), poly(glycolic acid) (PGA) [34], poly propylene fumarate [11], poly(ε-caprolactone), and poly phosphoester [35] to the PEG backbone chain; the ratio of the degradable group in the backbone can be tailored to fine-tune the degradation rate.

In addition to the modification of synthetic polymers, modification of natural polymers to render them photocrosslinkable has also been extensively reported. Some pertinent examples include hyaluronic acid (HA) [36], dextran [37], chondroitin sulphate (CS) [38], alginate [39], chitosan [40], and heparin [41]. Natural polymers such as HA, CS, and heparin are native components of body tissues and are cytocompatible, biodegradable, and play vital roles in complex cell signaling events and other biological mechanisms. Studies have shown that the degradation products of HA promote angiogenesis [42]. Leach et al. modified HA with glycidyl methacrylate and made photocrosslinkable HA hydrogels that were degradable by hyaluronidase (a mammalian enzyme secreted primarily by fibroblasts, macrophages, and endothelial cells in the body to remodel HA [8]). When the photocrosslinked HA hydrogels were implanted subcutaneously in rats, the hydrogels supported angiogenesis comparable to fibrin positive controls [36].

Much attention has focused on the incorporation of functional additives to photopolymerizable blends. Cell-adhesive proteins and peptides such as collagen, fibronectin, laminin, and arginine-glycine-aspartic acid (RGD) tripeptide sequences have been incorporated into hydrogels, which promote cell adhesion and infiltration. Acrylation of peptides and proteins was successfully performed by different research groups; in these studies the peptides were either crosslinked throughout the bulk of the hydrogel [25, 43] or in complex 3D patterns with spatial control to guide cellular adhesion and migration. Hubbell and colleagues [43] took a step forward and designed photopolymerizable PEG hydrogels with an engineered protein containing the cell-adhesive RGD peptide, the heparin-binding site, and a plasmin-degradation site. This hydrogel supported infiltration and migration of fibroblasts because of the presence of the RGD peptide and plasmin-degradation site, which resulted in serine protease-mediated matrix degradation favoring cell migration (Figure 15.1a, b). Hahn et al. [44] created internal, complex, 3D patterns of acrylated RGD peptide by two-photon absorption photolithography. 3D patterns were created by soaking the preswelled PEG-diacrylate hydrogels with acrylated, low molecular weight PEG-peptide monomer solution followed by crosslinking

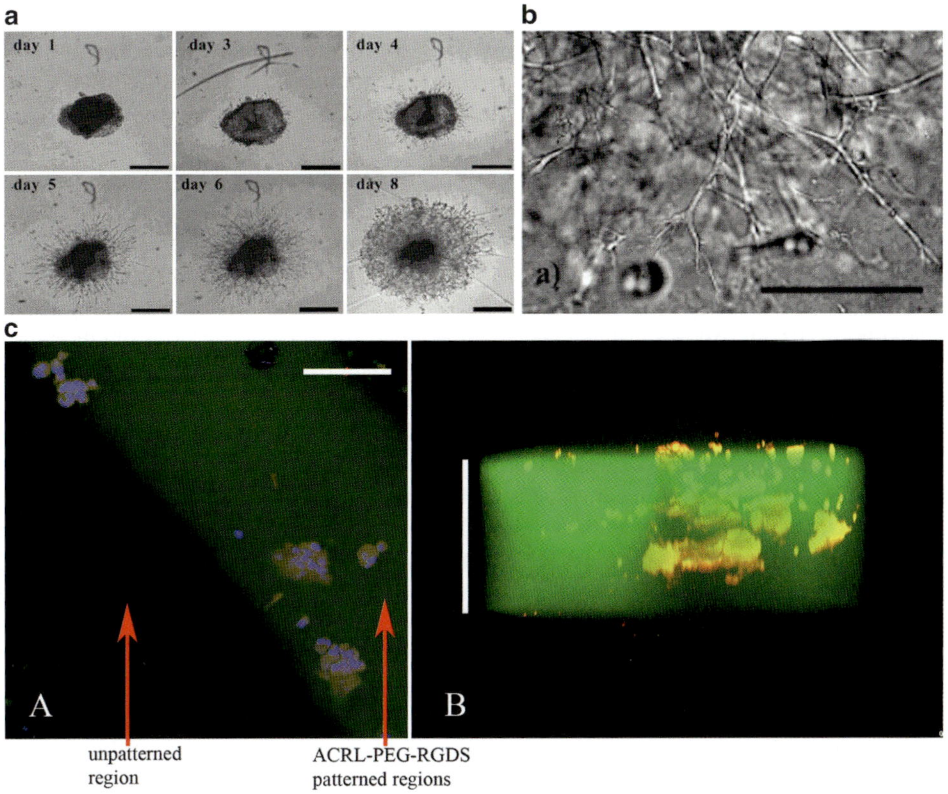

Figure 15.1. Cell migration confined to the RGDS-patterned channels of degradable PEG hydrogels. (**a**) Three-dimensional human fibroblast migration in PEG (6,000 MW) hydrogel grafted with recombinant protein at 1, 3, 4, 5, 6, and 8 days (scale bar = 400 μm). The recombinant protein had an RGD-binding site to promote cell adhesion and a plasmin-degradation site to facilitate matrix degradation. Reproduced from [42] with permission of the American Chemical Society, (**b**) three-dimensional human fibroblast migration at high magnification in the hydrogel described for (**a**) (scale bar = 100 μm). Reproduced from [42] with permission of the American chemical society, (**c**) similar study demonstrating cell migration confined to ACRL–PEG–RGDS-patterned regions of collagenase-degradable PEG hydrogels. Confocal cross-sectional image into the hydrogel, illustrating the confinement of migrating HT-1080 cells (cytoplasm: *orange;* nuclei: *blue*) to the fluorescently labeled ACRL–PEG–RGDS-patterned channels (*green*); scale bar = 100 μm (*left*). Side view of a 3D OsiriX volume rendering of a confocal image series (z-stack) of an ACRL-PEG-RGDS (*green*)-patterned, collagenase-degradable hydrogel into which HT-1080 cell clusters (*orange*) had migrated; scale bar = 550 μm (*right*). Reproduced from [43] with permission of Wiley.

with selective exposure to the desired sections; the result was patterned RGD peptide immobilized in the hydrogel. Cell infiltration in this hydrogels was observed when HT-1080 human fibrosarcoma cells were seeded on top of the gels, but was confined to the immobilized RGD domains after 4 weeks in vitro (Figure 15.1c).

15.4.2. Photografting

Tissue structural organization and cell functionalities can be significantly modulated with select polymer surface properties. For more than a decade, various surface modification strategies have been exploited for derivatization of polymer matrix surfaces with either biological or synthetic macromolecules; however, use of photografting for surface and bulk modification of preformed biomedical scaffolds and devices had not been explored until recently. Devices and scaffolds have been modified via covalent chemical modification techniques, e.g., via modification with self-assembled monolayers. This modification strategy resulted in high surface density of the desired macromolecules with improved retention of the photoinitiating moieties (photoiniferters) compared with the more commonly used non-covalent modification techniques. Photografting techniques, like atom-transfer radical polymerization and living radical polymerization (LRP) [45], can further improve the modification process by reducing the time required to achieve surface modification, enhancing graft stability, and also reducing biomolecule desorption [46, 47]. Other advantages of photografting include mild reaction conditions and the ability to pattern surfaces across user-defined regions of interest [44]. Matsuda et al. developed surface microprocessing technology and exploited phenylazido photochemistry to spatially pattern different polymers to precisely control cell function including cell adhesion, morphology, and migration kinetics [48–51]. Since then, many researchers have expanded the field by utilizing the concept developed by Otsu and Yoshida [52], called quasi-LRP, to spatially graft UV-reactive vinyl species on polymeric matrices. These studies involve use of dithiocarbamate (DTC)-based photoiniferters and phenylazido-based photoiniferters. Use of other PIs such as benzophenone [53] and 2,2-dimethoxy-2-phenylacetophenone (DMPA) [47] has also been reported.

By the late 1990s and early 2000s, Anseth and colleagues [47] broadened the possibilities of LRP chemistry to obtain novel constructs for biomedical applications. One of their early publications described improved detection of antigens using photografted antibodies on the surface of polymers [47]. Acrylated antibodies against IgG were first prepared using monoacrylated PEG-hydroxysuccinimide (ACRL-PEG-NHS) and photografted on polymer substrates prepared using urethane diacrylate and triethylene glycol diacrylate mixed with dithiocarbamide-based tetraethylthiuram disulfide (TED) and DMPA initiator (Figure 15.2a). Patterned photografting with either a single type of antibody or multiple antibodies (tagged with fluorescent moieties of various excitation wavelengths) was achieved using the LRP-based scheme and

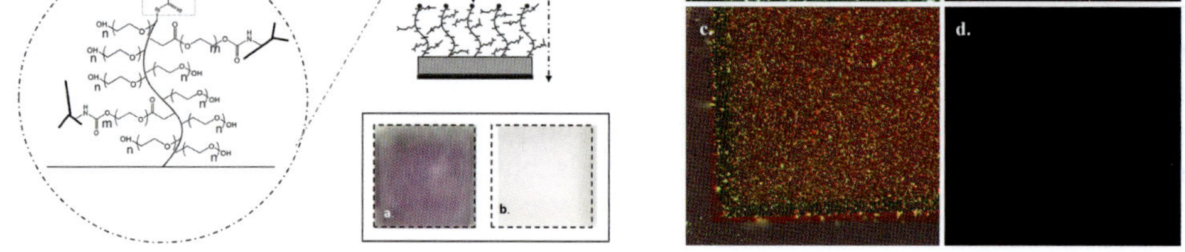

Figure 15.2. Photografting the antibody on a polymer surface using living radical photopolymerization chemistry, (**a**) antibody grafting scheme using controlled living radical photopolymerization chemistry to covalently graft antibody-containing polymer chains from a dithiocarbamate (DTC)-incorporated polymer surface, (**b**) multiple antibodies grafted simultaneously, each labeled with separate fluorophores. One antibody was labeled with AlexaFluor 488 and the other with AlexaFluor 594. A confocal micrograph of the surface square grafted (**a**) using a 488 nm FITC filter, (**b**) using a 594 nm Texas Red filter, (**c**) using both filters simultaneously at 488 nm and 594 nm, and (**d**) using a nonacrylated antibody. Reproduced from [46] with permission of the American chemical society.

standard photolithography techniques. The design facilitated control over numerous parameters including protein conformation and chain mobility, and nonspecific protein interactions, resulting in enhanced interaction between antigen and antibody (Figure 15.2b) [47].

The adaptability of the photografting procedure was further established with recent studies demonstrating precise control over spatial and temporal tethering of macromolecules. The surfaces of DTC-based polymer substrates were sequentially modified by photografting PEG, collagen type I, and a fluorogenic molecule (carboxyfluorescein diacetate) using TED/DMPA and LRP [46]. The objectives behind these approaches were first to prevent nonspecific binding of cells to the polymer substrate surface (by modifying the surface with PEG) and second to enhance cell-polymer surface interaction (by grafting cell adhesion-promoting collagen type I on the material substrate).

Regenerative tissue engineering utilizes both surface photografting of biomolecules and bulk modification of polymer scaffolds. Recent advancements in the field have shown immense potential in fabricating biochemical cues in 3D biocompatible platforms to promote cell functions such as adhesion, differentiation, and tissue regeneration. Luo and Shoichet fabricated a photolabile matrix for 3D photografting of an adhesive fibronectin peptide fragment that facilitated the guided orientation and migration of neurites in 3D [54]. Briefly, a photolabile agarose gel was obtained by modifying agarose with a cysteine-containing compound using a carbonyldiimidazole activation process [54]. When exposed to UV light, the preformed hydrogel scaffolds generated photolabile sulfhydryl moieties that further reacted with sulfhydryl-reactive peptide fragments upon exposure to a focused laser beam. This photografted matrix resulted in improved in vitro neurite guidance and orientation in 3D scaffolds.

Gomez et al. [30] patterned poly(dimethyl siloxane) (PDMS) surfaces with various extracellular matrix (ECM) molecules and physical cues to study the effect of these factors on axon initiation and elongation from neurons. For this reason, nerve growth factor (NGF) was immobilized on the surface of PDMS scaffolds using a polyallylamine-modified phenylazido group (PAA-azido) [30]. The PAA-azido was cast onto PDMS scaffolds thus creating photoinitiating moieties on the polymer scaffold to which NGF was later photo-immobilized using UV light. The effect of immobilized NGF on axonal initiation and elongation in hippocampal neurons was studied in the presence or absence of surface microtopography as a physical guidance cue. Further possibilities were explored by the same research group to tether NGF on the surface of polypyrrole to construct a biomimetic, electrically conducting polymer substrate for enhanced neurite outgrowth in PC12 cells [55].

Use of photografting has also been extended to microdevices and sensors. Spatial patterning can be used to modulate the properties of polymer matrices to respond to environmental factors such as temperature and pH [56]. Sebra et al. [45] developed an optically active, pH-sensitive PEG-tethered microfluidic device. These fluorescence-patterned devices, which were constructed using LRP onto DTC-functionalized polymer substrates, responded to pH changes by switching between fluorescent and nonfluorescent states; such constructs can be used as rapid pH sensors [45].

15.4.3. Advanced Methods

Photopolymerized hydrogels offer many advantages over hydrogels crosslinked using other techniques; among such advantages, temporal and spatial control are the most attractive properties. Further, photografting is often employed to graft the surface and bulk of polymer matrices with functional moieties in order to either promote cell functions such as cell adhesion [46] or reduce protein and cell interactions by creating, for example, antithrombotic and nonfouling surfaces [57]. In addition to traditional photopolymerization of hydrogels and photografting of surfaces and bulk materials, photofunctionalization can also be employed to design "smart materials" including stimuli-responsive materials and dual-crosslinked hydrogels. Stimuli-responsive materials respond to stimuli from their environment and act in an "intelligent" or "predetermined" manner. Photoinitiating systems can be used to create stimuli-responsive materials, which will sense and respond to a change in their microenvironment (such as pH, temperature, solvent type, and bioactive molecules).

As one example, Ehrick et al. [58] synthesized a stimuli-responsive hydrogel by conjugating a genetically engineered protein to the hydrogel substrate. Briefly, genetically engineered calmodulin (CaM) was photofunctionalized with allylamine and was photopolymerized using free radical polymerization by mixing CaM with the acrylamide hydrogel precursor solution. CaM exhibits different conformations in the presence of calcium, some peptides, and phenothiazine (antipsychotic) drugs. In the presence of these ligands, photocrosslinked CaM inside the hydrogel network underwent a conformation change that induced differential hydrogel swelling and thus biomolecule release (Figure 15.3). These kinds of stimuli-responsive hydrogels have potential applications in the field of drug delivery, microfluidics, and BioMEMS devices.

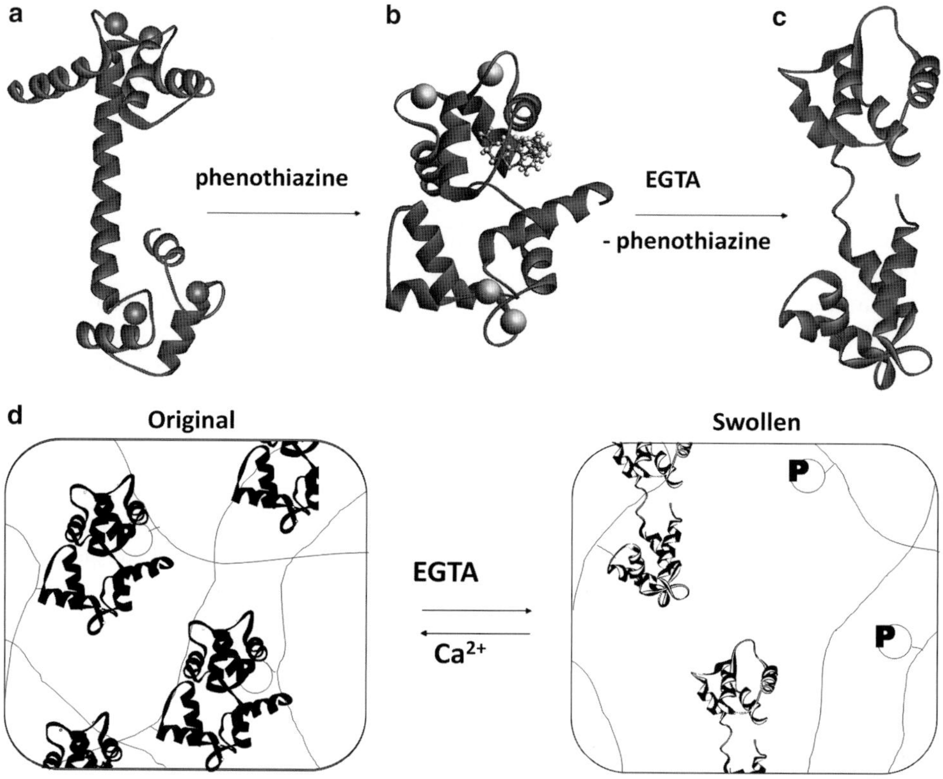

Figure 15.3. Schematic illustration of a stimuli-responsive hydrogel containing genetically engineered (CaM). CaM can have three different conformations: (**a**) dumbbell (spheres represent four bound Ca^{2+} ions), (**b**) phenothiazine-bound (Ca^{2+} and ball-and-stick structure for phenothiazine), and (**c**) native conformation in the absence of Ca^{2+}; (**d**) hydrogel swelling mechanism in response to EGTA. Original stage corresponds to CaM bound to phenothiazine. In the presence of EGTA, Ca^{2+} is removed from its binding sites in CaM and the noncovalent crosslinking is broken, creating expansion of the hydrogel network. Reproduced from [59] with permission of the Nature publishing group.

In another approach, Asoh et al. [59] used photopolymerization to fabricate stimuli-responsive hydrogels in which the hydrogel changed either its shape or volume when triggered by an external stimulus (i.e., temperature change). In this case, poly(*N*-isopropylacrylamide) hydrogels were formed by embedding silica particles in a photopolymerizable hydrogel precursor solution. Nanogradients of silica particles, or pores created by dissolving silica particles, were established in the hydrogel using electrophoresis and subsequent photopolymerization. These gradient hydrogels responded to environmental triggers, specifically, a temperature change, by changes in the hydrogel curvature and swelling behavior. The degree of curvature or bending of a hydrogel was a function of the gradient of silica particles or nanopores. The hydrogels acted as soft actuators and underwent stretching or bending in response to a temperature change.

Recently, some research groups have used photocrosslinking and other crosslinking methods to produce dual-crosslinked hydrogels known as "interpenetrating polymeric networks" (IPNs). This strategy usually involves creating two or more polymers that are crosslinked either simultaneously or sequentially by two different crosslinking methods; the result is the formation of two networks entangled within each other. Dual-crosslinked hydrogels have advantages over single-component gels because they exhibit properties of both polymers. Furthermore, properties of IPNs (such as degradation rate, tensile strength, and porosity) can be modulated by choosing different individual polymer networks [60]. IPN synthesis also results in strong and reinforced resultant scaffolds because of the presence of highly entangled networks. For example, Desai et al. [61] investigated the encapsulation of Islets of Langerhan cells in IPNs of alginate and PEG. Alginate was ionically crosslinked followed by PEG photocrosslinking resulting in the formation of dual-crosslinked gels, which were mechanically and chemically stable under physiological conditions.

Very recently, Schmidt and colleagues synthesized novel dual-crosslinked HA hydrogel scaffolds with photopatterned anisotropic swelling [62]. Dual-crosslinked hydrogels were obtained by a combination of chemical crosslinks and patterned photocrosslinks within a single hydrogel. The resulting dual-crosslinked hydrogel swelled anisotropically because of differential crosslink densities between the photopatterned and nonphotopatterned regions, thus permitting the hydrogel to contort and evolve a shape different from that of the unswollen hydrogel. Anisotropic swelling resulted in zonal distributions of crosslink density, water content, and viscoelasticity at the macro- and micro-scales in the hydrogel scaffold [62]. This dual-crosslinking technique represents a platform to fabricate tissue-engineering scaffolds exhibiting heterogeneity in structural and biomechanical properties.

15.5. Photofunctionalized Materials to Promote Cell Interactions for Tissue-Engineering Applications

Photofunctionalization has been used for a wide range of applications such as bone and cartilage regeneration [28], drug delivery devices [63, 64], blood vessel adhesives [65], adhesion prevention barriers [66–68], and cell encapsulation [69, 70]. Hubbell and colleagues were among the first to develop photopolymerized bioerodible hydrogels based on PEG diacrylate macromers that could be used for burn dressings, surgical adhesives, and as tissue barriers to improve healing response by preventing scar tissue formation [34]. Hubbell and colleagues further developed acrylated RGD peptide-functionalized hydrogels [71] using photoinitiated systems to render nonadhesive hydrogel surfaces cell adhesive for tissue resurfacing. Since then, a number of strategies have been adopted to create photofunctionalized hydrogels and scaffolds for numerous biomedical applications.

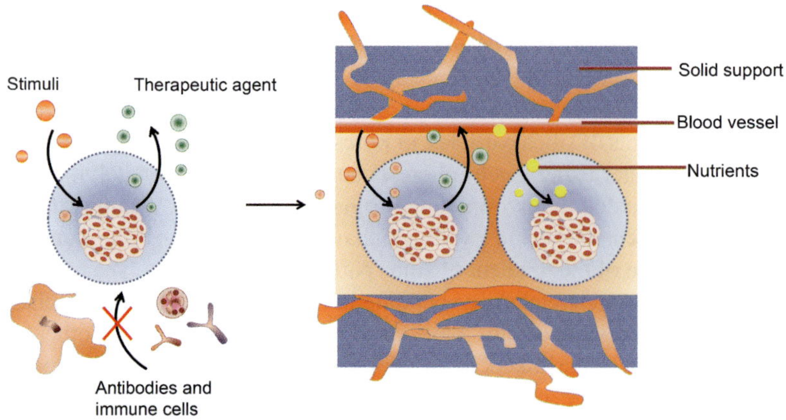

Figure 15.4. Schematic illustration of cell microencapsulation: (**a**) nutrients, oxygen, and stimuli diffuse across the membrane, whereas antibodies and immune cells are excluded, (**b**) prevascularized solid support system to facilitate optimal nutrition of the enclosed cells. Reproduced from [73] with permission of Nature publishing group.

Tissue adhesives offer an attractive alternative to conventional repair methods such as sutures for corneal perforations. Recently, Miki et al. developed a novel photopolymerized sealant based on HA for corneal lacerations [72], thus further extending the possibilities for the use of photopolymerized biomaterials in tissue engineering.

Vast opportunities for these photopolymerized biomaterials in tissue engineering can be envisioned by considering the latest developments using stem cells for tissue reconstruction. For instance, Elisseeff and colleagues developed a photopolymerizable poly(ethylene oxide) diacrylate hydrogel for bone marrow-derived mesenchymal stem cell (MSC) implantation using a transdermal photopolymerization technique for in vivo cartilage regeneration. The study successfully showed differentiation of MSCs into the chondrogenic lineage [19].

Cells can also be encapsulated in photocrosslinked microcapsules of polymer, which allows diffusion of cell-secreted biomolecules in the vicinity but isolates the cells from the host tissue and thus prevents immunorejection (Figure 15.4) [73]. In one study, Hubbell and colleagues fabricated microcapsules by photocrosslinking PEG to protect pancreatic islets cells from immunorejection [74].

15.5.1. Bone and Cartilage Tissue Engineering

In 2005, more than 500,000 patients received bone grafts in the United States; the number was over 2 million worldwide [75]. Autografts have been considered the "gold standard" clinically but are associated with complications such as a limited supply and the need for additional surgery, as well as pain and inflammation at the donor site. Allografts were proposed as alternatives to autografts, but these were also associated with limitations such as immune rejection and potential viral transmission [75]. Metal rods and screws have been widely used as bone implants for almost two decades but are also associated with several disadvantages. Metals provide immediate mechanical support but they do not integrate with the regenerating tissue, can cause loosening, and can be associated with fatigue loading, infection, and corrosion [76]. Ceramic-based bone grafts have also been used clinically. Since ceramics, however, are brittle and have low tensile loading, they are not suitable for implant material applications under bending or stress conditions [76].

Because of the inherent limitations associated with autografts, allografts, and traditionally used metal and ceramic implants, tissue-engineered bone substitutes have been investigated to meet the increasing clinical demand for bone implants. Bioresorbable PLGA screws and rods are now being used clinically as an alternative to metal implants [77, 78], but these polymers have some drawbacks such as relatively low mechanical strength (which limits their use to low load-bearing applications only), and frequent loss of stability during bone regeneration [79]. Although a number of strategies have been tried and different materials have been investigated to develop scaffolds for bone tissue regeneration, an ideal construct meeting all the requirements for an engineered bone scaffold is still not a reality.

Photopolymerized scaffolds have also been extensively researched for regeneration and restoration of bone and cartilage tissues [28, 80]. Photocrosslinking can result in strong crosslinked networks of polymeric chains imparting appropriate mechanical strength to orthopedic implants similar to that of natural, healthy bone [81, 82], as well as offering the possibility to entrap cell and protein components inside scaffolds. Recent studies using photopolymerized materials have shown tremendous potential in controlling the degradation rate of the implant and simultaneously maintaining mechanical integrity [83]. Since bone can regenerate on its own after injury, it is essential to create a resorbable scaffold that degrades at a rate synchronous with that of bone healing [84]. If the degradation rate is too slow, then the scaffold might hinder the growth of tissue or may obstruct cell–cell signaling; if the degradation rate is too fast, then the growing tissue might not be able to bear mechanical loads [14].

Photopolymerizable scaffolds, blends and fillers, and hydrogels have been researched for bone, cartilage, and anterior cruciate ligament reconstruction in animal models [12]. Photocrosslinked polyanhydrides were used to fabricate tailor-made bone constructs [14]. Compared with either traditional polyesters like poly(ε-caprolactone) or poly(α-hydroxy acids) and several permanent synthetic polymers [14], these biodegradable constructs demonstrated tunable biodegradation and improved mechanical strength because degradation by surface erosion preserved the polymer molecular weight. One such popular biomaterial is methacrylated sebacic acid, which has an elastic modulus similar to that of bone [14, 80]. Despite these efforts, mimicking actual bone integrity, complexity, mechanical properties, and chemical milieu is far from realization.

Several attempts have been made to incorporate native bone constituents such as hydroxyapatite (HAp), which can further improve the mechanical integrity of bone constructs and cell-tissue interactions [13, 85]. Li et al. [13] investigated the effect of incorporating HAp in poly(methacrylated pyromellitylimidoalanine) (PM[PMA-ala]). In this case, needle-shaped nanoconstructs were synthesized by photopolymerization of HAp with PM(PMA-ala). Incorporation of HAp by photocrosslinking with the PM(PMA-ala) blend resulted in significant improvement of the mechanical strength of the composite over that of PM(PMA-ala) alone. In addition, recent studies by Anseth's group showed that the mechanical strength of polyanhydrides can be improved by using fully functionalized tricarballylic acid, instead of sebacic acid, to increase the crosslinking density of the polyanhydrides, as well as using stiffer monomers to form these polymer photocrosslinked networks [14]. Alternatively, the uncontrolled degradation characteristics of polyesters can be overcome by using enzymatically cleavable acrylamide-based crosslinkers. The materials thus obtained are promising for various orthopedic applications, but detailed and specific application-based studies are needed.

Another major application of photopolymerized polymers is in cartilage tissue engineering. Articular cartilage in humans lacks blood vessels, a complex lymphatic system, and functional nerve endings [20, 86, 87]. Failure of native cartilage to repair itself after injury has made it one of the most widely researched tissue-engineering areas. Mimicking the complex cartilage infrastructure using synthetic scaffolds imposes a major challenge. During the 1990s, several research groups at Harvard and the Massachusetts institute of technology, led by Vacanti and colleagues [88], Langer and colleagues [89, 90], and Vunjak-Novakovic and colleagues [91], extensively explored the use of PGA- and PLA-based systems for developing 3D cartilage tissues. Chondrocytes, the cartilage cells, were cultured on PLA and PGA biomaterial constructs for 6 weeks in vitro and for 6 months in vivo. Survival and proliferation of chondrocytes was assessed at regular time intervals along with cartilage matrix production. Although cells cultured on PGA scaffolds produced sulfated glycosaminoglycans (GAG) and collagen II, this approach, however, had limitations since the cells were seeded on the top of the scaffold but were not dispersed in the 3D structure of the scaffold. In addition, PGA, which degrades by bulk erosion, released acidic degradation products causing inflammation and loss of the mechanical strength of the scaffold. Cells seeded on PLA (a hydrophobic material) produced minimal levels of sulfated GAG and collagen II; this outcome did support cartilage regeneration. Photopolymerized hydrogels of natural, as well as synthetic, materials, have emerged as potential candidates to overcome this limitation and have been successful in encapsulating chondrocytes, as well as maintaining cell viability, phenotypic characteristics, and protein-glycosaminoglycan secretion for several weeks to months [20, 92].

In situ fabrication of biopolymers using photopolymerization involves crosslinking of acrylated monomers in the presence of a PI exposed to UV light. Anseth et al. developed photopolymerizable, surface-erodible polyanhydride-based methacrylate monomers, using visible and UV light PIs, suitable for various orthopedic applications [28, 84, 93]. Further studies by this research group demonstrated the efficacy and biocompatibility of these methacrylated polyanhydride derivatives through in vivo studies performed by photopolymerizing the monomers in bone defects in tibia of mice and by subcutaneous scaffold implantations [94]. The cell inflammatory response and the thickness of fibrous capsule varied depending on the degradation rate and composition of the polymer scaffolds [94]. Langer and colleagues showed that a liquefied polymer mixture can be photopolymerized in situ by exposure to either UV or visible light transdermally [28]. In this case, photopolymerization was performed using an injectable poly(ethylene oxide)dimethacrylate prepolymer solution and chondrocyte mixture, administered subcutaneously in swine. The results provided evidence of development of neocartilage including ECM components such as GAG and collagen II. Transdermal photopolymerization is advantageous for cartilage tissue engineering since the liquid monomer solution was injected in vivo in a minimally invasive manner, photocrosslinked in situ acquiring the shape of the tissue defect, and permitted the cell-matrix integration. Hoshikawa et al. [95] used a photopolymerizable gelatin-based biomaterial scaffold and cultured chondrocytes in these hydrogels. Under these conditions, chondrocytes remained viable and retained their phenotypic characteristics [95]. Bowman and colleagues [96] developed photopolymerizable macromers based on PEG and poly(vinyl alcohol). Biodegradable hydrogel crosslinks were created using these multifunctional molecules for cartilage tissue-engineering applications. Chondrocytes were combined with the desired macromer/initiator solution, which was subsequently polymerized to form a hydrogel. The cells (encapsulated in the biodegradable hydrogel) produced cartilageneous matrix as evidenced by GAG production. The production of neocartilage was further enhanced by adding PLGA microparticles loaded with transforming growth factor (TGF)-β and insulin-like growth factor in the hydrogel [96].

Application of photoencapsulation of cells is not restricted to chondrocytes. Regeneration of cartilage tissue from MSCs offers an alternative and versatile tool for cartilage and bone tissue engineering [97, 98]. These cells can be differentiated into specific cell types. MSCs could potentially differentiate into chondrocytes in the presence of growth factors such as TGF-β [19]. Sharma et al. [19] developed an injectable photocrosslinkable PEG-based hydrogel encapsulating MSCs for cartilage regeneration. These researchers tested two groups of hydrogels, one with HA and TGF-β, and another with TGF-β only. Hydrogels with HA and TGF-β showed the highest expression of cartilage-specific genes such as collagen II and aggrecan. Histological staining also revealed the enhanced production of GAG and collagen II in the hydrogels containing HA and TGF-β [19].

Alhadlaq et al. [99] formulated articular condyles of the synovial joints using MSCs. Rat bone marrow-derived MSCs were encapsulated in a PEG-based photopolymerizable hydrogel to form an osteochondral assembly for total joint replacements. The rationale behind this study was that both chondrocytes and osteoblasts have the same progenitor parent stem cell (i.e., the adult MSC) and thus differentiation of MSCs into either osteoblasts or chondrocytes should lead to an articular condyle containing both cell types. For this purpose, two stratified, but integrated layers of hydrogels containing encapsulated MSCs were photopolymerized to form an osteochondral construct. After implantation, these osteochondral constructs retained their shape, produced cartilage-specific GAG and collagen II, and exhibited stratified osteogenesis and chondrogenesis [99].

15.5.2. Neural Tissue Engineering

Following nerve injury, major complications in the regeneration of the peripheral and central nervous systems include: axon orientation and guidance across injury sites; restoration of functional synaptic connections and remyelination; and, in the central nervous system, formation of a hostile microenvironment that includes formation of a glial scar, impeding axonal regeneration [36, 100]. Axonal guidance to target cells is important for the reestablishment of functional synaptic connections; a number of physical, chemical, and biological cues, which play important roles by interacting synergistically or individually with the cells, have been identified. Bioengineers and neuroscientists have made significant attempts to design tissue-engineered scaffolds that can mimic the intricate in vivo environment of the nervous system and, thus, study the interactions between different cues and neuronal cell function and neural tissue regeneration. A number of photofunctionalized materials including fibers [101], films [4], patterned surfaces [3, 30, 55], hollow conduits, matrix-filled conduits, and hydrogel scaffolds [102] have been explored as potential substrates to interact with neurons.

Photopolymerized hydrogels as axonal guidance matrices have been studied extensively in vitro for their ability to be fabricated with varying degrees of mechanical stiffness. For example, Gunn et al. [103] synthesized photopolymerized PEG hydrogels with varying degrees of mechanical stiffness and incorporated different types of cell-adhesive ligands for peripheral nerve regeneration applications. The study provided evidence that PC12 cells extended neurites both on RGD-immobilized and peptide sequence isoleucine–lysine–valine–alanine–valine-immobilized hydrogels, but the length of neurites was longer in RGD-immobilized hydrogels. Neurite length was also affected by the mechanical properties of the hydrogel; greater neurite extension was observed on the more flexible hydrogels [103].

Neurotrophic factors such as NGF, brain-derived neurotrophic factor (BDNF), ciliary neurotrophic factor (CTNF), neurotrophin (NT)-3, and fibroblast growth factor influence different neuronal cell functions including cell survival, differentiation, neurite orientation, and growth. Such growth factors were studied alone or in conjunction with other physical, chemical, and biological cues to promote neural cell response/function. Photoirradiation was used to immobilize growth factors on the surfaces of materials [30]; in addition, a photopolymerizable hydrogel was fabricated with encapsulated growth factors [26]. Langer and colleagues formulated unique photopolymerized biodegradable hydrogels and used them as delivery vehicles to release multiple neurotrophic factors such as CTNF, BDNF, and NT-3 [26]. The release kinetics of such growth factors was tailored by altering the crosslinking density, which in turn controlled the diffusion of molecules through the hydrogel network. In a following study [104], these researchers injected prepolymer solution with NT-3 into rat spinal cords and conducted photopolymerization in situ at the defect site. The hydrogels with the growth factor induced enhanced responses at both the functional level (improved recovery) and anatomical level (more axonal growth) [104]. Photocrosslinkable tubes of gelatinous fibers were produced and coated with laminin, fibronectin, and NGF to produce nerve conduits [101]. The extent of axonal regeneration in these tubes was comparable to that obtained on matrigel-coated tubes, suggesting their potential application as nerve guidance conduits.

More recently, fabrication of neural cell-encapsulated 3D hydrogel matrices have attracted attention from different research groups because such constructs mimic the native cellular environment and enable the encapsulated cells to secrete ECM components (such as collagen, HA, laminin, and fibronectin) and thus augment tissue regeneration. To this end, Mahoney and Anseth [105] encapsulated neural stem/progenitor cells in photopolymerizable PEG hydrogels for nerve regeneration and demonstrated that the rate of cell differentiation and cell spreading inside the hydrogel can be fine-tuned by controlling the hydrogel degradation rate. Furthermore, the hydrogel system tested supported stem cell survival and proliferation, and promoted the differentiation of the stem cells into glial cells and neurons.

Photofunctionalization has also been applied to materials beyond hydrogels. Polymer surfaces have been modified with growth factors using photoirradiation to promote neural cells growth. Gomez and Schmidt [4] combined chemical cues with electrical cues to stimulate neural cell functions. These researchers photografted NGF on the surface of polypyrrole, which is an electrically conducting polymer, and studied the synergistic influence of chemical and electrical cues on neural hippocampal cell polarization and axon elongation [4]. In another study, they studied the effect of electrical and topographical cues on neural cell polarization and neurite elongation and orientation [106]. To this purpose, these researchers fabricated microchannels on a polypyrrole surface, and further modified the polymer by photografting NGF [106]. Since these cues, individually or in combination, affected neural cell functions such as cell differentiation, adhesion, and growth, they could be advantageous in neural tissue-engineering applications aiming at improving axonal regeneration.

15.6. Conclusions

Photofunctionalization has been used in a number of biomedical studies for a variety of applications, mostly in photopolymerized hydrogels. Hydrogels can be photopolymerized in situ at the defect site, under spatial and temporal control; these conditions result in fabrication of complex scaffolds in a minimally invasive manner and allow scaffolds to conform to the shape of the tissue defect. The photocrosslinking density can be altered to fabricate scaffolds with different mechanical stiffnesses to match those of the tissues of interest. Photofunctionalization has also been widely used to modify, by grafting cell adhesive peptides, the surface properties of scaffolds so as to render them adhesive to cells, to make them non-biofouling by resisting protein and cell adhesion, and to graft bioactive molecules that promote specific cell functions pertinent to new tissue formation. Photoencapsulation of cells inside hydrogels provides a platform to study cellular mechanisms and tissue formation in a tissue-mimicking 3D micro-environment. Given the variety of modification strategies provided by photoinitiated systems, the biomaterials fabricated using these techniques will continue to be used and further developed as "designer scaffolds" in tissue-engineering applications.

Acknowledgments

We would like to thank and recognize Maeve Cooney for assistance with editing. We would like to thank Tera Sherrard for assistance with copyright permission and for obtaining high-resolution images. We would like to acknowledge Jeffrey A. Hubbell, Kristi S. Anseth, Sylvia Daunert, and Jose Luis Pedraz for contributions of figures used in this chapter. We would also like to thank Scott Zawko for his valuable feedback and input on this project.

References

1. Garagorri N, Fermanian S, Thibault R, Ambrose WM, Schein OD, Chakravarti S, Elisseeff J: Keratocyte behavior in three-dimensional photopolymerizable poly(ethylene glycol) hydrogels. *Acta Biomater* 2008, 4:1139–1147.
2. Carnahan MA, Middleton C, Kim J, Kim T, Grinstaff MW: Hybrid dendritic-linear polyester-ethers for in situ photopolymerization. *J Am Chem Soc* 2002, **124**:5291–5293.
3. Goldner JS, Bruder JM, Li G, Gazzola D, Hoffman-Kim D: Neurite bridging across micropatterned grooves. *Biomaterials* 2006, **27**:460–472.
4. Gomez N, Schmidt CE: Nerve growth factor-immobilized polypyrrole: bioactive electrically conducting polymer for enhanced neurite extension. *J Biomed Mater Res A* 2007, **81**:135–149.
5. Sakai Y, Matsuyama Y, Takahashi K, Sato T, Hattori T, Nakashima S, Ishiguro N: New artificial nerve conduits made with photocrosslinked hyaluronic acid for peripheral nerve regeneration. *Biomed Mater Eng* 2007, **17**:191–197.
6. Onuki Y, Nishikawa M, Morishita M, Takayama K: Development of photocrosslinked polyacrylic acid hydrogel as an adhesive for dermatological patches: involvement of formulation factors in physical properties and pharmacological effects. *Int J Pharm* 2008, **349**:47–52.
7. Moon JJ, Lee SH, West JL: Synthetic biomimetic hydrogels incorporated with ephrin-A1 for therapeutic angiogenesis. *Biomacromolecules* 2007, **8**:42–49.
8. Baier Leach J, Bivens KA, Patrick CW Jr, Schmidt CE: Photocrosslinked hyaluronic acid hydrogels: natural, biodegradable tissue engineering scaffolds. *Biotechnol Bioeng* 2003, **82**:578–589.
9. Kilambi H, Cramer NB, Schneidewind LH, Shah P, Stansbury JW, Bowman CN: Evaluation of highly reactive mono-methacrylates as reactive diluents for BisGMA-based dental composites. *Dent Mater* 2008, **25**:33–38.
10. Tilbrook DA, Clarke RL, Howle NE, Braden M: Photocurable epoxy-polyol matrices for use in dental composites I. *Biomaterials* 2000, **21**:1743–1753.
11. Suggs LJ, Mikos AG: Development of poly(propylene fumarate-co-ethylene glycol) as an injectable carrier for endothelial cells. *Cell Transplant* 1999, **8**:345–350.
12. Chen CH, Liu HW, Tsai CL, Yu CM, Lin IH, Hsiue GH: Photoencapsulation of bone morphogenetic protein-2 and periosteal progenitor cells improve tendon graft healing in a bone tunnel. *Am J Sports Med* 2008, **36**:461–473.
13. Li H, Chen Y, Xie Y: Photo-crosslinking polymerization to prepare polyanhydride/needle-like hydroxyapatite biodegradable nanocomposite for orthopedic application. *Mater Lett* 2003, **57**:2848–2854.
14. Young JS, Gonzales KD, Anseth KS: Photopolymers in orthopedics: characterization of novel crosslinked polyanhydrides. *Biomaterials* 2000, **21**:1181–1188.
15. Sawhney AS, Pathak CP, Hubbell JA: Modification of islet of langerhans surfaces with immunoprotective poly(ethylene glycol) coatings via interfacial photopolymerization. *Biotechnol Bioeng* 1994, **44**:383–386.
16. Fouassier JP: Photoinitiation, photopolymerization, and photocuring: fundamentals and applications. Munich, New York, Cincinnati: Hanser, 1995.
17. Peiffer RW: Applications of photopolymer technology. Washington, DC, American Chemical Society, 1996.
18. Salamone JC: *Polymeric materials encyclopedia*. Boca Raton: CRC, 1996.
19. Sharma B, Williams CG, Khan M, Manson P, Elisseeff JH: In vivo chondrogenesis of mesenchymal stem cells in a photopolymerized hydrogel. *Plast Reconstr Surg* 2007, **119**:112–120.
20. Bryant SJ, Nuttelman CR, Anseth KS: Cytocompatibility of UV and visible light photoinitiating systems on cultured NIH/3T3 fibroblasts in vitro. *J Biomater Sci Polym Ed* 2000, **11**:439–457.
21. Li Q, Wang J, Shahani S, Sun DD, Sharma B, Elisseeff JH, Leong KW: Biodegradable and photocrosslinkable polyphosphoester hydrogel. *Biomaterials* 2006, **27**:1027–1034.
22. Atsumi T, Murata J, Kamiyanagi I, Fujisawa S, Ueha T: Cytotoxicity of photosensitizers camphorquinone and 9-fluorenone with visible light irradiation on a human submandibular-duct cell line in vitro. *Arch Oral Biol* 1998, **43**:73–81.

23. Williams CG, Malik AN, Kim TK, Manson PN, Elisseeff JH: Variable cytocompatibility of six cell lines with photoinitiators used for polymerizing hydrogels and cell encapsulation. *Biomaterials* 2005, **26**:1211–1218.
24. Odian GG: *Principles of polymerization*. 4th edn. Hoboken, NJ: Wiley, 2004.
25. Park YD, Tirelli N, Hubbell JA: Photopolymerized hyaluronic acid-based hydrogels and interpenetrating networks. *Biomaterials* 2003, **24**:893–900.
26. Burdick JA, Ward M, Liang E, Young MJ, Langer R: Stimulation of neurite outgrowth by neurotrophins delivered from degradable hydrogels. *Biomaterials* 2006, **27**:452–459.
27. Fujisawa S, Kadoma Y, Masuhara E: Effects of photoinitiators for the visible-light resin system on hemolysis of dog erythrocytes and lipid peroxidation of their components. *J Dent Res* 1986, **65**:1186–1190.
28. Elisseeff J, Anseth K, Sims D, McIntosh W, Randolph M, Langer R: Transdermal photopolymerization for minimally invasive implantation. *Proc Natl Acad Sci U S A* 1999, **96**:3104–3107.
29. Guan J, Gao C, Feng L, Sheng J: Surface photo-grafting of polyurethane with 2-hydroxyethyl acrylate for promotion of human endothelial cell adhesion and growth. *J Biomater Sci Polym Ed* 2000, **11**:523–536.
30. Gomez N, Lu Y, Chen S, Schmidt CE: Immobilized nerve growth factor and microtopography have distinct effects on polarization versus axon elongation in hippocampal cells in culture. *Biomaterials* 2007, **28**:271–284.
31. Tessmar JK, Gopferich AM: Customized PEG-derived copolymers for tissue-engineering applications. *Macromol Biosci* 2007, **7**:23–39.
32. Andreopoulos FM, Deible CR, Stauffer MT, Weber SG, Wagner WR, Beckman EJ, Russell AJ: Photoscissable hydrogel synthesis via rapid photopolymerization of novel PEG-based polymers in the absence of photoinitiators. *J Am Chem Soc* 1996, **118**:6235–6240.
33. Cramer NB, Scott JP, Bowman CN: Photopolymerizations of thiol-ene polymers without photoinitiators. *Macromolecules* 2002, **35**:5361–5365.
34. Sawhney AS, Pathak CS, Hubbell JA: Bioerodible hydrogels based on photopolymerized poly(ethylene glycol)-co-poly(α-hydroxy acid) diacrylate macromers. *Macromolecules* 1993, **26**:581–587.
35. Dikovsky D, Bianco-Peled H, Seliktar D: The effect of structural alterations of PEG-fibrinogen hydrogel scaffolds on 3-D cellular morphology and cellular migration. *Biomaterials* 2006, **27**:1496–1506.
36. Leach JB, Bivens KA, Patrick CW, Schmidt CE: Photocrosslinked hyaluronic acid hydrogels: natural biodegradable tissue engineering scaffolds. *Biotechnol Bioeng* 2003, **82**:578–589.
37. van Dijk-Wolthuis WNE, Franssen O, Talsma H, van Steenbergen MJ, Kettenes-van den Bosch JJ, Hennink WE: Synthesis, characterization, and polymerization of glycidyl methacrylate derivatized dextran. *Macromolecules* 1995, **28**:6317–6322.
38. Li Q, Wang DA, Elisseeff JH: Heterogeneous-phase reaction of glycidyl methacrylate and chondroitin sulphate: mechanism of ring-opening-transesterificationcompetition. *Macromolecules* 2003, **36**:2556–2562.
39. Chou AI, Nicoll SB: Characterization of photocrosslinked alginate hydrogels for nucleus pulposus cell encapsulation. *J Biomed Mater Res A* 2008, in press.
40. Ono K, Saito Y, Yura H, Ishikawa K, Kurita A, Akaike T, Ishihara M: Photocrosslinkable chitosan as a biological adhesive. *J Biomed Mater Res* 2000, **49**:289–295.
41. Matsuda T, Magushi T: Preparation of vinylated polysaccharides and photofabrication of tubular scaffolds as potential use in tissue engineering. *Biomacromolecules* 2002, **3**:942–950.
42. West DC, Hampson IN, Arnold F, Kumar S: Angiogenesis induced by degradation products of hyaluronic acid. *Science* 1985, **228**:1324–1326.
43. Halstenberg S, Panitch A, Rizzi S, Hall H, Hubbell JA: Biologically engineered protein-graft-poly(ethylene glycol) hydrogels: a cell adhesive and plasmin-degradable biosynthetic material for tissue repair. *Biomacromolecules* 2002, **3**:710–723.
44. Hahn MS, Miller JS, West JW: Three-dimensional biochemical and biomechanical patterning of hydrogels for guiding cell behavior. *Adv Mater* 2006, **18**:2679–2684.
45. Sebra RP, Kasko AM, Anseth KS, Bowman CS: Synthesis and photografting of highly pH-responsive polymer chains. *Sens Actuators B Chem* 2006, **119**:127–134.
46. Sebra RP, Reddy SK, Masters KS, Bowman CN, Anseth KS: Controlled polymerization chemistry to graft architectures that influence cell-material interactions. *Acta Biomater* 2007, **3**:151–161.
47. Sebra RP, Masters KS, Bowman CN, Anseth KS: Surface grafted antibodies: controlled architecture permits enhanced antigen detection. *Langmuir* 2005, **21**:10907–10911.
48. Nakayama Y, Anderson JM, Matsuda T: Laboratory-scale mass production of a multi-micropatterned grafted surface with different polymer regions. *J Biomed Mater Res* 2000, **53**:584–591.
49. Nakayama Y, Matsuda T: Photo induced surface heparin immobilization. *ASAIO J* 1993, **39**:M754–M757.
50. Nakayama Y, Matsuda T, Irie M: A novel surface photo-graft polymerization method for fabricated devices. *ASAIO J* 1993, **39**:M542–M544.
51. Nakayama Y, Matsuda T: Surface macromolecular architectural designs using photo-graft copolymerization based on photochemistry of benzyl N,N-diethyldithiocarbamate. *Macromolecules* 1996, **29**:8622–8630.

52. Otsu T, Yoshida M: Role of initiator-transfer agent-terminator (iniferter) in radical polymerizations: polymer design by organic disulfides as iniferters. *Macromol Rapid Commun*1982, **26**:127–132.
53. Yang W, Ranby B: Bulk surface photografting process and its applications. II. Principal factors affecting surface photografting. *J Appl Polym Sci* 1996, **62**:545–555.
54. Luo Y, Shoichet MS: A photolabile hydrogel for guided three-dimensional cell growth and migration. *Nat Mater* 2004, **3**:249–253.
55. Gomez N, Chen S, Schmidt CE: Polarization of hippocampal neurons with competitive surface stimuli: contact guidance cues are preferred over chemical ligands. *J R Soc Interface* 2007, **4**:223–233.
56. Peng T, Cheng YL: PNIPAAm and PMAA co-grafted porous PE membranes: living radical co-grafting mechanism and multi-stimuli responsive permeability. *Polymer* 2001, **42**:2091–2100.
57. Wang P, Tan KL, Kang ET: Surface modification of poly(tetrafluoroethylene) films via grafting of poly(ethylene glycol) for reduction in protein adsorption. *J Biomater Sci Polym Ed* 2000, **11**:169–186.
58. Ehrick JD, Deo SK, Browning TW, Bachas LG, Madou MJ, Daunert S: Genetically engineered protein in hydrogels tailors stimuli-responsive characteristics. *Nat Mater* 2005, **4**:298–302.
59. Asoh TA, Kaneko T, Matsusaki M, Akashi M: Rapid and precise release from nano-tracted poly(N-isopropylacrylamide) hydrogels containing linear poly (acrylic acid). *Macromol Biosci* 2006, **6**:959–965.
60. Hasirci V, Lewandrowski K, Gresser JD, Wise DL, Trantolo DJ: Versatility of biodegradable biopolymers: degradability and an in vivo application. *J Biotechnol* 2001, **86**:135–150.
61. Desai NP, Sojomihardjo A, Yao Z, Ron N, Soon-Shiong P: Interpenetrating polymer networks of alginate and polyethylene glycol for encapsulation of islets of Langerhans. *J Microencapsul* 2000, **17**:677–690.
62. Zawko SA, Suri S, Trong Q, Schmidt CE: Photopatterned anisotropic swelling of dual-crosslinked hyaluronic acid hydrogels. *Acta Biomaterialia* 2008. doi: 10.1016/j.actbio.2008.09.012.
63. Zawko SA, Truong Q, Schmidt CE: Drug-binding hydrogels of hyaluronic acid functionalized with beta-cyclodextrin. *J Biomed Mater Res A* 2008, **87**:1044–1052.
64. Ward JH, Peppas NA: Preparation of controlled release systems by free-radical UV polymerizations in the presence of a drug. *J Control Release* 2001, **71**:183–192.
65. Dumanian GA, Dascombe W, Hong C, Labadie K, Garrett K, Sawhney AS, Pathak CP, Hubbell JA, Johnson PC: A new photopolymerizable blood vessel glue that seals human vessel anastomoses without augmenting thrombogenicity. *Plast Reconstr Surg* 1995, **95**:901–907.
66. West JL, Chowdhury SM, Sawhney AS, Pathak CP, Dunn RC, Hubbell JA: Efficacy of adhesion barriers. Resorbable hydrogel, oxidized regenerated cellulose and hyaluronic acid. *J Reprod Med* 1996, **41**:149–154.
67. West JL, Hubbell JA: Comparison of covalently and physically cross-linked polyethylene glycol-based hydrogels for the prevention of postoperative adhesions in a rat model. *Biomaterials* 1995, **16**:1153–1156.
68. West JL, Hubbell JA: Separation of the arterial wall from blood contact using hydrogel barriers reduces intimal thickening after balloon injury in the rat: the roles of medial and luminal factors in arterial healing. *Proc Natl Acad Sci U S A* 1996, **93**:13188–13193.
69. Elisseeff J, McIntosh W, Anseth K, Riley S, Ragan P, Langer R: Photoencapsulation of chondrocytes in poly(ethylene oxide)-based semi-interpenetrating networks. *J Biomed Mater Res* 2000, **51**:164–171.
70. Albrecht DR, Tsang VL, Sah RL, Bhatia SN: Photo- and electropatterning of hydrogel-encapsulated living cell arrays. *Lab Chip* 2005, **5**:111–118.
71. Hern DL, Hubbell JA: Incorporation of adhesion peptides into nonadhesive hydrogels useful for tissue resurfacing. *J Biomed Mater Res* 1998, **39**:266–276.
72. Miki D, Dastgheib K, Kim T, Pfister-Serres A, Smeds KA, Inoue M, Hatchell DL, Grinstaff MW: A photopolymerized sealant for corneal lacerations. *Cornea* 2002, **21**:393–399.
73. Orive G, Hernandez RM, Gascon AR, Calafiore R, Chang TM, De Vos P, Hortelano G, Hunkeler D, Lacik I, Shapiro AM, et al.: Cell encapsulation: promise and progress. *Nat Med* 2003, **9**:104–107.
74. Pathak CP, Sawhney AS, Hubbell JA: Rapid photopolymerization of immunoprotective gels in contact with cells and tissues. *J Am Chem Soc* 1992, **114**:8311–8312.
75. Marra KG: *Biodegradable polymers and microspheres in tissue engineering*. Boca Raton: CRC, 2005.
76. Salgado AJ, Coutinho OP, Reis RL: Bone tissue engineering: state of the art and future trends. *Macromol Biosci* 2004, **4**:743–765.
77. Fedorowicz Z, Nasser M, Newton JT, Oliver RJ: Resorbable versus titanium plates for orthognathic surgery. *Cochrane Database Syst Rev* 2007:CD006204.
78. Eppley BL, Morales L, Wood R, Pensler J, Goldstein J, Havlik RJ, Habal M, Losken A, Williams JK, Burstein F, et al.: Resorbable PLLA-PGA plate and screw fixation in pediatric craniofacial surgery: clinical experience in 1883 patients. *Plast Reconstr Surg* 2004, **114**:850–856; discussion 857.
79. Hutmacher DW, Schantz JT, Lam CX, Tan KC, Lim TC: State of the art and future directions of scaffold-based bone engineering from a biomaterials perspective. *J Tissue Eng Regen Med* 2007, **1**:245–260.

80. Anseth KS, Shastri VR, Langer R: Photopolymerizable degradable polyanhydrides with osteocompatibility. *Nat Biotechnol* 1999, **17**:156–159.
81. Timmer MD, Ambrose CG, Mikos AG: Evaluation of thermal- and photo-crosslinked biodegradable poly(propylene fumarate)-based networks. *J Biomed Mater Res A* 2003, **66**:811–818.
82. Timmer MD, Carter C, Ambrose CG, Mikos AG: Fabrication of poly(propylene fumarate)-based orthopaedic implants by photo-crosslinking through transparent silicone molds. *Biomaterials* 2003, **24**:4707–4714.
83. Timmer MD, Ambrose CG, Mikos AG: In vitro degradation of polymeric networks of poly(propylene fumarate) and the crosslinking macromer poly(propylene fumarate)-diacrylate. *Biomaterials* 2003, **24**:571–577.
84. Burkoth AK, Anseth KS: A review of photocrosslinked polyanhydrides: in situ forming degradable networks. *Biomaterials* 2000, **21**:2395–2404.
85. Li H, Chen Y, Xie Y: Nanocomposites of cross-linking polyanhydrides and hydroxyapatite needles: mechanical and degradable properties. *Mater Lett* 2004, **58**:2819–2823.
86. Mankin HJ: The reaction of articular cartilage to injury and osteoarthritis (second of two parts). *N Engl J Med* 1974, **291**:1335–1340.
87. Mankin HJ: The reaction of articular cartilage to injury and osteoarthritis (first of two parts). *N Engl J Med* 1974, **291**:1285–1292.
88. Cao Y, Vacanti JP, Paige KT, Upton J, Vacanti CA: Transplantation of chondrocytes utilizing a polymer-cell construct to produce tissue-engineered cartilage in the shape of a human ear. *Plast Reconstr Surg* 1997, **100**: 297–302; discussion 303–294.
89. Freed LE, Grande DA, Lingbin Z, Emmanual J, Marquis JC, Langer R: Joint resurfacing using allograft chondrocytes and synthetic biodegradable polymer scaffolds. *J Biomed Mater Res* 1994, **28**:891–899.
90. Freed LE, Marquis JC, Nohria A, Emmanual J, Mikos AG, Langer R: Neocartilage formation in vitro and in vivo using cells cultured on synthetic biodegradable polymers. *J Biomed Mater Res* 1993, **27**:11–23.
91. Freed LE, Langer R, Martin I, Pellis NR, Vunjak-Novakovic G: Tissue engineering of cartilage in space. *Proc Natl Acad Sci U S A* 1997, **94**:13885–13890.
92. Bryant SJ, Bender RJ, Durand KL, Anseth KS: Encapsulating chondrocytes in degrading PEG hydrogels with high modulus: engineering gel structural changes to facilitate cartilaginous tissue production. *Biotechnol Bioeng* 2004, **86**:747–755.
93. Burkoth AK, Burdick J, Anseth KS: Surface and bulk modifications to photocrosslinked polyanhydrides to control degradation behavior. *J Biomed Mater Res* 2000, **51**:352–359.
94. Poshusta AK, Burdick JA, Mortisen DJ, Padera RF, Ruehlman D, Yaszemski MJ, Anseth KS: Histocompatibility of photocrosslinked polyanhydrides: a novel in situ forming orthopaedic biomaterial. *J Biomed Mater Res A* 2003, **64**:62–69.
95. Hoshikawa A, Nakayama Y, Matsuda T, Oda H, Nakamura K, Mabuchi K: Encapsulation of chondrocytes in photopolymerizable styrenated gelatin for cartilage tissue engineering. *Tissue Eng* 2006, **12**:2333–2341.
96. Anseth KS, Metters AT, Bryant SJ, Martens PJ, Elisseeff JH, Bowman CN: In situ forming degradable networks and their application in tissue engineering and drug delivery. *J Control Release* 2002, **78**:199–209.
97. Bruder SP, N, Jaiswal NS, Ricalton JD, Mosca KH, Kraus S: Kadiyala 1998: Mesenchymal stem cells in osteobiology and applied bone regeneration. Clin Orthop Relat Res S247–S256.
98. Lutolf MP, Weber FE, Schmoekel HG, Schense JC, Kohler T, Muller R, Hubbell JA: Repair of bone defects using synthetic mimetics of collagenous extracellular matrices. *Nat Biotechnol* 2003, **21**:513–518.
99. Alhadlaq A, Elisseeff JH, Hong L, Williams CG, Caplan AI, Sharma B, Kopher RA, Tomkoria S, Lennon DP, Lopez A et al.: Adult stem cell driven genesis of human-shaped articular condyle. *Ann Biomed Eng* 2004, **32**:911–923.
100. Fawcett JW, Asher RA: The glial scar and central nervous system repair. *Brain Res Bull* 1999, **49**:377–391.
101. Gamez E, Ikezaki K, Fukui M, Matsuda T: Photoconstructs of nerve guidance prosthesis using photoreactive gelatin as a scaffold. *Cell Transplant* 2003, **12**:481–490.
102. Sarig-Nadir O, Seliktar D: Compositional alterations of fibrin-based materials for regulating in vitro neural outgrowth. *Tissue Eng Part A* 2008, **14**:401–411.
103. Gunn JW, Turner SD, Mann BK: Adhesive and mechanical properties of hydrogels influence neurite extension. *J Biomed Mater Res A* 2005, **72**:91–97.
104. Piantino J, Burdick JA, Goldberg D, Langer R, Benowitz LI: An injectable, biodegradable hydrogel for trophic factor delivery enhances axonal rewiring and improves performance after spinal cord injury. *Exp Neurol* 2006, **201**:359–367.
105. Mahoney MJ, Anseth KS: Three-dimensional growth and function of neural tissue in degradable polyethylene glycol hydrogels. *Biomaterials* 2006, **27**:2265–2274.
106. Gomez N, Lee JY, Nickels JD, Schmidt CE: Micropatterned polypyrrole: a combination of electrical and topographical characteristics for the stimulation of cells. *Adv Funct Mater* 2007, **17**:1645–1653.

16

Hydrogel Nanocomposites in Biology and Medicine: Applications and Interactions

Nitin S. Satarkar, Ashley M. Hawkins, and J. Zach Hilt

Hydrogel nanocomposites are a new class of biomaterials that have recently attracted a lot of attention for applications in medical and pharmaceutical areas. The nanocomposites may consist of various types of nanoparticles, such as clay, ceramic, metallic, or metal oxides dispersed in a hydrogel matrix. The hydrogel nanocomposites have been investigated for various biological applications including drug delivery, tissue engineering, antimicrobial materials, and thermal therapy. In particular, different techniques to control the drug release rate from the nanocomposite matrix have been highlighted. In this chapter, various tissue-engineering areas are discussed, including bone, articular cartilage, and cornea repair. Biological interactions with nanocomposites, including cell adhesion and pertinent cytotoxicity studies, are also discussed.

Abbreviations

ALP	alkaline phosphate
AMF	alternating magnetic field
DC	direct current
HA	hydroxyapatite
HEMA	hydroxyethylmethacrylate
HepG2	human hepatoma cells
pHEMA	poly(hydroxyethylmethacrylate)
IR	infrared
LCST	lower critical solution temperature
MTT	3-(4,5-dimethylthiazol-2-yl)-2,5-diphenyltetrazolium bromide
n-HAp	nanohydroxyapatite
PNIPAAm	poly(*N*-isopropylacrylamide)
PVA	poly(vinyl alcohol)
RC	remote controlled
RGR	relative growth rate

N.S. Satarkar, A.M. Hawkins, and J.Z. Hilt • Department of Chemical and Materials Engineering, University of Kentucky, Lexington, KY 40506, USA

D.A. Puleo and R. Bizios (eds.), *Biological Interactions on Materials Surfaces*,
DOI 10.1007/978-0-387-98161-1_16,

SPH	super porous hydrogel
SPHC	super porous hydrogel composite
TCP	tricalcium phosphate

16.1. Introduction

Hydrogels are three-dimensional polymeric networks with the ability to swell several times their dry weight by absorption of water and other biological fluids. Hydrogels are currently considered for numerous biomedical and pharmaceutical applications including drug-delivery devices, contact lenses, tissue-engineering scaffolds, biosensors, sutures, and components of microfluidic devices [1–4]. Responsive hydrogels are a class of hydrogels with swelling properties dependent on environmental factors like pH, temperature, ionic strength, and presence of a particular molecule [5–7]. Temperature- and pH-responsive hydrogels have been widely investigated for drug delivery, sensors, and actuator types of applications [8–11]. For example, poly(*N*-isopropylacrylamide) (PNIPAAm) is a negative temperature-responsive hydrogel with a lower critical solution temperature (LCST) of 34°C [12]. PNIPAAm is in swollen state below LCST and is collapsed (less swollen) above LCST. Appropriate choice of the PNIPAAm hydrogel composition allows tuning the LCST close to body temperature to dictate when collapse will occur [13]. This aspect has allowed for PNIPAAm systems to be used for temperature-triggered drug release and in microfluidic devices in other temperature regimes [14, 15].

Hydrogel nanocomposites are obtained by incorporating different types of nanoparticles, such as metallic, clay, and ceramics, into a hydrogel matrix. The particulate material can be incorporated in hydrogels either by mixing the particles in monomer solution followed by polymerization, or particles can be precipitated in the hydrogel matrix after polymerization [16]. Hydrogel nanocomposites have unique properties such as improved mechanical strength, stimuli-responsive behavior, biological interactions, optical properties, and ability of remote actuation [17–19]. For example, high water content and elasticity of hydrogels sometimes leads to inferior mechanical performance and limits their applications. Incorporation of particulate material like nanophase hydroxyapatite (HA) has improved the mechanical properties and made the nanocomposites attractive materials for tissue-engineering applications [20]. In addition to mechanical properties, drug release, and stimuli-responsive behavior, hydrogels can also be altered by nanoparticles for specific applications [21–23]. Additionally, hydrogels can be made responsive to external stimuli such as either magnetic field or light by incorporation of stimuli-specific nanoparticulates in the hydrogel network [24–26].

In this chapter, we primarily focus on biological applications of matrix-type hydrogel composites consisting of inorganic nanoparticulate materials. Applications of hydrogel nanocomposites in drug delivery, tissue engineering, and other therapeutic areas are discussed. Here, we have highlighted various nanocomposites based on magnetic iron oxide [27], clays [28, 29], gold, silver [30], HA [31–37], and tricalcium phosphate (TCP) [21, 35, 38]. Applications of various nanocomposite systems with selected references are summarized in Table 16.1. In the final section of this chapter, biological interactions of hydrogel nanocomposites are discussed with emphasis on cytotoxicity studies [39]. Hydrogel-polymer nanocomposites and hydrogel nanocomposites in particle form are not within the scope of this chapter, and thus, readers are directed to other literature references for more information [30, 40].

Table 16.1. Summary of applications of various nanocomposite systems with selected references.

General application	Specific application	Particulate	Polymer	References
Drug release	Remote control	Iron oxide (Fe_3O_4)	Poly(*N*-isopropylacrylamide), gelatin, poly(vinyl alcohol)	[24, 41–46]
		Gold nanoshells	Poly(*N*-isopropylacrylamide)-co-poly(acrylamide)	[25, 47, 48]
	Enhanced drug-release profile	Hydrotalcite clay	Poly(acrylic acid-co-poly[ethylene glycol] methyl ether acrylate)	[49]
		Mica anionic clay	Poly(*N*-isopropylacrylamide)	[50]
		Hydroxyapatite	Acrylamide	[32]
Tissue engineering	Bone tissue repair	Hydroxyapatite	Poly(vinyl alcohol), poly(vinyl alcohol)/collagen, bacterial cellulose, HYAFF 11, poly(vinyl alcohol)/gelatin	[31, 33, 35, 36]
		Tricalcium phosphate	Oxidized alginate/gelatin, alginate, HYAFF 11	[21, 38, 35]
		Biphasic calcium phosphate	Si-hydroxypropylmethylcellulose	[51]
	Articular cartilage repair	Bioglass	Poly(vinyl alcohol)	[20]
		Hydroxyapatite	Poly(vinyl alcohol)	[37]
	Cornea repair	Hydroxyapatite	Poly(vinyl alcohol)/collagen	[34]
		Clay	Poly(*N*-isopropylacrylamide)	[52]
	Cell adhesion applications	Fumed silica	Poly(hydroxyethylmethacrylate)	[53]
		Silver	Acrylic acid	[54]
Other therapeutic applications	Antimicrobial applications	Silver	Poly(*N*-isopropylacrylamide)-co-poly(sodium acrylate), poly(vinyl pyrrolidone)-acrylamide IPN, poly (acrylamide-co-acrylic acid)	[55–58]
	Thermal therapy	Iron oxide (Fe_3O_4)	Poly(*N*-isopropylacrylamide), bis(ethylhexyl) sodium sulfosuccinate, poly(vinyl alcohol)	[24, 39, 59–62]

16.2. Hydrogel Nanocomposites for Drug-Delivery Applications

The field of controlled drug delivery continues to be one of the key areas of current research in pharmaceutics and medicine. Due to unique and tailorable properties, hydrogels have been extensively studied as materials for drug-delivery applications [3, 11, 63]. For example, the hydrophilicity and porosity of the hydrogel network can be controlled by the choice of the monomer(s) and the crosslinker. Manipulation of network design allows for the development of drug-delivery systems with a desired release profile for a given drug [64]. Incorporation of nano- and micro-scale materials in the hydrogel matrix allows additional control on the network properties (e.g., potential for remote actuation). Various types of hydrogel nanocomposites have been developed in the past few years, and their swelling and drug-release properties have been extensively studied. Here, we review selected key drug-delivery applications of hydrogel nanocomposites.

16.2.1. Hydrogel Nanocomposites for Remote-Controlled Drug Release

In many applications, drug-delivery systems are designed to release a drug at a constant rate for long periods of time. In other situations, therapeutics such as peptides and hormones often require pulsatile release to match the natural release profile of such chemical compounds in the body. In the design of pulsatile drug-delivery systems, devices have been created that are preprogrammed, self-regulating depending on the presence of a specific molecule, and externally actuated with specific stimuli [48, 65–67]. The design of externally actuated pulsatile drug-delivery systems can be tailored by the choice of nanoparticulates, the hydrogel matrix, and the external stimulus. Inherent advantage of these systems over other approaches is that the drug-release profile can be altered after implanting the device in the body.

The idea of remotely actuating polymer composites was first pursued by Langer and coworkers [68, 69]. Magnetic steel beads were embedded in an ethylene-vinyl polymer matrix; release of macromolecules was regulated by application of a low-frequency oscillating magnetic field. A magnetic polymer composite containing insulin was implanted subcutaneously in diabetic rats, and effective control of glucose levels was demonstrated with an oscillating magnetic field [70]. Remote control of drug release has attracted a lot of attention, and hydrogel nanocomposites responsive to various stimuli like alternating magnetic field (AMF) [59], direct current (DC) magnetic fields [46], and light [47] have been investigated. Some of the studies on remote-controlled (RC) release are highlighted in the sections that follow.

Modulation of Drug Release with AMF

Magnetic fields are attractive in RC medical applications because they can be transmitted through biological tissues for a range of frequencies and intensities; for this reason, AMF-responsive systems are attracting much attention. For example, Satarkar and Hilt incorporated 20–30 nm Fe_3O_4 particles in a temperature-responsive PNIPAAm hydrogel matrix [59]. Application of AMF at 300 kHz heated the magnetic particles (due to Neel and Brownian relaxations), in turn, heating the entire nanocomposite matrix [71]. The effect of AMF on different particle loadings was examined; it was observed that the temperature of the hydrogel matrix could be remotely increased above LCST, leading to collapse. This phenomenon was used to obtain pulsatile release of chemical compounds from hydrogel discs on application of on–off cycles of AMF, as shown in Figure 16.1a [24]. Pulsatile drug release was demonstrated for methylene blue, pyrocatechol violet, and vitamin B_{12} as model drug molecules. Figure 16.1b shows the release rate and cumulative release profile of vitamin B_{12} in response to on–off AMF pulses [24].

In another study, Fe_3O_4 nanoparticles were incorporated in a gelatin hydrogel matrix, and the pulsed release of vitamin B_{12} was observed on application of a high-frequency magnetic field [42]. The effect of magnetic particle size (10–250 nm) was also studied to obtain maximum pulse effect [42]. It was speculated that pulsed release was obtained due to heating of the magnetic particles leading to heating, and resultant shrinking, of gelatin network. The underlying mechanism for this phenomenon in ferrogels is not completely understood.

Modulation with DC Magnetic Field

Magnetic hydrogels have also been synthesized by mixing Fe_3O_4 particles of different sizes in a poly(vinyl alcohol) (PVA) matrix using a freeze–thaw method [44, 46]. Chen and colleagues showed that application of a DC magnetic field to a PVA nanocomposite resulted in accumulation of drug around the ferrogel; this situation was designated as the "closed" configuration of pores.

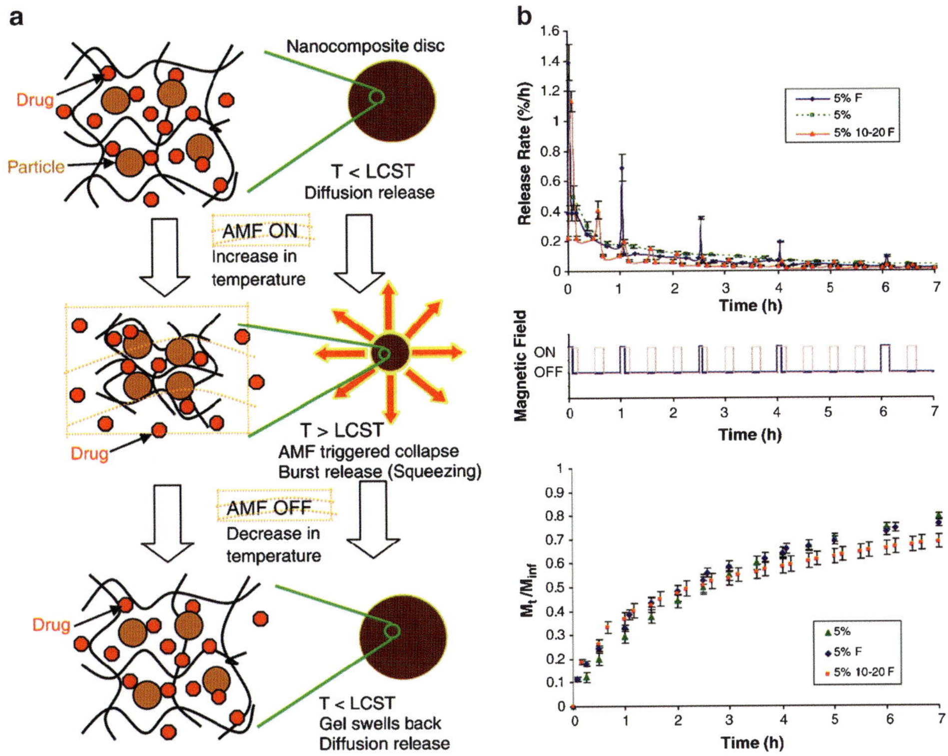

Figure 16.1. (**a**) Schematic showing the effect of on-off cycles of AMF on the release of chemical compounds from magnetic nanocomposites of PNIPAAm, (**b**) vitamin B_{12} release from magnetic nanocomposites of PNIPAAm on pulsed application of AMF. The symbol "%" represents particle loading by weight. *F* represents application of AMF. M_t represents cumulative mass released at time *t*. M_{inf} represents cumulative mass released over 48 h. Reprinted with permission from [24].

When the field was switched off, the pores opened and the drug was released, as shown in Figure 16.2a [46]. Studies with different sized particles showed that hydrogels with larger sized particles exhibited better actuation profiles due to stronger saturation magnetization and smaller coercive forces. Overall, the release profiles showed reduced drug release on application of AMF. Further studies with the system tested composites with various particle loadings in an attempt to achieve the optimum magnetic-responsive behavior. Various system parameters like permeability coefficient, partition coefficient, space restriction, and magnetization were investigated [45]. A similar effect with a DC magnetic field was observed on drug release from magnetic nanocomposites of gelatin hydrogels [41]. In another study, nanocomposites were fabricated by in situ synthesis of magnetic particles using a coprecipitation method [43]. Vitamin B_{12} was used as a model drug for the release study; reduced drug release was obtained by application of the magnetic field, as shown in Figure 16.2b.

Modulation of Drug Release with Light

Near-infrared (near-IR) light is an alternative method to achieve remote actuation of materials and devices. Of particular interest is near-IR light with wavelengths between 800 and 1,200 nm, which can be transmitted through tissue for short (e.g., a few centimeters)

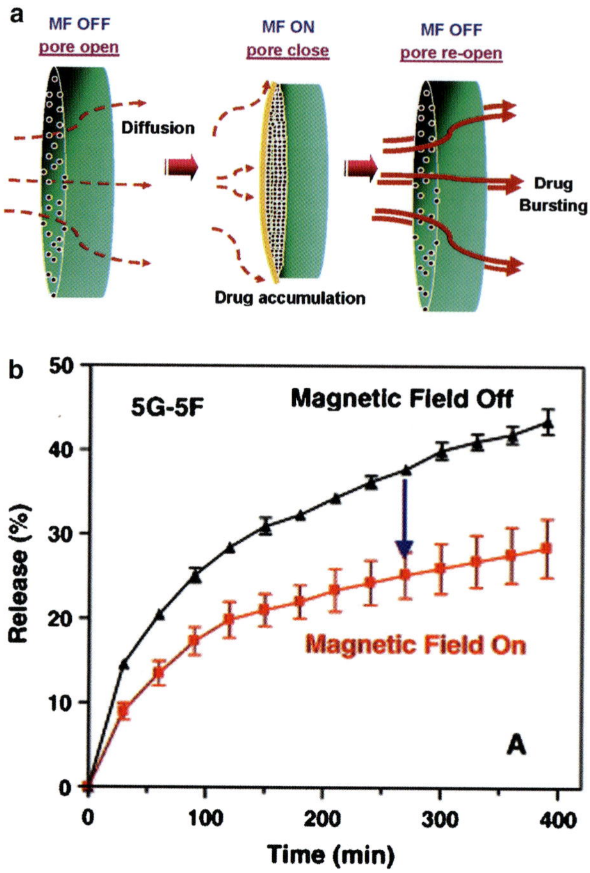

Figure 16.2. (**a**) Schematic illustration of the mechanism of the "closed" configuration of ferrogels due to aggregation of Fe_3O_4 nanoparticles under "on" magnetic fields, a condition that causes the porosity of ferrogels to decrease, (**b**) vitamin B_{12} release profiles with and without applying a magnetic field. Reprinted with permission from [43, 46].

distances with very little attenuation. Nanoshells have been developed to obtain selective absorbance of light in the near-IR region with potential applications for contrast agents, drug delivery, photothermal ablation, and immunoassays [72]. For example, gold nanoshells consisting of optically active nanoparticles with a dielectric core surrounded by a thin layer of gold were fabricated. The optical absorbance properties of these nanoparticles can be tuned by appropriate choice of the geometry and composition of the nanoshells. West and coworkers incorporated nanoshells in PNIPAAm-co-acrylamide hydrogels and subjected the composite to near-IR light. The nanoshells selectively absorbed the near-IR light and generated heat, which led to heating of the composite. The generated heat increased the composite temperature above the LCST, leading to hydrogel collapse. The extent of such a collapse could be tailored by varying the nanoshell concentration and the strength of the laser light. The results of this study (Figure 16.3) showed release of insulin and lysozyme from SiO_2–gold composite hydrogels on irradiation by a laser of wavelength 808 nm [47]. This approach demonstrated RC drug delivery with different drugs and RC activation of valves in a microfluidic device [25, 47, 73].

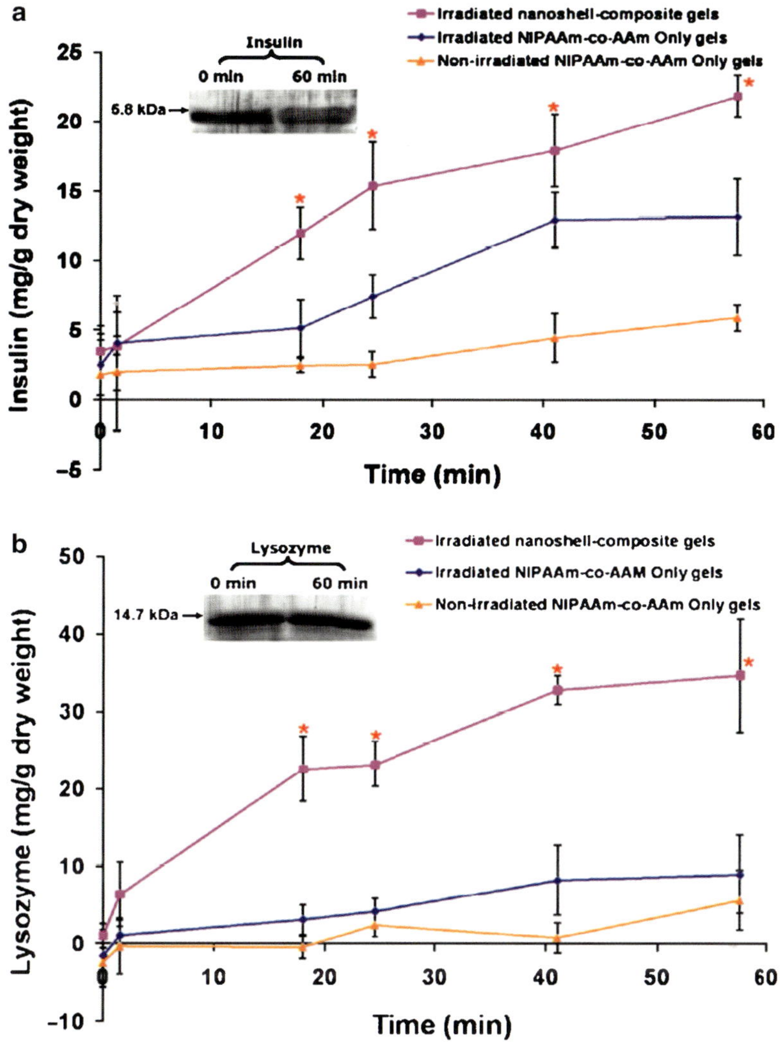

Figure 16.3. Release of (**a**) insulin and (**b**) lysozyme from PNIPAAM-co-acrylamide hydrogels as a function of time; *mg/g dry weight* represents milligrams of drug released per gram of polymer dry weight. Reprinted with permission from [47].

16.2.2. Hydrogel–Clay Nanocomposites for Enhanced Drug-Release Profile

In addition to remote actuation, nanoparticles can also be utilized to manipulate transport properties of hydrogel networks. Clay is one of the most widely studied filler materials that affect the hydrophilicity of hydrogel networks in order to modulate drug release. Haraguchi and co-workers synthesized nanocomposite hydrogels with water-swellable inorganic clay [74]. Clay nanoparticles acted as active components of the network structure and allowed additional control of the crosslink density as well as of the inter-crosslinking molecular weight independently. Incorporation of clay into hydrogel matrices resulted in composites with unique properties including high mechanical strength, better optical properties, and controlled response to stimuli [29, 75–78]. Hectorite, hydrotalcite, montmorillonite, and synthetic mica are some of the examples of clay incorporated in hydrogels. The mechanism of formation of clay–PNIPAAm composites was studied in order to investigate the reasons for the unique nanocomposite properties [79]. These studies revealed that polymerization proceeded on the clay particles, which act as multifunctional crosslinking agents.

Lee and Chen prepared nanocomposites of poly(acrylic acid-*co*-poly[ethylene glycol] methyl ether acrylate) hydrogels with hydrotalcite clay [49]. The results showed that clay incorporation affected slightly the pH swelling response of the hydrogels; moreover, the gels became more hydrophilic with increased clay content (Figure 16.4a). Hydrotalcite clay has a positive charge on its surface, while acrylic acid networks have a negative charge. Hence, incorporation of various amounts of clay could control the charge of the nanocomposite. "Drug"-release studies were conducted using model chemical compounds with different charge properties to study the effects of the "drug"–composite interaction. Figure 16.4b

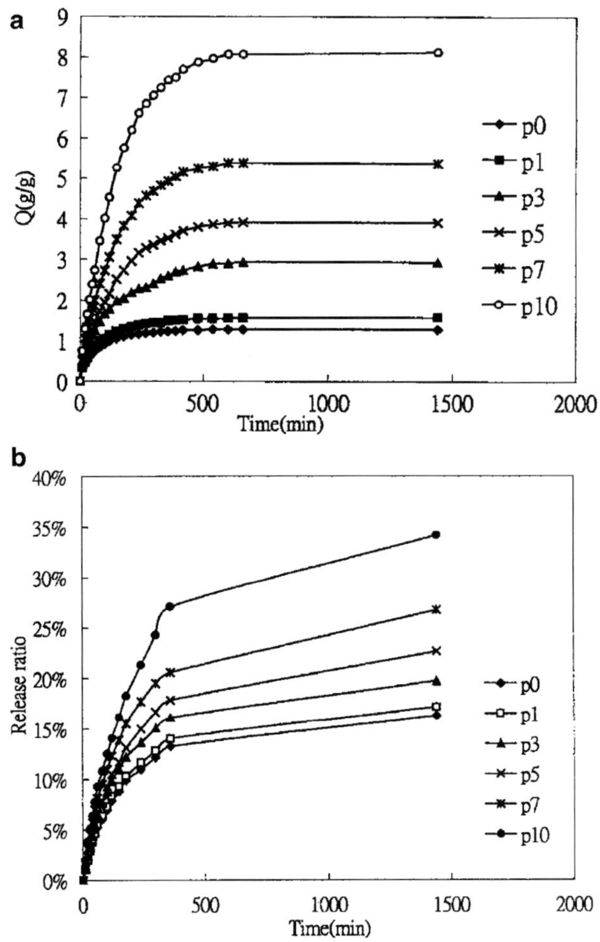

Figure 16.4. (**a**) Swelling ratio of acrylic acid-based hydrogels with time for different hydrotalcite clay loadings at 25°C in saline solution, (**b**) time course of crystal violet release profiles from a nanocomposite in saline solution at 32°C. "*p*" indicates wt% of hydrotalcite clay in the hydrogel tested. Reprinted with permission from [49].

shows that the release profile of crystal violet, a cationic model chemical compound, can be altered significantly by clay loading [49].

In another study, the charge properties of PNIPAAm were adjusted by using anionic clay (mica) and/or a cationic monomer (trimethyl [acrylamido propyl] ammonium iodide). Various properties for the resultant nanocomposite, including temperature-responsive swelling and drug-release behavior, were studied. In that case, "drug"-release behavior depended on the charge properties of the clay and the chemical compounds tested [50].

16.3. Hydrogel Nanocomposites for Tissue-Engineering Applications

Composite hydrogels and their application in tissue engineering is a relatively new research area with extensive growth in the last few years. The objective of tissue engineering is to repair or replace damaged and diseased tissue, thus increasing tissue and organ function and enhancing the quality of life for the patient [23]. Usually, tissue-engineering platforms use a synthetic material as the scaffold, which directs molecular signals and native cells to carry out the desired tissue regeneration. These scaffolds must provide the mechanical support necessary for the intended application and direct cell function pertinent to new tissue formation, usually through the incorporation of growth factors. Composite hydrogels with properties tailored for specific applications can be synthesized. The following sections highlight the recent research endeavors and achievements in this field.

16.3.1. Hydrogel Nanocomposites for Bone Tissue Engineering

Large bone segments can be lost or damaged through injury, infection, and tumor removal; the most reliable forms of treatment currently available have high success rates but many disadvantages [80]. The most common technique used is autografting, which involves removing bone from another anatomical site and placing it into the damaged site [81]. Though successful, this method has its limitations in the amount of bone that can be extracted and the morbidity associated with the donor site. If either a human cadaver or animal donor are used, the issue of immune response becomes an additional concern [80, 81]. It is important to note that natural bone has an inherent nanocomposite structure containing an organic component (primarily, type I collagen), and an inorganic component, HA (Figure 16.5) [81, 82]. Hydrogel nanocomposites have been designed to meet some of the specifications of natural bone and have been tested for tissue-engineering applications.

HA Composite Applications

The most important criterion for success of tissue-engineered bone implants is the ability to induce new bone formation and provide a tight bond between the implant and the new bone. For this reason, bioactive or osteoconductive components have been incorporated into the hydrogel. For example, composites have been formulated that contain calcium phosphates [83], such as HA ($Ca_{10}[PO_4]_6[OH]_2$) which is found intrinsically in bone [81, 84]. Incorporation of HA promotes the osteoconductivity of the composite and may also result in mechanical properties close to those of bone. Demirtas and others incorporated 25-45 μm HA into a super porous hydrogel (SPH) and formulated a SPH composite (SPHC) with an HA concentration of 1,000 mg HA per 1,000 μL monomeric mixture (Figure 16.6) [32].

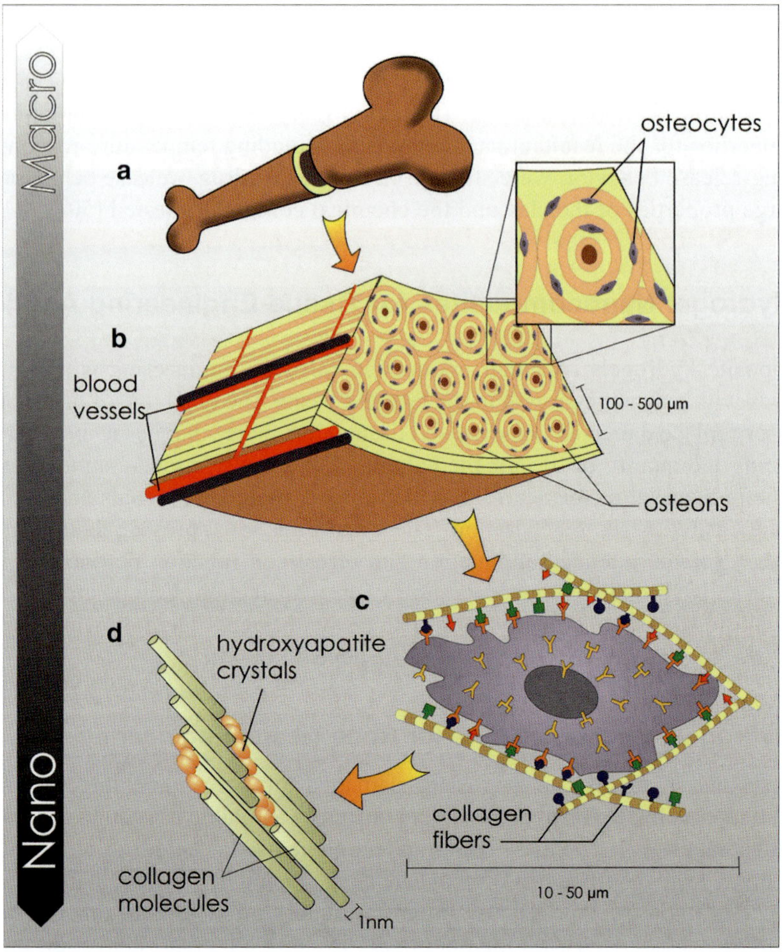

Figure 16.5. Hierarchical organization of bone over different length scales. Bone has a strong calcified outer compact layer (**a**), which encloses many cylindrical Haversian systems, or osteons (**b**). Bone cells express various cell membrane receptors with which they adhere to specific binding sites (**c**) of the well-defined nanoarchitecture of the surrounding extracellular matrix (**d**). Reprinted with permission from [82].

These hydrogel composites underwent swelling studies to determine the optimum HA concentration at which the pores of the construct remained open for nutrient and metabolic waste transport. Incorporation of HA significantly increased the mechanical properties of the construct: the compressive moduli increased from 0.63 ± 0.04 N/mm^2 for the SPH to 6.59 ± 0.35 N/mm^2 for the SPHC. The cytocompatibility of SPHC was confirmed using fibroblasts in vitro and SPHC did not release cytotoxic compounds. Other research groups formulated composites in order to enhance the material strength of composites such as hydrogels. Degirmenbasi et al. incorporated nanohydroxyapatite (n-HAp) into either PVA, collagen, or PVA/collagen composites [31]. The viscoelastic properties of these constructs were significantly increased by the incorporation of the n-HAp, and were further enhanced through cryogenic treatments.

The Hutchens group explored using polymers as a means of simply arranging the HA to mimic that found in natural bone and to induce new tissue formation [33]. Bacterial cellulose

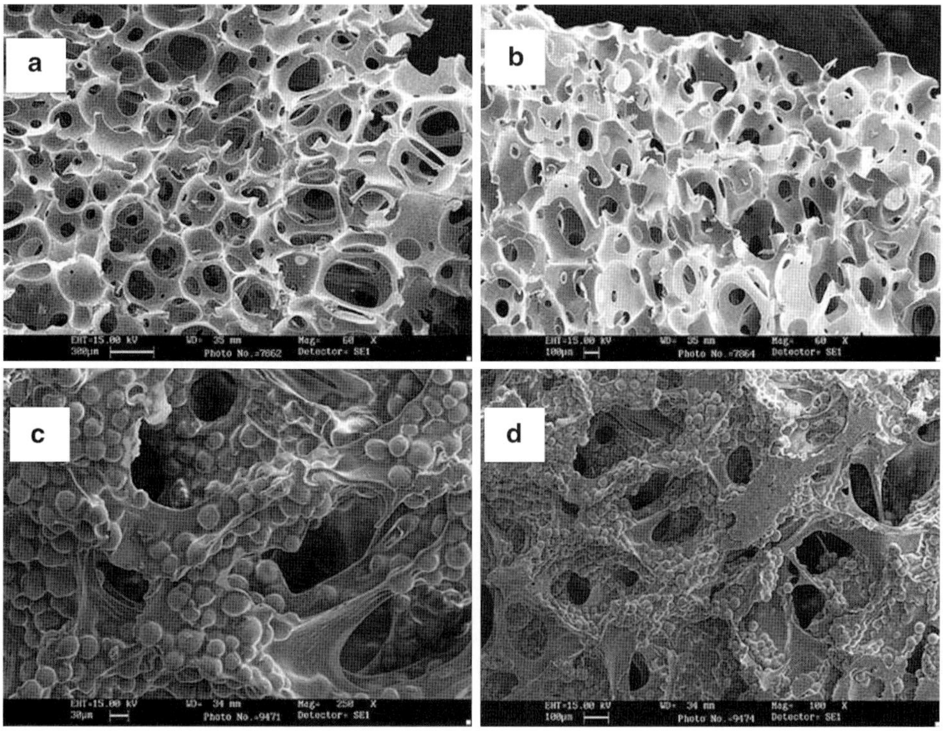

Figure 16.6. Scanning electron micrographs of: (**a**) the SPH surface (original magnification: ×60), (**b**) SPH cross-sections (original magnification: ×60), (**c**) SPHC surface (original magnification: ×250), and (**d**) SPHC cross-section (original magnification: ×100). Reprinted with permission from [32].

was used because it is biocompatible and has a degradation rate that could be altered chemically. This research group chose to use calcium-deficient HA because it would both promote bone growth and degrade over time as new bone tissue was formed. The results of this study provided evidence that the system tested promoted new bone formation. Other researchers incorporated HA into HYAFF 11 hydrogel networks (which are composed mainly of hyaluronan) and reported increased mechanical properties pertinent to application in bone tissue engineering [35].

TCP Composite Applications

TCP is used to increase the osteoconductivity of in bone tissue-engineering constructs [84]. The Cai group formulated a novel hydrogel composite of oxidized alginate, gelatin, and TCP and studied its properties in vitro [21]. Because this gel exhibited controllable degradation properties (based on the degree of oxidation of the alginate), this substrate is suitable for tissue engineering and drug-release applications. As evidenced by cell viability (Figure 16.7a) and alkaline phosphatase activity (Figure 16.7b), these gels were cytocompatible.

Other composites were formulated using TCP and growth factors that are osteoinductive stimuli [38]. The Luginbuehl group used a composite of alginate gel and TCP containing poly(lactic-co-glycolic acid) microspheres, which released the insulin-like growth factor (IGF)-I. These composites exhibited increased stiffness and decreased gelling times with

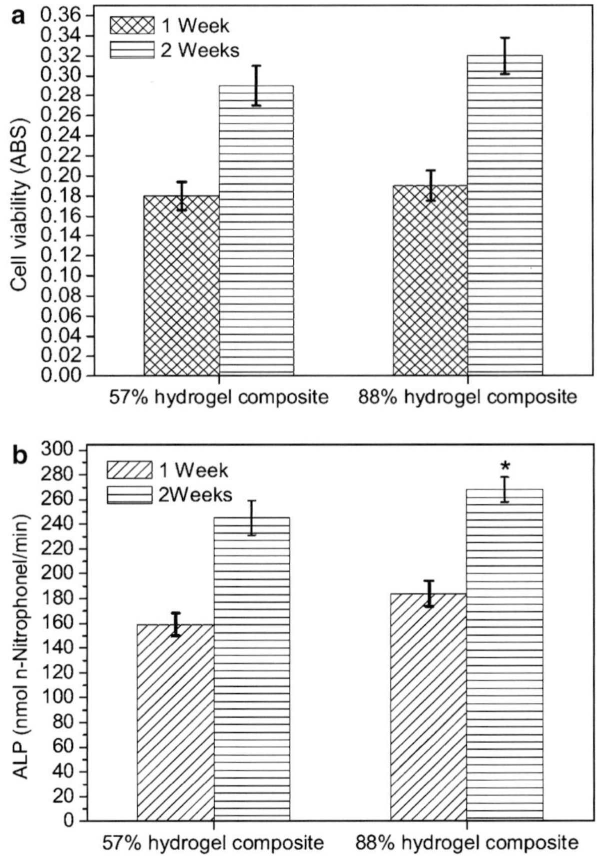

Figure 16.7. (**a**) Formazan absorbance was used as an index of viability of osteoblasts (using the MTT assay) cultured either on control substrates or encapsulated into alginate hydrogel composites of different oxidation degrees indicated as percentage ($n = 6$), (**b**) Alkaline phosphatase activity of osteoblasts cultured either on substrates or encapsulated into the alginate hydrogel composite of different oxidation degrees ($n = 6$). The hydrogel composites were prepared using equal volumes of 20% oxidized sodium alginate solution/15% gelatin solution and TCP in the presence of 0.1 M borax. The weight ratio of the three components correspond to the ratio 4/3/1. Reprinted with permission from [21].

increasing TCP content as well as cytocompatibility. The Sanginario group incorporated TCP into HYAFF 11 hydrogels and reported increased mechanical properties [35].

The Trojani group fabricated nanocomposites using biphasic calcium phosphate particles (a combination of HA and TCP) in a Si-hydroxypropylmethylcellulose matrix [51]. Incorporation of bone marrow stromal cells on these substrates promoted new bone formation in vivo. These constructs were implanted beneath the dorsal skin and intramuscularly in the hind leg of mice. After 8 weeks of implantation, new bone had fully filled seven of the ten original implants; the remaining implants had significant bone growth. The seven implants with new bone had an average of 30.2 ± 5.2 multinucleated cells/mm^2 and 18.2 ± 3.5 vessels/mm^2, indicating that they induced bone growth. Such constructs have potential for bone-engineering applications.

16.3.2. Hydrogel Nanocomposites for Articular Cartilage Tissue Engineering

Adult cartilage defects and damage have become an ever-increasing problem as the human population ages and as individuals live longer and more active lives. There are several forms of cartilage in the human body; the focus of most engineering research, however, has been on articular cartilage because of patient needs and the potential of providing the most benefits to patient care [85]. Articular cartilage is a form of hyaline cartilage that is present in diarthroidial joints, covering the areas of two bones that come in contact during daily activity; for normal function in the body, cartilage must be able to withstand both compressive and tensile forces. Natural articular cartilage is composed of type II collagen fibers and a high concentration of proteoglycans, giving it the needed mechanical integrity [86]. Cartilage can be damaged by injury as well as by several degenerative and rheumatic diseases [86]. These problems are further compounded by the inability of the cartilage tissue to recover and repair itself [85, 86].

Due to the fact that cartilage is attached to areas of bone contact, it is important that the engineered tissue bind tightly to bone. To aid in this effort, many composite materials include HA because of its osteoconductive properties. Wang and others fabricated a composite hydrogel from HA, PVA, and gelatin [36]. This research group found the system to be very porous and water-swellable, thus aiding in both the nutrient/waste transport and the dampening of vibrations. After immersion in simulated body fluid, a mineral (similar to that of bone apatite) was deposited on the surface and pore walls of the composite hydrogel; this result indicated that the composite could be osteoconductive.

Bioglass has been used extensively in tissue engineering because of its good bioactivity [20]. In addition, bioglass degradation products promote upregulation of seven families of genes, thus affecting subsequent cell responses to the construct [87]. Xu and others worked specifically with a PVA/bioglass composite for articular cartilage repair in order to increase the mechanical strength of the implant [20] and reported that the bioglass percent content in the gel was directly correlated to the elastic compression modulus of the material (Figure 16.8).

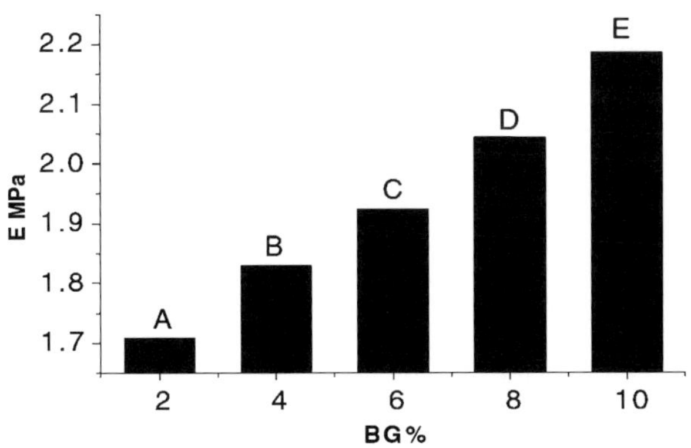

Figure 16.8. Effect of Bioglass content on the elastic compression modulus of 25 wt% PVA composite hydrogels. Reprinted with permission from [20].

16.3.3. Hydrogel Nanocomposites for Cornea Applications

Corneal disease, which may result in blindness, can be the end result of injury or disease of the eye. Artificial corneas have been developed to restore vision to affected patients; the success rate of these interventions, however, is low due to undesirable binding between the host tissue and the keratoprosthesis [34, 88]. Nanocomposite hydrogels are designed and synthesized to have the desired mechanical and bioactive properties to meet the requirements of being an artificial cornea fringe. Several research groups have used HA in test constructs (because of its bioactivity) and enhanced the bond between the construct and biological tissue [34, 37, 89]. Xu and others formulated a hydrogel of PVA with incorporated n-HAp [37]. Incorporation of HA increased the mechanical properties up to the point to 10 wt% n-HAp loading (Table 16.2). Increased HA content further increased the compressive strength, but the tensile strength and elongation of the composite decreased. To determine the ability to bind to cells, the researchers conducted cell attachment and proliferation assays. Compared with pertinent results obtained on the nanocomposite, rabbit corneal fibroblast adhesion was greater on the composite fringe; this result indicated an advantage of using the bioactive composite.

Earlier work by Kobayashi et al. used a system with collagen immobilized on the PVA surface [34]. Although not a nanocomposite (the dimensions of the HA used ranged from 0.3 to 4 μm), the degradation rate could be controlled by the degree of crystallinity of the HA. Chick embryo keratocyte-like cells adhered strongly, demonstrating that this composite had potential for strong adhesion to the natural tissue (Figure 16.9); this composite, therefore, has potential as a biomaterial for this tissue-engineering application [34].

16.3.4. Hydrogel Nanocomposites for Cell Adhesion Applications

Adhesion of anchorage-dependent cells and subsequent functions (such as proliferation, etc.) on material surfaces is very important for many tissue-engineering applications. Haraguchi and others developed a clay nanocomposite that allows the culture of cells on the composite surface and later removal of cell sheets by simply changing the temperature [52]. These researchers created PNIPAAm/clay composites and reported that incorporation of clay improved the mechanical and swelling properties compared with the pertinent properties of PNIPAAm gels alone. In addition, human hepatoma cells (HepG2), fibroblasts, and human umbilical vein endothelial cells proliferated only on the surfaces of gels of intermediate clay concentrations. The nanocomposite gels also exhibited temperature sensitivity that allowed HepG2 and fibroblast cell sheet removal without the addition of trypsin; by simply lowering the medium temperature from 37°C (a hydrophobic domain) to 10–20°C (a hydrophilic domain),

Table 16.2. Summary of the mechanical properties of PVA and n-HAp/PVA materials ($n = 5$).

Construct (10 wt% PVA)	Tensile strength (kPa)	Elongation-at-break (%, $\times 10^{-2}$)	Compressive strength (kPa)
n-HA/PVA (0 wt% n-HA)	1.05 ± 0.06	170 ± 14	0.12 ± 0.01
n-HA/PVA (5 wt% n-HA)	1.17 ± 0.09	120 ± 9	0.16 ± 0.02
n-HA/PVA (10 wt% n-HA)	1.70 ± 0.10	210 ± 18	0.30 ± 0.02
n-HA/PVA (20 wt% n-HA)	1.21 ± 0.08	195 ± 16	0.41 ± 0.04

Reprinted with permission from [37]

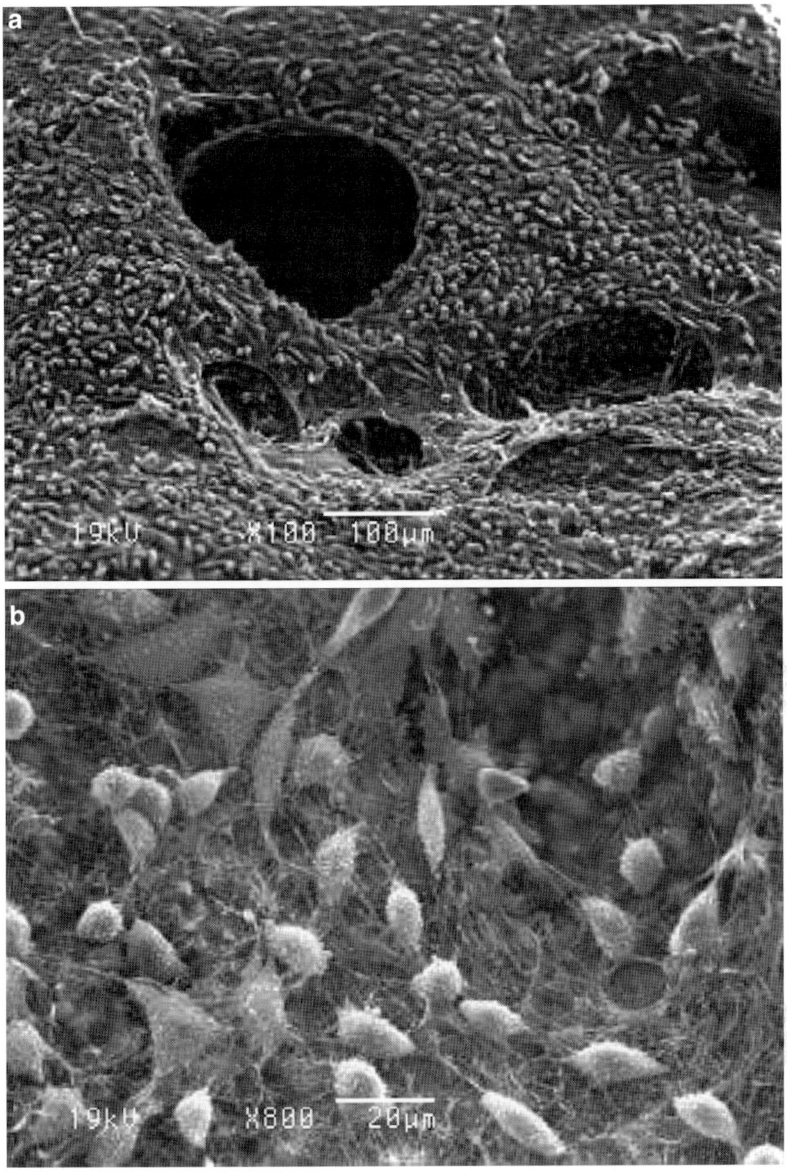

Figure 16.9. Scanning electron micrographs of keratocyte-like cells adhering on PVA–collagen–hydroxyapatite 7 days after cell seeding. (**a**) keratocyte-like cells fully covered the surface of the composite. Original magnification: ×100, (**b**) keratocyte-like cells were well-spread on the composite surface. Original magnification: ×800. Reprinted with permission from [34].

the cell sheets were easily detached from the gel surface [52]. The properties imparted to the hydrogels tested (as a result of the incorporation of clay) make this composite system of potential interest for a variety of applications in biomedicine.

In another interesting study, Schiraldi and others formulated a composite of hydroxyethyl-methacrylate (HEMA) with incorporated fumed silica nanoparticles as filler [53] and reported that increased percentage of silica decreased the swelling behavior, but increased the mechanical

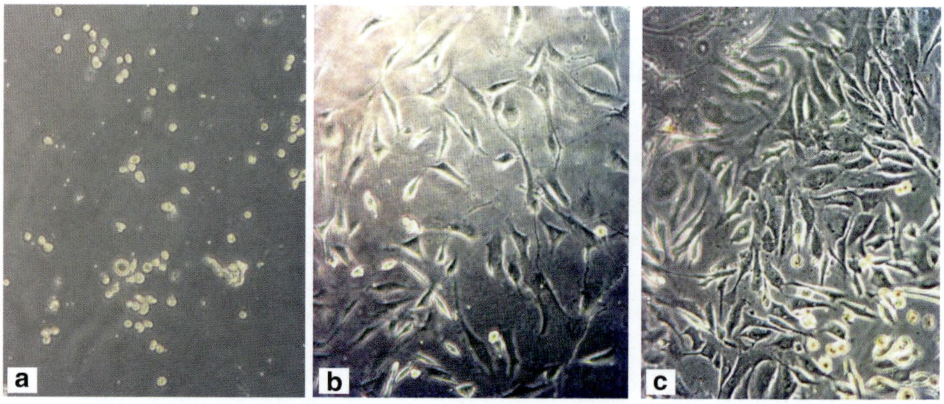

Figure 16.10. Phase contrast micrographs of human osteoblasts cultured on pHEMA (**a**), on pHEMA + 10% fumed silica (**b**), and pHEMA + 23% fumed silica (**c**) for 24 h. Original magnification: ×100. Reprinted with permission from [53].

properties, of the constructs. Most importantly, incorporation of silica in the poly(hydroxyethyl methacrylate) (pHEMA) gels increased the adhesion of human osteoblasts from 5% (on plain pHEMA) to 70% (on the composite) of the polystyrene control (Figure 16.10).

16.4. Hydrogel Nanocomposites for Other Therapeutic Applications

In addition to drug delivery and tissue engineering, hydrogel nanocomposites are also attractive for various other therapeutic applications, such as antimicrobial materials and thermal therapy. For example, antimicrobial activity of hydrogels can arise from the incorporation of silver nanoparticles. Thermal therapy for cancer treatment is based on the concept of heating magnetic nanoparticles remotely by magnetic fields. This section briefly addresses the aforementioned two applications.

16.4.1. Antimicrobial Applications

Metallic nanoparticles are toxic to microorganisms. Silver nanoparticles exhibit antimicrobial effects by binding to microbial DNA, preventing bacterial replication, and also causing inactivation of bacteria function [58, 90]. Different methods of incorporating silver nanoparticles in hydrogels have been reported in the literature. Nanocomposites can be synthesized by incorporation of particles prior to polymerization, during polymerization, or after polymerization of hydrogels [30, 91]. Incorporation of nanoparticles after polymerization often involves synthesis of hydrogel systems, swelling gels in a silver salt, followed by reduction to form nanoparticles in the matrix (Figure 16.11a). This method has been adopted to synthesize nanocomposites of various types of hydrogel systems, including poly(acrylamide-co-acrylic acid), poly(NIPAAm-co-sodium acrylate), and poly(vinyl pyrrolidone) interpenetrated with poly(acrylamide) [55–58]. The resultant hydrogel nanocomposites demonstrated excellent antibacterial effects when tested using *Escherichia coli* [55, 57]. In fact, the

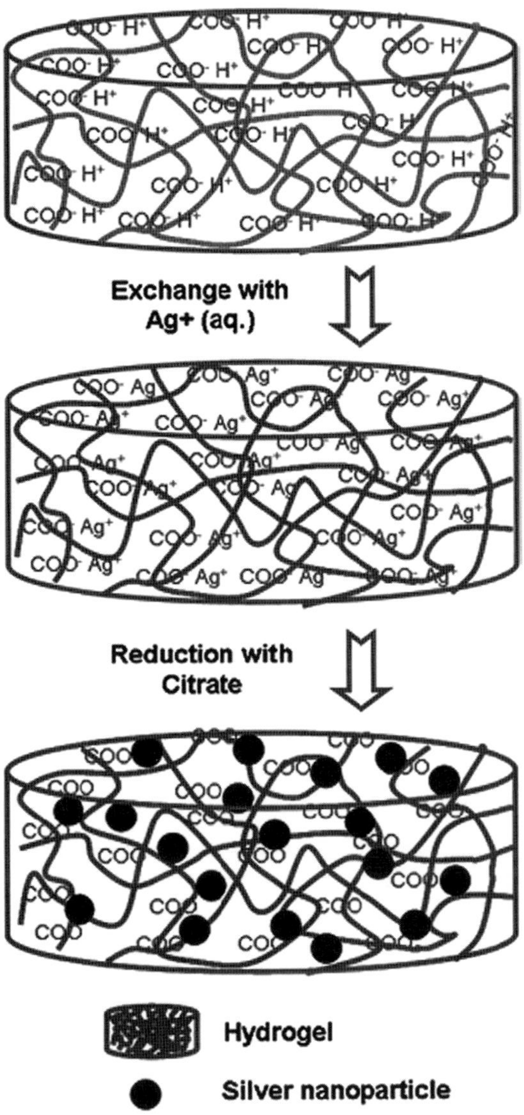

Figure 16.11. (**a**) Schematic illustration of the formation of silver nanoparticles in a swollen acrylamide-co-acrylic acid hydrogel network. Note: *Ag + (aq.)* represents aqueous solution of silver ions

antibacterial activity of these composites was enhanced by reducing the nanoparticle size and increasing the nanoparticle loading [58].

Lee and Tsao first synthesized silver nanoparticles, mixed them with an acrylic acid-based monomer solution, and then obtained nanocomposite films by photopolymerization [54]. The effect of particle concentration on various properties including swelling kinetics, gel strength, electrical conductivity, and antimicrobial activity was studied. Increased nanoparticle concentration resulted in increased antimicrobial activity of the construct (Figure 16.11b) [54].

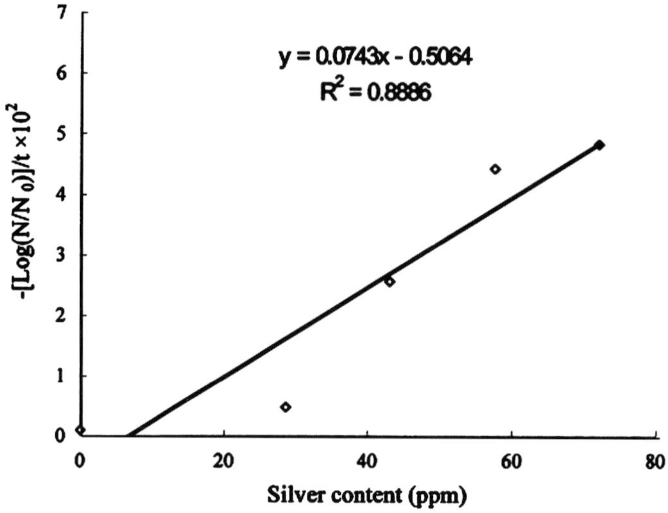

Figure 16.11. (continued) (**b**) initial rate of bacterial inactivation ($-\log[N/N_0]/t \times 10^2$) as a function of the silver content in the acrylic acid-based hydrogels. Reprinted with permission from [54, 58].

16.4.2. Thermal Therapy Applications

In the last few years, hyperthermia, which uses heat as a therapy, has attracted much interest in the treatment of cancer cells. It is interesting to note that hyperthermia is actually one of the oldest cancer therapies known [92]. Recent advances in thermal therapy systems (e.g., those based on nanoparticle systems) allow better control of the spatial and temporal delivery of thermal energy. As a result, there has been renewed interest in investigating the effect of hyperthermia in combination with radiation, chemotherapy, immunotherapy, and drug-delivery modalities [93–95].

Hilt and coworkers incorporated superparamagnetic Fe_3O_4 nanoparticles in PNIPAAm hydrogels and applied an AMF. Heating of the magnetic particles led to heating the hydrogel matrix [24, 39, 59]. The particle loading in the hydrogel matrix was varied to examine the resultant heating effect. Figure 16.12 shows the surface temperatures of dry PNIPAAm discs in response to 297 kHz frequency and 2.98 kA m^{-1} strength [59]. An increase in particle loadings increased the heating effect; the resulting temperature profile was also governed by heat transfer in the material. This system can potentially be useful for hyperthermia treatment [24, 39, 59].

In other studies, iron oxide particles of different sizes were incorporated in different hydrogel systems and an AMF was applied to achieve remote heating. The effect of variation of the magnetic field strength and particle loadings on the resultant heating was also studied [60–62]. When a magnetic organogel was subjected to the AMF, the specific absorption rate increased linearly with increasing ferrite content in the composite [61]. On the other hand, studies with PNIPAAm-based magnetic composites showed that increasing the AMF strength increased the rate of heating as well as the resultant temperatures [60].

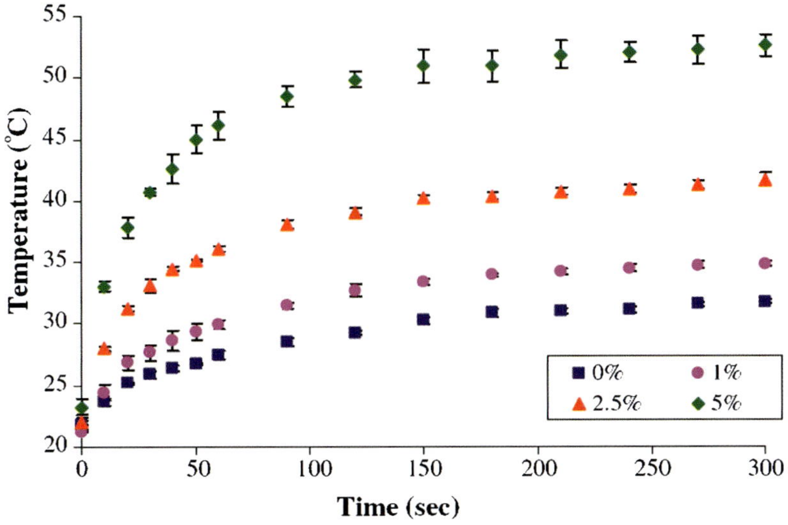

Figure 16.12. Temperature increase of nanocomposites with varying particle loadings subjected to AMF. The symbol "%" represents particle loading by weight in the PNIPAAm-based nanocomposite. Reprinted with permission from [59].

16.5. Hydrogel Nanocomposites and Biological Interactions

Hydrogel nanocomposites resemble soft tissue and typically are known for their good biocompatibility. Hydrogels are currently used in various biomedical applications such as contact lenses, drug-delivery devices, and implants for tissue regeneration [2, 96]. Nanoparticulates in their raw form, however, exhibit toxicity in certain situations [97–99]. For applications of hydrogel nanocomposites in contact with biological systems (e.g., as biomedical implants), it is essential to test the effects of such synthetic formulations on surrounding body tissues. Typically, evaluation of the biocompatibility of materials includes both in vitro and in vivo tests. In vitro analysis consists of a number of techniques including direct and indirect contact testing.

Direct contact testing exposes cells to the composite material and evaluates subsequent cell function (including death). Superporous hydrogel composites formulated by the Demirtas group were tested by direct contact with fibroblasts in vitro [32]. After 8 h of contact with the acrylamide-HA-based gels, there was neither cell death nor morphological changes in the fibroblasts tested. For this reason, these researchers concluded that the system was cytocompatible [32]. Meenach et al. [39] conducted cytotoxicity tests on magnetic particles and magnetic–NIPAAm composites using NIH 3T3 murine fibroblasts (Figure 16.13). Fibroblast viability in contact with the magnetic nanoparticles was approximately 75%. Cell viability on nanocomposites with 3 wt% magnetic particles and various crosslinking densities was about 90%. It appeared that the nanocomposite hydrogel matrix shielded the nanoparticles and retained cell viability.

To test the cytocompatibility of the HA/PVA/gelatin composite, Wang et al. evaluated the relative growth rate (RGR) in the extraction fluid (medium that had been in contact with the composite for 24 h) and the toxicity grade of the composite using 3-(4,5-dimethylthiazol-2-yl)-2,5-diphenyltetrazolium bromide (MTT) analysis [36]. The high RGR value indicated that the composite was cell compatible, and the toxicity grade result of 1 demonstrated that the composite was nontoxic. Xu et al. also used the MTT assay to determine that their n-HAp/PVA composite did not adversely affect cells [37]. Montmorillonite clay–chitosan

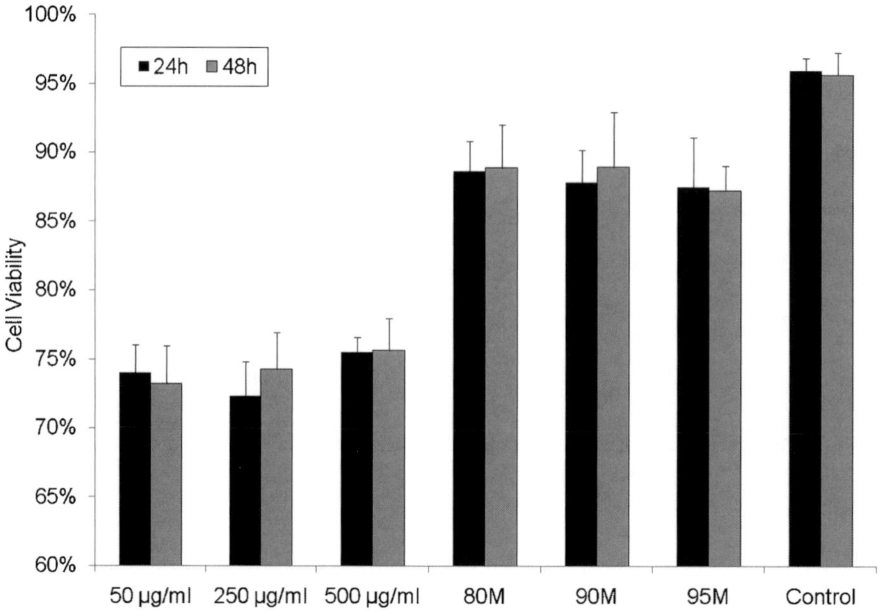

Figure 16.13. Percent viability of fibroblasts exposed to nanoparticles (at 50, 250, and 500 µg/ml), in magnetic hydrogel composites at varying crosslinking and control materials (polystyrene). Note: The numbers *80, 90,* and *95* indicate mol% of NIPAAm in NIPAAm-tetra (ethylene glycol) dimethacrylate hydrogels. *M* indicates the presence of 3 wt% Fe_3O_4 particles. Reprinted with permission from [39].

composite films were evaluated by Wang et al. for drug release and biocompatibility; MTT analysis showed no apparent cytotoxicity [100]. The Schiraldi group also determined the viability of cells on pHEMA-fumed silica gels [53] and reported that their systems were nontoxic for the murine fibroblasts and Raji cells tested.

Other cell function studies investigated cell proliferation on scaffold material. Kobayashi et al. reported that the PVA/HA/collagen composites supported strong cell adhesion [34]. The Xu group used a MTT assay to demonstrate that, compared with PVA alone, the n-HAp/PVA composite supported cell proliferation [37].

The Luginbuehl group determined alkaline phosphate (ALP) activity and reported that alginate/TCP/IGF-I composites induced bone growth in in vitro experiments [38]. The Cai group also used the ALP assay to demonstrate that their composite promoted enhanced bone formation [21]. The Wang group examined the compatibility of their HA/PVA/gelatin composite in rats; they reported an initial inflammatory response and fibrous capsule formation, which, after 12 weeks, grew into the pores of the implant [36].

In this section, biological interactions with hydrogel nanocomposites were reviewed. To date, the cytotoxicity of hydrogel nanocomposites has been studied in vitro; most of the studies indicated cytocompatibility of the nanocomposites. However, in vivo studies in animal models, which are a prerequisite for clinical studies, are yet to be investigated in detail.

16.6. Concluding Remarks

In this chapter, the new and rapidly emerging field of hydrogel nanocomposites was introduced. Hydrogel nanocomposites containing different types of nanoparticulates including iron oxide, clay, silver, gold, HA, and TCP were highlighted. The enhanced and unique properties of these nanocomposite biomaterials were emphasized along with some of the methods used for their preparation. The primary focus of this chapter was on the biological applications of hydrogel nanocomposites containing inorganic nanoparticulate materials.

Applications of hydrogel nanocomposites in drug delivery, tissue engineering, antimicrobial materials, and thermal therapy were discussed. In particular, the drug-delivery section discussed modulation of drug release from nanocomposites induced by different external stimuli such as AMF, DC magnetic fields, and near-IR light. The tissue-engineering section focused on applications of hydrogel nanocomposites in bone, articular cartilage, and cornea regeneration. These novel materials have the potential to greatly impact the biomedical field.

References

1. Lowman AM, Peppas NA. Hydrogels. In: Mathiowitz E, editor. Encyclopedia of Controlled Drug Delivery. New York: Wiley, 1999. pp. 397–418.
2. Peppas NA, Hilt JZ, Khademhosseini A, Langer R. Hydrogels in biology and medicine: from molecular principles to bionanotechnology. Adv Mater 2006;18:1345–1360.
3. Peppas NA, Bures P, Leobandung W, Ichikawa H. Hydrogels in pharmaceutical formulations. Eur J Pharm Biopharm 2000;5:27–46.
4. Hoffman AS. Hydrogels in biomedical applications. Adv Drug Deliv Rev 2002;43:3–12.
5. Chaterji S, Kwon IK, Park K. Smart polymeric gels: redefining the limits of biomedical devices. Prog Polym Sci 2007;32(8–9):1083–1122.
6. Ulijn RV, Bibi N, Jayawarna V, Thornton PD, Todd SJ, Mart RJ, et al. Bioresponsive hydrogels. Materials Today 2007;10(4):40–48.
7. Miyata T, Asami N, Uragami T. A reversibly antigen-responsive hydrogel. Nature 1999;399(6738):766–769.
8. Hilt JZ, Gupta AK, Bashir R, Peppas NA. Ultrasensitive Biomems sensors based on microcantilevers patterned with environmentally responsive hydrogels. Biomed Microdevices 2003;5(3):177–184.
9. Dong L, Jiang H. Autonomous microfluidics with stimuli-responsive hydrogels. Soft Matter 2007;3(10): 1223–1230.
10. Beebe DJ, Moore JS, Bauer JM, Yu Q, Liu RH, Devadoss C, et al. Functional hydrogel structures for autonomous flow control inside microfluidic channels. Nature 2000;404(6778):588–590.
11. Qiu Y, Park K. Environment-sensitive hydrogels for drug delivery. Adv Drug Deliv Rev 2001;53(3):321–339.
12. Schild HG. Poly(N-isopropylacrylamide): experiment, theory, and application. Prog Polym Sci 1992;17: 163–249.
13. Yoshida R, Sakai K, Okano T, Sakurai Y. Modulating the phase transition temperature and thermosensitivity in N-isopropylacrylamide copolymer gels. J Biomater Sci Polym Ed 1994;6:585–598.
14. Hoffman AS. Applications of thermally reversible polymers and hydrogels in therapeutics and diagnostics. J Control Release 1987;6(1):297–305.
15. Agarwal AK, Dong L, Beebe DJ, Jiang H. Autonomously-triggered microfluidic cooling using thermo-responsive hydrogels. Lab Chip 2007;7(3):310–315.
16. Goldberg M, Langer R, Jia X. Nanostructured materials for applications in drug delivery and tissue engineering. J Biomater Sci Polym Ed 2007;18:241–268.
17. Frimpong RA, Hilt JZ. Hydrogel nanocomposites for intelligent therapeutics In: Peppas NA, Hilt JZ, Thomas JB, editors. Nanotechnology in Therapeutics: Current Technology and Applications. Norfolk: Horizon Scientific, 2007. pp. 241–256.
18. Thomas V, Namdeo M, Mohan YM, Bajpai SK, Bajpai M. Review on polymer, hydrogel and microgel metal nanocomposites: a facile nanotechnological approach. J Macromol Sci, Part A: Pure Appl Chem 2008;45(1): 107–119.
19. Gattas-Asfura KM, Zheng Y, Micic M, Snedaker MJ, Ji X, Sui G, et al. Immobilization of quantum dots in the photo-cross-linked poly(ethylene glycol)-based hydrogel. J Phys Chem B 2003;107(38):10464–10469.
20. Xu H, Wang YJ, Zheng YD, Chen XF, Ren L, Wu G, et al. Preparation and characterization of bioglass/polyvinyl alcohol composite hydrogel. Biomed Mater 2007;2(2):62–66.
21. Cai K, Zhang J, Deng LH, Yang L, Hu Y, Chen C, et al. Physical and biological properties of a novel hydrogel composite based on oxidized alginate, gelatin and tricalcium phosphate for bone tissue engineering. Adv Eng Mater 2007;9(12):1082–1088.

22. Lee W-F, Fu Y-T. Effect of monmorillonite on the swelling behavior and drug-release behavior of nanocomposite hydrogels. J Appl Polym Sci 2003;89:3652–3660.
23. Biondi M, Ungaro F, Quaglia F, Netti PA. Controlled drug delivery in tissue engineering. Adv Drug Deliv Rev 2008;60(2):229–242.
24. Satarkar NS, Hilt JZ. Magnetic hydrogel nanocomposites for remote controlled pulsatile drug release. J Control Rel 2008;130(3):246–251.
25. Sershen SR, Westcott SL, Halas NJ, West JL. Temperature-sensitive polymer-nanoshell composites for photothermally modulated drug delivery. J Biomed Mater Res 2000;51(3):293–298.
26. Zrínyi M. Intelligent polymer gels controlled by magnetic fields. Colloid Polym Sci 2000;278(2):98–103.
27. Filipcsei G, Csetneki I, Szilagyi A, Zrinyi M. Magnetic field-responsive smart polymer composites. Adv Polym Sci 2007;206:137–189.
28. Okada A, Usuki A. Twenty years of polymer-clay nanocomposites. Macromol Mater Eng 2006;291(12): 1449–1476.
29. Haraguchi K. Nanocomposite hydrogels. Curr Opin Solid State Mater Sci 2007;11:47–54.
30. Bajpai SK, Mohan YM, Bajpai M, Tankhiwale R, Thomas V. Synthesis of polymer stabilized silver and gold nanostructures. J Nanosci Nanotechnol 2007;7:2994–3010.
31. Degirmenbasi N, Kalyon DM, Birinci E. Biocomposites of nanohydroxyapatite with collagen and poly(vinyl alcohol). Colloids Surf B Biointerfaces 2006;48(1):42–49.
32. Demirtas TT, Karakecili AG, Gumusderelioglu M. Hydroxyapatite containing superporous hydrogel composites: synthesis and in vitro characterization. J Mater Sci Mater Med 2008;19(2):729–735.
33. Hutchens SA, Benson RS, Evans BR, O'Neill HM, Rawn CJ. Biomimetic synthesis of calcium-deficient hydroxyapatite in a natural hydrogel. Biomaterials 2006;27(26):4661–4670.
34. Kobayashi H, Kato M, Taguchi T, Ikoma T, Miyashita H, Shimmura S, et al. Collagen immobilized PVA hydrogel-hydroxyapatite composites prepared by kneading methods as a material for peripheral cuff of artificial cornea. Mater Sci Eng C 2004;24(6–8):729–735.
35. Sanginario V, Ginebra MP, Tanner KE, Planell JA, Ambrosio L. Biodegradable and semi-biodegradable composite hydrogels as bone substitutes: morphology and mechanical characterization. J Mater Sci Mater Med 2006;17(5):447–454.
36. Wang MB, Li YB, Wu JQ, Xu FL, Zuo Y, Jansen JA. In vitro and in vivo study to the biocompatibility and biodegradation of hydroxyapatite/poly(vinyl alcohol)/gelatin composite. J Biomed Mater Res A 2008;85A(2):418–426.
37. Xu FL, Li YB, Deng YP, Xiong J. Porous nano-hydroxyapatite/poly(vinyl alcohol) composite hydrogel as artificial cornea fringe: characterization and evaluation in vitro. J Biomater Sci Polym Ed 2008;19(4):431–439.
38. Luginbuehl V, Wenk E, Koch A, Gander B, Merkle HP, Meinel L. Insulin-like growth factor I-releasing alginate-tricalciumphosphate composites for bone regeneration. Pharm Res 2005;22(6):940–950.
39. Meenach SA, Anderson AA, Suthar M, Anderson KW, Hilt JZ. Biocompatibility analysis of magnetic hydrogel nanocomposites based on poly(N-isopropylacrylamide) and iron oxide. J Biomed Mater Res A, In Press.
40. Nayak S, Lyon LA. Soft nanotechnology with soft nanoparticles. Angew Chem Int Ed 2005;44(47):7686–7708.
41. Liu T-Y, Hu S-H, Liu K-H, Liu D-M, Chen S-Y. Preparation and characterization of smart magnetic hydrogels and its use for drug release. J Magn Magn Mater 2006;304:e397–e399.
42. Hu S-H, Liu T-Y, Liu D-M, Chen S-Y. Controlled pulsatile drug release from a ferrogel by a high-frequency magnetic field. Macromolecules 2007;40(19):6786–6788.
43. Hu S-H, Liu T-Y, Liu D-M, Chen S-Y. Nano-ferrosponges for controlled drug release. J Control Release 2007;121(3):181–189.
44. Reséndiz-Hernández PJ, Rodríguez-Fernández OS, García-Cerda LA. Synthesis of poly(vinyl alcohol)-magnetite ferrogel obtained by freezing-thawing technique. J Magn Magn Mater 2008;320(14):e373–e376.
45. Liu T-Y, Hu S-H, Liu K-H, Liu D-M, Chen S-Y. Study on controlled drug permeation of magnetic-sensitive ferrogels: effect of Fe3O4 and PVA. J Control Release 2008;126(3):228–236.
46. Liu TY, Hu SH, Liu DM, Chen SY. Magnetic-sensitive behavior of intelligent ferrogels for controlled release of drug. Langmuir 2006;22(14):5974–5978.
47. Bikram M, Gobin AM, Whitmire RE, West JL. Temperature-sensitive hydrogels with SiO2-Au nanoshells for controlled drug delivery. J Control Release 2007;123(3):219–227.
48. Sershen SR, West JL. Implantable, polymeric systems for modulated drug delivery. Adv Drug Deliv Rev 2002;54:1225–1235.
49. Lee W-F, Chen Y-C. Effect of hydrotalcite on the physical properties and drug-release behavior of nanocomposite hydrogels based on poly[acrylic acid-co-poly(ethylene glycol) methyl ether acrylate] gels. J Appl Polym Sci 2004;94(2):692–699.
50. Lee W-F, Tsao K-T. Effect of intercalant content of mica on the various properties for the charged nanocomposite poly(N-isopropyl acrylamide) hydrogels. J Appl Polym Sci 2007;104(4):2277–2287.

51. Trojani C, Boukhechba F, Scimeca JC, Vandenbos F, Michiels JF, Daculsi G, et al. Ectopic bone formation using an injectable biphasic calcium phosphate/Si-HPMC hydrogel composite loaded with undifferentiated bone marrow stromal cells. Biomaterials 2006;27(17):3256–3264.
52. Haraguchi K, Takehisa T, Ebato M. Control of cell cultivation and cell sheet detachment on the surface of polymer/clay nanocomposite hydrogels. Biomacromolecules 2006;7(11):3267–3275.
53. Schiraldi C, D'Agostino A, Oliva A, Flamma F, De Rosa A, Apicella A, et al. Development of hybrid materials based on hydroxyethylmethacrylate as supports for improving cell adhesion and proliferation. Biomaterials 2004;25(17):3645–3653.
54. Lee W-F, Tsao K-T. Preparation and properties of nanocomposite hydrogels containing silver nanoparticles by ex situ polymerization. J Appl Polym Sci 2006;100(5):3653–3661.
55. Murali Mohan Y, Lee K, Premkumar T, Geckeler KE. Hydrogel networks as nanoreactors: a novel approach to silver nanoparticles for antibacterial applications. Polymer 2007;48(1):158–164.
56. Murali Mohan Y, Premkumar T, Lee K, Geckeler KE. Fabrication of silver nanoparticles in hydrogel networks. Macromol Rapid Commun 2006;27(16):1346–1354.
57. Murthy PSK, Murali Mohan Y, Varaprasad K, Sreedhar B, Mohana Raju K. First successful design of semi-IPN hydrogel-silver nanocomposites: a facile approach for antibacterial application. J Colloid Interface Sci 2008;318(2): 217–224.
58. Thomas V, Yallapu MM, Sreedhar B, Bajpai SK. A versatile strategy to fabricate hydrogel-silver nanocomposites and investigation of their antimicrobial activity. J Colloid Interface Sci 2007;315(1):389–395.
59. Satarkar NS, Hilt JZ. Nanocomposite hydrogels as remote controlled drug delivery systems. Acta Biomater 2008;4:11–16.
60. Ang KL, Venkatraman S, Ramanujan RV. Magnetic PNIPA hydrogels for hyperthermia applications in cancer therapy. Mater Sci Eng C 2007;27(3):347–351.
61. Babincová M, Leszczynska D, Sourivong P, Cicmanec P, Babinec P. Superparamagnetic gel as a novel material for electromagnetically induced hyperthermia. J Magn Magn Mater 2001;225(1–2):109–112.
62. Lao LL, Ramanujan RV. Magnetic and hydrogel composite materials for hyperthermia applications. J Mater Sci Mater Med 2004;15:1061–1064.
63. Hoare TR, Kohane DS. Hydrogels in drug delivery: progress and challenges. Polymer 2008;49(8):1993–2007.
64. Lin C-C, Metters AT. Hydrogels in controlled release formulations: network design and mathematical modeling. Adv Drug Deliv Rev 2006;58(12–13):1379–1408.
65. Kikuchi A, Okano T. Pulsatile drug release control using hydrogels. Adv Drug Deliv Rev 2002;54:53–77.
66. Kost J, Langer R. Responsive polymeric delivery systems. Adv Drug Deliv Rev 2001;46:125–148.
67. Bussemer T, Otto I, Bodmeier R. Pulsatile drug-delivery systems. Crit Rev Ther Drug Carrier Syst 2001;18(5): 433–458.
68. Edelman ER, Kost J, Bobeck H, Langer R. Regulation of drug release from polymer matrices by oscillating magnetic fields. J Biomed Mater Res 1985;19(1):67–83.
69. Hsieh DS, Langer R, Folkman J. Magnetic modulation of release of macromolecules from polymers. Proc Natl Acad Sci U S A 1981;78(3):1863–1867.
70. Kost J, Wolfrum J, Langer R. Magnetically enhanced insulin release in diabetic rats. J Biomed Mater Res 1987;21:1367–1373.
71. Vaishnava PP, Tackett R, Dixit A, Sudakar C, Naik R, Lawes G. Magnetic relaxation and dissipative heating in ferrofluids. J Appl Phys 2007;102:063914.
72. Hirsch L, Gobin A, Lowery A, Tam F, Drezek R, Halas N, et al. Metal nanoshells. Ann Biomed Eng 2006;34(1):15–22.
73. Sershen SR, Mensing GA, Ng M, Halas NJ, Beebe DJ, West JL. Independent optical control of microfluidic valves formed from optomechanically responsive nanocomposite hydrogels. Adv Mater 2005;17(11): 1366–1368.
74. Haraguchi K, Takehisa T. Nanocomposite hydrogels: a unique organic-inorganic network structure with extraordinary mechanical, optical, and swelling/de-swelling properties. Adv Mater 2002;14(16):1120–1124.
75. Ma J, Xu Y, Zhang Q, Zha L, Liang B. Preparation and characterization of pH- and temperature-responsive semi-IPN hydrogels of carboxymethyl chitosan with poly (N-isopropyl acrylamide) crosslinked by clay. Colloid Polym Sci 2007;285(4):479–484.
76. Xiang Y, Peng Z, Chen D. A new polymer/clay nano-composite hydrogel with improved response rate and tensile mechanical properties. Eur Polym J 2006;42(9):2125–2132.
77. Kokabi M, Sirousazar M, Hassan ZM. PVA-clay nanocomposite hydrogels for wound dressing. Eur Polym J 2007;43(3):773–781.
78. Haraguchi K, Li H-J. Control of the coil-to-globule transition and ultrahigh mechanical properties of PNIPA in nanocomposite hydrogels. Angew Chem Int Ed 2005;44(40):6500–6504.

79. Haraguchi K, Li HJ, Matsuda K, Takehisa T, Elliott E. Mechanism of forming organic/inorganic network structures during in-situ free-radical polymerization in PNIPA-clay nanocomposite hydrogels. Macromolecules 2005;38(8):3482–3490.
80. Cancedda R, Giannoni P, Mastrogiacomo M. A tissue engineering approach to bone repair in large animal models and in clinical practice. Biomaterials 2007;28(29):4240–4250.
81. Stevens MM. Biomaterials for bone tissue engineering. Mater Today 2008;11(5):18–25.
82. Stevens MM, George JH. Exploring and engineering the cell surface interface. Science 2005;310(5751): 1135–1138.
83. Yuan HP, Van den Doel M, Li SH, Van Blitterswijk CA, De Groot K, De Bruijn JD. A comparison of the osteoinductive potential of two calcium phosphate ceramics implanted intramuscularly in goats. J Mater Sci Mater Med 2002;13(12):1271–1275.
84. Ge ZG, Jin ZX, Cao T. Manufacture of degradable polymeric scaffolds for bone regeneration. Biomed Mater 2008;3(2):22001.
85. Chung C, Burdick JA. Engineering cartilage tissue. Adv Drug Deliv Rev 2008;60(2):243–262.
86. Schulz RM, Bader A. Cartilage tissue engineering and bioreactor systems for the cultivation and stimulation of chondrocytes. Eur Biophys J 2007;36(4–5):539–568.
87. Pereira MM, Jones JR, Orefice RL, Hench LL. Preparation of bioactive glass-polyvinyl alcohol hybrid foams by the sol-gel method. J Mater Sci Mater Med 2005;16(11):1045–1050.
88. Legeais JM, Renard G. A second generation of artificial cornea (Biokpro II). Biomaterials 1998 ;19(16):1517–1522.
89. Sinha A, Das G, Sharma BK, Roy RP, Pramanick AK, Nayar S. Poly(vinyl alcohol)-hydroxyapatite biomimetic scaffold for tissue regeneration. Mater Sci Eng C 2007;27(1):70–74.
90. Sondi I, Salopek-Sondi B. Silver nanoparticles as antimicrobial agent: a case study on E. coli as a model for Gram-negative bacteria. J Colloid Interface Sci 2004;275(1):177–182.
91. Saravanan P, Padmanabha Raju M, Alam S. A study on synthesis and properties of Ag nanoparticles immobilized polyacrylamide hydrogel composites. Mater Chem Phys 2007;103(2–3):278–282.
92. Hildebrandt B, Wust P, Ahlers O, Dieing A, Sreenivasa G, Kerner T, et al. The cellular and molecular basis of hyperthermia. Crit Rev Oncol Hematol 2002;43(1):33–56.
93. Coffey DS, Getzenberg RH, DeWeese TL. Hyperthermic biology and cancer therapies: a hypothesis for the "Lance Armstrong effect". JAMA 2006;296(4):445–448.
94. Horsman MR, Overgaard J. Hyperthermia: a potent enhancer of radiotherapy. Clin Oncol 2007;19(6):418–426.
95. Meyer DE, Shin BC, Kong GA, Dewhirst MW, Chilkoti A. Drug targeting using thermally responsive polymers and local hyperthermia. J Control Release 2001;74(1–3):213–224.
96. Vihola H, Laukkanen A, Valtola L, Tenhu H, Hirvonen J. Cytotoxicity of thermosensitive polymers poly(N-isopropylacrylamide), poly(N-vinylcaprolactam) and amphiphilically modified poly(N-vinylcaprolactam). Biomaterials 2005;26(16):3055–3064.
97. Wick P, Manser P, Limbach LK, Dettlaff-Weglikowska U, Krumeich F, Roth S, et al. The degree and kind of agglomeration affect carbon nanotube cytotoxicity. Toxicol Lett 2007;168(2):121–131.
98. Müller K, Skepper JN, Posfai M, Trivedi R, Howarth S, Corot C, et al. Effect of ultrasmall superparamagnetic iron oxide nanoparticles (Ferumoxtran-10) on human monocyte-macrophages in vitro. Biomaterials 2007;28(9): 1629–1642.
99. Kirchner C, Liedl T, Kudera S, Pellegrino T, MunozJavier A, Gaub HE, et al. Cytotoxicity of colloidal CdSe and CdSe/ZnS nanoparticles. Nano Lett 2005;5(2):331–338.
100. Wang X, Du Y, Luo J. Biopolymer/montmorillonite nanocomposite: preparation, drug-controlled release property and cytotoxicity. Nanotechnology 2008;19(6):065707.

17

Protein and Cell Interactions with Nanophase Biomaterials

Courtney M. Creecy, David A. Puleo, and Rena Bizios

This chapter provides an overview of current knowledge regarding protein and cell interactions with nanophase materials, that is, materials with feature sizes of less than 100 nm in (at least) one dimension. Nanophase biomaterials promote select and specific interactions with biomolecules and, subsequently, support enhanced mammalian cell functions pertinent to new tissue formation. Although some advances have been made in elucidating aspects of these processes, many questions regarding fundamental processes controlling protein and cell interactions with nanophase biomaterials remain unanswered. Advantages and limitations of these material formulations are highlighted in this chapter; challenges and needs for further research that will elucidate the cellular and molecular processes, as well as pertinent underlying mechanisms, are outlined. These issues must be addressed and resolved before the full potential and promise of these novel biomaterial formulations can be realized.

Abbreviations

CO_2	carbon dioxide
DMEM	Dulbecco's modified Eagle medium
FBS	fetal bovine serum
HA	hydroxyapatite
PBS	phosphate-buffered saline
PLA	poly(L-lactic acid)
PMMA	poly(methyl methacrylate)
RGD	arginine–glycine–aspartic acid

C.M. Creecy and R. Bizios • Department of Biomedical Engineering, The University of Texas at San Antonio, San Antonio, TX 78249, USA
D.A. Puleo • Center for Biomedical Engineering, University of Kentucky, Lexington, KY 40506, USA

D.A. Puleo and R. Bizios (eds.), *Biological Interactions on Materials Surfaces*,
DOI 10.1007/978-0-387-98161-1_17,

17.1. Introduction

Cells of mammalian tissues are sensitive to their surrounding milieu: they exist and function by responding to cues from extracellular spatial domains and structures as well as from biochemical (such as chemical composition of their surroundings and soluble, diffusible factors) and biophysical (such as mechanical) stimuli. In the past, research provided important information regarding pertinent phenomena but was limited regarding the level or scale at which these investigations were conducted, specifically, macro- and microlevels. In comparison, the relevant feature size in physiological tissues (such as bone) and of biomolecules are on the order of nanometers. For example, the soluble plasma protein fibrinogen has a length of 40–50 nm and a width of 4–10 nm, while collagen fibers in the extracellular matrix are approximately 60 nm in diameter. For a long time, such scale could not be reproducibly duplicated in vitro.

The advent of nanotechnology made available nanophase biomaterials with surface features in the nanometer level, that is, on the order of those present in the physiological environment. These developments inspired new technologies over a broad spectrum of biomedical applications [1]. In the case of biomaterials for implant devices, where success is often determined by integration into surrounding tissues and formation of new functional tissue, pertinent research sought to elucidate interactions on nanophase biomaterials at the cellular/molecular level.

This chapter provides an overview of current knowledge regarding protein and cell interactions with nanophase materials, which by definition have feature sizes with (at least) one dimension of less than 100 nm [2]. The advantages and limitations of these material formulations are highlighted; challenges and needs for further research that will elucidate the cellular and molecular processes, as well as pertinent underlying mechanisms of critical importance at the tissue/implant interface are outlined in the sections that follow.

17.2. Protein Interactions with Nanophase Materials

Since proteins are the first components of the biological milieu to rapidly interact with biomaterials that come in contact with body fluids and tissues [3], the presence and function of macromolecules and the subsequent cell/tissue interactions they mediate at the surfaces of devices play crucial roles in the long-term performance and fate of implants. Recognition of this fundamental aspect of events at tissue/biomaterial interfaces has motivated extensive studies of protein type, structure, and function on biomaterial surfaces. Reviews of current knowledge of the experimental and theoretical aspects of the behavior of proteins at solid/liquid interfaces are provided in other chapters (Chaps. 1–4) in this book as well as in numerous reports in the scientific literature (e.g., Ref. [3]).

Because of previously exclusive availability of conventional materials, it was inevitable that past studies addressed, and elucidated, various aspects of protein interactions with those surfaces. The broad scope of potential applications (such as functional nanomaterials, biosensing, drug delivery, and imaging) for nanophase materials in the emerging field of nanobiotechnology has prompted examination of protein interactions on such substrates; Chap. 5 of this book and recent reviews [4, 5] provide details appropriate to those applications. The present chapter will focus on protein interactions on nanophase materials that are pertinent to implant materials.

17.2.1. Protein Adsorption

Compared with events observed on planar, conventional material surfaces, adsorption, the first stage of interaction between a protein and a solid biomaterial surface, is different on planar nanophase materials of the same chemical composition. The earliest such evidence was provided by seminal observations and results from in vitro studies. The total amount (μg/cm^2) of protein adsorbed from a supernatant solution of 10% fetal bovine serum (FBS; which contains often unknown amounts of various proteins) in phosphate-buffered saline (PBS; pH 7.4) on nanophase alumina (24-nm grain size) and on nanophase hydroxyapatite (HA) (67-nm grain size) in a 37°C, humidified, 5% CO_2/95% air environment for 4 h was significantly ($p < 0.01$) higher than that obtained on conventional alumina and HA (167- and 179-nm grain size, respectively) under the same experimental conditions [6]. Adsorption of individual proteins (from 5 μg each single-component solutions in PBS) in a 37°C, humidified, 5% CO_2/95% air environment for 4 h was selective: while adsorption of denatured collagen (285 kDa) and vitronectin (70 kDa) was significantly ($p < 0.01$) enhanced on nanophase alumina and HA (24- and 67-nm grain size, respectively), albumin (66 kDa) and fibronectin (440 kDa) adsorption was similar on these nanophase and their respective conventional ceramics (167- and 179-nm grain size, respectively); in contrast, adsorption of laminin (900 kDa) was lower ($p < 0.01$) on nanophase than on the respective conventional alumina and HA [6]. Preferential adsorption of vitronectin (65–78 kDa), but not of albumin (66 kDa), immunoglobulin G (heavy and light chain, 110 and 52 kDa, respectively), α_1-antitrypsin (50 kDa), inter-α-trypsin inhibitor family heavy chain-related protein (120 kDa), and immunoglobulin G (162 kDa), was observed on nanophase versus conventional (grain sizes 24 and 167 nm, respectively) alumina in a 37°C, pH 7.4, humidified, 5% CO_2/95% air environment for 4 h [7].

In summary, these in vitro studies provided evidence that the nanophase features (and not the chemistry) of the ceramic substrates tested modulated the observed interaction of biomolecules; most importantly, planar nanophase ceramics promoted adsorption of select proteins (but not of a specific molecular weight). Although various hypotheses (including selective deposition of proteins at surface discontinuities [8] and conformational changes of the adsorbed protein molecules [7]) have been proposed, the underlying mechanisms of protein interactions with nanophase materials remain unknown.

Surface Properties of Nanophase Materials

In addition to the properties of adsorbing proteins, the properties of the planar surface modulate and control biomolecule interactions at the tissue/implant interface.

A number of other material surface properties, some of which are similar to those of conventional substrates but others are unique to nanophase materials, may affect protein adsorption: (1) chemistry [9]; (2) increased surface area [9]; (3) wettability [9]; (4) more grain boundary area [10]; (5) higher proportion of edge sites [10]; (6) altered topography (roughness) [9]; and (7) modified charge [11].

To date, the number of studies that addressed the aforementioned aspects of nanomaterials has been small; in addition, the scope of such studies largely has been limited to phenomenological investigations of cell interactions but not the details of the underlying mechanisms of the observed events. Correlation with surface roughness (which was proposed by several researchers as the material property responsible for subsequent interactions and activity/functions of components of the biological milieu) is not the fundamental aspect that dictates protein interactions with nanophase materials. Undoubtedly, further research is needed to investigate, for example, energetics at the grain boundaries [12], determine the kinetics of protein adsorption and strength of protein binding on nanomaterials, establish the priority of importance of pertinent physicochemical aspects (such as hydrophobic, electrostatic, van der Waals, and other surface energy forces) in these events, as well as elucidate the mechanisms that govern protein interactions on nanophase biomaterials. These endeavors need coordinated theoretical and experimental studies whose results will establish the credentials of this promising field of biomaterials.

Protein Conformation

Adsorption of proteins on conventional solid substrates is accompanied by changes in the structure of these biomolecules. Unfolding to accommodate further bonding with the substrate surface leads to conformational changes, which may promote either enhanced or attenuated bioactivity (including protein denaturation). Undoubtedly, this aspect of protein interaction with solid material surfaces is instrumental in subsequent responses of surrounding cells and tissues with implant biomaterials and devices.

The earliest study to explore the role of protein conformation on planar nanophase biomaterials focused on vitronectin. In a competitive environment, this protein, which is known for its specific mediation of cell adhesion on solids, adsorbed in the highest concentration on nanophase (23-nm grain size) alumina [7]. Vitronectin adsorption on the nanophase alumina tested was correlated to calcium-mediated mechanisms. Surface-enhanced Raman scattering revealed a different unfolding pattern, which enhanced the bioactivity of vitronectin, on the nanophase substrate. The underlying mechanism of this result involved exposure of cell-adhesive sequences, primarily, arginine-glycine-aspartic acid (RGD), in the protein [7].

The aforementioned results provided the first, but admittedly very interesting, insight into the mechanisms underlying the interactions of proteins (and subsequently of cells) on nanophase substrates that are pertinent to biomedical implant applications. Additional theoretical and experimental studies are needed, however, in order to definitively determine and, thus, establish the mechanisms of protein interactions with nanophase biomaterials.

Protein Interactions with Nanoparticles and Carbon Nanotubes

The interactions of proteins and biomolecules with nanophase substrates have been the focus of recent nanobiotechnological developments aimed at applications in medical imaging, biosensors, and drug delivery [4, 5]. Although these developments are not directly related to biomaterials designed, fabricated, and used in implantable devices, they represent important basic science studies that may provide insights, and thus contribute, to current understanding of fundamental aspects of planar nanophase materials with grain sizes and/or features of similar dimensions.

Pertinent in vitro studies used special formulations of nanophase materials (such as nanoparticles and carbon nanotubes) in suspension and addressed the role of specific dimensions, geometries, and material surface properties on modulating and controlling the structure, stability, and bioactivity of interacting proteins. These studies provided evidence that the curvature of nanosurfaces affects the adsorbed protein structure, stability, and activity. Specifically, high surface curvature of small (4 nm in diameter) silica nanoparticles promoted retention of the native structure and function of adsorbed enzymes, specifically chicken egg lysozyme (14.6 kDa) [13] and human carbonic anhydrase [14]. In contrast, adsorption of these enzymes on larger (100 nm) diameter particles affected the secondary structure and correlated with greater loss in the α-helix content of the biomolecule [13, 14]. Similar results regarding retention of the secondary structure and activity were obtained when these enzymes were adsorbed onto single-walled carbon nanotubes [15].

Furthermore, the type and structure of the adsorbing protein are affected by the chemistry and curvature of the substrate nanoparticles. In this respect, bovine serum albumin retained its native structure but bovine serum fibrinogen denatured to a greater extent on similar small (radii less than 30 nm) silica spheres; both proteins tested were more denatured on hydrophobic than on hydrophilic surfaces [16]. Control of chymotrypsin structure and function was achieved by judiciously tailoring the surface chemistry of cadmium selenide nanoparticles and ranged from no binding, to inhibition with denaturation, to inhibition with retention of structure [17]. Surface-functionalized gold nanoparticles were designed and used to selectively bind either cytochrome c or cytochrome c peroxidase but to inhibit cytochrome c peroxidase activity and thus disrupt protein–protein interactions [18]. Enhanced protein stability in harsh environments (i.e., high temperature) was also achieved when the biomolecules were adsorbed on single-walled carbon nanotubes [19]; this is a desirable property for achieving not only interaction of specific proteins with nanomaterials but also maintaining stability and selectivity of function and bioactivity.

In summary, the aforementioned studies regarding protein interactions with nanoparticles and carbon nanotubes in suspension provide a roadmap for similar studies that need to be conducted in order to establish the fundamental scientific principles of protein interactions with nanophase implant materials.

In the case of planar materials, one study addressed the extent of surface coverage by nanoparticles and of the nanoparticle diameters on the amount, type, distribution, conformation, and bioactivity (related to subsequent adhesion of mammalian cells) of adsorbed proteins. For this purpose, native oxide-coated silicon surfaces were decorated with silica nanoparticles of either 4-, 20-, or 100-nm diameter and up to 100% surface coverage by the respective particles [20]. Adsorption of vitronectin and fibronectin increased with increased particle diameter on surfaces with 80% coverage by particles. In these cases, the secondary structure of fibronectin was similar on the 4- and 20-nm-diameter particles but exhibited decreased β sheet content and increased unordered structure on the 100-nm-diameter particles. The structure of vitronectin exhibited similar extent of structure loss on all particle sizes tested. These aspects of the adsorbed adhesive proteins were correlated to the patterns of osteoblast and fibroblast adhesion (in serum-free media) on these specially decorated surfaces preadsorbed with either vitronectin or fibronectin. Osteoblast adhesion on the 4- and 20-nm, but not on the 100-nm, diameter particles was enhanced when fibronectin, but not when vitronectin, was the preadsorbed protein [20]. In summary, this study brought attention to the importance of controlling the substrate material properties (such as nanoscale feature size and uniformity as well as distribution on, and surface coverage of, planar substrates) to promote protein interactions that induce subsequent cell functions.

17.3. Cell Interactions with Nanophase Materials

The advent of methods to reproducibly fabricate nanostructured materials has provided substrates with nanometer topography similar to that of native, physiological tissues. The unexplored potential of these novel materials formulations to control, direct, and/or modulate the type and extent of molecular and cellular interactions at the tissue/implant interface has motivated research endeavors that sought to examine biological responses to nanophase materials as an index of their cytocompatibility, capability of integration in surrounding tissues, and promotion of tissue regeneration around implants.

17.3.1. Adhesion of Bone Cells on Nanophase Materials

The biocompatibility of conventional ceramic materials and their strong bonding to bone following implantation in the body is well established [21–25]. This outcome is the result of the excellent cytocompatibility of these biomaterials: osteoblasts (the bone-forming cells) adhere to and function exceptionally well on conventional ceramics [9].

The first in vitro study to explore bone cell adhesion on nanophase and conventional ceramics investigated rat calvarial osteoblasts cultured in Dulbecco's modified Eagle medium (DMEM) supplemented with 10% FBS in a 37°C, humidified, 5% CO_2/95% air environment for durations ranging from 0.5 to 4 h. It revealed significantly ($p < 0.05$) greater osteoblast adhesion on nanophase alumina (49- and 23-nm grain size) and titania (97- and 32-nm grain size) than on their respective conventional counterparts (177-nm and 2,120-nm grain sizes for alumina and titania, respectively) for all time intervals tested [9]. Osteoblast adhesion on alumina for 4 h increased significantly ($p < 0.05$) as the grain size of the substrate decreased (from 49 to 23 nm for alumina and from 32 to 20 nm for titania); maximal ($p < 0.01$) adhesion was observed on the 20-nm grain size alumina [9]. Compared with results obtained on conventional (177-nm grain size) HA, osteoblast adhesion (under the aforementioned experimental conditions) was significantly ($p < 0.01$) increased on the nanophase (67-nm grain size) HA substrates [6]. These results provided evidence that nanophase characteristics (specifically the grain size), but not the chemistry, of the ceramic substrates tested were the key parameters in the observed adhesion of osteoblasts. Moreover, a threshold grain size (between 49 and 67 nm for alumina and between 32 and 56 nm for titania) was associated with the observed preferential adhesion of osteoblasts on the nanophase substrates tested [9]. The trend of increased osteoblast adhesion with decreasing ceramic grain size established a criterion that can be used to design and fabricate biomaterials with predetermined enhanced cytocompatibility for orthopedic/dental applications.

The aforementioned and continuation studies revealed a unique characteristic of nanophase ceramics that is unrepresented in their conventional counterparts: selectivity. In contrast to the enhanced adhesion of rat calvarial osteoblasts on nanophase alumina (45- and 24-nm grain size), titania (39-nm grain size), and HA (67-nm grain size), adhesion of rat skin fibroblasts (ubiquitous mammalian cells primarily associated with soft tissues) decreased with decreasing grain size on all substrates tested. Furthermore, while bovine arterial endothelial cell (cells on the innermost layer of blood vessel networks present in vascularized tissues) adhesion decreased with decreasing grain size on alumina and titania, it was unaffected on HA [6]. It should be noted that, compared with results obtained with osteoblasts, adhesion of fibroblasts was lower on all alumina and titania and on nanophase HA, while adhesion of endothelial cells was far less on all ceramic substrates tested. Maximal adhesion of fibroblasts was observed on the conventional (179-nm grain size) HA [6].

Similar trends of enhanced and selective adhesion of rat calvarial osteoblasts were observed on composites of nanophase ceramics (alumina, titania, and HA) with polymers, specifically poly(L-lactic acid) (PLA) and poly(methyl methacrylate) (PMMA) [26]. In this study, osteoblast adhesion to the 50/50 (weight percent) nanophase alumina/PLA substrates (which exhibited significantly [$p < 0.05$] greater bending moduli than either the brittle ceramic alone or the composites with similar loading of conventional alumina) was similar to that observed on the 100% nanophase ceramic substrate; in contrast, rat skin fibroblast adhesion to the nanophase composites was either comparable to, or lower than, that observed on the PLA and PMMA composites with all conventional ceramics tested. Similar trends of cell adhesion were observed on composites of the other ceramics and polymers tested in that study [26]. The important contribution of this study was the evidence that, on composites containing above 50/50 (weight percent) loading, the nanoceramic content mediated select and enhanced adhesion of the bone-forming cells that was similar to that observed on the respective 100% ceramics. Maintenance of the select and specific protein and cell interactions obtained on 100% nanoceramics combined with improvement of mechanical properties of the composites constitute attractive characteristics of these novel material formulations [26]. Further studies are needed, however, to optimize pertinent material and mechanical parameters of nanomaterials and their composites with polymers.

In addition to contributions in aspects of cellular biology, the cell adhesion findings on nanophase biomaterials are important in that they identified a novel approach to obtain proactive, implant biomaterials, that is, materials designed and formulated to promote select, specific, and timely responses for surrounding cells and tissues in the complex and competitive physiological milieu. Because adhesion is the first and, for this reason, the most crucial function for the survival (and a prerequisite for subsequent functions) of anchorage-dependent cells, which are the constituents of connective tissue, this aspect remains the greatest promise and potential of nanophase biomaterials used as implants in humans and animals. It should be stated, however, that the long-term in vivo success of implant materials and devices requires other cell functions pertinent to new tissue formation that follow the extremely important, but relatively short-term, cell adhesion process; these aspects are discussed in Sects. 17.3.3 and 17.3.4 of this chapter.

17.3.2. Mechanism of Cell Adhesion on Nanophase Materials

A major clue regarding the mechanism underlying the aforementioned enhanced, but selective, adhesion of osteoblasts on nanophase ceramics (Sect. 17.3.1) was provided by the results of in vitro cell adhesion experiments *in the absence of FBS* in a 37°C, humidified, 5% CO_2/95% air environment for 4 h [6]. Without serum, cell adhesion was minimal and independent of the type of cell (namely, rat calvarial osteoblasts, rat skin fibroblasts, and bovine arterial endothelial cells), substrate chemistry (i.e., alumina, titania, and HA), and the substrate material phase (i.e., conventional and nanophase) tested [6]. These results were in stark contrast to those obtained under similar experimental conditions but in the presence of 10% FBS as described in Sect. 17.3.1.

Serum, a standard supplement in cell culture media, contains proteins whose identity and concentrations are usually not disclosed by the commercial vendors of these supplements. Some vendors market certified varieties of sera that have been biochemically analyzed, but even then, only select biomolecules have been measured; generally such analyses do not include the important adhesive proteins. Since serum is necessary for the survival and function of cultured cells, its use (at 5–15 vol%) has become standard practice. In the case of cell adhesion (and because proteins mediate and modulate adhesion of anchorage-dependent cells on substrates), the contents of the serum supplement play a crucial role in the results obtained on nanophase materials.

Further investigation provided evidence that, compared with results obtained on conventional (167- and 179-nm grain size) substrates, the respective nanophase alumina (24-nm grain size) and HA (67-nm grain size) adsorbed significantly ($p < 0.01$) higher amounts from the protein "cocktail" contained in 10% FBS [6]. Individual proteins (5 μg protein in PBS) exhibited different adsorption trends: albumin and fibronectin adsorption was similar on all substrates tested, while laminin adsorbed preferentially on conventional substrates and was maximal on conventional alumina. In contrast, greater amounts of denatured collagen and vitronectin adsorbed on nanophase alumina and HA. Unequivocal evidence that these proteins mediated the observed cell adhesion was provided by further examination of osteoblast adhesion (in DMEM *without serum* in a 37°C, humidified, 5% CO_2/95% air environment for 4 h) on nanophase (24-nm grain size) alumina on which various individual proteins had been preadsorbed; in this case, enhanced osteoblast adhesion was observed on substrates pretreated with 0.5 or 5 μg/mL of either vitronectin or fibronectin [6].

In addition to establishing the role of proteins in osteoblast adhesion on nanophase ceramics, these results identified the proteins that are the major mediators in this cell function, and provided an explanation for the observed enhanced and selective adhesion of osteoblasts on these substrates. Vitronectin and collagen, which adsorbed preferentially and mostly on the nanophase substrates tested, are known mediators of osteoblast adhesion [27–30].

In addition to the type and concentration of specific proteins, the underlying mechanism in the process of cell adhesion involves appropriate conformation of the mediating biomolecules. In this respect, adsorption of vitronectin on nanophase (24-nm grain size) alumina induced increased calcium-mediated unfolding of these molecules to expose cell-adhesive epitopes, such as RGD sequences, which are the ligands of receptors on cell membranes of osteoblasts and other cells [7].

These in vitro studies were the first to elucidate key aspects of the underlying mechanisms of adhesion, a most important requirement for anchorage-dependent cell survival and a prerequisite for subsequent cell functions on nanophase materials.

17.3.3. Other Functions of Bone Cells on Nanophase Materials

In addition to adhesion (a short-term cell function on the order of 2 to 4 h), other long-term (i.e., on the order of days to weeks) cell functions pertinent to the formation of tissue were investigated. Compared with results obtained on the respective conventional formulations, rat calvarial osteoblast proliferation was significantly ($p < 0.01$) increased on nanophase alumina (24-nm grain size), titania (39-nm grain size), and HA (67-nm grain size) after both 3 and 5 days of culture [31]. Alkaline phosphatase activity, which is considered a marker for the osteoblastic phenotype, was significantly ($p < 0.01$) enhanced on the nanophase substrates tested at 21 and 28 days. During these longer time intervals, accumulation of calcium, an index for the inorganic, mineral phase of bone, also increased for osteoblasts cultured on all nanophase substrates tested [31]. Since calcium accumulates on and in collagen, an index of the organic phase of bone, in the extracellular matrix produced by osteoblasts, this result is indirect evidence of enhanced collagen deposition by the cells cultured on the nanophase ceramics.

These in vitro studies provided evidence that, compared with their conventional counterparts, nanophase ceramics promote interactions and support functions of osteoblasts that are pertinent to new bone tissue formation.

17.3.4. Adhesion and Function of Other Cells on Nanophase Materials

The seminal in vitro studies discussed in Sects. 17.3.1–17.3.3 were followed by numerous, but similar, investigations of chondrocyte [32], endothelial cell [33], fibroblast [34], neural cell [35], and human mesenchymal stem cell [36] interactions (primarily adhesion and proliferation) with various nanophase substrates, including ceramics [37], metals [38], polymers [34], carbon nanofibers [39], and composites [32].

Unquestionably, the later studies expanded the repertoire of cell type and material substrate tested. Taken collectively, these in vitro studies confirmed the conclusions of the pioneering investigations; specifically, the importance of the nanoscale grain/feature size in the observed cell interactions pertinent to new tissue formation. These reports, however, did not provide further insights into the underlying mechanism(s). It is possible that either the same type of adhesive proteins adsorb on various nanophase materials or that the induced conformational changes of the protein molecules reveal binding sites recognized by many types of cells. This explanation is supported by the fact that not only do biomolecules have many, and different, bioactive domains, but that such ligands are recognized by various receptors on the membranes on many types of cells.

The similarity in adhesion trends reported for cells from various species, anatomical locations, tissues, and types requires careful review and appropriate resolution. The cases of results that at first glance appeared contradictory were settled when the identity and role of the mediating adhesive proteins were taken into consideration [40]. A most important aspect of the performance of nanophase materials was specificity, for example, enhanced adhesion of osteoblasts but decreased adhesion of fibroblasts with decreasing grain size of the materials substrate [6]. Loss of such discriminating capability among cells of the same tissue eliminates one of the most attractive and promising aspects of nanophase biomaterials for orthopedic/dental applications. In contrast, the report that adhesion of *Staphylococcus epidermidis* bacteria on nanophase zinc oxide and titania substrates is attenuated [41] implies potential control, if not prevention, of infection at implants sites and represents another very attractive aspect of these biomaterial formulations for clinical applications.

17.4. Concluding Remarks

Nanotechnology has provided novel material formulations whose surfaces promote select and specific interactions with biomolecules as well as functions of mammalian cells pertinent to new tissue formation. Although significant advances have been made in elucidating some aspects of these processes, the field is at an early stage of development; many questions regarding fundamental processes controlling protein and cell interactions with nanophase biomaterials remain unanswered, and their resolution is needed before the full potential and promise of these novel biomaterial formulations can be realized. These include, for example, unambiguous identification of the material surface properties that control and drive adsorption of protein molecule assemblies over time scales from seconds through hours and days, and determination of the details of underlying biological/molecular mechanisms responsible for cell responses, including receptor-ligand interactions and subsequent signal transduction pathways within the cells. Theoretical analyses and development of pertinent models that correlate to, expand, and predict the bio-related but (so far) exclusively experimental studies using nanophase material are needed to establish and extend this novel field of biomaterials. Last, but not least, the in vivo performance of nanophase materials must be determined. It remains unknown whether the interesting findings from the in vitro cell culture models discussed in this chapter will be manifested at the tissue level in the complex physiological milieu in vivo; moreover, the biocompatibility of nanophase materials, specifically addressing potential adverse effects (e.g., "nanotoxicity"), needs to be established.

Undoubtedly, nanophase materials have the potential to make seminal contributions to current and future developments of biomaterials, which could revolutionize medical diagnostics and therapeutic methodologies as well as have enormous impact on medical implant devices and tissue-engineering applications.

References

1. Sniadecki N, Desai R, Ruiz S, Chen C. Nanotechnology for cell–substrate interactions. Annals of Biomedical Engineering. 2006;34(1):59–74.
2. Siegel RW. Creating nanophase materials. Scientific American. 1996;275(6):74–79.
3. Horbett T. Role of adsorbed proteins in tissue response to biomaterials. In: Ratner B, Hoffman A, Schoen A, Lemmons J, editors. Biomaterials Science: An Introduction to Materials in Medicine, 2nd ed. San Diego, CA: Elsevier Academic Press 2004. pp. 237–246.
4. Asuri P, Bale SS, Karajanagi SS, Kane RS. The protein-nanomaterial interface. Current Opinions in Biotechnology. 2006;17(6):562–568.
5. Kane RS, Stroock AD. Nanobiotechnology: protein-nanomaterial interactions. Biotechnology Progress. 2007;23(2):316–319.
6. Webster TJ, Ergun C, Doremus RH, Siegel RW, Bizios R. Specific proteins mediate enhanced osteoblast adhesion on nanophase ceramics. Journal of Biomedical Materials Research. 2000;51(3):475–483.
7. Webster TJ, Schadler LS, Siegel RW, Bizios R. Mechanisms of enhanced osteoblast adhesion on nanophase alumina involve vitronectin. Tissue Engineering. 2001;7(3):291–301.
8. Andersson A-S, Olsson P, Lidberg U, Sutherland D. The effects of continuous and discontinuous groove edges on cell shape and alignment. Experimental Cell Research. 2003;288(1):177–188.
9. Webster TJ, Siegel RW, Bizios R. Osteoblast adhesion on nanophase ceramics. Biomaterials. 1999;20(13):1221–1227.
10. Klabunde KJ, Stark J, Koper O, Mohs C, Park DG, Decker S, et al. Nanocrystals as stoichiometric reagents with unique surface chemistry. The Journal of Physical Chemistry. 1996;100(30):12142–12153.
11. Luck M, Paulke BR, Schroder W, Blunk T, Muller RH. Analysis of plasma protein adsorption on polymeric nanoparticles with different surface characteristics. Journal of Biomedical Materials Research. 1998;39(3):478–485.
12. Kasemo B, Department of Physics, Chalmers University of Technology, Gothenburg, Sweden. Personal communication to RB, 2003.
13. Vertegel AA, Siegel RW, Dordick JS. Silica nanoparticle size influences the structure and enzymatic activity of adsorbed lysozyme. Langmuir. 2004;20(16):6800–6807.
14. Lundqvist M, Sethson I, Jonsson B-H. Protein adsorption onto silica nanoparticles: conformational changes depend on the particles' curvature and the protein stability. Langmuir. 2004;20(24):10639–10647.
15. Karajanagi SS, Vertegel AA, Kane RS, Dordick JS. Structure and function of enzymes adsorbed onto single-walled carbon nanotubes. Langmuir. 2004;20(26):11594–11599.
16. Roach P, Farrar D, Perry CC. Surface tailoring for controlled protein adsorption: effect of topography at the nanometer scale and chemistry. Journal of the American Chemical Society. 2006;128(12):3939–3945.
17. Hong R, Fischer NO, Verma A, Goodman CM, Emrick T, Rotello VM. Control of protein structure and function through surface recognition by tailored nanoparticle scaffolds. Journal of the American Chemical Society. 2004;126(3):739–743.
18. Bayraktar H, Ghosh PS, Rotello VM, Knapp MJ. Disruption of protein-protein interactions using nanoparticles: inhibition of cytochrome c peroxidase. Chemical Communications (Cambridge, England). 2006(13):1390–1392.
19. Asuri P, Karajanagi SS, Yang H, Yim T-J, Kane RS, Dordick JS. Increasing protein stability through control of the nanoscale environment. Langmuir. 2006;22(13):5833–5836.
20. Ballard JD. Investigation of cell adhesion to silica nanoparticle-decorated surfaces and the associated protein-mediated mechanisms. Troy, NY: Rensselaer Polytechnic Institute; 2005.
21. Davies JE, Matsuda T. Extracellular matrix production by osteoblasts on bioactive substrata in vitro. Scanning Microscopy 1988;(23):1445–1452.
22. Hayashi H, Uchida A, Hamada H, Yoshikawa H, Shinto Y, Ono K. Alumina ceramic prostheses for bone tumor surgery. Archives Of Orthopaedic And Trauma Surgery. 1992;(1121):1–4.
23. Matsuda T, Davies JE. The in vitro response of osteoblasts to bioactive glass. Biomaterials. 1987;84):275–284.
24. Sautier J, Nefussi J, Forest N. Ultrasonic study of bone formation on synthetic hydroxyapatite in osteoblast cultures. Cells and Materials. 1991;1:209–217.

25. Toni A, Lewis CG, Sudanese A, Stea S, Calista F, Savarino L, et al. Bone demineralization induced by cementless alumina-coated femoral stems. The Journal of Arthroplasty. 1994;9(4):435–444.
26. McManus AJ, Doremus RH, Siegel RW, Bizios R. Evaluation of cytocompatibility and bending modulus of nanoceramic/polymer composites. Journal of Biomedical Materials Research. Part A. 2005;72(1):98–106.
27. Ayad S, Boot-Handford R, Humpries M, Kadler K, Shuttleworth A. The extracellular matrix facts book. San Diego, CA: Academic Press Inc; 1994.
28. Gronowicz G, McCarthy MB. Response of human osteoblasts to implant materials: integrin-mediated adhesion. Journal of Orthopaedic Research. 1996;14(6):878–887.
29. Moursi AM, Globus RK, Damsky CH. Interactions between integrin receptors and fibronectin are required for calvarial osteoblast differentiation in vitro. Journal of Cell Science. 1997;110 (Pt 18):2187–2196.
30. Schneider G, Burridge K. Formation of focal adhesions by osteoblasts adhering to different substrata. Experimental Cell Research. 1994;214(1):264–269.
31. Webster TJ, Ergun C, Doremus RH, Siegel RW, Bizios R. Enhanced functions of osteoblasts on nanophase ceramics. Biomaterials. 2000;211(7):1803–1810.
32. Kay S, Thapa A, Haberstroh KM, Webster TJ. Nanostructured polymer/nanophase ceramic composites enhance osteoblast and chondrocyte adhesion. Tissue Engineering. 2002;8(5):753–761.
33. Khang D, Lu J, Yao C, Haberstroh KM, Webster TJ. The role of nanometer and sub-micron surface features on vascular and bone cell adhesion on titanium. Biomaterials. 2008;29(8):970–983.
34. Vance RJ, Miller DC, Thapa A, Haberstroh KM, Webster TJ. Decreased fibroblast cell density on chemically degraded poly-lactic-co-glycolic acid, polyurethane, and polycaprolactone. Biomaterials. 2004;25(11):2095–2103.
35. Webster TJ, Waid MC, McKenzie JL, Price RL, Ejiofor JU. Nano-biotechnology: carbon nanofibres as improved neural and orthopaedic implants. Nanotechnology. 2004;15(1):48–54.
36. Dulgar-Tulloch A, Bizios R, Siegel R. Human mesenchymal stem cell adhesion and proliferation in response to ceramic chemistry and nanoscale topography. Journal of Biomedical Materials Research. Part A (published online, 18 June 2008).
37. Balasundaram G, Sato M, Webster TJ. Using hydroxyapatite nanoparticles and decreased crystallinity to promote osteoblast adhesion similar to functionalizing with RGD. Biomaterials. 2006;27(14):2798–2805.
38. Webster TJ, Ejiofor JU. Increased osteoblast adhesion on nanophase metals: Ti, Ti6Al4V, and CoCrMo. Biomaterials. 2004;25(19):4731–4739.
39. Price RL, Ellison K, Haberstroh KM, Thomas J. Webster. Nanometer surface roughness increases select osteoblast adhesion on carbon nanofiber compacts. Journal of Biomedical Materials Research 2004;70A(1):129–138.
40. Dulgar-Tulloch A. The effects of nanophase ceramic materials on the cellular functions of human mesenchymal stem cells. Troy, NY: Rensselaer Polytechnic Institute; 2005.
41. Colon G, Ward BC, Thomas J. Webster. Increased osteoblast and decreased *Staphylococcus epidermidis* functions on nanophase ZnO and TiO. Journal of Biomedical Materials Research. Part A. 2006;78A(3):595–604.

18

Inflammatory Response to Implanted Nanostructured Materials

Kristy M. Ainslie, Rahul G. Thakar, Daniel A. Bernards, and Tejal A. Desai

Nanostructured materials are implanted for dynamic application in therapies such as drug delivery, tissue engineering, biosensing, and imaging. The interaction between nanostructured materials and the tissues of the body can be used to alter cellular attachment and motility, detect analytes in vivo, and simulate tissue organization. These interactions may also lead to a variety of adverse immune responses, such as acute inflammation and fibrous encapsulation. Nanostructured surfaces have been shown to attenuate the acute inflammatory response of immune cells to a material, as well as change the normal encapsulation response that generally results in implants being quarantined by a wall of cells. Additionally, many of these novel materials improve the time course of wound healing in vivo. In general, nanostructured materials are equally or less immunogenic than a planar surface of the same material. However, much like the basis of material biocompatibility, further studies need to be completed on a per material basis to properly evaluate the inflammation potential of novel nanomaterials.

Abbreviations

CX3CL1	chemokine ligand 1
CXCL2	chemokine ligand 2
CXCL3	chemokine ligand 3
CXCL10	chemokine ligand 10
EBL	electron beam lithography
ECM	extracellular matrix
GM-CSF	granulocyte-macrophage colony-stimulating factor
IL-6	interleukin six
IL-8	interleukin eight
MIP-1α	macrophage inflammatory protein 1 alpha
MIP-1β	macrophage inflammatory protein 1 beta

K.M. Ainslie, R.G. Thakar, D.A. Bernards, and T.A. Desai • Department of Physiology, Division of Bioengineering, University of California–San Francisco, San Francisco, CA 94158, USA

D.A. Puleo and R. Bizios (eds.), *Biological Interactions on Materials Surfaces*,
DOI: 10.1007/978-0-387-98161-1_18,

MCP-1	monocyte chemotactic protein-1
MSC	marrow stromal cells
PCL	poly(caprolactone)
PDMS	poly(dimethylsiloxane)
PHBV	poly(3-hydroxybutyrate-co-3-valerate)
PLGA	poly(lactide-*co*-glycolide)
PMMA	poly(methylmethacrylate)
PTFE	poly(tetrafluoroethylene)
RANTES	regulated upon activation, normal T-cell expressed and secreted
ROS	reactive oxygen species
TNF-α	tumor necrosis factor alpha
TGF-β	transforming growth factor-beta

18.1. Introduction

Cells exist in a universe unseen by the unaided eye. This universe operates on the microscale with critical events affecting cellular behavior transpiring at the nanoscale. Because of the cellular world's dependency on nanoscale events, application of nanotechnology to the biomedical arena has tremendous potential for development of new therapeutic modalities. Besides cell signaling through chemical cytokines, signal transduction is also mediated by cellular interactions with the extracellular matrix (ECM). The ECM is the defining feature of connective tissue in animals and is composed of polysaccharides and proteins, such as collagen, laminin, and fibronectin [1]. The elements of the ECM are nanometer in scale, further highlighting the importance of nanotechnology and cellular interaction.

The impact of nanostructured materials can be observed in the orthopedic, neural, urinary and cardiovascular systems. Adhesion of bone and bone precursor cells on orthopedic implants is of importance. More than 30,000 orthopedic implant surgeries are revised each year [2]; these repeated operations could be reduced with increased implant fixation and cellular adherence. Bone and cartilage precursor cells exhibited increased adhesion on nanomaterials [3]. Marrow stromal cells (MSC) cultured on nanoporous alumina membranes displayed increased adhesion, as well as increased ECM formation [3]. When MSCs were cultured on nanostructured titanium nanotubes, a similar phenomenon was observed. Osteoblasts cultured on silica nanowires also displayed improved proliferation, adhesion, and matrix formation [4]. Additionally, chondrocytes cultured on nanoporous poly(lactide-*co*-glycolide) (PLGA) scaffolds displayed enhanced cellular function [5]. Besides orthopedic implants, nanostructured materials have shown importance in neuronal cell growth. Axon guidance formation is significant in the tissue engineering of nerve cells. PC12 (rat neural cells) displayed similar cellular adhesion and proliferation on nanostructured alumina membranes as on flat tissue culture polystyrene, displaying the ability of these sensitive cells to be maintained on such material surfaces. Aligned, patterned nanofibrous scaffolds provide both physical and biochemical cues to induce, enhance, and guide neurite outgrowth and skin cell migration [6]; in addition, these materials are capable of guiding morphogenesis of tissues with anisotropic structure, such as cardiac muscle, blood vessel, tendon, and ligament [7]. The high incidence of bladder cancer and urinary incontinence results in the need for tissue engineered bladders. On PLGA and poly(urethane) nano-roughened surfaces, bladder smooth muscle cell adhesion increased [8]. Cellular alignment is of high importance in the tightly-organized vascular wall. Vascular endothelial cells aligned on nanostructured titanium surfaces [9].

The aforementioned examples illustrate that cells respond to the contact guidance cues presented by nanostructured materials.

As nanostructured devices are increasingly designed and proposed for in vivo applications, the inflammatory response of these materials must be evaluated in conjunction with the classical definition of biocompatibility: the ability of a device to perform a specific application given an appropriate host response [10, 11]. Oftentimes the biocompatibility of a synthetic material is determined without an inflammatory response. Measurement of non-immune cell viability, proliferation, and/or morphology misrepresents the entire picture of biocompatibility because the inflammatory response to the target material is completely neglected. To date, a limited number of studies report the inflammatory response to nanostructured materials. Understanding the construction of nanostructured materials, as well as the inflammatory response, will aid in applications of nanotechnology to the biomedical arena.

18.2. Fabrication Techniques

A variety of techniques exist to fabricate material surfaces with nanostructures. Nanostructured materials are developed through several fabrication techniques that allow for elements such as fibers, pores, and through-channels, which are on the size scale of elements of the ECM. Because of the unique properties of nanostructured surfaces, the techniques used to form non-particle nanostructures will first be reviewed.

18.2.1. Nanofibers

Polymer fibers can be produced that are less than 100 nm in diameter, through techniques like electrospinning and jet-blowing. Electrospinning is accomplished by applying a potential between a metallic needle and a conducting collection plate and can be tailored to produce from micro- to nano-scale fiber diameters depending on the collection plate. Using electrospun collagen nanofibers is an efficient and inexpensive approach, which has been successfully applied to tissue engineering [12]. Electrospinning for tissue engineering, however, is not limited to collagen nanofibers. Biocompatible polymers with therapeutic potential for implant applications, such as poly(L-lactide) or PLGA, can also form nanofibers via electrospinning [13]. High molecular weight polymers, such as poytetrafluoroethylene (PTFE), can be jet-blown into fibers at high pressures with inert gases such as argon and nitrogen (Figure 18.1a) [14]. In general, nanofibers mimic ECM fibrils in scale and provide structure to guide tissue morphogenesis and remodeling.

18.2.2. Electrochemical Methods

Electrochemistry, the injection or removal of electronic charge to induce chemical reactions, is a popular approach to achieve structured surfaces in inorganic materials. Through relatively simple methods, it is possible to create surfaces with a variety of micro- and nano-structures, such as titanium nanotubes [15] (Figure 18.1b), nanoporous silicon [16] and metal alloy nanowires [17–20]. The majority of these structures result from an interplay between growth and etch rates at the reactive surface, which depend on the composition of the reaction bath and applied electrochemical currents. Electrochemistry underlies several fabrication techniques, including electropolishing, anodization, and electrodeposition.

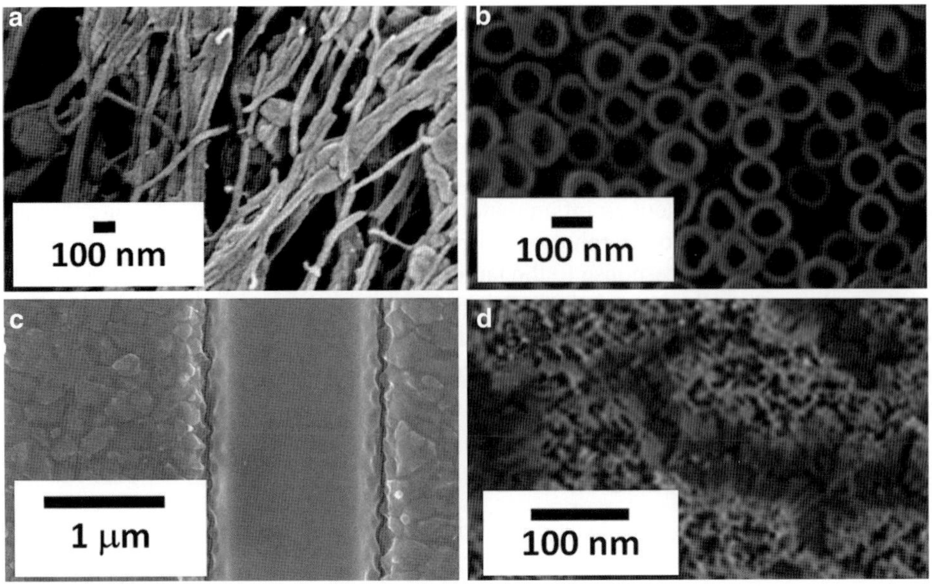

Figure 18.1. (**a**) Nanofibrous poly(tetrafluoroethylene) jet-blown onto a glass coverslip [14], (**b**) titanium nanotubes fabricated through electrochemical anodization [68], (**c**) silicon nanoporous membrane developed through micromachining methods including photolithography and chemical etching [81], (**d**) polycaprolactone (PCL) nanowires produced by micromolding of hot PCL into an aluminum membrane [21].

In the process of anodization the native oxide of a material is increased by electrochemical deposition; this approach can be used to generate structured surfaces. The surface of interest is typically attached to an electrode and placed in an acid solution with a counter electrode. The formulation of the reaction solution influences the rate of oxidation and specific dissolution of the oxide layer, as well as the type of structures formed. This process can be used to make nanoporous alumina with 10–150 nm diameter pores [16] and titanium nanotubes with diameters on the order of 60 nm [15]. The depth of these nano-features varies, but the typical range is 10–100 microns. These micro- and nano-structured materials can then be used as molds for a method like electrodepostion.

Electrodeposition occurs when a galvanic cell is used in reverse, with the plated surface as the cathode and the metal to be plated as the anode. With a potential applied between the two electrodes, metal ions from solution are reduced and deposited on the target surface. For example, this procedure can be used to create nanowires when an aluminum nanoporous membrane is used as the active substrate. Metal ions from solution are deposited within the membrane and fill the pores when given sufficient growth time. When the anodized membrane is filled, it can be subsequently etched, resulting in metal nanowires [18, 20]. Electrochemically processed micro- and nano-structured surfaces have been used as templates for further microstructures, such as nanowires [18, 20, 21].

18.2.3. Lithographic Techniques

Lithographic techniques for patterning at the micro- and nano-scale have predominantly been developed by the semiconductor processing field and have become increasingly popular for use in the biological sciences. Photolithography, the predominant micropatterning technique used industrially, is easily adapted for patterning features comparable to cell size. In this process, a photoactive polymer is selectively exposed to light through a mask, leaving a pattern in the polymer that can be developed and used in further processing. Projection photolithography is the current standard for high resolution lithography and can be used to obtain features down to 50 nm in high quality systems (Figure 18.1c). Unfortunately, due to the equipment and mask fabrication costs, nanoscale patterning is limited to high throughput manufacturing and is not conducive to the study of biological systems.

In addition to photolithography, electron beam lithography (EBL) has been used to obtain high resolution patterning. EBL scans a highly focused electron beam across an electron sensitive polymer (typically poly(methylmethacrylate) (PMMA)) to define patterns. Research-level EBL systems reproducibly resolve features of 20 nm; however, this technique requires serial exposure of patterns and operation under high vacuum, leading to low throughput. In general, nanoscale lithography is typically not used in biological applications because of its complexity and expense, as well as the wide range of ordered nanostructured materials available using other processing techniques [22–27].

More recently the use of nanoscale colloids was demonstrated as another technique to obtain nanoscale features. This approach uses the ability of colloids to form ordered patterns, which can then be etched to define highly ordered nanoscale patterns. Instead of using a photopatterned polymer, films of self-assembled colloids are used as an etch mask. This approach achieved features down to 20 nm but is limited to self-assembled structures of colloids on surfaces [28].

18.2.4. Molding and Embossing

Similar to how electrochemical deposition is used to fill nanoscale templates, molds can be used to transfer patterns to materials that are otherwise difficult to structure at the desired size scale. For example, beeswax micromolding has been carried out since the time of the Egyptians [29]. In place of natural polymers, synthetic polymers, such as poly(dimethylsiloxane) (PDMS), are used to transfer patterns from a master mold. The initial master substrate can be fabricated using a variety of methods, including electrochemical fabrication and lithography. Roughing the master is sometimes required to promote adhesion of the mold polymer when planar features are less than 20 nm [29]. A master pattern can be replicated through techniques like injection molding, hot embossing, and melt or solution casting. Plastics casting is commonly used for microfabricated implants. By casting a bio-friendly polymer, like PDMS, patterns can be subsequently transferred to a secondary polymer, such as biodegradable PLGA or poly(caprolactone) (PCL). For PDMS, feature sizes on the order of several hundred nanometers have been reported [30]. Micro- and nano-structured thin films of PDMS, PLGA, and PCL have been used for tissue engineering applications in a variety of tissues [31–34].

Molding can also be accomplished with electrochemically fabricated surfaces. Tao et al. [21] used commercially available aluminum oxide membranes to mold PCL nanowires (Figure 18.1d). PCL was melt-processed to fill alumina membranes, which were then removed by etching with sodium hydroxide. By varying the contact time (up to 60 min at 130°C) of the molten polymer with the alumina membrane, nanowire lengths of 2.5–27.0 microns were fabricated. Nanowire diameter did not vary with contact time but did vary with temperature, from 168 nm at 65°C to 196 nm at 130°C. The nanowires did degrade over a period of 7 weeks, as measured by scanning electron microscopy and release of a fluorescent molecule [21].

The techniques briefly reviewed in this section can be used to develop a variety of nanostructured devices for applications in the biomedical sciences. The interaction of these devices with cells and tissues allow for dynamic treatments or measurements for therapeutic applications. To properly function in vivo, a device fundamentally needs to fulfill its design parameters but must also avoid a significant immune response. The immune response to several nanostructured materials is presented in Sect. 1–3.

18.3. Immune Response to Implanted Nanostructured Materials

Nanostructured materials are used in vivo for dynamic therapies, such as drug delivery, tissue engineering, biosensing, and imaging. Nanowires and other non-particulate planar nanostructured materials have been proposed to alter cellular attachment and motility [35–38] detect analytes in vivo [39, 40] simulate tissue organization [7, 41–43] and serve as an alternative stent topography [14]. The interaction between nanostructured materials and elements of the body can be beneficial but also lead to a variety of adverse immune responses ranging from bacterial infection to tumorigenesis. A primary concern with increased use of nanomaterials is safety, as it relates to possible toxicity [44–47]. Numerous studies have linked inhaled nanoparticles to airway injury or disease such as coronary disease [44, 48–50]. The reported toxicity of nanoparticles brings into question the immunogenicity of non-particulate nanomaterials. Numerous studies have focused on the inflammation response to nanoparticles and carbon nanotubes, but little attention has focused on the immune response to nanostructured materials, such as nanowires, nanoporous membranes, and nanotubular surfaces. Recently, an increasing number of studies have characterized the inflammatory response of nanostructured materials both in vitro and in vivo. To understand the immune response to nanostructured materials, first the response to flat and nano-particulate materials will be reviewed.

18.3.1. Inflammatory Response to Biomaterials

The innate immune system is considered the primary response to a sterile foreign implanted material. In contrast to the adaptive immune system that responds to pathogen fragments presented on the surface of antigen presenting cells, the innate immune system responds through specific signal (or pattern) recognition [51–53]. The cells of the innate and adaptive immune system function cooperatively: one system can initiate the response to the other [51–53]. For the purposes of this chapter, however, innate functions will be highlighted since cells of that system are fundamental to the body's response to implants.

The immune system response to implanted biomaterials is depicted by the flow chart in Figure 18.2. Phagocytic cells, such as macrophages and neutrophils, comprise the primary response to the innate immune system [10, 54]. Upon implantation, proteins from the existing ECM and blood immediately adhere to the surface of the implant [11]. Inherent tissue damage caused by surgical implantation as well as protein adsorption and conformations at the material surface can lead to further immune response [10, 54, 55]. Tissue damage activates an immune response through release of "danger signals". Danger signals are proteins (released by cells that die through necrosis, viral infection, or stress), which are biochemical alarms that activate the immune system. According to the danger model, apoptotic cells do not release danger signals and do not promote an immune response [56]. Danger signals activate immune cells and promote immune cell chemotaxis (directed cellular migration induced by a cytokine concentration gradient), increased cell numbers at the site of implantation, and

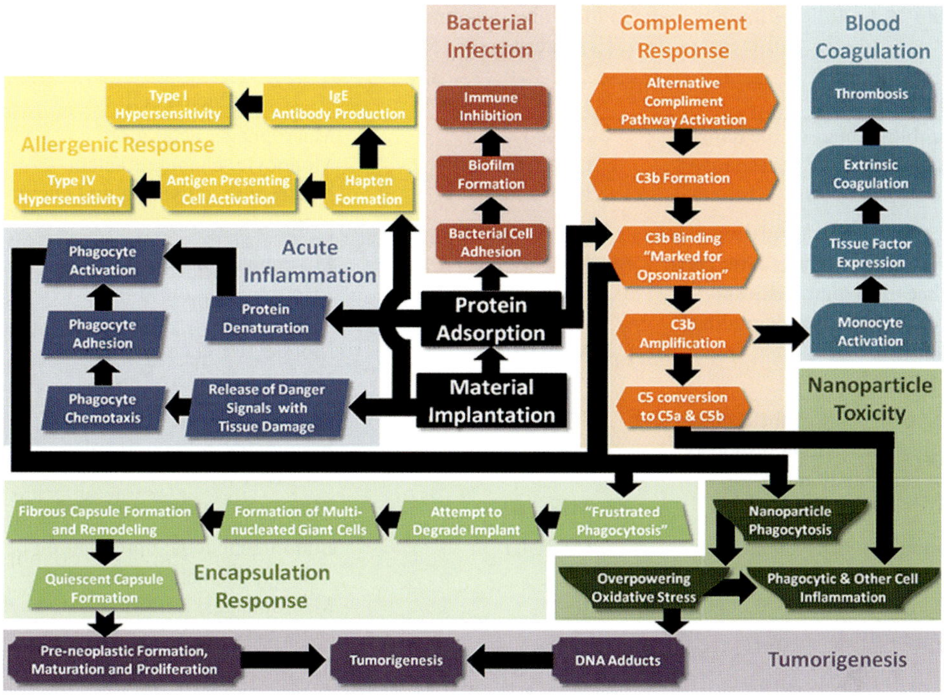

Figure 18.2. Schematic of potential immune responses to implanted planar and nanoparticulate materials. A cascade of possible immunological responses are depicted as initiated by material implantation and protein adsorption and are represented in the middle of the flow chart [10, 44, 55]. Most arrows indicate responses that have been well characterized; an incomplete arrow () represents a tentative immune response that is not yet well established.

activation of immune cells. Chemotaxis and activation of immune cells result in the body's inflammation response to implanted biomaterials.

Several cell types have functional roles in the inflammatory response, as listed in Table 18.1. Damage to fibroblasts, endothelial cells, and mast cells results in the release of danger signals initiating an inflammatory response. These signals can cue other cells to migrate to the area of implantation through chemotaxis or activate immune cells to further propagate the signal.

Immune cell chemotaxis primarily involves phagocytic cells. Initially neutrophils migrate to the site of inflammation, and as the short-lived cells die off, they are replaced by monocytes that can differentiate into macrophages or antigen-presenting dendritic cells. Dendritic cells stimulate T-cells by presenting antigens as part of the adaptive immune response. Cytokine signals from neutrophils and macrophages up-regulate adhesion molecules at the endothelial cell surface. These adhesion molecules (I-CAM for phagocytes and selectins for endothelial cell surfaces) facilitate the extravasation of additional phagocytes to the implant site. Once the phagocytes arrive at the implant site, they adhere to the material surface [51, 53]. For objects larger than a few microns, phagocytes are incapable of engulfing them and undergo "frustrated phagocytosis" [10, 54]. Frustrated phagocytosis is the result of the phagocyte inability to engulf a foreign object larger than itself [57]. Because the phagocytes cannot remove the foreign object by engulfment, they attempt to degrade the material surface [58], and the subsequent response leads to formation of a capsule that separates the implant from the tissue surrounding it. In order to form a continuous capsule, phagocytic

Table 18.1. Role of immune cells in the acute inflammation and encapsulation response to implanted materials [51, 53, 80].

Cell: Role	Cytokines	Affected cell: result
Fibroblasts: part of the danger-signal cascade in response to implantation	MCP-1	Macrophage: activation
	CXCL10, MIP-1α	Monocyte: chemoattractant
	IL-8	Neutrophil: activation and chemoattractant
Macrophages: phagocytic cells that form giant cells and release material degrading species	MCP-1, TNF-α	Macrophages: activation
	IL-6, GM-CSF	Macrophages: multi-nucleated cell formation
	MIP-1α & β	Monocytes: chemoattractant
	IL-8	Neutrophils: activation
Mast cells: part of the danger-signal cascade in response to implantation	TNF-α	Macrophages: activation Endothelial cells: upregulation of ICAM
	MIP-1α	Monocytes: chemoattractant
Monocytes: differentiate into either macrophages or dendritic cells upon chemical stimulation	CX3CL1	Endothelial cells: upregulation of leukocyte adhesion molecules
	IL-6	Macrophages: multi-nucleated cell formation
	MCP-1, TGF-β	Macrophages: activation and IL-1 production
	CXCL10	Monocytes: activation
Neutrophils: phagocytes, first responders to implanted material and marker of acute inflammation	MIP-1β	Monocytes: chemoattractant
Vascular endothelial cells: part of the danger-signal cascade in response to implantation	CX3CL1	Endothelial cells: upregulation of leukocyte adhesion molecules
	CXCL10, MIP-1β, Eotaxin, RANTES Fractalkine	Monocytes: chemoattractant
	RANTES	Monocytes: promote chronic inflammation
	IL-8, CXCL1, CXCL2, CXCL3	Neutrophils: activation and chemoattractant

cells fuse into multinucleated giant cells. These cells combine with ECM components and fibroblasts to form a cellular capsule [10, 54]. Within a month after implantation, a cell capsule can encase the implant, with a thickness that ranges from tens to hundreds of microns [59]. Once the implant is successfully surrounded, the capsule generally reaches a quiescent phase [10].

The wound healing response is characterized by three steps: inflammation, proliferation, and remodeling [10]. Inflammation brings monocytes and other phagotcytic cells to the implant site in an attempt to dilute and clear damaged tissues and cells. Once the area is clear of such debris, the body can begin to repair itself through cell proliferation and tissue remodeling. Proliferation of endothelial cells promotes vascularization at the implant site to augment damaged capillaries. For tissue surrounding the implant that cannot be correctly reinstated, fibroblastic tissue can turn into scar tissue. Multinucleated giant cells, formed as a result of frustrated phagocytosis, become surrounded by a fibrous capsule that isolates the implant [10, 54]. The fibrous capsule contains ECM components such as collagen, proteoglycans, and fibroblasts [54].

The body's inflammatory response to biomaterials has an acute short-term phase and a chronic long-term phase [10, 54]. Initially, acute inflammation is observed when phagocytic cells adhere to the foreign surface and become activated. Neutrophils migrate to the material surface in response to the release of danger signals associated with tissue damage [54]. These short-lived cells then induce trafficking (chemotaxis) of other monocytes to the surface. Monocytes can then differentiate into macrophages or dendritic cells [51, 53, 54]. Activation of macrophages through denatured protein or cytokine-signaling characterizes the acute response of the innate system to an implanted material surface [55]. Histological observations, such as edema and redness of tissue, are seen clinically with acute inflammation. Without excision of the implanted biomaterial, acute inflammation may progress to chronic inflammation [51].

Chronic inflammation takes many forms (Figure 18.2) and has been observed with both non-structured surfaces and nano-particles, such as metal particles [44, 47, 60, 61] and carbon nanotubes [46]. The toxicity of nanoparticles and carbon nanotubes has been linked to activation of phagocytic cells and resulting nanoparticle engulfment. Concentrations of nanoparticles of 1 mg/mL or greater can adversely affect tissues [62, 63]. When exposed to toxic levels of nanoparticles, phagocytic cells produce reactive oxygen species (ROS) that may cause either cell death or tumorigenesis. While nanoparticles can be engulfed by phagocytic cells, planar materials induce frustrated phagocytosis [10, 54]. With frustrated phagocytosis, cells initiate an encapsulation response, by which the implant is walled-off from surrounding tissue. Additionally, activated macrophages release products, such as ROS, in an attempt to degrade the material surface [58]. Macrophages then fuse together to wall off the implant from the surrounding tissue [10, 54, 64]. An encapsulation response rarely results in tumorigenesis [10, 65].

In addition to the encapsulation response, some of the immune responses depicted in Figure 18.2 are less common than acute inflammation but can still occur at the implant surface. A link between nanoparticle feature size and complement (C3b) binding has been reported, indicating the potential importance of C3b in nanostructured implants. Another critical immune response is tumorigensis. This response was observed in cases of chronic nanoparticle inhalation and mammoplasty. With concentrated chronic nanoparticle inhalation, oxidative stress overwhelms the cells the nanoparticles are in contact with; the resulting DNA damage can lead to tumorgenic cell formation, similar to clinical examples following asbestos exposure and mesothelioma formation [10].

18.3.2. Acute Inflammation and Encapsulation in Response to Nanostructured Biomaterials

A nanostructured device is defined as a planar-based material with at least one feature less than 100 nm in dimension. Materials such as nanowires, nanoporous metal oxides, nanotubular surfaces, and some microfabricated surfaces are all nanostructured surfaces. Nanoparticles fail to satisfy this definition since they lack a planar base and are purely spherical in nature. Additionally, the biocompatibility of nanoparticles has been thoroughly reviewed [44, 46, 47, 60, 61].

In Vitro Studies

The inflammatory response studied in vitro typically investigates the interaction of a single cell type (Table 18.1) with the nanostructured surface. Analysis of various inflammatory, connective tissue, and vascular cell responses, such as release of cytokines, morphology, and proliferation, may be indicative of an inflammatory response the material surface might generate in vivo. There is considerable evidence both in vitro and in vivo (presented in Tables18.2 and 18.3) that nanostructures can alter the inflammatory response compared to

Table 18.2. In vitro acute and chronic inflammation response to nanomaterials.

Material	Dimensions	Conclusion	Reference
Short PCL nanowires	196 nm diameter, 2.6 μm length	The short nanowire PCL surface induced decreased cytokine expression compared to conventional PCL, but the overall inflammation response was not significantly less	[68]
Long PCL nanowires	196 nm diameter, 27 μm length	The long PCL nanowire surface was significantly less inflammatory than the conventional PCL	
Titania nanotubes	79 nm diameter, 0.4 μm length	The nanotubular surface was significantly less inflammatory than the conventional titanium surface	
Short silicon oxide nanowires	40 nm diameter, 6–8 μm length	The short silicon oxide nanowires were significantly less inflammatory than the conventional glass surface	
Long silicon oxide nanowires	40 nm diameter, 30–40 μm length	The long silicon oxide nanowires induced a similar inflammatory response as the conventional glass surface.	
Nanoporous silicon	2–50 nm diameter pores	The nanoporous silicon surface induced a similar inflammatory response as the conventional silicon surface.	[66]
Nanochanneled silicon	100 nm channels	The nanochanneled silicon surfaces were equally inflammatory as the conventional silicon surface.	
Nanofibrous PTFE	20–30 nm width, 3–4 mm length	The nanofibrous PTFE surface was not immunogenic and might attenuate foreign body giant cell formation.	[14]
Ni-Co-Fe nanowire	75 nm diameter, 10 μm length	Compared to cell responses on conventional surface of the same material, macrophages, fibroblasts and endothelial cells on the nanowire surface, were smaller and less viable. The nanowire surface was more inflammatory.	[74]
Nanopitted PMMA	120 nm diameter, 100 nm depth	The chemokines and complement activation involved in inflammation were downregulated on the nanostructured surfaces compared to conventional PMMA.	[71]
Nanocolumnar PMMA	120 nm diameter, 100 nm high		
PHBV with hydroxyapatite	60 × 20 nm^2 crystals	Hydroxyapatite crystals in PHBV induced the lowest expression of pro-inflammatory signaling compared to all surfaces tested, including the non-nanostructured PHBV control.	[70]
PHBV with calcined hydroxyapatite	80 × 25 nm^2 crystals	PHBV with calcined hydroxyapatite crystals induced lower pro-inflammatory signaling than PHBV.	
PHBV with b-tricalcium phosphate	100–400 nm in diameter crystals	PHBV with b-tricalcium phosphate crystals induced lower pro-inflammatory signaling than PHBV.	

(continued)

Table 18.2. (Continued)

Material	Dimensions	Conclusion	Reference
Nanoporous alumina	20–200 nm in diameter	Nanoporous membranes with collagen I, fibrinogen, or blood serum proteins had limited influence on neutrophil activation. Higher initial ROS levels were induced on the non-coated 20 nm pore-size membrane compared to results obtained on the 200 nm membrane. On the 200 nm membrane, phagocyte activation time was longer ROS started decreasing later.	[67]
3-Hydroxybutyrate-co-3-hydroxyvalerate nanopits	100 nm depth, 120 nm diameter	Nanopits and nanocylinders induced decreased pro-inflammatory cytokine release compared to conventional surfaces of the same polymer.	[69]
PCL nanocylinders	160 nm height, 100 nm diameter		
Polystyrene-co-poly(4-bromo-styrene) Nanohills	95 nm in height	No differences were noted in the inflammation response observed on the nanohills and conventional controls.	[73]
Silicon nanogroves	2 and 10 μm width, 30–282 nm deep	Macrophages on nanochannels exhibited increased phagocytic activity.	[76]

Table 18.3. In vivo acute and chronic inflammation response to nanomaterials. The in vivo studies reported here were performed by subcutaneous implantation in the murine models.

Material	Dimensions	Conclusion	Reference
Titania nanoparticles in gelatin	20–60 nm diameter	TNF-α and IL-6 serum levels were similar to control at day 7 and 14 post implantation. The wound healing response observed for the nanostructured matrix at 12 days post implantation was superior to that of the control. Complete healing was observed only in the case of the nanostructured matrix at day 18 post implantation.	[77]
Titania nanotubes	79 nm diameter, 0.4 μm length	Lack of fibrous scar formation after 4 weeks post implantation.	[42]
Nanopits of 3-hydroxybutyrate-co-3-hydroxyvalerate	100 nm depth, 120 nm diameter	Nanopits induced increased fibrous capsule formation compared to PCL nanocylinders. The capsule, however, was not significantly different from that formed around the conventional polymer surface.	[69]
PCL nanocylinders	160 nm height, 100 nm diameter	Compared to nanopits, nanocylinders induced increased cellularity of the fibrous capsule and vascular density.	
Nanoporous titania	150 nm	After 7 and 12 days post implantation, the fibroblast layers were significantly fewer on the nanoporous implant surface than on conventional titanium.	[78]

conventional material surfaces. Most of the research regarding the inflammatory response to nanostructured materials concludes that structured surfaces are equally [14, 66–68] or less [68–70] inflammatory than a conventional surface of the same material, with one study of toxic nanowires finding increased inflammation [19].

Titania nanotubes significantly reduced the amount of inflammatory cytokines released from human monocytes adhered to the material surface. The nanotube surface was significantly less inflammatory, resulting in an increased rate of apoptosis of adherent monocytes due to lack of activation [68]. In the same study, long (27 microns in length) PCL nanowires and short (6–8 μm in length) silica nanowires were significantly less inflammatory than their conventional counterparts. Both nanostructured materials induced significantly down-regulated inflammatory cytokine expression from adherent human monocytes compared to a conventional surface of the same material [68]. In the Ainslie et al. [68] study, short PCL and long silica nanowires were also studied. Interestingly, the length of the nanowires led to different results for the two different materials, silicon oxide and PCL: the short nanowires were least inflammatory in the silicon oxide series (compared to the long nanowires) and vice versa for the PCL substrates. This disparity indicated that the inflammatory response in vitro is independent of the aspect ratio. Compared to nanowires, a reduced inflammatory response was also reported for several nanostructured materials with decreased aspect ratios. While the aforementioned nanowires had an aspect ratios on the order of 100, PMMA [71] and PCL [69] nanocolumns (cylinders) had aspect ratios on the order of 1. The inflammatory response to PMMA and PCL nanocolumns for the nanostructured surface was less than the conventional control. For a surface with nanopits of the same aspect ratio, the inflammatory response was also reduced compared to that in the case of the non-structured control [69, 71]. Furthermore, when low-aspect ratio crystals were embedded in a matrix to form a composite, reduced inflammatory response was observed [70]. Nanosized crystals of three different materials (hydroxyapatite, calcined hydroxyapatite, and tricalcium phosphate) embedded in poly(3-hydroxybutyrate-co-3-valerate) (PHBV) induced reduced inflammatory signaling compared to neat PHBV [70].

The independence of the inflammatory response from the aspect ratio of the nanostructured material would indicate that material properties (e.g., surface energy and surface charge) play a role in the inflammatory response. Indeed, it has been hypothesized that increased hydrophilicity, as a result of increased surface roughness [72], leads to decreased protein adsorption on Ni–Co–Fe nanowires [19]. By altering protein adsorption, subsequent cell adherence is certainly modulated and thus the cell-based inflammatory response is influenced. Additionally, surface charge may play a role. Among nanoparticles, cationic particles are more likely to induce inflammatory reactions than anionic particles and neutral species [61]. However, due to limitations in measuring surface charge of nanostructured materials, no data are available for the charge effects on these nanostructured surfaces.

A smaller subset of nanostructured materials induces an inflammatory response that is similar to that of the respective conventional material. A comparison of nanoporous silicon [66] and alumina [67] to their respective conventional surfaces reported that the activation of immune cells was similar. Similarly, the immune response to nanochanneled silicon was similar to that induced by conventional silicon [66]. The inflammatory response induced by short PCL and long silicon oxide nanowires was similar to that observed on conventional PCL and silicon, respectively [68]. Additionally, nanohills of polystyrene-co-poly(4-bromostyrene) induced a similar inflammatory response compared to a conventional surface of the same co-polymer [73]. Macrophages did not significantly upregulate their inflammatory response when chemically inert and biocompatible PTFE was investigated as a nanofibrous mat [14].

The results presented so far in this chapter provide evidence that nanostructured materials have the attractive property of being equal or less inflammatory than their respective conventional counterparts.

In contrast to the aforementioned studies that show equivalent or decreased inflammatory response to nanostructured materials at least one study found that nanostructures led to increased inflammation [74]. Nanowires of magnetostrictive and cytotoxic Ni–Co–Fe proved to be more inflammatory than a conventional surface of the same material [74]. Two of the three alloy metals, nickel and colbalt, are considered cytotoxic [75]. In contrast to nanostructures created from relatively inert materials, nanostructures of these toxic elements (Ni and Co), led to an increased inflammatory response.

Evaluation of the phagocytic activity alonc is inconclusive for increased macrophage activation, but it can be indicative of an increased inflammatory response. Increased phagocytic activity can be induced by the topography of material surfaces [76]. Inert silicon was used to form nanogrooves that induced an increased phagocytic activity in macrophages. Nanostructures fabricated using typically non-inflammatory materials produce non-inflammatory surfaces; in contrast when toxic elements are used to generate nanostructured materials, the inflammatory response increases. Increased interfacial surface area may result in increased release of toxic ions, such as Co and Ni, which induce elevated activation of inflammatory cells.

In Vivo Studies

Of the few in vivo studies performed on the inflammatory and wound healing response to nanomaterial implants, most report the thickness of the fibrotic capsule to indicate degree of immune activation (Table 18.3) [42, 69, 77]. Similar to observations made in the in vitro studies, the in vivo studies reported that nanostructures led to a decreased inflammatory response compared to the respective conventional surfaces. Murine peritoneal implantation of a nanotubular titania surface lacked fibrous scar formation at the implant site [42]. When nanoporous titania was implanted subcutaneously in a murine model, the number of fibroblast layers that was observed around the nanostructured implant (an indication of capsule thickness) was significantly less compared to results obtained on conventional titanium [78]. When nanocylindrical PCL, was implanted subcutaneously in a murine model, increased number of fibroblasts and increased vascular density were observed. PCL nanopits induced increased fibrous capsule formation but the inflammatory response was similar to that observed with conventional PCL. The wound healing response was enhanced with PCL nanocylinders compared to that observed with the inverted nanopit structures [69]. Improved wound healing response in a subcutaneous murine model was also observed when a composite material of titania nanoparticles in gelatin was used. Accelerated wound healing was induced by the nanostructured composite compared to that observed with the a conventional control of the same material [77].

In summary, in vivo studies indicated that not only was the inflammatory response attenuated in the presences of nanostructured materials, but that the wound healing process was accelerated. For applications such as biosensing, drug delivery, and tissue engineering, which require close proximity of the vasculature for analyte sensing, therapeutics delivery, and delivery of nutrients to newly formed tissue, accelerated wound healing and angiogenesis are most beneficial.

Nanostructured surfaces induce an immune response distinct from that of nanoparticles. Champion *et al.* concluded that, when the particle volume is greater than a typical cell (with a characteristic dimension of around 15 microns) macrophages do not engulf such particles [57, 79]. The largest nanostructured dimensions of the materials listed in Tables 18.2 and 18.3 are well under 15 microns, but since cells can not engulf the continuous material, frustrated phagocytosis results. In addition, nanoparticles with a volume larger than the typical cell dimension induce macrophage spreading [57, 79]. In contrast, other studies reported that, instead of macrophage spreading on nanostructured surfaces, these cells reduce in area [14, 66, 68, 74]. Besides phagocytosis of foreign materials and cell spreading, activated macrophages also generate ROS when exposed to concentrated nanoparticles [44]. ROS formation, however, was not significant when monocytes were incubated with nanostructured materials [44].

Ainslie *et al.* reported that monocytes in contact with a variety of nanostructured materials did not produce either ROS or superoxides [66, 68]. Lack of engulfment, cell spreading, and ROS formation indicates that, like conventional biomaterials, nanostructured materials, induce frustrated phagocytosis in vitro. In vivo data (Table 18.3) provides additional evidence that nanostructured materials, like conventional materials, induce frustrated phagocytosis that can result in limited fibrous capsule formation. Both in vitro and in vivo research has indicated that nanostructured materials do not provoke an immune response similar to that induced by nanoparticles, but rather induce a frustrated phagocytosis response similar to that associated with conventional implant biomaterials.

18.4. Concluding Remarks

Nanotechnology is becoming intricately linked with the study of the cell in its microenvironment. Nanomaterials, like nanofibers and self-assembled peptides, are emerging as promising alternatives for three-dimensional in vivo-like microenvironments. Surfaces with nanostructured features, like wires, tubes and pits, provide contact guidance and enable design and fabrication of novel implantable tissue engineered scaffolds. Such technologies pave the way for future discoveries as well as the continually merging of medicine, biology, and engineering.

Nanostructured surfaces both attenuate the acute inflammation response of immune cells to implanted materials, and affect the encapsulation response. Some general conclusions can be ascertained from the studies reported in the literature:

- The inflammatory response around nanostructured materials is primarily an innate system immune response.
- The immune response is independent of the nanostructure aspect ratio.
- Nanostructures induce decreased inflammation, perhaps due to their increased surface energy affecting protein adsorption and, therefore, subsequent cell adhesion.
- Nanomaterials that release toxic molecules interact adversely with surrounding tissue in vivo.
- The time course of wound healing is decreased in the presence of nanostructures.
- Whereas nanoparticles may induce toxicity, nanostructured materials induce "frustrated phagocytosis" responses by macrophages. This response can lead to an altered encapsulation response.

Like other aspects of biocompatibility, further studies are needed on a per-material-basis to properly evaluate the inflammation potential of nanomaterials. In addition to understanding how nanostructured materials interact with the human body, insight into these materials and their interactions with the immune system may aid in developments like novel vaccines and tolerance therapies.

References

1. Yurchenco PD, Birk DE, Mecham RP, editors. Extracellular matrix assembly and structure. San Diego: Academic Press, Inc.; 1994.
2. Liu HA, Webster TJ. Nanomedicine for implants: A review of studies and necessary experimental tools. Biomaterials. 2007;28(2):354–69.
3. Swan EEL, Popat KC, Grimes CA, Desai TA. Fabrication and evaluation of nanoporous alumina membranes for osteoblast culture. J Biomed Mater Res Part A. 2005;72A(3):288–95.
4. Popat KC, Daniels RH, Dubrow RS, Hardev V, Desai TA. Nanostructured surfaces for bone biotemplating applications. J Orthop Res. 2006;24(4):619–27.
5. Park GE, Pattison MA, Park K, Webster TJ. Accelerated chondrocyte functions on NaOH-treated PLGA scaffolds. Biomaterials. 2005;26(16):3075–82.
6. Patel S, Kurpinski K, Quigley R, Gao H, Hsiao BS, Poo MM, et al. Bioactive nanofibers: Synergistic effects of nanotopography and chemical signaling on cell guidance. Nano Lett. 2007;7(7):2122–28.
7. Huang NF, Patel S, Thakar RG, Wu J, Hsiao BS, Chu B, et al. Myotube assembly on nanofibrous and micropatterned polymers. Nano Lett. 2006;6(3):537–42.
8. Thapa A, Miller DC, Webster TJ, Haberstroh KM. Nano-structured polymers enhance bladder smooth muscle cell function. Biomaterials. 2003;24(17):2915–26.
9. Lu J, Rao MP, MacDonald NC, Khang D, Webster TJ. Improved endothelial cell adhesion and proliferation on patterned titanium surfaces with rationally designed, micrometer to nanometer features. Acta Biomaterialia. 2008;4(1):192–201.
10. Anderson JM, Cook G, Costerton B, Hanson SR, Hensten-Pettersen A, Jacobsen N, et al. Host reactions to biomaterials and their evalulation. In: Ratner BD, Hoffmann AS, Schoen FJ, Lemons JE, editors. Biomaterials science. 2nd ed. San Francisco: Elsevier Academic; 2004.
11. Williams DF. General concepts of biocompatibility. In: Black J, Hastings G, editors. Handbook of biomaterial properties. New York: Chapman & Hall; 1998.
12. Matthews JA, Boland ED, Wnek GE, Simpson DG, Bowlin GL. Electrospinning of collagen type II: A feasibility study. J Bioactive and Compatible Polymers. 2003;18(2):125–34.
13. Zong X, Bien H, Chung CY, Yin L, Fang D, Hsiao BS, et al. Electrospun fine-textured scaffolds for heart tissue constructs. Biomaterials. 2005;26(26):5330–38.
14. Ainslie KM, Bachelder EM, Borkar S, Zahr AS, Sen A, Badding JV, et al. Cell adhesion on nanofibrous polytetrafluoroethylene (nPTFE). Langmuir. 2007; 23(2):747–54.
15. Grimes CA. Synthesis and application of highly ordered arrays of TiO2 nanotubes. J Mater Chem. 2007;17(15):1451–57.
16. Reimhult E, Kumar K, Knoll W. Fabrication of nanoporous silicon nitride and silicon oxide films of controlled size and porosity for combined electrochemical and waveguide measurements. Nanotechnology. 2007;18(27).
17. Jung M, Lee S, Jhon YM, Mho SI, Cho JW, Woo D. Nanohole arrays with sub-30 nm diameter formed on GaAs using nanoporous alumina mask. Jpn J Appl Phys Part 1. 2007;46(7A):4410–12.
18. Sharma G, Pishko MV, Grimes CA. Fabrication of metallic nanowire arrays by electrodeposition into nanoporous alumina membranes: effect of barrier layer. J Mater Sci. 2007;42(13):4738–44.
19. Ainslie KM, Sharma G, Dyer MA, Grimes CA, Pishko MV. Attenuation of protein adsorption on static and oscillating magnetostrictive nanowires. Nano Lett. 2005;5(9):1852–56.
20. Sharma G, Grimes CA. Synthesis, characterization, and magnetic properties of FeCoNi ternary alloy nanowire arrays. J Mater Res. 2004;19(12):3695–703.
21. Tao SL, Desai TA. Aligned arrays of biodegradable poly(epsilon-caprolactone) nanowires and nanofibers by template synthesis. Nano Lett. 2007;7(6):1463–68.
22. Cheng J, Fortina P, Surrey S, Kricka LJ, Wilding P. Microchip-based devices for molecular diagnosis of genetic diseases. Mol Diagn. 1996;1(3):183–200.
23. Flaim CJ, Chien S, Bhatia SN. An extracellular matrix microarray for probing cellular differentiation. Nat Methods. 2005;2(2):119–25.
24. Hudson TJ, Stein LD, Gerety SS, Ma J, Castle AB, Silva J, et al. An STS-based map of the human genome. Science. 1995 22;270(5244):1945–54.
25. Lockhart DJ, Dong H, Byrne MC, Follettie MT, Gallo MV, Chee MS, et al. Expression monitoring by hybridization to high-density oligonucleotide arrays. Nat Biotechnol. 1996;14(13):1675–80.
26. McIntyre PM. Microfabrication technology for DNA sequencing. Trends Biotechnol. 1996;14(3):69–73.
27. Westin L, Xu X, Miller C, Wang L, Edman CF, Nerenberg M. Anchored multiplex amplification on a microelectronic chip array. Nat Biotechnol. 2000;18(2):199–204.

28. Norman J, Desai T. Methods for fabrication of nanoscale topography for tissue engineering scaffolds. Ann Biomed Eng. 2006;34(1):89–101.
29. Madou M. Fundamentals of microfabrication: The science of miniaturization. 2nd ed. New York, NY: CRC; 2002.
30. Lee H, Lee BP, Messersmith PB. A reversible wet/dry adhesive inspired by mussels and geckos. Nature 2007;448(7151):338–41.
31. Seunarine K, Gadegaard N, Tormen M, O Meredith D, O Riehle M. Wilkinson CDW. 3D polymer scaffolds for tissue engineering. Nanomedicine. 2006;1(3):281–96.
32. Sarkar S, Dadhania M, Rourke P, Desai TA, Wong JY. Vascular tissue engineering: Microtextured scaffold templates to control organization of vascular smooth muscle cells and extracellular matrix. Acta Biomaterialia. 2005;1(1):93–100.
33. Sarkar S, Lee GY, Wong JY, Desai TA. Development and characterization of a porous micro-patterned scaffold for vascular tissue engineering applications. Biomaterials. 2006;27(27):4775–82.
34. Ainslie KM, Desai TA. Lab Chip. 2008;8(11):1864–78.
35. Curtis AS, Gadegaard N, Dalby MJ, Riehle MO, Wilkinson CD, Aitchison G. Cells react to nanoscale order and symmetry in their surroundings. IEEE Trans Nanobioscience. 2004;3(1):61–5.
36. Dalby MJ, Riehle MO, Sutherland DS, Agheli H, Curtis AS. Changes in fibroblast morphology in response to nano-columns produced by colloidal lithography. Biomaterials. 2004;25(23):5415–22.
37. Gallagher JO, McGhee KF, Wilkinson CD, Riehle MO. Interaction of animal cells with ordered nanotopography. IEEE Trans Nanobioscience. 2002;1(1):24–8.
38. Yim EK, Reano RM, Pang SW, Yee AF, Chen CS, Leong KW. Nanopattern-induced changes in morphology and motility of smooth muscle cells. Biomaterials. 2005;26(26):5405–13.
39. Cui Y, Wei Q, Park H, Lieber CM. Nanowire nanosensors for highly sensitive and selective detection of biological and chemical species. Science. 2001;293(5533):1289–92.
40. Lin H, Datar RH. Medical applications of nanotechnology. Natl Med J India. 2006;19(1):27–32.
41. Kim DHK, P.; Suh, KY, Choi SK, Lee SH, Kim B. Modulation of adhesion and growth of cardiac myocytes by surface nanotopography. 27th Annual International Conference of the Engineering in Medicine and Biology Society. 2005:4091–4.
42. Popat KC, Leoni L, Grimes CA, Desai TA. Influence of engineered titania nanotubular surfaces on bone cells. Biomaterials. 2007;28(21):3188–97.
43. Schindler M, Nur EKA, Ahmed I, Kamal J, Liu HY, Amor N, et al. Living in three dimensions: 3D nanostructured environments for cell culture and regenerative medicine. Cell Biochem Biophys. 2006;45(2):215–27.
44. Nel A, Xia T, Madler L, Li N. Toxic potential of materials at the nanolevel. Science. 2006;311(5761):622–7.
45. Tsuji JS, Maynard AD, Howard PC, James JT, Lam CW, Warheit DB, et al. Research strategies for safety evaluation of nanomaterials, part IV: Risk assessment of nanoparticles. Toxicol Sci. 2006;89(1):42–50.
46. Lam CW, James JT, McCluskey R, Arepalli S, Hunter RL. A review of carbon nanotube toxicity and assessment of potential occupational and environmental health risks. Crit Rev Toxicol. 2006;36(3):189–217.
47. Xia T, Kovochich M, Brant J, Hotze M, Sempf J, Oberley T, et al. Comparison of the abilities of ambient and manufactured nanoparticles to induce cellular toxicity according to an oxidative stress paradigm. Nano Lett. 2006;6(8):1794–807.
48. Inoue K, Takano H, Yanagisawa R, Sakurai M, Ichinose T, Sadakane K, et al. Effects of nano particles on antigen-related airway inflammation in mice. Respir Res. 2005;6.
49. Sayes CM, Wahi R, Kurian PA, Liu YP, West JL, Ausman KD, et al. Correlating nanoscale titania structure with toxicity: A cytotoxicity and inflammatory response study with human dermal fibroblasts and human lung epithelial cells. Toxicol Sci. 2006;92(1):174–85.
50. Yamawaki H, Iwai N. Mechanisms underlying nano-sized air-pollution-mediated progression of atherosclerosis – Carbon black causes cytotoxic injury/inflammation and inhibits cell growth in vascular endothelial cells. Circ J. 2006;70(1):129–40.
51. Paul WE. Fundamental immunology. 4th ed. New York: Raven; 1994.
52. Mitchell RN. Innate and adaptive immunity: The immune response to foreign material. In: Ratner BD, Hoffmann AS, Schoen FJ, Lemons JE, editors. Biomaterials science: An introduction to materials in medicine. 2nd ed. San Francisco: Elsevier; 2004.
53. Janeway C, Travers P, Walport M, Shlomchik M. Immunobiology. 5th ed. New York: Garland Science; 2001.
54. Anderson JM. Soft tissue response. In: Black J, Hastings G, editors. Handbook of biomaterial properties. New York: Chapman & Hall; 1998.
55. Tang L, Eaton JW. Natural responses to unnatural materials: A molecular mechanism for foreign body reactions. Mol Med. 1999;5(6):351–8.
56. Gallucci S, Lolkema M, Matzinger P. Natural adjuvants: endogenous activators of dendritic cells. Nat Med. 1999;5(11):1249–55.

57. Champion JA, Mitragotri S. Role of target geometry in phagocytosis. Proc Natl Acad Sci U S A. 2006;103(13):4930–4.
58. McNally AK, Anderson JM. Foreign body-type multinucleated giant cell formation is potently induced by alpha-tocopherol and prevented by the diacylglycerol kinase inhibitor R59022. Am J Pathol. 2003;163(3):1147–56.
59. Padera RF, Colton CK. Time course of membrane microarchitecture-driven neovascularization. Biomaterials. 1996;17(3):277–84.
60. BeruBe K, Balharry D, Sexton K, Koshy L, Jones T. Combustion-derived nanoparticles: Mechanisms of pulmonary toxicity. Clin Exp Pharmacol Physiol. 2007;34(10):1044–50.
61. Dobrovolskaia MA, McNeil SE. Immunological properties of engineered nanomaterials. Nat Nanotechnol. 2007;2(8):469–78.
62. Barlow PG, Clouter-Baker A, Donaldson K, Maccallum J, Stone V. Carbon black nanoparticles induce type II epithelial cells to release chemotaxins for alveolar macrophages. Part Fibre Toxicol. 2005;2:11.
63. Davoren M, Herzog E, Casey A, Cottineau B, Chambers G, Byrne HJ, et al. In vitro toxicity evaluation of single walled carbon nanotubes on human A549 lung cells. Toxicol In Vitro. 2007;21(3):438–48.
64. Anderson JM, Defife K, McNally A, Collier T, Jenney C. Monocyte, macrophage and foreign body giant cell interactions with molecularly engineered surfaces. J Mater Sci Mater Med. 1999;10(10/11):579–88.
65. Tanaka Y, Morishima I, Kikuchi K. Invasive micropapillary carcinomas arising 42 years after augmentation mammoplasty: A case report and literature review. World J Surg Oncol. 2008;6:33.
66. Ainslie KM, Tao SL, Popat KC, Desai TA. In vitro immunogenicity of silicon based micro- and nano-structured surfaces. ACS Nano. 2008;2(5):1076–84.
67. Karlsson M, Tang L. Surface morphology and adsorbed proteins affect phagocyte responses to nano-porous alumina. J Mater Sci Mater Med. 2006;17(11):1101–11.
68. Ainslie KM, Tao SL, Popat KC, Daniels H, Hardev V, Grimes CA, Desai TA. J Biomed Mater Res A. 2008 5 [Epub ahead of print].
69. Giavaresi G, Tschon M, Daly JH, Liggat JJ, Sutherland DS, Agheli H, et al. In vitro and in vivo response to nanotopographically-modified surfaces of poly(3-hydroxybutyrate-co-3-hydroxyvalerate) and polycaprolactone. J Biomater Sci Polym Ed. 2006;17(12):1405–23.
70. Cool SM, Kenny B, Wu A, Nurcombe V, Trau M, Cassady AI, et al. Poly(3-hydroxybutyrate-co-3-hydroxyvalerate) composite biomaterials for bone tissue regeneration: In vitro performance assessed by osteoblast proliferation, osteoclast adhesion and resorption, and macrophage proinflammatory response. J Biomed Mater Res A. 2007;82(3):599–610.
71. Dalby MJ, Gadegaard N, Herzyk P, Agheli H, Sutherland DS, Wilkinson CD. Group analysis of regulation of fibroblast genome on low-adhesion nanostructures. Biomaterials. 2007;28(10):1761–9.
72. Bico J, Tordeux C, Quere D. Rough wetting. Europhys Lett. 2001;55(2):214–20.
73. Dalby MJ, Marshall GE, Johnstone HJ, Affrossman S, Riehle MO. Interactions of human blood and tissue cell types with 95-nm-high nanotopography. IEEE Trans Nanobioscience. 2002;1(1):18–23.
74. Ainslie KM, Bachelder EM, Sharma G, Grimes C, Pishko MV. Macrophage cell adhesion and inflammation cytokines on magnetostrictive nanowires. Nanotoxicology. 2007;1(4):279–90.
75. Little MC, Gawkrodger DJ, MacNeil S. Chromium- and nickel-induced cytotoxicity in normal and transformed human keratinocytes: an investigation of pharmacological approaches to the prevention of Cr(VI)-induced cytotoxicity. Br J Dermatol. 1996;134(2):199–207.
76. Wojciak-Stothard B, Curtis A, Monaghan W, MacDonald K, Wilkinson C. Guidance and activation of murine macrophages by nanometric scale topography. Exp Cell Res. 1996 15;223(2):426–35.
77. Peng CC, Yang MH, Chiu WT, Chiu CH, Yang CS, Chen YW, et al. Composite nano-titanium oxide-chitosan artificial skin exhibits strong wound-healing effect-an approach with anti-inflammatory and bactericidal kinetics. Macromol Biosci. 2008 9;8(4):316–27.
78. Areva S, Paldan H, Peltola T, Narhi T, Jokinen M, Linden M. Use of sol-gel-derived titania coating for direct soft tissue attachment. J Biomed Mater Res A. 2004;70(2):169–78.
79. Champion JA, Katare YK, Mitragotri S. Particle shape: A new design parameter for micro- and nanoscale drug delivery carriers. J Control Release. 2007;121(1–2):3–9.
80. Lemaire I, Yang H, Lafont V, Dornand J, Commes T, Cantin MF. Differential effects of macrophage- and granulocyte-macrophage colony-stimulating factors on cytokine gene expression during rat alveolar macrophage differentiation into multinucleated giant cells (MGC): Role for IL-6 in type 2 MGC formation. J Immunol. 1996;157(11):5118–25.
81. Desai TA, Chu WH, Tu JK, Beattie GM, Hayek A, Ferrari M. Microfabricated immunoisolating biocapsules. Biotechnol Bioeng. 1998;57(1):118–20.

19

Collagen I-Coated Titanium Surfaces for Bone Implantation

Marco Morra, Clara Cassinelli, Giovanna Cascardo, and Daniele Bollati

Biological interactions at the tissue/implant material interface can be modulated by surface-linked cell-signalling biological molecules. Collagen type I, the main extracellular matrix protein of bone tissue, has been widely investigated in biomolecular surface modification of bone-contacting titanium implant devices. Literature reports on the biological effects of collagen-based coatings are, however, often contradictory. From a biomolecular surface-engineering perspective, a possible explanation is that the definition "collagen-coated surface" encompasses widely different molecular and supramolecular structures: adsorbed collagen, covalently linked collagen, crosslinked collagen, fibrillar versus monomeric collagen, and many other variation of this theme. Relevant details are not always described and proper surface characterization is often lacking. This chapter attempts to build up a rational frame of reference to describe surface modification of implant devices by collagen type I from a surface chemistry point of view, as well as to discuss relevant implications for process design.

Abbreviations

AFM	atomic force microscopy
Arg–Gly–Asp	arginine–glycine–aspartic acid
BMTiS	biochemical modification of titanium surfaces
Co	cobalt
DAE	double acid etched
ECM	extracellular matrix
EDC	1-ethyl-3-(3-dimethylaminopropyl)-carbodiimide
NHS	*N*-hydroxysuccinimide
OC	osteocalcin
OP	osteopontin
PBS	phosphate-buffered saline
PDGF	platelet-derived growth factor

M. Morra, C. Cassinelli, G. Cascardo, and D. Bollati • Nobil Bio Ricerche s.r.l, V. Valcastellana 26, 14037 Portacomaro (AT), Italy

D.A. Puleo and R. Bizios (eds.), *Biological Interactions on Materials Surfaces*,
DOI 10.1007/978-0-387-98161-1_19,

PDL	periodontal ligament
PEG	poly(ethylene glycol)
RGD	arginine–glycine–aspartic acid
RT-PCR	real-time polymerase chain reaction
Ti	titanium
Ti6Al4V	titanium/aluminum/vanadium alloy
ToF-SIMS	time-of-flight static secondary ion mass spectroscopy
UHV	ultra-high vacuum
XPS	X-ray photoelectron spectroscopy

19.1. Introduction

Titanium has gained wide acceptance for load-bearing, bone-contacting devices since the pioneering work of Branenmark. Nowadays, Ti implant devices are used for a variety of applications; most of the techniques in use are evidence-based and predictable. Because interactions at the bone-implant interface are recognized as the key to osteointegration and the literature on titanium surfaces and interfaces is extensive [1–4], numerous approaches focusing on the surface modification of titanium to further improve clinical results and extend the spectrum of biomedical applications have been developed. Despite its significant success as a biomaterial, Ti and its surface are still actively investigated for applications such as dental implantology, where the need exists to address difficult clinical settings, e.g., an intended implant site compromised because of poor bone quality. Examples of poor bone quality include, for instance, low bone density in the case of highly cancellous bone and insufficient quantity of bone (in the case of the width of the alveolar ridge). In the case of artificial vertebral discs, Cunningham recently suggested that the most important and challenging aspect for the success of orthopaedic implants is to achieve osseointegration at the bone-metal interface while preserving the necessary biomechanical properties of motion [5]. In both dental implantology and orthopaedic applications, increased life expectancy of human subjects poses new challenges: novel surface treatment of implant devices are needed to improve bone density locally, to accelerate healing, and to perform with dependability during integration even in old or pathologic bone for long periods of time.

Traditionally, the approach to surface modification of titanium has been based on the control of surface topography [3], on physicochemical [6] and inorganic approaches, and on the use of ceramic coatings [7]. Presently, a significant research effort is aimed at the biochemical modification of titanium surfaces (BMTiS); these strategies were described by Puleo and Nanci [8] as follows: "Biochemical surface modification endeavours to utilize current understanding of the biology and biochemistry of cellular function and differentiation. Much has been learned about the mechanisms by which cells adhere to substrates, and major advances have been made in understanding the role of biomolecules in regulating differentiation and remodelling of cells and tissues, respectively. The goal of biochemical surface modification is to immobilize proteins, enzymes, or peptides on biomaterials for the purpose of inducing specific cell and tissue responses or, in other words, to control the tissue implant interface with molecules delivered directly to the interface.... In contrast to calcium phosphate coatings, biochemical surface modification utilizes critical organic components of bone to affect tissue response."

Among the organic components of bone, collagen is most important and of special interest: the extracellular matrix (ECM) of bone contains approximately 85% type I collagen.

Cell interactions with ECM components (such as adhesive proteins and growth factors) are crucial in the regulation of the osteoblast phenotype. Through the amino acid sequence Arg–Gly–Asp (RGD), collagen mediates adhesion of cells of direct relevance to orthopaedic applications [9]. Collagen also plays an important role in the osteoblastic differentiation of bone marrow cells and controls various aspects of their progression along the osteogenic pathway [10–12]. Collagen also interacts with other biomolecules and growth factors, providing the cooperative signalling described by Sampath and Reddi in their report on the bioactivity of the bone morphogenetic proteins: "Combining the extract with the collagenous matrix rendered the recombined matrix osteoinductive… this allows to enunciate the concept of a collaboration between a soluble signal and the insoluble ECM substratum to initiate new bone formation" [13]. Collagen also exhibits a strong procoagulant (hemostatic) activity and activates platelets in a unique way [14]; this aspect is of special interest because of the importance of blood-implant interactions in the early stages of bone healing [15–17]. For the aforementioned reasons, collagen is used in dental and orthopaedic surgery as osteogenic and bone-filling material [18, 19] to promote regeneration of bone defects. In the spirit of BMTiS [8], implant devices covered with a surface layer of collagen could make the properties of this interesting biomolecule available directly at the site of implantation. Indeed, as reviewed in Sect. 19.2, the in vitro and in vivo response to collagen-coated, bone-contacting materials has been addressed in a number of studies. Some of the literature reports are conflicting and even in vitro results are sometime contradictory. Improvements were observed in some but not in other studies. This situation has been referred to as "the grand challenge of molecular surface chemistry" [20]. As for any other surface-modification process involving immobilization of biomolecules, the general definition "collagen-coated surface" encompasses many possible surface structures: relevant details concerning molecular interactions, such as surface density, stability in aqueous media, resistance to hydrolysis and degradation, conformational freedom, presentation of active epitopes, and supramolecular assembly, are directly related to the way the biomolecules are linked to the surface, that is, to the specific surface-engineering process used.

This chapter tries to establish a rationale for the description of structure and properties of collagen-coated Ti surfaces by focusing on some of the most important surface parameters involved. The chapter includes an initial short review of existing data on the subject. Then, several key aspects of surface modification of Ti implants by collagen type I are described; relevant topics (such as the respective merits of monomeric and fibrillar collagen, how much collagen should be present on the surface, how to detect it, and whether it should be crosslinked, covalently bonded, or simply adsorbed), are discussed. The final section summarizes the main points presented in the chapter.

19.2. Literature Reports Regarding Collagen-Coated Ti Surfaces

19.2.1. In Vitro Studies of Collagen-Coated Ti Surfaces

Collagen coatings, applied by simple adsorption on material surfaces, have long been used in cell culture studies. Jikko and coworkers [10] addressed the role of collagen and its adhesive peptide epitopes on osteoblast differentiation. Osteogenic precursor cells adhere to collagen-coated Ti surfaces. The underlying mechanisms involve cell responses to cues from their extracellular environment such as adhesive proteins and diffusible molecules (growth factors). Early papers published in the scientific literature on surface modification of Ti by proteins were mostly concerned with improving adhesion of soft tissues on implants in order to achieve stable seals and avoid bacterial invasion at the bone-contacting regions. For this purpose, Lowenberg and coworkers [21] coated porous titanium/aluminum/vanadium alloy (Ti6Al4V) discs with either collagen or collagen containing platelet-derived growth factor (PDGF). Compared with respective controls, these constructs promoted significantly higher attachment and orientation indices of gingival fibroblasts in vitro; the orientation index was further enhanced on the PDGF-containing collagen coatings. It was concluded that attachment and orientation of cells on Ti can be modified by the application of biological molecules, suggesting a useful application in clinical implantology. Matsumura and coworkers developed and applied their methodology for surface immobilization of collagen on Ti implants [22–24]. Briefly, collagen was linked to poly(ethylene-co-vinyl alcohol) coatings on Ti implants. Human periodontal ligament (PDL) cells adhered on this "hybrid" material; these results suggested that these materials could promote PDL regeneration on dental implants.

Geissler and coworkers [25] investigated adhesion of rat calvarial osteoblasts to uncoated and collagen type I-coated titanium over a period of 24 h. Collagen was passively adsorbed, with and without fibrillogenesis, on Ti6Al4V disks. In the presence of fetal calf serum, the collagen type I coatings accelerated initial adhesion of osteoblasts; cell adhesion, however, was similar on all surfaces tested 1 h after cell seeding. Investigation of the underlying mechanisms using soluble peptides indicated that osteoblasts adhered to collagen type I-coated materials via integrin β1 but did interact with the RGD domain; in contrast, osteoblast adhesion to the uncoated titanium alloy was mediated by RGD sequences but not via integrin β1. This result deserves some further consideration and reflection. It is sometimes claimed that one of the objectives of coating implant devices with ECM proteins is to provide a surface more suitable for cell adhesion. This claim is not necessarily the case: like many other rigid, high-energy materials, unmodified Ti surface, in general, is a very good substrate for cell adhesion [26]. A variable extent of surface contamination of Ti, in particular from low-energy hydrocarbons [27], is often the reason behind conflicting reports of cell adhesion results observed on coated versus noncoated Ti. Cell adhesion to the ECM is coupled to signal transduction; in fact, most adhesion receptors function as signalling molecules. Engagement of different receptors leads to a wide variety of intracellular signals, often acting synergistically with signals derived from activation of growth factor receptors. Thus, material coatings that mimic the ECM are not simply a scaffold for cell adhesion, but may also provide cues that regulate cell responses such as differentiation and various cell functions pertinent to new tissue formation. An example is shown in Figure 19.1. Figure 19.1 shows a fluorescence micrograph illustrating SaOS-2 cell adhesion onto machined Ti and double acid etched (DAE) Ti, with and without a covalently linked layer of fibrillar collagen type I, at a surface density of less than 1 $\mu g/cm^2$ [28]. Compared with results obtained on uncoated surfaces, the collagen coating did not affect significantly the cell density. Specifically, reduced cell proliferation and impairment of cell spreading were observed on rough DAE surfaces [29]. The flat phenotype and the high cell proliferation rate on machined Ti is associated with lower differentiation of osteoblast cells; this in vitro observation provides an explanation for the superior clinical performances of DAE (compared with machined) surfaces [29]. As illustrated in Figure 19.1, both cell density and morphology were similar on the collagen-coated surfaces. While, at least in the present case, cell number and morphology is apparently controlled solely by surface topography (and not by surface chemistry), specific markers such as alkaline phosphatase activity was enhanced on the two collagen-coated surfaces tested [28]. Moreover, in vivo tests in rabbits confirmed an acceleration of the osteointegration process of collagen-coated (versus uncoated) DAE Ti implants. These results provided evidence that ECM-based signalling at the biomaterial interface can lead to enhanced cell function.

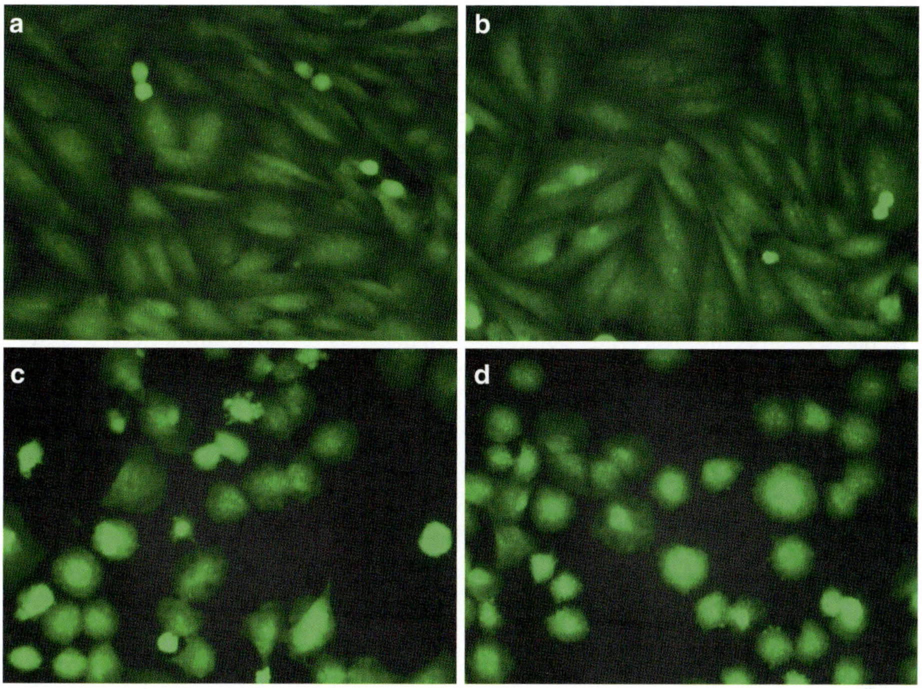

Figure 19.1. Fluorescence micrographs of SaOS-2 osteoblast cells after 84 h of culture on: (**a**) machined Ti; (**b**) collagen-coated, machined Ti; (**c**) double acid-etched (DAE) Ti; and (**d**) collagen-coated DAE Ti. The collagen type I coating was obtained by covalent linking of adsorbed fibrillar collagen onto Ti aminated by deposition from allylamine plasma. Original magnification, ×200.

The effect of collagen coating on titanium on the initial attachment of human gingival fibroblasts was investigated by Nagai and coworkers [30]. Scanning electron microscopy was used to investigate the morphological changes of cultured human gingival fibroblasts on four different surfaces, i.e., noncoated, mirror-polished titanium, collagen-coated titanium, noncoated tissue culture polystyrene, and collagen-coated polystyrene. Collagen coating of titanium effectively promoted initial cell attachment, thus improving subsequent function of human gingival fibroblasts.

Roehlecke and coworkers [31] investigated the function (including cell attachment, spreading, cytoskeletal organization, focal contact formation, proliferation, and expression of a differentiated phenotype) of primary osteoblasts cultured on both fibrillar collagen-coated and tropocollagen-coated Ti6A14V. In comparison with results obtained on uncoated titanium alloy, collagen-coated alloy enhanced spreading and resulted in rapid formation of focal adhesions and their associated stress fibers. Osteoblast proliferation was enhanced and the intracellular expression of osteopontin was upregulated on collagen-coated samples. In contrast, Becker and coworkers [32], who evaluated proliferation, differentiation, and mineralization of osteoblasts on type I collagen-coated Ti6Al4V, reported no extensive effect on cell proliferation, the activity of alkaline phosphatase activity, collagen synthesis, calcium accumulation in the ECM, and upregulation of the messenger RNA levels for collagen I alpha1, osteopontin, osteocalcin, matrix metalloproteinase, and tissue inhibitor of matrix metalloproteinase. In that study, adsorption of collagen was performed at 25°C for 15 min in phosphate buffer at pH 7, yielding a surface density of 3–6 μg/cm^2. The results obtained in the Becker and coworkers study [32] suggest that collagen coating alone is not sufficient to accelerate differentiation of rat calvarial osteoblasts on Ti6Al4V.

Similarly, Van den Dolder and coworkers [33] reported lack of significant effects of various protein coatings on biomaterials. The objective of that study was to evaluate the effects of fibronectin and collagen I coatings on titanium fiber mesh on the proliferation and osteogenic differentiation of rat bone marrow cells. Three treatment groups were investigated and compared with uncoated titanium fiber meshes: Ti meshes were coated with (1) fibronectin, (2) with collagen I, and (3) first with collagen I and subsequently with fibronectin. Rat bone marrow cells were cultured for 1, 4, 8, and 16 days on the coated titanium fiber meshes and uncoated control meshes. DNA quantification analysis revealed similar cell proliferation on all surfaces tested. When antibodies against fibronectin and collagen I integrins were used, significant reduction in cell proliferation was observed on the uncoated titanium meshes, as well as on meshes coated with collagen, and meshes coated with collagen and fibronectin. The different coatings also did not affect the alkaline phosphatase activity of the cells seeded on the coated meshes. Altogether, these results provided evidence that neither a fibronectin nor a collagen I coating stimulated the differentiation of rat bone marrow cells seeded on titanium fiber meshes.

Bierbaum and coworkers [34, 35] used coatings of adsorbed collagen I, collagen III, and fibronectin and investigated the effect of fibrillogenesis and fibril morphology on adsorption. Increasing the collagen type III density (between 40 and 60 μg/cm^2) resulted in decreased fibril diameter, but in no significant changes in adsorption. Interestingly, the amount of fibronectin bound to the heterotypic fibrils depended on fibrillogenesis parameters such as ionic strength and concentration of phosphate, and varied with the percentage of integrated type III collagen. A second paper by the same research group addressed responses of primary osteoblastic cells from rat calvariae [35]. Differences in alkaline phosphatase activity and collagen synthesis were observed when cells were cultured on the various collagen coatings tested. In addition, cell shape and morphology was affected by the thickness and nature of the collagen layer.

Kim and coworkers investigated the stability of collagen I coatings (obtained by adsorption), and the role of fibrillogenesis and of crosslinking on osteoblast (MG63 cell line) function [36]. The degree of assembly varied with the incubation time; the differently assembled collagen was coated on Ti. Some samples were crosslinked by carbodiimide. In that study, both fibrillar assembly and crosslinking improved the stability of collagen. Cell responses improved significantly when collagen fibrils were used and their assembly degree was increased. These researchers proposed that some of the conflicting results that had been reported in the literature regarding in vitro tests of adsorbed collagen could be accounted by differences in the degree of the assembly of collagen.

Müller and coworkers [37, 38] investigated the covalent bonding of fibrillar collagen to Ti and Co alloys using surface analytical approaches. Silane chemistry was used to introduce different densities of amino groups to electrochemically oxidized alloy surfaces. Type I calf skin collagen fibrils were either adsorbed to oxidized (but nonsilanized) metal surfaces, or covalently linked to aminated surfaces (via its aspartic and glutamic acid residues) and 1-ethyl-3-(3-dimethylaminopropyl)-carbodiimide (EDC)/*N*-hydroxysuccinimide (NHS) chemistry. Using the Sirius Red method, the amount of collagen detected on the material surface ranged from 0.36 to 1.9 μg/cm^2. Stability testing in collagenase and MG-63 cell adhesion and proliferation experiments provided evidence of enhanced cell functions on the coated, over the native, metal surfaces.

Recently, Douglas and coworkers expanded the biomimicry strategy by binding the small leucine-rich proteoglycans decorin and biglycan (which are chemical compounds expressed in bone together with collagen) to fibrils during fibrillogenesis in vitro [39–41]. These researchers reported that collagen type II bound significantly more biglycan and decorin than the other collagens tested. Therefore, osteoblast (from rat calvaria and from the human knee) functions on titanium surfaces coated with collagen fibrils with or without decorin and biglycan (either adsorbed or crosslinked) were examined in vitro. The fibrils were adsorbed to Ti and crosslinked. The results showed that the small proteoglycans bound to collagen II fibrils on the titanium surfaces promoted proliferation of, and collagen synthesis by, primary osteoblasts [39–41].

19.2.2. In Vivo Studies on Collagen-Coated Ti Surfaces

Reports on in vivo performance of collagen-coated Ti implants have been published since 2002 [42]. Rammelt and coworkers [43] conducted a study involving 3–6 $\mu g/cm^2$ type I collagen coatings on titanium rods. Coated and uncoated pins were inserted into the tibia of adult rats and histology and immunohistochemistry were performed at 1, 2, 4, 7, 14, and 28 days. At day 4, higher numbers of cathepsin D-positive mononucleated cells in granulation tissue were observed at the interface with collagen-coated (compared with the uncoated) implants. Osteoblast numbers increased around the collagen-coated pins at days 4 and 7. Increased early bone remodelling around titanium pins on the collagen coatings was observed. However, the results of direct bone contact and newly formed bone present around the collagen-coated and uncoated pins were similar 28 days postimplantation [43].

The first literature report providing in vivo evidence of enhanced osteointegration by surface modification of Ti implants by collagen was published in 2003 [44]. In this case, type I collagen, not reconstituted in fibrils, was covalently linked to a layer of polyacrylic acid grafted on Ti surfaces; the substrate samples were sterilized by exposure to ethylene oxide. Compared with results obtained on the control Ti implants, significant increase of bone growth and bone-to-implant contact was observed on the collagen-coated Ti 4 weeks postimplantation in the femurs of rabbits. The histomorphometric results were later supported by additional findings from the same in vivo model, of significant improvement of mechanical parameters such as peri-implant bone microhardness [45], and push-out force [46]. While the aforementioned data were obtained on "machined" Ti surfaces, a more recent publication by the same group of researchers reported the advantages of collagen coatings over more topographically sophisticated titanium surfaces. The combination of topography and biochemical cues is indeed an interesting development for modifying the materials used in implant devices. In one study, collagen type I coatings were applied to titanium implants nanoporous surface topography that had been treated with galvanostatic anodization [47]. Compared with respective controls, improved bone-to-implant contact and bone in-growth (the latter having a $p < 0.056$ value) was observed 4 weeks postimplantation when these materials were implanted in trabecular (cancellous) bone at the epiphyses of rabbit distal femurs. Figure 19.2 compares histological sections from control and collagen-coated implant interfaces with the surrounding biological tissues; these figures illustrate the very low density that is typical of trabecular bone but, most importantly, the more abundant bone regeneration on the collagen-coated Ti. The large interconnected pores between the struts and sheets of trabecular bone tissue were filled by marrow that provided mesenchymal cells for bone regeneration. Cell differentiation was induced by biochemical cues, both soluble and insoluble, in the cell microenvironment. Insoluble cues resulted in cell adhesion to ECM proteins, such as collagen [2].

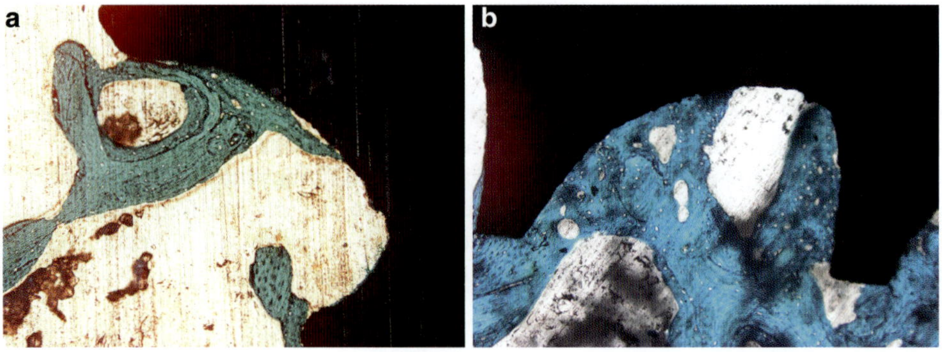

Figure 19.2. Histology examples of sections obtained from implants in the rabbit femur trabecular bone 4 weeks postimplantation. The implant materials were: (**a**) nanoporous Ti; and (**b**) collagen-coated nanoporous Ti. The surface-modification process used was described in reference [47]. Stains: fast green and acid fuchsin; original magnification, ×10. The histology was performed by Drs. Milena Fini, Gianluca Giavaresi, and Roberto Giardino, Istituti Ortopedici Rizzoli, Bologna, Italy.

Another study [28] compared the response of rabbit bone to Ti implants roughened by double acid etching, with and without a covalently linked surface layer of type I collagen fibrils (surface density about 1 $\mu g/cm^2$). Osteointegration of collagen-coated implants was significantly enhanced 2 weeks postimplantation but was similar to that observed on the uncoated controls 4 weeks postimplantation; at that time, the osteointegration process was complete.

Schliephake and coworkers [48] compared Ti with Ti coated with collagen type I (containing two different concentrations of immobilized RGD) implanted in the mandible of foxhounds. The collagen layer was crosslinked and had a surface density of 44 $\mu g/cm^2$. After 1 month, and compared with uncoated implants, the volume density of new bone was significantly higher around all implants with collagen and collagen plus RGD coatings. Both new bone volume density and bone implant contact on collagen-coated and collagen plus RGD-coated implants were similar at 1 month postimplantation, but compared with TI alone surfaces, they were significantly higher at 3 months postimplantation.

Bernhardt and coworkers [49] reported differences in the responses to collagen I versus collagen III, the latter inducing faster bone apposition in the less dense trabecular bone. That research group attributed the observed difference to the different functions induced by the two collagen types. Specifically, collagen I is critical in the signalling cascade that leads to the expression of the mature osteoblast phenotype and to mineralization of the ECM; this type of collagen is particularly effective in settings of high tissue and cell density. Collagen type III, on the other hand, affects early phenomena, such as cell proliferation and matrix deposition; these aspects are of critical importance in cases of low bone and cell density.

Several recent publications by the aforementioned groups of investigators [50–54] addressed the in vivo effects of collagen type I and other biomolecule coatings. Collagen in fibrillar form, at a surface density in the range of 15–50 $\mu g/cm^2$, was used in one of those studies. Quoting from one of these publications [50]: "The addition of ECM components significantly enhances bone remodelling in the early stages of bone healing around Ti implants, eventually leading to increased new bone formation at the implant surface after 4 weeks."

Svehla and coworkers [55] used a skeletally mature bicortical, bilateral ovine tibia model, but, in contrast to previous findings, observed no effect on bone in-growth on fibrillar atelopeptide and poly(ethylene glycol) (PEG) crosslinked collagen coatings layered directly onto porous, sintered beads. The implants tested were either injected with 60 mg/mL bovine skin fibrillar collagen, 60 mg/mL bovine skin fibrillar collagen (crosslinked with 0.5% disuccinimidylglutarate PEG), or left without collagen. At 4 weeks, woven bone was present within the pores and remodelled with time. Although bone in-growth into the implants increased with time, the results were similar on all substrates tested.

19.3. Design of Collagen-Coated Biomaterial Surfaces

Interest in collagen as a biomimetic coating on biomaterials has generated an extensive literature describing in vitro and in vivo studies of collagen-coated bone-contacting implant devices. Most of these results support the hypothesis that the material surfaces that contain ECM motifs affect bone tissue response at those interfaces and improve host response; to date, however, it is difficult to define precisely and quantitatively the advantages of such approaches. In this respect, the results of studies that no significant improvements were obtained under pertinent conditions need to be reconciled.

Undoubtedly, the intrinsic complexity of collagen(s) [56, 57] and of its(their) biochemical behavior plays a role in the differences documented in the scientific literature. To date, the published results do not assist the surface bioengineer, who is confronted with the task to design suitable processes and seeks answers to even simple questions such as: "how much collagen should be coated on an implant material surface to improve bone response?" In the studies mentioned so far in the present chapter, the surface density of collagen ranged from 0.36 to 80 μg/cm^2. What do these values mean from the point of view of material surface properties? What does this information mean for the mechanisms controlling interfacial responses at the cell and tissue levels? To date, there are no definitive answers to such questions.

Besides collagen surface density, a number of other aspects are pertinent; the most important among them are the following:

- the role of supramolecular arrangement of collagen (specifically, "monomeric" versus "fibrillar");
- characterization of the collagen-coated surfaces; and
- collagen surface chemistry (specifically, the role of crosslinking and of covalent bonding versus simple adsorption).

The rest of this chapter is written from a practical surface-engineering perspective and highlights the most important critical aspects that should be addressed in the design and fabrication of collagen-based coatings for biomaterials. This discussion will be centred on collagen type I (henceforth, referred to as "collagen") not only because it is by far the most investigated compound of this chemical family, but also because of its practical importance (since it is the most readily available type in the human body) and of its impact on implant devices.

19.3.1. Collagen Coatings on Ti Implants: Relevant Parameters

Strategies for surface modification of implant devices must deal with the supramolecular hierarchy of the class of proteins known as "collagens" [57]. The molecules of most collagens self-associate to form a higher-order structure, such as fibrils and networks; such supramolecular structures are responsible for the mechanical and binding properties that are characteristic of, and critical to, the function of these biomolecules. The collagen molecule comprises three polypeptide chains (α-chains), which form a unique triple-helical structure.

In collagen type I, two of the three α chains are identical, thus type I collagen is designated $[\alpha 1(I)]_2\alpha 2(I)$. For the three chains to wind into a triple helix, they must have a glycine (the smallest amino acid) at every third residue along each chain. Each of the three chains, therefore, has the repeating structure "Gly–Xaa–Yaa," in which Xaa and Yaa can be any amino acid but are frequently either proline or hydroxyproline.

For most studies, collagen is dissolved in an acidic aqueous solution and, then, reconstituted in fibrils. The fibril-forming collagen molecules consist of an uninterrupted triple helix of approximately 300 nm in length and 1.5 nm in diameter, which contains about 1,000 residues and is flanked by short extrahelical telopeptides [57, 58]. The telopeptides, which do not have a repeating "Gly–Xaa–Yaa" structure and do not adopt a triple-helical conformation, account for 2% of the molecule and are critical for fibril formation. The 300-nm-long collagen molecule is often (even if somehow improperly) called "monomeric" collagen, to distinguish it from the assembled fibrillar form. Even if assembly does not involve any polymerization, the term "monomeric" is commonly used to refer to the unassembled collagen molecules. Unassembled, monomeric porcine collagen, applied to a mica surface from a diluted aqueous solution, is shown in the atomic force microscopy (AFM) image of Figure 19.3. The calculated length of the monomers in Figure 19.3 is 308 ± 8 nm; this result is in good agreement with published X-ray results [59, 60].

Collagen molecules spontaneously self-assemble into fibrils, a higher-order structure. To quote Yamauchi, this phenomenon "is mainly an entropy-driven process through the loss of solvent molecules from the surface.... This implies that all information necessary to initiate the fibrillogenesis is intrinsic to the collagen molecules themselves" [61]. From the perspective of molecular function, Birk and Bruckner state: "Collagen assembly yields highly multimeric suprastructural aggregates, thereby converting protomeric, mainly inactive/nonfunctional, molecules into their active/functional state" [62]. In nature, the collagen fibrils are stabilized by covalent crosslinking, a process initiated via oxidative deamination of specific lysine and hydroxylysine residues by lysyl oxidase. Fibrillogenesis in the lab is usually obtained by

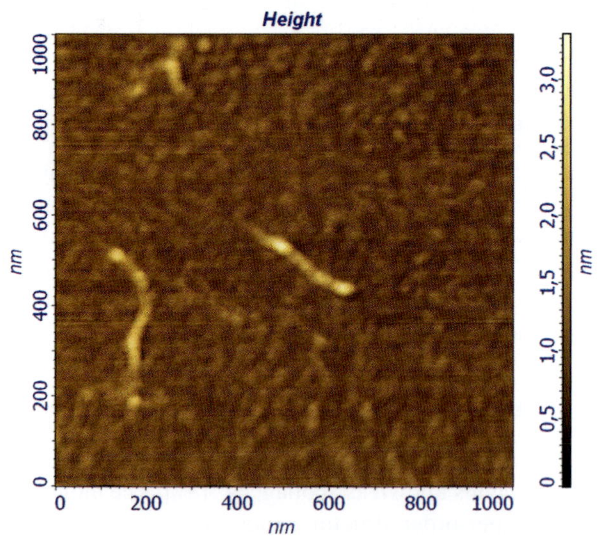

Figure 19.3. AFM image of porcine type I collagen "monomers" on mica in air. The length of the molecules is about 300 nm; the width is greater than the expected 1.5 nm, due to artefactual enlargement by the cantilever tip (see also Ref. [60]). This image was obtained in the vibrational mode of AFM.

neutralizing the acidic monomeric collagen solution (usually at acetic pH) by either phosphate or similar buffers. Ionic strength, the specific nature of ions in solution, and the presence of other biomolecules are parameters that can affect the kinetics of fibril formation, size, and function. A number of published reports described cofibrillation of collagen with other extracellular components [39–41, 50–52].

Further assembly results in the characteristic D-periodic cross–striated fibrils (where D ≈ 67 nm, is the characteristic axial periodicity of collagen, explained by the classic gap-overlaps model [56–58]). AFM images of collagen fibrils on mica are shown in Figure 19.4. The exquisite precision of the nature-driven self-assembling processes is exemplified by the three-dimensional view of a collagen fibril (about 1-μm long), shown in Figure 19.5.

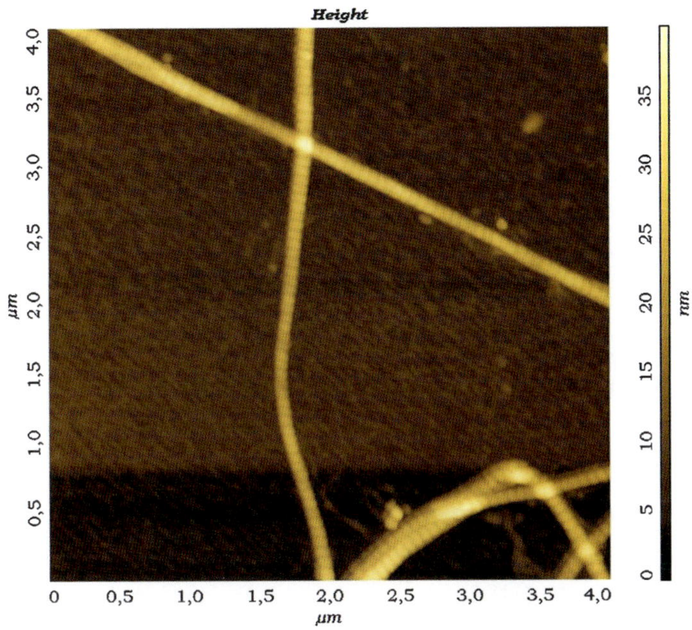

Figure 19.4. AFM image of porcine type I collagen fibrils on mica in air. The measured length of the D-periodicity is 66 ± 4 nm. This image was obtained in the vibrational mode of AFM.

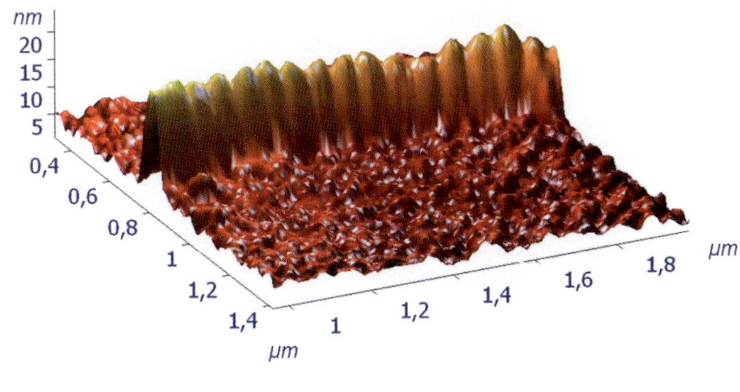

Figure 19.5. Three-dimensional AFM view of porcine type I collagen fibrils on mica in air. The image shows the periodic structure due to gap and overlaps of the assembled collagen molecules.

The calculated periodicity from this, and similar images, is 66 ± 4 nm; this result is in good agreement with values reported in the literature. It should be noted that it is within the gap (holes) or the fibrillar structure that the mineral crystals of bone first appear.

Like most proteins, collagen irreversibly adsorbs at aqueous-solid interfaces; this characteristic can be used to modify surfaces of implant materials. AFM images of monomeric and fibrillar collagen adsorbed to mica surfaces are shown in Figure 19.6. In the case of

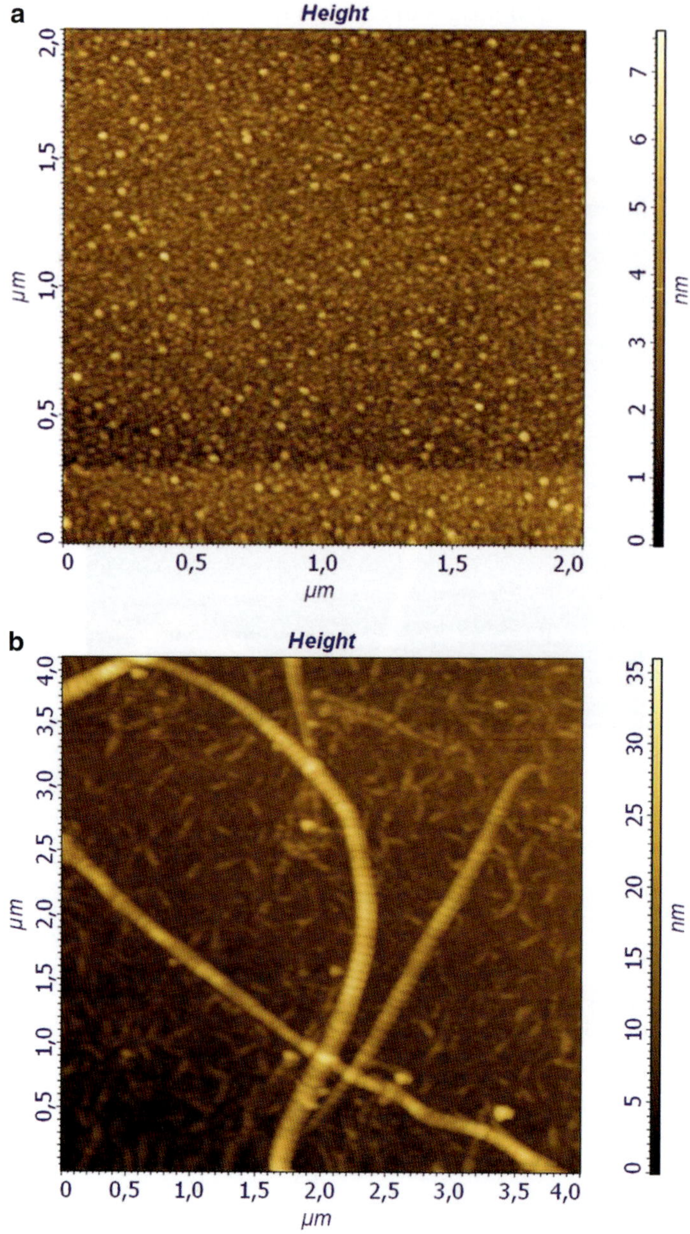

Figure 19.6. AFM images of (**a**) monomeric collagen adsorbed onto mica; and (**b**) fibrillar collagen adsorbed onto mica. In both cases the surface density of collagen was less than 1 mg/cm^2.

Table 19.1. Surface composition of the monomeric and fibrillar collagen samples (shown in Figure 19.6) as determined by XPS. The results are expressed as atomic %.

Collagen sample	O	C	N	Other (<1%)
Adsorbed, monomeric	18.0	68.8	13.2	Na, Cl, Mg, Si
Adsorbed, fibrillar	18.3	68.3	13.4	Si, Cl, Na
Powder (reference)	17.9	68.9	13.2	Si, Cl, Na

monomeric collagen, the overlayer was dense and fully packed, with maximum heights of a few nanometers over the atomically flat mica surface. In the case of fibrillar collagen, the overlayer contained fibrils with the typical collagen periodicity (the vertical scale in this case was in tens of nanometers). In both cases, X-ray photoelectron spectroscopy (XPS) analysis (Table 19.1) revealed the typical stoichiometry of collagen, confirming that:

- In the case of monomeric collagen, homogeneity in the vertical direction indicated that the adsorbed collagen layer was thicker than the XPS sampling depth (about 8 nm). This, in turn, means that fully packed collagen molecules were not adsorbed "side on" (as they were obtained under high-dilution conditions in Figure 19.3) on the substrate surface.
- In the case of fibrillar collagen, homogeneity in the lateral direction indicated that even the regions of the substrate surface that were not covered by fibrils were coated by adsorbed either monomeric collagen or small fibrils of collagen.

Consideration of the identical stoichiometry of the monomeric and fibrillar collagen (Table 19.1) and of the different nanostructures of these collagen forms (Figure 19.6) underscores that supramolecular arguments must be taken into account in the description of collagen-coated surfaces.

Another interesting point for consideration is that literature reports stated collagen surface concentrations ranging from 0.36 to 80 μg/cm^2. How does this range of values agree with the scenario of Figure 19.6? Considering the size of monomeric collagen molecules, theoretical calculations predict a fully packed monolayer a few hundredths of a microgram to a few micrograms per square centimeter. The lower limit of this range is for a hypothetical side-on adsorption (with the 300-nm long-axis parallel to the substrate surface) of collagen molecules, while the upper limit is for a hypothetical end-on adsorption (with 300-nm long-axis normal to the substrate surface); an average molecular weight of 100 Da for about 1,000 residues of the collagen molecule was assumed. In this respect, Horbett indicated that, regardless of the initial concentration of the protein in solution and the duration of the adsorption reaction, the amount of collagen present in a monolayer on the surface of a substrate is in the 0.1–0.5 μg/cm^2 range [63]. The wide range of reported collagen densities on material surfaces and the several tens of micrograms per square centimeter mentioned in most literature reports underline a very significant aspect of surface modification by collagen: to date, surface-modification approaches occur in two different modes, specifically, "adsorption" and "coating." While the term "adsorbed collagen" is often used to refer to collagen densities up to a few tens of micrograms, "true" collagen adsorption, as illustrated in Figure 19.6, occurs because of protein-surface interactions. In order to work in the adsorption regime, the collagen solution must be displaced from the substrate surface at the end of the process, by dilution displacement, which washes away unbound or loosely bound protein while avoiding exposure of the surface to air [64]. In fact, it is possible in this case to avoid transfer of surface-active protein at the solid–air interface when extracting the sample from the protein solution.

On the other hand, the same phenomenon can be purposely exploited, as in common dip-coating procedures, working in the coating regime and increasing the amount of collagen on the substrate surface up to several tens of micrograms per square centimeter. These two modes are well defined in some of the aforementioned publications: Müller and coworkers [38] described a carefully characterized process and reported that "adsorptive binding to oxidized titanium surfaces resulted in 0.36 μg/cm² (0.26–0.44) surface bound collagen." Schliephake and coworkers states: "Afterwards, the collagen layer thickness was increased by dip-coating… leading to a layer density of 44 ± 5 μg/cm² on the implant surface" [48].

Examples of results obtained starting from the same collagen solution but applying either the adsorption or the coating methods are shown in Figure 19.7. Since the fibril density on the adsorbed sample (Figure 19.7a) is much lower, approaching cells will encounter significantly different nanotopography, mechanical properties, and surface-induced signalling. These conditions affect both the structure of, and the biological events that take place at, the tissue-implant interface.

The amount of collagen present on material surfaces has been measured using various methods. The most widely used analytical technique for quantification of collagen on implant materials is ultraviolet-visible spectroscopy. The widely used Sirius Red approach [36, 38, 65] exposes the collagen-coated substrate surfaces to a 0.1% dye solution in saturated picric acid at room temperature in the dark for 1 h. After rinsing with 1 mL of 0.01 N HCl (until the effluent is colorless), the substrates are treated with 0.1 N NaOH (to elute the red dye). Quantitative results are obtained by reading the light absorbance at 540 nm, and converting that data to collagen concentrations using a pertinent standard curve.

The limit of detection and limit of quantification of the Sirius Red method are in the range of 1 μg/mL [66]. Taking into consideration the typical size of the substrate samples and the experimental procedure, this value translates into about 1 μg collagen per cm² substrate surface area. This evaluation approach can be used for collagen-coated surfaces, where the surface density is above that expected for monolayer assemblies. On the other hand, the collagen densities obtained on material surfaces in the adsorption regime are at the border line of the sensitivity range of this approach. Collagen detection following adsorption requires material surface-sensitive techniques such as XPS.

In addition to rendering significantly different surface structures that are crucial for subsequent biological interactions, adsorption and coating affect differently the stability of proteins because they are engendered by intermolecular forces. Adsorbed proteins are held at the material surfaces by interfacial forces. While protein adsorption is virtually irreversible in aqueous solutions, adsorbed proteins can be exchanged or displaced in protein-containing fluids. Since the thick protein layer obtained by coating substrates is more loosely held in situ, it is less stable in aqueous solutions. The publication by Kim and coworkers presented an interesting discussion of the stability of thick collagen coating in cell coculture medium [36].

In principle, the stability of collagen coatings can be improved by either crosslinking or covalent bonding to the material surface. Carboxyl and amino groups are the moieties most often used in pertinent crosslinking reactions to promote bonding to functionalized surfaces. In this respect, –COOH and $–NH_2$ coupling reactions promoted by water-soluble carbodiimide (EDC) and NHS are the most commonly used strategy. Due to the presence of both functionalities in collagen, crosslinking always occurs together with covalent bonding to functionalized material surfaces; in fact, it is difficult to determine definitively whether covalent bonding to the surface occurred.

Crosslinking of collagen with EDC/NHS is well known in the biomedical devices industry, where it is widely used to prepare gels and scaffolds. It is known that both size and

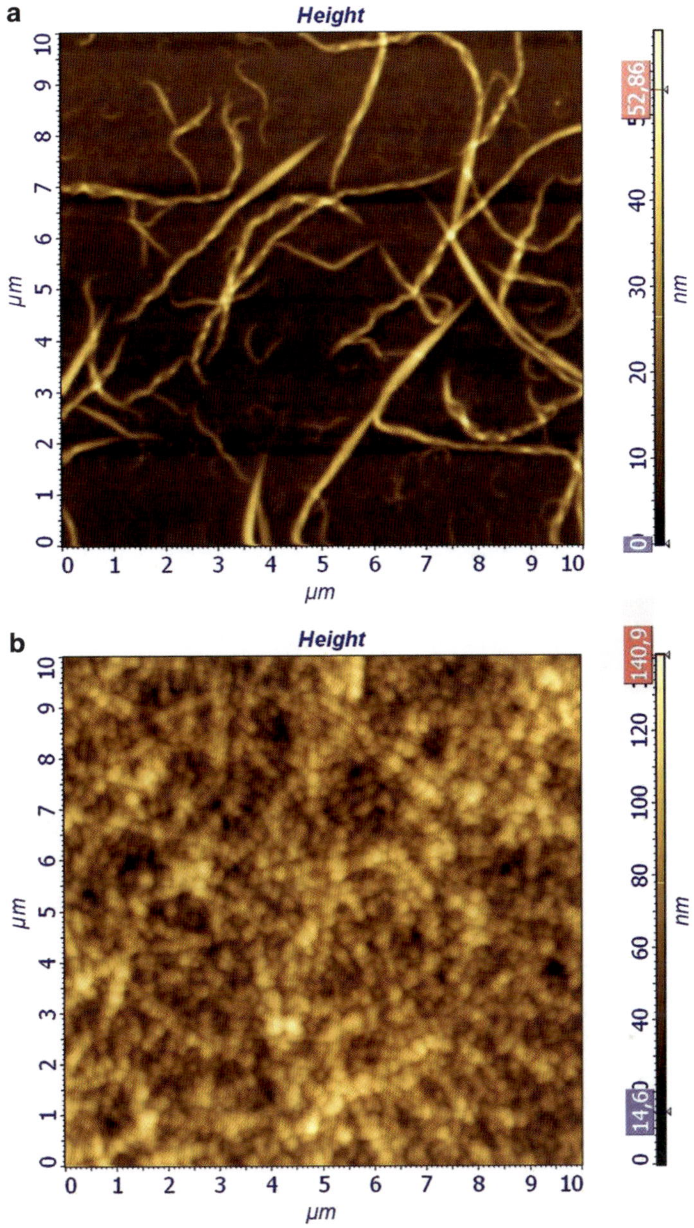

Figure 19.7. AFM images of the results of the classical dilution-displacement method.(a) fibrillar collagen adsorption onto mica The collagen surface density was less than 1 μg/cm^2. (b) in this case the substrate sample was removed from the collagen solution without dilution-displacement, washed and dried. In this case, the surface density of collagen was several μg/cm^2

chemistry of collagen constructs is affected by crosslinking procedures. In the case of surface-adsorbed and surface-coated collagen, it must be emphasized that, depending on the crosslinker concentration, the size and length of the collagen fibrils are affected. As an example, the effect of increasing EDC concentration on fibrillar collagen adsorbed onto aminated mica is shown in Figure 19.8. The size and height of the fibrils are affected by the process, as reported in Table 19.2; the resultant major modification of the surface structure could have implications on subsequent function of the surrounding cells and tissues. Another interesting aspect is illustrated by Figure 19.9. Crosslinking led to the loss of a significant amount of surface-coated collagen. Such loss decreased with decreasing surface density; in this case, protein was in the adsorption regime, and the molecules were strongly held in situ.

The surface stoichiometry of the material samples of Figure 19.8 was analyzed by XPS and found to be identical. This result indicated that the uppermost 5–8 nm was coated by

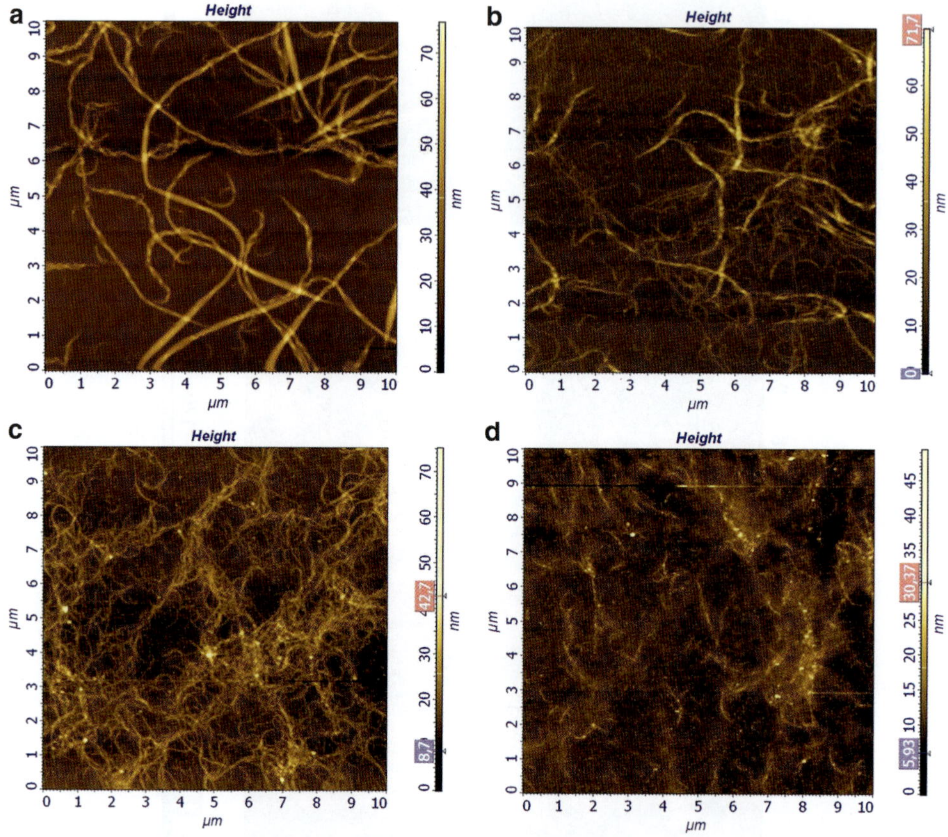

Figure 19.8. AFM images of the results of EDC-NHS crosslinking of collagen fibrils adsorbed onto mica. The collagen tested was as follows: (**a**) untreated; (**b**) 0.011 M EDC, 0.013 M NHS; (**c**) 0.023 M EDC, 0.026 M NHS; and (**d**) 0.036 M EDC, 0.043 M NHS. The measurements of fibril length and height are reported in Table 19.2.

Table 19.2. Typical size of collagen fibrils of noncrosslinked and crosslinked collagen on mica. These measurements (maximum–minimum of 20 random measurements) were made on the samples shown in Figure 19.8.

Sample	Length (μm)	Height (nm)	Diameter (nm)
a (uncrosslinked)	3–10	20–40	150–250
b	2–5	10–25	80–160
c	0.5–4	10–18	60–100
d	0.5–2	2–8	<60

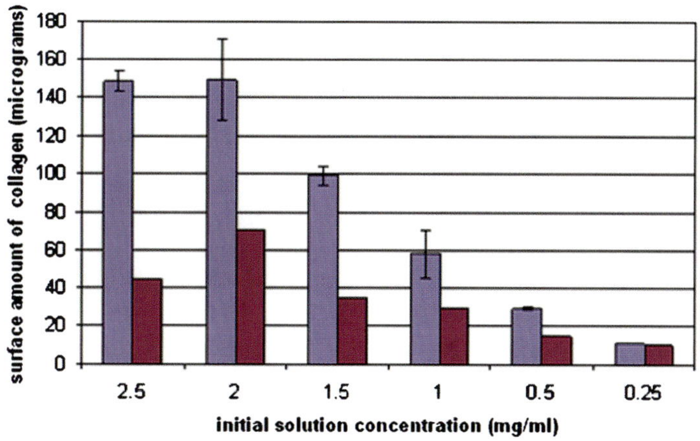

Figure 19.9. Surface amount of collagen before (*blue bars*) and after (*red bars*) crosslinking with EDC-NHS. The collagen coating was applied using solutions with different collagen concentrations. The amount of collagen on the substrate surface was determined from the weight increase of each sample using a microbalance (1×10^{-6} g sensitivity). The nominal surface area of each disk was about 1.5 cm^2.

stoichiometric collagen. Modification of the relative amount and of the nature of the present functional groups can be determined by either deconvolution of the high-resolution C1s peak [45] or by the application of UHV techniques, such as time-of-flight static secondary ion mass spectroscopy (ToF-SIMS) [67–70], which are more sensitive to molecular details.

Discussion of the stability of collagen-coated implant devices is not complete without consideration of its resistance to collagenase, an enzyme present in wounds, which is expected to be present in the peri-implant microenvironment. Both Kim and coworkers [36] and Müller and coworkers [38] address the stability of collagen coatings (with and without EDC/NHS crosslinking and covalent bonding) onto the material surface to the collagenase challenge. In these cases, noncrosslinked collagen proved sensitive to the enzymatic attack. Crosslinking greatly enhanced collagen resistance to enzymatic degradation; moreover, covalent linking of the adsorbed collagen onto an aminated substrate further improved the collagen resistance [38]. Results from our group showed that crosslinking, even at low EDC/NHS concentration, was necessary to preserve the collagen surface layer. Figure 19.10 illustrates the underlying mica substrate surface (on which fibrillar collagen had been adsorbed) before and after treatment with a 2.5 U/mL collagenase solution at 37°C for 3 h.

19.3.2. Collagen Coating of Ti Implants: Summary

A number of molecular and supramolecular parameters of collagen-coated implant devices were identified and discussed because they affect the structure of material surfaces and, thus, may modulate subsequent biological responses at the tissue–implant interface. Is it possible to derive guiding principles for the design of material surfaces from those parameters and the results reported so far in the scientific literature?

Most publications addressing the biological role of fibrillar versus monomeric collagen come from medical settings. While very informative regarding cell biology, they do not provide sufficient details pertinent to material surface characteristics. As an example, Fassett and coworkers provided evidence that various aspects of hepatocyte responses depended on whether the type I collagen was in fibrillar or monomeric form: hepatocytes adherent to monomeric

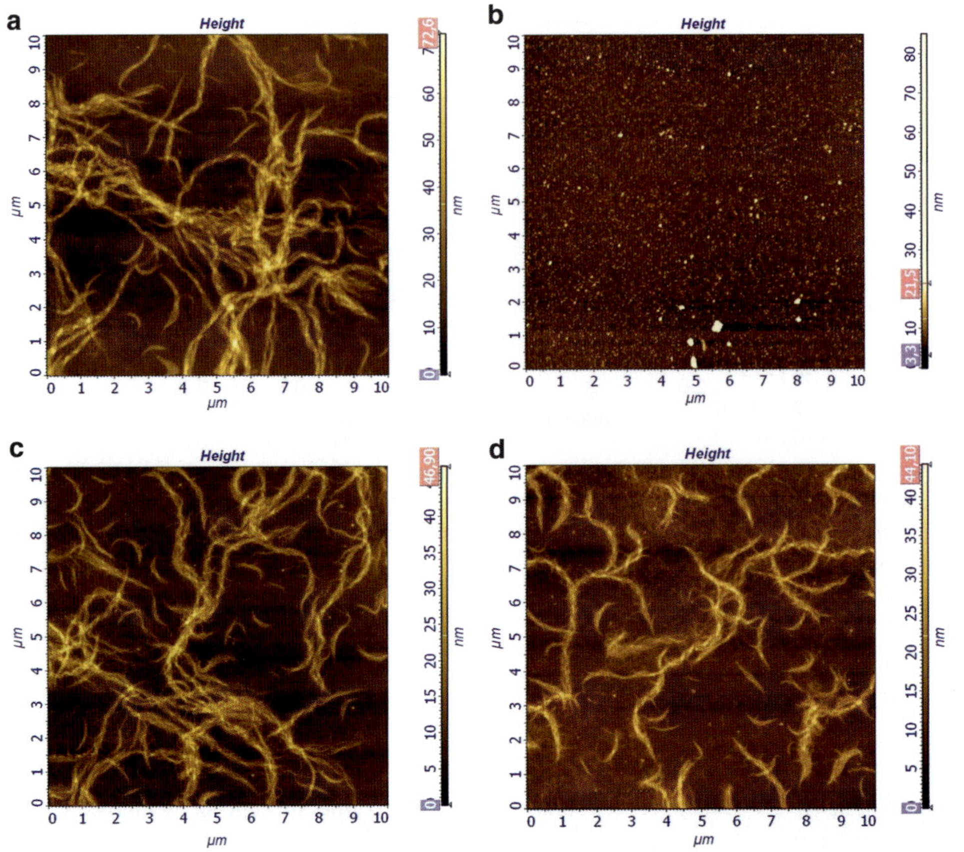

Figure 19.10. AFM images illustrating the results of exposing collagen adsorbed onto mica to the enzymatic activity of a collagenase solution. (**a**) Collagen adsorbed, before exposure to collagenase; (**b**) collagen adsorbed, after exposure to collagenase; (**c**) crosslinked collagen, before exposure to collagenase; and (**d**) crosslinked collagen, after exposure to collagenase. These experiments were performed using a 2.5 U/mL collagenase solution at 37°C for 3 h.

collagen spread and proliferated, whereas those adherent to fibrillar collagen remained rounded and growth arrested [71]. The substrate surfaces used in this study were prepared by coating from aqueous solutions; however, no information regarding the surface density of the protein was given. Monomeric and fibrillar collagen are referred to as "film" and "gels," respectively; such descriptions suggest the possibility of widely different pertinent mechanical properties. Henriet and coworkers showed that human M24met melanoma cells remained responsive to proliferation-regulatory signals that resulted from contact with type I collagen and that the effect on cell proliferation depended on the physical structure of the collagen used: on fibrillar collagen, M24met cells were growth arrested, while on nonfibrillar collagen, the cells entered the cell cycle [72]. A possible explanation for this result is that these cell proliferation-regulatory effects involve contact between type I collagen and the collagen-binding integrin $\alpha 2\beta 1$ on the cell membrane; this integrin was restricted in the presence of fibrillar collagen. Other details of the experimental procedures may have contributed to this outcome; for example, 1 mg/mL fibrillar collagen in phosphate-buffered saline (PBS) was used to form a thick gel in tissue culture dishes at 37°C for at least 1 h before the cells were plated. Nonfibrillar collagen was prepared using a 1 mg/mL solution in tissue culture dishes at 37°C overnight. All collagen preparations were washed twice with PBS before the addition of cells. These descriptions do not provide important details regarding collagen surface density; in many such studies the surface concentration of collagen (in milligrams per milliliter) is given with reference to the concentration of the coating solution. Considering how rinsing, for example, may affect the interfacial environment (Figure 19.6), this detail is not really informative. Most importantly, thick gels of fibrillar collagen could affect cell responses not just by supramolecular assembly, but also via their mechanical properties. Compared with the morphology acquired by cells plated on rigid substrates, cells on gels often do not spread; this aspect was addressed satisfactorily by Bierbaum and coworkers [35, 73, 74].

In the publication by Kim and coworkers, adhesion and proliferation of MG-63 osteoblast-like cells was studied on unassembled and assembled, with and without crosslinking collagen-coated Ti (40–80 μg/cm^2) [36]. In this case, the fibrillar assembly of collagen promoted enhanced cell attachment and proliferation. Crosslinking had a significant effect on cell function only in the case of the unassembled collagen coating.

Our research group evaluated the effect of adsorbed monomeric, adsorbed fibrillar (surface density < 1 μg/cm^2), and coated fibrillar (surface density > 10 μg/cm^2) collagen on glass on the differentiation of human mesenchymal cells, with and without EDC/NHS crosslinking [75]. The cells were cultured in basal medium (i.e., in the absence of osteogenic differentiation chemical stimuli such as dexamethasone, glycerolphosphate, and ascorbic acid) [76] in order to evaluate the effect of insoluble cues and interfacial interactions on cell differentiation. In this study, the effect of collagen crosslinking proved more important than either the surface density or the supramolecular assembly of collagen. Figure 19.11 illustrates

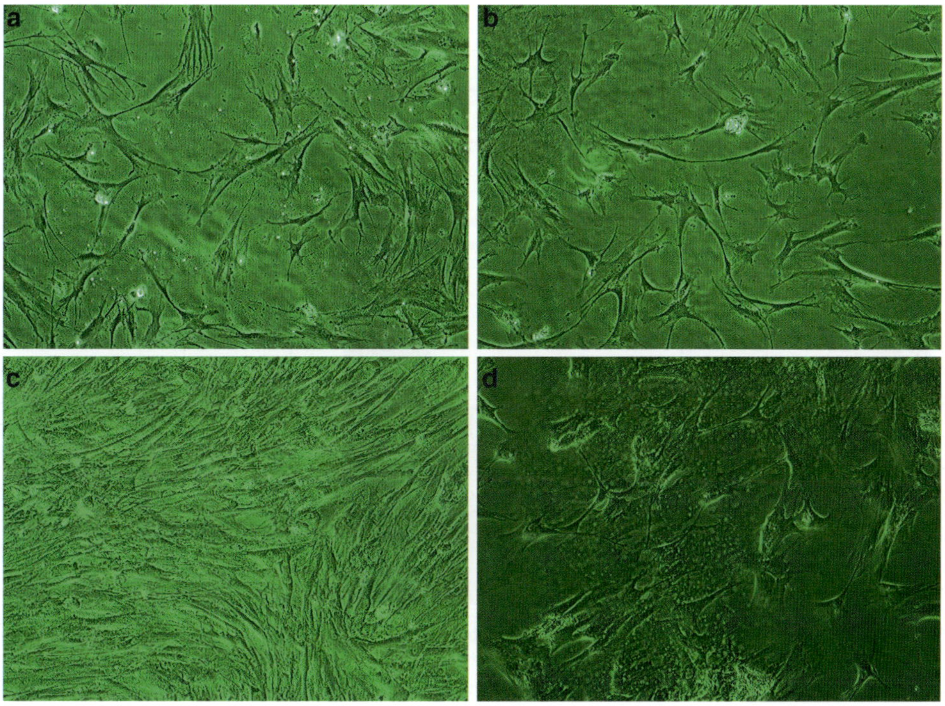

Figure 19.11. Light micrographs of human mesenchymal cells on the following substrates: (**a**) and (**c**) monomeric collagen adsorbed onto glass; (**b**) and (**d**) monomeric collagen adsorbed onto glass and crosslinked (0.023 M EDC, 0.026 M NHS). Micrographs (**a**) and (**b**) were taken after 24 h while micrographs (**c**) and (**d**) were taken after 16 days of cell culture. Original magnification, ×100.

the shape of cells on monomeric collagen with and without crosslinking after 24 h and 16 days of culture. The morphology of cells was similar on all substrates tested at 24 h. After 16 days of culture, however, the typical morphology of nondifferentiated mesenchymal cells was still observed only on monomeric, noncrosslinked collagen [76]. The hallmark of osteogenic differentiation, specifically, transition from spindle-like to cuboidal morphology and a reduced proliferation rate, was observed on crosslinked collagen. This result was confirmed by gene expression analysis using real-time polymerase chain reaction (RT-PCR) determination of osteocalcin (OC) and osteopontin (OP) gene expression after 24 days of culture (Figure 19.12). In agreement with other reports, OC was neither significantly different nor upregulated on the various substrate tested [77, 78]. In contrast, and compared with results obtained on the noncrosslinked substrates, OP was significantly upregulated on the crosslinked, adsorbed, monomeric collagen monolayer. Compared with the respective noncrosslinked pair, the results on every crosslinked collagen tested were enhanced; the values measured on the adsorbed fibrillar, not crosslinked, collagen layer were similar to those for monomeric collagen (data not shown). Even if these results were obtained on collagen-coated glass, it is of interest to notice that the similar upregulation of OP, without significant increase of OC, was detected by RT-PCR analysis of human alveolar bone-derived cells cultured on Ti surfaces coated with a covalently linked layer of monomeric collagen [78]. In addition to emphasizing the importance of stabilization of the collagen surface layer, Figure 19.12 suggests a possible positive effect of surface density (or of three-dimensional structures) with increasing concentrations of collagen on substrate surfaces.

What are, then, the guiding principles for surface engineering of implant devices by collagen [79]. Based on the results of the lastly mentioned study alone, crosslinking is the

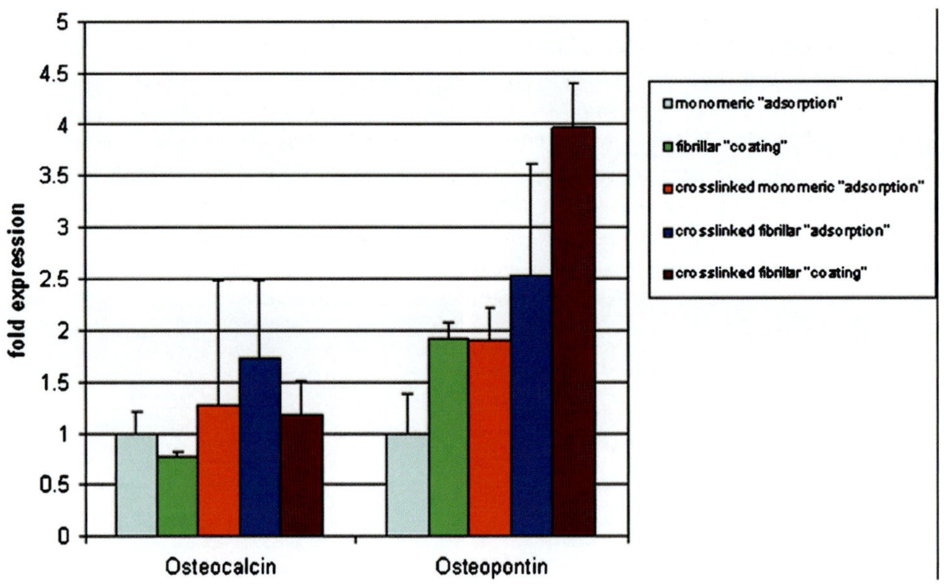

Figure 19.12. RT-PCR results showing expression of osteocalcin and osteopontin genes by human mesenchymal cells cultured in basal medium without proosteogenic-soluble chemical stimuli for 24 days. These cells were cultured on glass coated with adsorbed monomeric collagen, adsorbed fibrillar collagen (both at a surface density of less than 1 mg/cm^2) and coated with fibrillar collagen (surface density greater than 10 mg/cm^2), both not crosslinked and EDC/NHS crosslinked.

most important parameter to guide protein coatings of implant materials. On the other hand, the prolonged presence of a full-density collagen layer, assembled or not, is the minimum affecter of mesenchymal cell function. This conclusion is based on in vivo results of studies by our research group [28, 44–47], and it is in agreement with the Schliephake and coworkers results [48]. The comparatively high surface density of collagen (even in the absence of crosslinking) reported by Rammelt and coworkers [50] and Berhardt and coworkers [49] could provide interfacial collagen long enough to affect the wound-healing mechanisms. On the other hand, the low (3–6 $\mu g/cm^2$) surface density of uncrosslinked collagen used in the first reports by Rammelt and coworkers could possibly be enough to promote enhanced bone remodelling, without, however, achieving significantly increased bone volume [42, 43]. It should be noted that this is only a speculation, since there is no direct in vivo comparison to determine (all other parameters being the same) whether assembly or increased surface density can further improve the results obtained with the crosslinked collagen monolayers; a similar comment is applicable to all other aspects discussed in the present chapter.

19.4. Conclusions

Coating titanium surfaces with ECM molecules, in particular with collagen, accelerates bone healing and increases new bone formation in vivo. Adhesion of mesenchymal cells on select ECM molecules promotes their differentiation along the osteogenic pathway ([12]; Figures 19.11 and 19.12). Despite conflicting literature reports, many studies observed, and reported, enhanced osteointegration in various animal models [44–54].

Details of the mechanisms of interfacial interactions leading to improved tissue responses on collagen-coated Ti surfaces need to be further elucidated. A key aspect in elucidating biological interactions on materials surfaces is an improved understanding and thorough characterization of materials surfaces. This aspect is very important in the case of biochemical surface modifications [80], whose aim is to introduce defined chemical structures that are ligands of their corresponding cell membrane receptors. Since supramolecular assembly of biological molecules is strictly related to protein function, this aspect adds another level of complexity to material surface characterization. Because the cost of research endeavours required to reveal the subtleties of biological responses to bioengineered implant surfaces rises exponentially, special attention should be paid to critical aspects of pertinent experimental methodology.

The research activities discussed in this chapter may lead to establishing guiding principles for the design of surface-engineering approaches for collagen-coated implant devices. It is, however, clear that questions still outnumber answers and that a significant, concerted effort is still needed to achieve satisfactory understanding the chemistry and application of surfaces modified with type I collagen.

Acknowledgments

Most of this work was performed under the program "Coating bioattivi per dispositivi a contatto con osso" Legge 598/94 art. 11 – Ricerca, Regione Piemonte.

The significant contribution of Dr. Eng. Ilaria Cannella to the analytical work (which was part of her thesis work) is acknowledged.

References

1. Davies JE, Ed, The bone-biomaterial interface, Toronto, University of Toronto Press, 1991
2. Davies JE, Ed, Bone Engineering, Toronto, Em Squared, 2000
3. Brunette DM, Tengvall P, Textor M, Thomsen P, Eds, Titanium in Medicine, Berlin, Springer, 2001
4. Ellingsen JE, Lyngstadaas SP, Eds, Bio-Implant Interface, CRC, Boca Raton, 2003
5. Cunningham BW, Basic scientific considerations in total disc arthroplasty, Spine J, 2004;4:219S–230S
6. Rupp F, Scheideler L, Olshanska N, de Wild M, Wieland M, Geis-Gerstorfer J, Enhancing surface free energy and hydrophilicity through chemical modification of microstructured titanium implant surfaces, J Biomed Mater Res, 2006;76A:323–334
7. Cooper LF, Zhou Y, Takebe J, Guo J, Abron A, Holmen A, Ellingsen JE, Fluoride modification effects on osteoblast behavior and bone formation at TiO2 grit-blasted c.p. titanium endosseous implants, Biomaterials, 2006;27:926–936
8. Puleo DA, Nanci A, Understanding and controlling the bone-implant interface, Biomaterials, 1999;20:2311–2321
9. Lebaron RG, Athanasiou KA, Extracellular matrix cell adhesion peptides: functional applications in orthopedic materials, Tissue Eng, 2000;6:85–104
10. Jikko A, Harris SE, Chen D, Mendrick DL, Damsky CH, Collagen integrin receptors regulate early osteoblast differentiation induced by BMP-2, J Bone Miner Res, 1999;14:1075–1083
11. Mizuno M, Fujisawa R, Kuboki Y, Type I collagen-induced osteoblastic differentiation of bone-marrow cells mediated by collagen-alpha2beta1 integrin interaction, J Cell Physiol, 2000;184:207–213
12. Salasznyk RM, Williams WA, Boskey A, Batorsky A, Plopper GE, Adhesion to vitronectin and collagen I promotes osteogenic differentiation of human mesenchymal stem cells, J Biomed Biotechnol, 2004;1:24–34
13. Sampath TK, Reddi AH, Dissociative extraction and reconstitution of extracellular matrix components involved in local bone differentiation, Proc Natl Acad Sci U S A 1981;78:7599–7603
14. Heemskerk JWM, Wust WMJ, Feijge MAH, Reutelingsperger CMP, Lindhout T, Collagen but not fibrinogen surfaces induce bleb formation, exposure of phosphatidylserine, and procoagulant activity of adherent platelets: evidence for regulation by protein tyrosine kinase-dependent Ca2 + responses, Blood, 1997;90:2615–2625
15. Park JY, Gemmell CH, Davies JE, Platelet interactions with titanium, modulation of platelet activity by surface topography, Biomaterials, 2000;22:2671–2682
16. Davies JE, Housseini MM, Histodynamics of endosseous wound healing. In: Davies JE, Ed, Bone Engineering, Toronto, Em Squared, 2000;1–14
17. Gemmell CH, Park JY, Initial blood interactions with endosseous implant materials. In: Davies JE, Ed, Bone Engineering, Toronto, Em Squared, 2000;108–117
18. Panduranga Rao K, Recent developments of collagen based biomaterials for medical applications and drug delivery systems, J Biomater Sci Polym Ed, 1995;7:623–645
19. Gungormus M, Kaya O, Evaluation of the effect of heterologous type I collagen on healing of bone defects, J Oral Maxillofac Surg, 2002;60:541–545
20. Somorjai G, Biointerfaces: The Grand Challenge of Molecular Surface Chemistry, invited talk at the 25th Anniversary of NESAC/BIO, University of Washington, Seattle, WA, August 24–27, 2008
21. Lowenberg BF, Pilliar RM, Aubin JE, Sodek J, Melcher AH, Cell attachment of human gingival fibroblasts in vitro to porous-surfaced titanium alloy discs coated with collagen and platelet-derived growth factor, Biomaterials, 1988;9:302–309
22. Matsumura K, Hyon SH, Nakajima N, Peng C, Tsutsumi S, Surface modification of poly(ethylene-co-vinyl alcohol) (EVA). Part I. Introduction of carboxyl groups and immobilization of collagen, J Biomed Mater Res, 2000;50:512–517
23. Peng C, Tsutsumi S, Matsumura K, Nakajima N, Hyon SH, Morphologic study and syntheses of type I collagen and fibronectin of human periodontal ligament cells cultured on poly(ethylene-co-vinyl alcohol) (EVA) with collagen immobilization, J Biomed Mater Res, 2001;54:241–246
24. Matsumura K, Hyon SH, Nakajima N, Peng C, Iwata H, Tsutsumi S, Adhesion between poly(ethylene-co-vinyl alcohol) (EVA) and titanium, J Biomed Mater Res, 2002;60:309–315
25. Geissler U, Hempel U, Wolf C, Scharnweber D, Worch H, Wenzel K, Collagen type I-coating of Ti6Al4V promotes adhesion of osteoblasts, J Biomed Mater Res, 2000;51:752–760
26. Baier RE, Meyer AE, Future directions in surface preparation of dental implants, J Dent Ed, 1988;52:788–791
27. Cassinelli C, Morra M, Bruzzone G, Carpi A, Di Santi G, Giardino R, Fini M, Surface chemistry effects of topography modification of titanium dental implants surfaces: 1. In vitro experiments, Int J Oral Maxillofacial Implants, 2003;18:45–62
28. Morra M, Cassinelli C, Cascardo G, Bollati D, Rodriguez Y, Baena R, Multifunctional implant surfaces: surface characterization and bone response to acid-etched Ti implants surface-modified by fibrillar collagen I, J Biomed Mater Res, A, submitted

29. Boyan BD, Lossdörfer S, Wang L, Zhao G, Lohmann CH, Cochran DL, Schwartz Z, Osteoblasts generate an osteogenic microenvironment when grown on surfaces with rough microtopographies, Eur Cell Mater, 2003;6:22–27
30. Nagai M, Hayakawa T, Fukatsu A, Yamamoto M, Fukumoto M, Nagahama F, Mishima H, Yoshinari M, Nemoto K, Kato T, In vitro study of collagen coating of titanium implants for initial cell attachment, Dent Mater J, 2002;21: 250–260
31. Roehlecke C, Witt M, Kasper M, Schulze E, Wolf C, Hofer A, Funk RW, Synergistic effect of titanium alloy and collagen type I on cell adhesion, proliferation and differentiation of osteoblast-like cells, Cells Tissues Organs, 2001;168:178–187
32. Becker D, Geissler U, Hempel U, Bierbaum S, Scharnweber D, Worch H, Wenzel KW, Proliferation and differentiation of rat calvarial osteoblasts on type I collagen-coated titanium alloy, J Biomed Mater Res, 2002;59:516–527
33. Van den Dolder J, Bancroft GN, Sikavitsas VI, Spauwen PH, Mikos AG, Jansen JA, Effect of fibronectin- and collagen I-coated titanium fiber mesh on proliferation and differentiation of osteogenic cells, Tissue Eng, 2003;9:505–515
34. Bierbaum S, Beutner R, Hanke T, Scharnweber D, Hempel U, Worch H, Modification of Ti6Al4V surfaces using collagen I, III, characteristics of the adsorbed matrix, J Biomed Mater Res A, 2003;67:421–430
35. Bierbaum S, Hempel U, Geissler U, Hanke T, Scharnweber D, Wenzel KW, Wenzel KW, Modification of Ti6AL4V surfaces using collagen I, III, and fibronectin. II. Influence on osteoblast responses, J Biomed Mater Res A, 2003;67:431–438
36. Kim HW, Li LH, Lee EJ, Lee SH, Kim HE, Fibrillar assembly and stability of collagen coating on titanium for improved osteoblast responses, J Biomed Mater Res A, 2005;75:629–638
37. Müller R, Abke J, Schnell E, Macionczyk F, Gbureck U, Mehrl R, Ruszczak Z, Kujat R, Englert C, Nerlich M, Angele P, Surface engineering of stainless steel materials by covalent collagen immobilization to improve implant biocompatibility, Biomaterials, 2005;26:6962–6972
38. Müller R, Abke J, Schnell E, Scharnweber D, Kujat R, Englert C, Taheri D, Nerlich M, Angele P, Influence of surface pretreatment of titanium- and cobalt-based biomaterials on covalent immobilization of fibrillar collagen, Biomaterials, 2006;27:4059–4068
39. Douglas T, Heinemann S, Hempel U, Mietrach C, Knieb C, Bierbaum S, Scharnweber D, Worch H, Characterization of collagen II fibrils containing biglycan and their effect as a coating on osteoblast adhesion and proliferation, J Mater Sci Mater Med, 2008;19:1653–1660
40. Douglas T, Hempel U, Mietrach C, Viola M, Vigetti D, Heinemann S, Bierbaum S, Scharnweber D, Worch H, Influence of collagen-fibril-based coatings containing decorin and biglycan on osteoblast behavior, J Biomed Mater Res A, 2008;84:805–816
41. Douglas T, Heinemann S, Mietrach C, Hempel U, Bierbaum S, Scharnweber D, Worch H, Interactions of collagen types I and II with chondroitin sulfates A-C and their effect on osteoblast adhesion, Biomacromolecules, 2007;8:1085–1092
42. Rammelt S, Schulze E, Wolf E, Scharnweber D, Holch M, Worch H, Zwipp H, Immunohistochemical characterization of the bone implant interface of collagen coated titanium pins in the undecalcified rat-tibia, Eur J Trauma, 2002;28:125–126
43. Rammelt S, Schulze E, Bernhardt R, Hanisch U, Scharnweber D, Worch H, Zwipp H, Biewener A, Coating of titanium implants with type-I collagen, J Orthop Res, 2004;22:1025–1034
44. Morra M, Cassinelli C, Cascardo G, Cahalan P, Cahalan L, Fini M, Giardino R, Surface engineering of titanium by collagen immobilization. Surface characterization and in vitro and in vivo studies, Biomaterials, 2003;24:4639–4654
45. Morra M, Cassinelli C, Meda L, Fini M, Giavaresi G, Giardino R, Surface analysis and effects on interfacial bone microhardness of collagen-coated titanium implants: a rabbit model, Int J Oral Maxillofac Implants, 2005;20:23–30
46. Morra M, Cassinelli C, Fini M, Giardino R, Enhanced osteointegration by biochemical surface modification: covalent linking of collagen I to intervertebral metal disk surface, Eur Cell Mater, 2005;10(suppl. 3):6
47. Morra M, Cassinelli C, Cascardo G, Mazzucco L, Borzini P, Fini M, Giavaresi G, Giardino R, Collagen I-coated porous titanium surfaces: mesenchymal cell adhesion and in vivo evaluation in trabecular bone implants, J Biomed Mater Res A, 2006;78:449–458
48. Schliephake H, Scharnweber D, Dard M, Sewing A, Aref A, Roessler S, Functionalization of dental implant surfaces using adhesion molecules, J Biomed Mater Res Part B Appl Biomater, 2005;73B:88–96
49. Bernhardt R, van den Dolder J, Bierbaum S, Beutner R, Scharnweber D, Jansen J, Beckmann F, Worch H, Osteoconductive modifications of Ti-implants in a goat defect model: characterization of bone growth with SR μCT and histology, Biomaterials, 2005;26:3009–3019
50. Rammelt S, Illert T, Bierbaum S, Scharnweber D, Zwipp H, Schneiders W, Coating of titanium implants with collagen, RGD peptide and chondroitin sulfate, Biomaterials, 2006;27:5561–5571
51. Stadlinger B, Pilling E, Mai R, Bierbaum S, Berhardt R, Scharnweber D, Eckelt U, Effect of biological implant surface coatings on bone formation, applying collagen, proteoglycans, glycosaminoglycans and growth factors, J Mater Sci Mater Med, 2008;19:1043–1049

52. Stadlinger B, Pilling E, Huhle M, Mai R, Bierbaum S, Scharnweber D, Kuhlisch E, Loukota R, Eckelt U, Evaluation of osseointegration of dental implants coated with collagen, chondroitin sulphate and BMP-4: an animal study, Int J Oral Maxillofac Surg, 2008;37:54–59
53. Rammelt S, Heck C, Bernhardt R, Bierbaum S, Scharnweber D, Goebbels J, Ziegler J, Biewener A, Zwipp H, In vivo effects of coating loaded and unloaded Ti implants with collagen, chondroitin sulfate, and hydroxyapatite in the sheep tibia, J Orthop Res, 2007;25:1052–1061
54. Stadlinger B, Pilling E, Huhle M, Mai R, Bierbaum S, Bernhardt R, Scharnweber D, Kuhlisch E, Hempel U, Eckelt U Influence of extracellular matrix coatings on implant stability and osseointegration: an animal study, J Biomed Mater Res B Appl Biomater, 2007;83:222–231
55. Svehla M, Morberg P, Bruce W, Walsh WR, No effect of a type I collagen gel coating in uncemented implant fixation, J Biomed Mater Res B Appl Biomater, 2005;74:423–428
56. Hay ED, Ed, Cell Biology of the Extracellular Matrix, 2nd edition, New York, Plenum, 1991
57. Brinckmann J, Notbohm H, Muller PK, Eds, Primer in structure, processing and assembly, In: Topics in Current Chemistry: Collagen, Berlin, Springer, 2005
58. Kadler KE, Holmes DF, Trotter JA, Chapman JA, Collagen fibril formation, Biochem J, 1996;316:1–11
59. Hodge AJ, In: Ramachandran GN, Ed, Treatise on Collagen, Vol. 1, New York, Academic, 1967;185:205–213
60. Gale M, Pollanen MS, Markiewicz P, Goh MC, Sequential assembly of collagen revealed by atomic force microscopy, Biophys J, 1995;68:2124
61. Yamauchi M, Collagen Biochemistry: An Overview, New York, WSP, 2003;93–112
62. Birk DE, Bruckner P, Collagen suprastructures, In: Brinckmann J, Müller PK, Notbohm H, Eds, Topics in Current Chemistry: Collagen, Berlin, Springer, 2005; 245:185–206
63. Horbett TA, Principles underlying the role of adsorbed plasma-proteins in blood interactions with foreign materials, Cardiovasc Pathol, 1993;2:S137–S148
64. Bergstrom K, Holmberg K, Safrani A, Hoffmann AS, Edgell MJ, Kozlowski A, Hovanes BA, Milton Harris J, Reduction of fibrinogen adsorption on PEG-coated polystyrene surfaces, J Biomed Mater Res, 1992;26:779–790
65. Zhu Y, Chan-Park MB, Density quantification of collagen grafted on biodegradable polyester: its application to esophageal smooth muscle cell, Anal Biochem, 2007;363:119–127
66. Cannella I, Sviluppo di un protocollo di caratterizzazione di un dispositivo da impianto in osso con modifica superficiale biomolecolare, Biomedical engineering thesis, Politecnico of Turin, 2008
67. Martin SM, Schwartz JL, Giachelli CM, Ratner BD, Enhancing the biological activity of immobilized osteopontin using a type-I collagen affinity coating, J Biomed Mater Res, 2004;70A:10–19
68. Lhoest JB, Wagner MS, Tidwell CD, Castner DG, Characterization of adsorbed protein films by time of flight secondary ion mass spectrometry, J Biomed Mater Res, 2001;57:432–440
69. Wagner MS, Castner DG, Characterization of adsorbed protein films by time of flight secondary ion mass spectrometry with principal component analysis, Langmuir, 2001;17:4649–4660
70. Wagner MS, Tyler BG, Castner DG, Interpretation of static time of flight secondary ion mass spectra of adsorbed proteins films by multivariate pattern recognition, Anal Chem, 2002;74:1824–1835
71. Fassett J, Tobolt D, Hansen LK, Type I, collagen structure regulates cell morphology and EGF signaling in primary rat hepatocytes through cAMP-dependent protein kinase A, Mol Biol Cell, 2006;17:345–356
72. Henriet P, Zhong ZD, Brooks PC, Weinberg KI, DeClerck YA, Contact with fibrillar collagen inhibits melanoma cell proliferation by up-regulating p27KIP1, Proc Natl Acad Sci U S A, 2000;97:10026–10031
73. Halliday NL, Tomasek JJ, Mechanical properties of the extracellular matrix influence fibronectin fibril assembly in vitro, Exp Cell Res, 1995;217:109–117
74. Roskelley CD, Desprez PY, Bissell MJ, Extracellular matrix dependent tissue-specific gene expression in mammary epithelial cells requires both physical and biochemical signal transduction, Proc Natl Acad Sci U S A 1994;91:12378–12382
75. Morra M, Cassinelli C, Cascardo G, Bollati D, manuscript in preparation
76. Cheng SL, Yang JW, Rifas L, Zhang SF, Avioli LV, Differentiation of human bone marrow osteogenic stromal cells in vitro: induction of the osteoblast phenotype by dexamethasone, Endocrinology, 1994;134:277–286
77. Freitas de Assis A, Beloti MM, Crippa GE, de Oliveira PT, Morra M, Luiz Rosa A, Development of osteoblastic phenotype in human alveolar bone-derived cells grown on collagen type I-coated titanium surface, Clin Oral Implant Res, submitted
78. Guo L, Kawazoe N, Hoshiba T, Tateishi T, Chen G, Zhang X, Osteogenic differentiation of human mesenchymal stem cells on chargeable polymer-modified surfaces, J Biomed Mater Res, Published Online: 28 Jan 2008
79. Ratner B, New ideas in biomaterials science – a path to engineered biomaterials, J Biomed Mater Res, 1993;27:837–850
80. Morra M, Biomolecular modification of implant surfaces, Expert Rev Med Devices, 2007;4:361–372

20

Prevention of Postsurgical Adhesions: A Biomaterials Perspective

John M. Medley and Thomas D. Dziubla

One of the most common complications following abdominal surgery is the formation of postsurgical adhesions (PSAs). While often asymptomatic, these formations can result in numerous complications, including pain, infertility, and bowel obstructions, and can increase the difficulty of future surgeries. Various strategies have been developed to reduce the incidence of these adhesions, yet the most successful to date have been the use of physical barriers composed of biocompatible materials. Yet, despite the success of these materials, PSAs remain a significant clinical challenge.

This chapter provides a summary of the field from a biomaterials perspective, an overview of the mechanism of PSA formation and a review of the technologies developed in attempting to reduce their occurrences, including the limitations that exist with each of these strategies. At the end of this chapter, a summary of the analytical methods used in assessing adhesion barrier performance and their limitations, and a call for more quantitative analytical materials performance characterization are presented.

Abbreviations

CMC	carboxymethylcellulose
ePTFE	expanded poly(tetrafluoroethylene)
FDA	US Food and Drug Administration
FGM	fibrin gel matrix
HA	hyaluronic acid
IP	intraperitoneal
MW	molecular weight
ORC	oxidized regenerated cellulose
PEG	poly(ethylene) glycol
PLG	poly(L-glutamate)
PLL	poly(L-lysine)
PSA	postsurgical adhesion
PTFE	poly(tetrafluoroethylene)
QCM-D	quartz crystal microbalance with dissipation

J.M. Medley and T.D. Dziubla • Department of Chemical and Materials Engineering, University of Kentucky, Lexington, KY, 40506, USA

D.A. Puleo and R. Bizios (eds.), *Biological Interactions on Materials Surfaces*,
DOI 10.1007/978-0-387-98161-1_20,

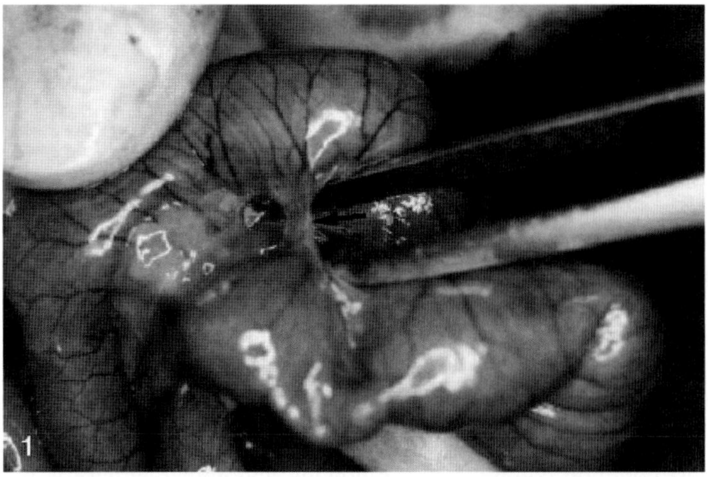

Figure 20.1. An example of intestinal postsurgical adhesion (PSA) formation. In a rat model, sections of cecum and ileum were abraded using dry gauze until hemorrhage formed. Fourteen days after surgery, adhesion formation was visualized. Figure reproduced with permission from Kutlay et al. [3].

20.1. Introduction

Each year in the United States, millions of people submit themselves to abdominal surgery for a variety of reasons, including gynecological reconstruction, tumor removal, and Cesarean sections [1, 2]. While the success rate and recovery times for the initial surgeries are continually improving, complications frequently occur. One major complication of abdominal surgery, occurring in 65–97% of patients, is the formation of tissue adhesion. This acellular, collagen-rich tissue often forms when an organ is damaged due to inadvertent desiccation or trauma during surgery. During healing, this damaged tissue often becomes permanently attached to adjacent tissues by the formation of a fibrous scar, as shown in Figure 20.1. While the nearly ubiquitous formation of adhesions is not generally problematic, numerous problems can arise that require subsequent physician care. Many patients experience abdominal or pelvic pain, intestinal obstructions, infertility, and increased difficulty in subsequent surgical procedures. In addition to the immeasurable costs in patient pain and suffering, an estimated 440,000 adhesiolysis procedures are performed in the United States each year [4, 5].

20.2. Postsurgical Adhesion (PSA) Formation

The mechanism of PSA formations has been shown to follow the same basic pathway as intraperitoneal wound healing. The cells, proteins, and tissues involved in this mechanism, which have been described in numerous review articles, are shown graphically in Figure 20.2 and can be summarized as a sequence of eight steps, as stated in Figure 20.3 [7–9]. When this process initiates on an open surface, typical wound healing is observed, including the associated scar formation. If an opposing surface is bridged by the fibrin gel matrix (FGM) (see Figure 20.3, step 6), the resulting scar tissue can permanently connect these tissues, resulting in the formation of PSAs.

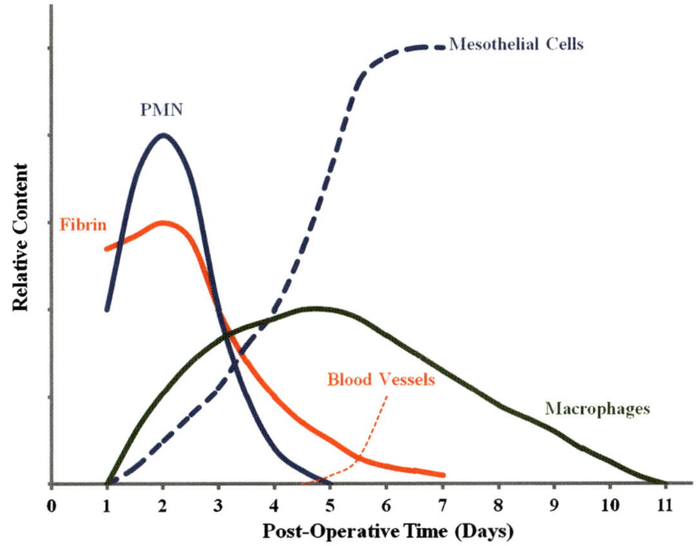

Figure 20.2. Stages and time course of the wound-healing process. Initially, a high concentration of *fibrin* is observed in the peritoneal fluid. Polymorphonuclear leukocytes (*PMNs*) are observed within approximately 2 days after tissue trauma, followed by *macrophages*. Reperitonealization occurs with the appearance of *mesothelial cells* within the first 5 days, and vascularization begins within 1 week following tissue damage. Adapted from [6].

1. Damage to the peritoneal mesothelium exposes an acellular, denuded surface, known as the "basement membrane", which consists primarily of type I collagen, type II collagen, fibronactin, and laminin [1, 2].
2. As the vasculature constricts, homestasis is obtained and the coagulation cascade begins with the deposition of fibrin [1].
3. The quantities of fluid and plasma proteins increase near the site of injury and inflammatory exudate develops witin the peritoneal fluid [7].
4. The composition of this exudate, initially consisting primarily of neutrophils, changes. If there is no infection, it becomes rich in macrophages over the course of 24 hours [1, 7].
5. Ephthelial cells are deposited on the surface of the injury by a process called "Island implantation" and re-epithelialization of the peritoneal surface proceeds [7].
6. The fibrin gel matrix (FGM), considered the "principal player in adhesion formation", consists of fibronectin, hyaluronic acid, glucosaminoglycans and proteoglycans, and forms 72–108 hours after surgery [1, 7].
7. The FGM is remodeled and stregthened, for approximately 1 month [1, 8].
8. The FGM is slowly replaced by fibrous collagen molecules and vascularized in a process controlled, at least partially, by macrophages [1].

Figure 20.3. Timeline of events occurring during postsurgical tissue adhesion formation.

When such adhesions form within the body, numerous complications are possible. If the adhesions affect the intestines, bowel obstructions can result [10–12]. In female patients, adhesions can obstruct the fallopian tubes and result in reduced fertility [10, 13, 14]. Following surgery, many patients experience abdominal pain, which is believed to be a result of reduced mobility of the organs within the abdominal cavity [10, 13, 15, 16]. Finally, subsequent surgeries can be complicated as tissue adhesions may interfere with the surgeon's access to organs [17, 18].

The costs associated with PSA in the USA are substantial. Approximately 300,000 hospital readmissions annually are either directly or indirectly related to the formation of PSA; the annual financial expenditure for pertinent treatment costs exceeds $1.3 billion [4, 7]. Based on the significant financial costs and the immeasurable cost in terms of patient discomfort, the need to either eliminate or reduce the incidence and severity of PSA is apparent.

20.3. Methods of PSA Prevention and Control

Based on the well-characterized PSA formation mechanism, researchers and physicians have developed numerous approaches to prevent their formation. These approaches can be classified into four general categories: modification of surgical technique, pharmaceutical approaches, liquid instillates, and adhesion barriers. While none of these techniques can eliminate the incidence of PSA completely, each approach has demonstrated encouraging results, with research continuing along all four directions.

20.3.1. Modification of Surgical Technique

Since PSAs are known to form as the result of trauma to tissue, it is logical that the number of, and severity of, PSAs can be reduced by reducing the trauma to tissues during surgery. Simple changes, such as implementing the use of nonpowdered surgical gloves, can reduce desiccation of the peritoneum and have a positive impact. During open abdominal surgery, the peritoneum is often sutured closed to aid healing. This practice increases the length of surgery and, since the peritoneum is known to heal rapidly via island implantation, it has recently been questioned [19–21]. Several studies have recently been conducted to assess the necessity of suturing the peritoneum and its affect on the formation of PSA; the results have been mixed [1, 2, 14, 19–21]. The presence of suture materials can induce an exacerbated wound-healing response, due to the foreign-body reaction, and promote PSA formation [22].

Another common surgical procedure, installation of prophylactic drainage tubes following abdominal surgeries, has also been subject of investigation [23]. Often performed as an effort either to remove excess serum and blood (to reduce infection) or to provide an early warning of surgical complications, the necessity of this technique has been questioned [24–27]. In fact, the presence of drainage tubes can increase the likelihood of infection following abdominal surgery and has been shown to increase the incidence of PSA [26].

Recent investigations suggest that increasing the oxygenation of tissues in the abdominal cavity may affect the biochemistry of the wound-healing process. Reduced oxygen levels in tissue (e.g., due to decreased blood flow) can reduce the ratio of tissue plasminogen activator to plasminogen activator inhibitor-1 present in that tissue [28]. Since plasmin is necessary to break down the FGM (see Figure 20.3, step 6), this reduced oxygen level may contribute to increased PSA formation. Oxygenating tissues during surgery has demonstrated some success at reducing the impact of PSA [28].

Since PSA is a result of trauma to the internal organ surfaces, reductions in the incision size, shortening the time of surgery, and reducing the overall invasiveness of the procedure should result in a reduction of PSA. Indeed, laparoscopic surgical methods have resulted in a major reduction in the incidence and severity of PSA [29–31]. Unfortunately, not all surgeries, such as gynecological reconstruction or emergency procedures, are amendable to laparoscopic methods. In addition, even though the incidence of PSA is reduced, laparoscopy can result in some tissue trauma, with PSAs still a likely outcome. Implementation of improved surgical procedures, while effective at reducing the incidence and severity of adhesions, does not eliminate the risk of PSA formation. A significant clinical need still exists for alternative interventions to inhibit PSA formation.

20.3.2. Pharmaceutical Interventions

Since PSA formation is a product of uncontrolled wound healing, by pharmacologically intervening in this process, it may be possible to reduce the effect of pathways leading to tissue adhesions. However, implementation of these techniques has been frustrated by the complex biochemical nature of the wound-healing process. For example, while it may be desirable to modify the healing process at the site of basement membrane exposure, biological approaches can potentially disrupt the beneficial healing throughout the body, resulting in more severe postsurgical complications.

Initial attempts to employ pharmaceutical agents focused on the reduction of tissue trauma. Local application of antiinflammatory medications (i.e., ibuprofen, steroids, antihistamines, etc.) has been used to reduce inflammation at the surgical site. Interestingly, these therapies have demonstrated some efficacy during preclinical trials at reducing the formation of PSA [32, 33]. This result is likely due to the known link between excessive inflammation and fibrosis [34].

Similarly, attempts at mitigating oxidative stress at the wound site have also been employed to suppress the formation of PSA. While generation of oxygen radicals (typically from the leukocyte respiratory burst) [35, 36] is a necessary part of the wound-healing cascade, an excessive amount is known to activate collagen-producing cells, resulting in exaggerated levels of fibrosis [37, 38]. It is logical to expect, then, that administration of antioxidant molecules would suppress formation of collagen-rich tissues [39]. Melatonin, vitamin E, and catalase have all shown some promise in reducing the severity of PSA formation [39–42].

Heparin, a naturally occurring anticoagulant, interrupts various steps of the coagulation cascade and inactivates thrombin. Used to treat numerous pathologies and to create biologically passivated surfaces, heparin has been investigated as an agent to reduce PSA formation. The results of these investigations, however, have been disappointing [43].

The most beneficial pharmaceutical interventions have employed fibrinolytic agents to reduce PSA formation. These enzymes, occurring naturally in the body, are important in the degradation of blood clots. As a result of trauma caused by surgery, production of these enzymes is suppressed [44, 45]. It was proposed that formation of PSA can be prevented by accelerating the degradation of fibrin deposits with locally delivered fibrinolytic agents to break down the FGM. This adhesive FGM is responsible for the formation of the initial deleterious union of adjacent tissue surfaces. Early investigations involved treatment with bacterially derived streptokinase, a protein that catalyzes formation of plasmin, which, in turn, breaks down the fibrin mesh of blood clots [46]. Increased plasmin activation proved beneficial in both animals and in humans and resulted in the commercialization of "Varidase" in the 1970s [47–49]. Additional investigations used the plasminogen activator enzyme, urokinase, and plasmin directly via intraperitoneal injection; in these cases, mixed results and potentially immunogenic responses were observed [50, 51]. Despite these inconsistencies, direct management of fibrin, the primary adhesive protein, represents an attractive PSA-prevention strategy. Recent studies, focusing on the use of recombinant tissue plasminogen activator demonstrated significant potential to successfully address PSA formation [45].

Oral delivery of drugs to reduce PSA formation has recently been investigated [52]. Chymases, naturally occurring proteases, stimulate accumulation of neutrophils and other inflammatory cells at the trauma site; these enzymes cleave procollagen into collagen fibrils, a primary step in the formation of the collagen-rich tissue bridge. It was hypothesized that, by inhibiting/slowing the formation of collagen fibrils, the FGM could break down before PSA formation can occur. A recent study employed oral delivery of NK3201, a novel chymase inhibitor, and showed significant reduction in chymase activity and, thereby, suppressed PSA formation in a hamster model [52].

20.3.3. Liquid Instillates

During surgery, saline solution is typically instilled into the abdominal cavity to accomplish a number of goals. In addition to maintaining cleanliness (by removing blood and other biological debris), this solution also maintains hydration of exposed tissues. In an effort to reduce PSA formation, several alternate liquid instillates have been investigated.

Hyaluronic acid (HA), whose chemical structure is shown in Table 20.1, is a naturally occurring biological polymer consisting of a long chain of disaccharide units with a molecular weight that can exceed 20 million daltons. HA is a principal component of the extracellular matrix of tissues throughout the human body. HA can readily be broken down into its principle sugar molecules and be metabolized within the body; these degradation units, both monomeric and oligomeric, are potent signals for macrophage and cell recruitment, angiogenesis, and the presumed nonscarring, fetal-like, wound-healing response [53–56]. Because of these properties, the use of HA has been investigated as a means to reduce PSA formation. If instilled prior to surgery, HA can effectively coat the organs in the abdominal cavity and provide protection against surgical trauma and tissue desiccation [57, 58]. HA reduces the tissue adhesion score to 0.9 ± 1.1 versus 2.0 ± 0.8 obtained from the controls in a mouse model and has been approved for use by the US Food and Drug Administration (FDA) [57].

Several chemical modifications of HA have been formulated in order to improve its in vivo performance. By slowing HA degradation kinetics, it is believed that additional protection against PSA formation can be obtained. HA crosslinked with ferric ion reduced adhesion scores by 59% in clinical trials and led to the commercial product “Intergel” [59, 60]. This product was withdrawn from the market after reports of additional complications, including pain and a foreign-body response attributed to high iron content [43, 61]. By esterifying the hydroxyl groups on one HA chain with the carboxyl groups on another HA chain, it is possible to form an autocrosslinked HA derivative (HAx). This material forms a hydrogel and provides surface tissue protection for a longer duration. Tests in a rabbit model demonstrated the reduction in the adhesion score from 3.0 for controls to 0.5 in the treatment groups [62]. Clinical trials have also demonstrated reductions in adhesion scores of approximately 60% versus controls [63, 64]. In addition, it is not clear whether these esterified HA monomers possess the same degree of chemotacticity as native HA and, thereby, alter the rate of cellular infiltration and wound healing at the surgery site [65].

The most effective HA-derived liquid instillate employs crosslinked HA with another biologically derived polymer, carboxymethylcellulose (CMC). CMC, whose structure is

Table 20.1. Structures of chemical compounds used for the prevention of PSA.

Compound	Structure
(a) Hyaluronic Acid (HA)	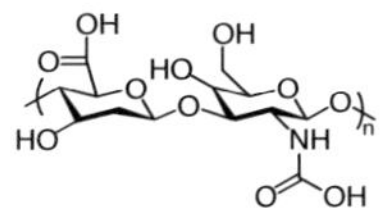
(b) Carboxymethyl Cellulose (CMC)	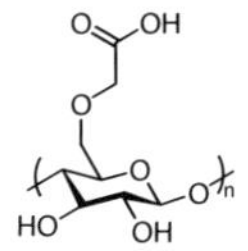
(c) Phospholipid	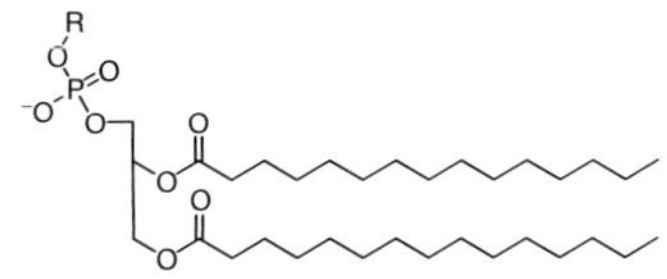
(d) "Adept" Icodextrin	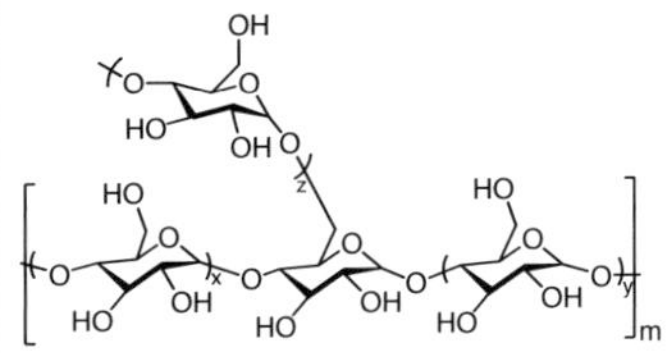
(e) Polyethylene Glycol (PEG)	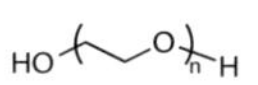
(f) Poly (L-Lysine)	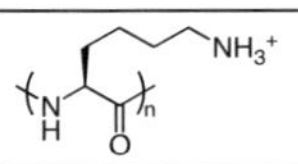
(g) Poly (L-Glutamate)	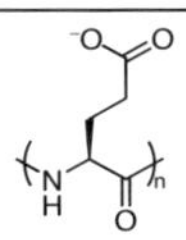
(h) Polyethylene (PE)	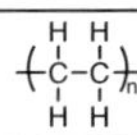
(i) Poly (Tetrafluoroethylene) (PTFE)	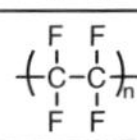
(j) Cellulose	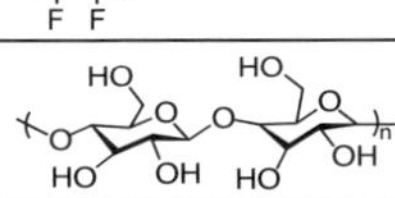
(k) "SprayGel" System	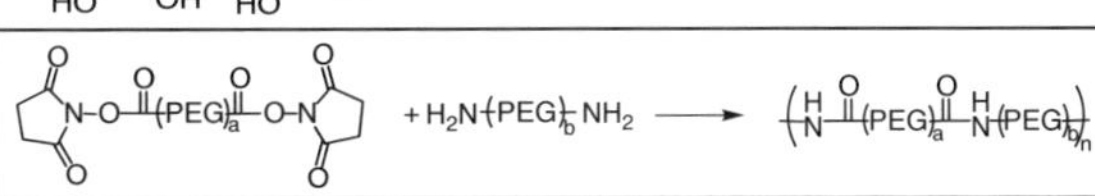

shown in Table 20.1, is a nontoxic material, used as a thickening agent in many food and drug products. CMC is not degraded in the body as rapidly as HA [62, 66, 67]. After partial oxidation, HA can be chemically crosslinked with CMC; this chemical combination of HA with CMC, which is commercially available as "Sepracoat," results in an instillate with a residence time in the abdominal cavity of approximately 7 days [59, 68–70]. This material has demonstrated significant protection against PSA formation in animal models but failed to convince the FDA panel of its clinical effectiveness against preventing adhesions in human patients [71].

Another category of liquid instillates is based on the chemistry of phospholipids. These materials, whose general chemical structure is shown in Table 20.1, are amphiphilic: they have a polar, hydrophilic head group and a hydrophobic tail. This amphiphilicity accounts for the surfactant properties exhibited by these materials; they adhere to hydrophilic surfaces and form a protective boundary–lubricating layer with their hydrophobic tails extending from the surface [72]. Several amphiphilic lipids (phospholipids, sphingolipids, and galactolipids), have been tested to reduce PSA formation and demonstrated significant potential to reduce PSA formation [73–75]. In a rabbit model, use of phospholipids reduced the area of adhesion formation from 691 to 191 mm^2 [73].

Another class of solutions has been designed to provide protection in the abdominal cavity without pretreatment via hydroflotation. The saline solutions typically used for irrigation during surgical procedures fall into this category. "Adept," a commercial product based on icodextrin, a polymeric starch, whose chemical structure is shown in Table 20.1, has been used for irrigation and rinsing during surgery [76]. After surgery, the polymeric starch is absorbed into the lymphatic system very slowly due to its high molecular weight [77]. Because of its high concentration, this material increases the osmotic pressure in the abdominal cavity and maintains an increased level of fluid [78]. The presence of this excess fluid results in tissue separation and has been shown to decrease PSA formation [77, 79–82]. In the rabbit dual uterine horn model, "Adept" increased the number of adhesion-free sites to 50% versus the 5% obtained for the untreated controls [77].

Another material that has been employed in efforts to reduce PSA formation is poly(ethylene) glycol (PEG). This material, often referred to as polyethylene oxide, was proven biocompatible in a number of applications and is under investigation for a number of additional uses [83]. PEG, whose chemical structure is shown in Table 20.1, consists of a series of repeating ether units with either hydroxyl or methoxy terminations at both ends. This structure renders PEG amphoteric and resistant to protein adsorption. Soluble in many organic solvents, PEG is highly hydrophilic and can form a hydrated layer with a large excluded volume and high steric stabilization [83, 84]. PEG surfaces present few binding sites for protein interactions; for this reason, adsorption of proteins, and subsequent adhesion of cells, is strongly inhibited. Injection of PEG into the peritoneal cavity has been moderately successful in preventing PSA formation [5]. Yet, because of their antiadhesive nature, free PEG chains poorly adhere to surfaces. Attempts to increase the protective ability of PEG by crosslinking with CMC into a viscous gel have achieved a 42% reduction in the American Fertility Society adhesion score versus the relevant controls [85].

A recent attempt to prevent PSA formation employed sequential use of solutions containing charged polypeptides [86]. A solution containing poly(L-lysine) (PLL) was instilled into the peritoneal cavity followed by a solution of poly(L-glutamate) (PLG) 5 min later. Positively charged PLL, whose chemical structure is shown in Table 20.1, adhered to damaged tissues. Negatively charged PLG, whose chemical structure is also shown in Table 20.1, associated with the positively charged PLL on the tissue surface and formed an insoluble film at the site of injury. In animal studies employing a mouse model, this system reduced the incidence of PSAs from 49.5% in the control animals to 8.81% in the treated animals [86, 87]. Further implementation of these materials/solutions may be hampered because of toxicity concerns associated with the highly charged polyamines employed [86].

20.3.4. Adhesion Barriers

The most effective methods that reduce the incidence and severity of PSA formation provide physical separation of wounded from surrounding tissues. By isolating damaged or potentially damaged surfaces from each other during the wound-healing process, normal healing without the formation of PSAs was achieved [2, 7]. Because of the simplicity and efficacy of this approach, numerous barrier systems have been investigated.

The first adhesion barrier system investigated consisted of grafts of peritoneal tissue aimed at providing protection to areas of damaged peritoneum. The presence of necrotic graft material, however, promoted, rather than inhibited, PSA formation in animal trials [43]. Other attempts to protect tissues by stretching the nearby peritoneal membrane also proved ineffective [43]. Decreased blood flow in this distorted tissue resulted in decreased plasmin activation and enhanced the formation of the FGM [43].

Another material that has been investigated for use as an adhesion barrier is poly(tetrafluoroethylene) (PTFE) [88]. Because this material is known to resist adhesion to most substances, PTFE is a natural candidate for use as a PSA barrier. PTFE, whose structure is shown in Table 20.1, consists of a straight chain of carbon atoms onto which fluorine atoms are fully substituted. The fluorine atoms present an inert, electron-rich surface that prevents strong interactions with other materials. PTFE is readily available as an expanded mesh, commonly referred to as ePTFE, under the trade name "Gore-Tex Surgical Membrane" [89, 90]. The presence of PFTE between damaged tissue surfaces prevents the FGM from bridging these tissues and, thus, prevents PSA formation [4, 91–94]. In animal tests employing a porcine model, PTFE reduced the adhesion score to 0.14 ± 0.12 versus the observed 1.33 ± 0.41 for controls [94].

The major drawbacks of PTFE result from its high degree of inertness. Since PTFE is so effective at preventing interactions with biological tissues, it typically must be sutured into place in the body [95]. In addition, PTFE is not biodegradable. Ideal for permanent prosthetic applications, the use of PTFE as an adhesion barrier can be problematic [92]. After the body has healed from the trauma induced by surgery, typically within about 2 weeks, the presence of the PTFE membrane at the site of trauma is undesirable. A second surgery, then, is typically necessary to remove this barrier material and ensure that unwanted complications are prevented [93].

Several other barrier membranes have been investigated that overcome this limitation of PTFE. Designed to be readily biodegradable, such membranes often do not need suturing into position and, thus, subsequent removal. After the desired protection has been rendered, these membranes break down and are metabolized by the body. Numerous biodegradable materials have been investigated for use as adhesion barriers, including cellulose derivatives, HA derivatives, and PEG-based materials [95–100].

A barrier composed of oxidized regenerated cellulose (ORC) was successfully commercialized under the trade name "Interceed." Cellulose, a naturally occurring polysaccharide, whose structure is shown in Table 20.1, is partially oxidized to improve aqueous solubility. This biodegradable material has a gauze-like consistency and can be applied directly to damaged tissues in order to prevent PSA. When properly applied, this material has shown a high degree of efficacy in reducing PSA formation in surgical procedures. There are, however, several limitations to the clinical use of "Interceed" [95, 100]. Because it is thrombogenic, contact with blood greatly increases the likelihood of subsequent surrounding tissue adhesion; because of this property, complete homeostasis must be achieved prior to in situ application of "Interceed." In addition, application must be limited to a single layer of material in all areas; multiple layers may promote adhesions [43]. These difficult handling requirements render "Interceed" unsuitable for laparoscopic procedures and limit its general applicability to more invasive laparotomic procedures.

Barrier membranes based on HA crosslinked with CMC are available under the trade name "Seprafilm." This material is applied as rigid sheets that, once hydrated, form a protective gel on the abdominal tissue surface and prevent PSA formation. With a chemical composition nearly identical to "Sepragel," this material is readily biodegraded in vivo within about 1 week. Provided that a surgeon has knowledge and access to all areas of tissue damage, "Seprafilm" can be very effective at preventing PSA formation [95–97]. In animal studies employing a rat model, the total adhesion score was reduced to 4.15 ± 0.58 from 7.21 ± 0.52 obtained for the respective controls [70]. A meta-analysis of clinical data also demonstrated "Seprafilm" to be effective at lowering the incidence of PSAs in abdominal surgeries [101]. While this study mistakenly included a non-"Sepracoat" HA/CMC material in its analysis, it reported that patients receiving the "Sepracoat" were 8.71 times more likely ($p < 0.01$) to have no adhesions than the non-"Sepracoat" control group ($p < 0.01$) [101]. While the rigid structure of "Seprafilm" has traditionally prevented its use in laparoscopic surgery, newer techniques involving the use of rolled films has enabled limited laparoscopic use [102].

Since the number of procedures performed by laparoscopic surgery is continually increasing, the ability to apply adhesion barriers via this method is increasingly important. A material recently developed to address this clinical need is the two-component system commercially available as "Spraygel." This system consists of a double-barrel syringe used to apply the contents to damaged tissue via laparoscopy. Each barrel contains a reactive PEG precursor solution. Upon mixing, an amine group from one component reacts with *N*-hydroxylsuccinimide ester in the other component to form a crosslinked hydrogel, as shown in Table 20.1. This crosslinked material degrades in vivo into PEG units that are rapidly removed from the body by renal clearance [98, 99]. Since "SprayGel" is readily applicable in laparoscopic procedures, the major limitation for its use in PSA is the surgeon's awareness of, and access to, damaged tissues. In clinical trials, this material showed significant efficacy because it reduced the mean adhesion tenacity score by 64.7% versus pertinent values obtained for the untreated controls [103].

The aforementioned adhesion barriers have proven efficacious in reducing PSA formation in patients; for this reason, these commercially available products are enjoying success in the marketplace. Even with these methods, however, PSA formation is still occurring in many patients. For instance, in a 2007 meta-analysis of "Seprafilm" efficacy, of the 371 patients in the adhesion barrier treatment group (combined over several studies), 196 still had developed a PSA [101]. With a prevention rate of only ~48%, multiple hypotheses have emerged to explain the persistent occurrence of PSA. If the surgeon is unaware of damaged tissue or if the damaged site is inaccessible, current barrier methods may not be effective and tissue adhesion formation remains a likely outcome. To overcome these deficiencies, many new strategies for the prevention of PSA are currently being investigated. Some of these investigations focus on combinations of techniques that have demonstrated success in the past. For example, various hydrogels have been loaded with either antiinflammatory, antibiotic, or fibrinolytic agents, including budesonide, paclitaxel, and streptokinase [33, 41, 104–106]. In addition, changes to pertinent surgical techniques are under investigation [26, 28, 52]. A brief summary of the most widely investigated PSA-prevention methods is given in Table 20.2.

20.4. PSA Evaluation Methods

Perhaps one of the biggest impediments to the rapid development of PSA-prevention strategies, especially to the testing of barrier materials, is the lack of a standard and widely used method to evaluate and compare pertinent performance. The situation is complicated because various animal models are used and each clinical trial is conducted independently. Due to the

Table 20.2. Summary of PSA-prevention methodologies.

	Technology	Application method	Biodegradability	Level of tunable properties	Level of success	References
Modification of surgical technique	Nonsuturing of damaged peritoneum	N/A	N/A	N/A	Significant decrease in PSA formation, but adoption of this practice by surgeons problematic	[1, 2, 14, 19–21]
	Use of laparoscopic procedures	N/A	N/A	N/A	Significant decrease in PSA formation, but: Application of technique is limited to certain surgeries Precludes use of many other adhesion-reduction techniques	[29–31]
	Other technique changes (nonpowdered gloves, nonuse of drainage tubes, increased tissue oxygenation)	N/A	N/A	N/A	Significant decrease in PSA formation, but adoption of this practice by surgeons is problematic	[1, 2, 26, 28]
Pharmaceutical interventions	Antiinflammatories	IP injection	Yes	Dose, timing, and selection of the therapeutic agent	Moderate success because of the associated risk of side effects	[32, 33]
	Antioxidants	IP injection	Yes	Dose can be altered	Very successful Vitamin E interrupts several parts of the adhesion formation mechanism	[40, 41]
	Fibrinolytic agents	IP injection	Yes	Dose can be altered	Moderately successful but has potential for immunogenic response and reduced hemostasis	[10, 44, 45, 47–49]
	Chymase inhibitors	Oral delivery	Yes	Dose and timing of the therapeutic agent can be altered	Moderately successful	[52]

(continued)

Table 20.2. (continued)

	Technology	Application method	Biodegradability	Level of tunable properties	Level of success	References
Peritoneal instillates	HA and its derivatives	IP injection	Yes	MW and dose of HA can be optimized	Moderately successful	[57–64, 69, 70]
	Phospholipids	IP injection	Yes	Dose can be optimized	Very effective The zwitterionic nature of the phospholipids allows them to bind onto the epithelial cell membranes and provide lubrication	[72–75]
	Hydroflotation initiators ("adept")	IP injection	Yes	MW and dose of the starch molecules can be optimized	Moderately successful	[76–82]
	Poly(ethylene glycol)	IP injection	Yes	MW and concentration of PEG solution can be optimized	Moderately successful: Significant decrease in number and collagen content of PSAs	[5, 83, 84]
Adhesion barriers	PTFE ("Gortex Surgical Membrane," "Preclude")	Film placed directly on the damaged tissue	No	Physical properties of the mesh can be controlled	Moderately successful: Nondegradable; it remains in site and must be removed with another surgical procedure Requires suture fixation Incompatible with laparoscopy	[88, 92–95]
	ORC ("Interceed")	Film placed directly on the damaged tissue	Yes	MW, degree of oxidation, and the physical properties of the film can be controlled	Moderately successful: Thrombogenic Difficult to handle Incompatible with laparoscopy	[43, 95, 100]

Crosslinked HA gel ("Seprafilm")	Film placed directlyon the damaged tissue	Yes	MW and degree of the crosslinking can be optimized	Very successful: Adheres to the peritoneum Degrades Is not procoagulant But causes: Severe inflammation Abscess formation Must be placed directly on the damaged tissue	[95–97]
In vivo polymerization of PEG ("Spray Gel")	In vivo spray polymeriza-tion	Yes	MW and degree of polymerization can be optimized	Very effective; can be employed with laparoscopic procedures	[98, 99, 103]

N/A not applicable; *IP* intraperitoneal; *MW* molecular weight; *HA* hyaluronic acid; *ORC* oxidized regenerated cellulose

Table 20.3. Summary of the animal models used in PSA testing.

Tissue(s) involved	Animal model	Pros	Cons	References
Abdominal wall	Mouse	Measurements of both extent and type of adhesion	Semiquantitative scoring	[28]
	Rat	Quantitative analysis of the adhesion area		[86, 87]
	Rabbit	Surgical damage carefully controlled	Semiquantitative scoring	[73, 74]
Intestinal wall	Rat		Semiquantitative scoring	[104, 105]
Intestine/ abdominal wall	Rabbit	Surgical damage carefully controlled		[75]
	Rabbit		Semiquantitative scoring	[33, 62, 106, 112]
Uterus	Hamster		Semiquantitative scoring	[52]
	Pig	Internal control possible by treating two uterine horns separately	Semiquantitative scoring	[98]
	Rabbit	Internal control possible	Semiquantitative scoring	[33, 99]
	Rat	Internal control possible Quantitative analysis of the adhesion area		[41, 82, 93]
Uterus reformation	Mouse	Multiple adhesion	Semiquantitative scoring	[57]

complex interplay of the many facets of the wound-healing process, inflammatory cascades, and anatomical aspects, in vivo studies are an integral part of PSA evaluation. By its nature, however, animal testing is very complex, time consuming, and limited in its ability to provide useful data comparisons among studies, which use different species. In addition, large-scale screening of materials is not typically possible because of the associated high cost, time requirements, and ethical issues. Furthermore, in most animal models, failure of PSA treatment strategy does not always provide useful information regarding the underlying mechanism of failure. Methods of assessment have been primarily based on semiquantitative animal models, including mouse, rat uterine horn, rabbit, and pig; these models are summarized in Table 20.3. The results of such studies are typically scored based on visual observations (e.g., no adhesion, filmy avascular adhesion, vascularized adhesion, collagen-rich adhesion) in an attempt to provide meaningful comparisons. Because of the large variability in the test methods and in data analyses, it is generally not possible to conduct meaningful comparisons between independent investigations. For these reasons, the results of material performance are often either confounded or obscured by the differences in the models and scoring systems employed.

Furthermore, there has been significant lack of translational relevance from in vivo PSA treatment models to clinical application and success. A strong need exists to revisit the fundamental aspects of tissue adhesion formation and obtain more information regarding the underlying mechanisms for both the success and failure of each treatment. Recent attempts to address these issues have focused on standardizing the methods of testing and quantifying tissue adhesion formation. A grading system, described in Table 20.4, was developed to

Table 20.4. Details of the adhesion scoring system. Each animal is semiquantitatively graded for the extent, severity, and degree of PSA formation. The sum of these scores (0–10) is the total adhesion score [107].

Score	Extent of adhesion	Severity	Degree
0	None	No adhesions	No adhesions
1	1–25%	Filmy, avascular	Adherent tissue separation with gentle traction
2	26–50%	Vascular and/or opaque	Tissue layer separation requires moderate traction
3	51–75%	Cohesive attachment	Tissue layer separation requires sharp dissection
4	76–100%		

evaluate three characteristics (specifically, extent, severity, and degree) of adhesion formation; application of these three measures assigns a total adhesion score [107]. While still semiquantitative, this grading system provides some quantitative information and allows comparisons to be made between the results of various studies.

Most recently, complete quantitative investigations and in vitro models are being developed to allow for a rational optimization of potential treatments prior to large-scale animal tests. Investigators have reported preliminary results indicating that tissue adhesions can be reproducibly formed in a porcine model and analyzed using a tensile test system [108]. This tensile strength analysis method provides bulk modulus data of the formed scar tissue, allowing direct quantitative assessment of the quality and maturity of the adhesion and does not rely alone upon the physical appearance of the tissue [108].

Finally, in vitro methods are being developed that address-specific aspects of the tissue adhesion events, providing rapid evaluation of chemophysical methods in preventing PSA. For example, for treatments that attempt to inhibit the initial adhesion between damaged tissue surfaces, a quartz crystal microbalance with dissipation (QCM-D) method has been described [109]. This system allows the sequential flow of molecules across a model surface, allowing for the evaluation of solution-applied barriers. The substrate resonates at a characteristic frequency, typically 5 MHz. As molecules from the solution interact with this substrate, the resonance frequency decreases by a small amount (about 17.7 ng/cm^2) [110, 111]. Using a surface, which mimics the FGM, it is possible to monitor the interactions of soluble, self-assembled barrier materials with the tested substrates (e.g., fibrin, hyaluronan, collagen I, laminin) and relate pertinent aspects (e.g., strength of adhesion, durability of material, viscoelasticity) to in vivo phenomena such as wound-healing rate and tissue adhesion formation. By monitoring the change in resonant frequency and energy loss exhibited as the substrate and solution are allowed to interact, the QCM-D method allows rapid, quantitative analysis of these assembling interactions and resistance of subsequent adhesion events with these surfaces. Since adsorption of proteins to surfaces is a known prerequisite for subsequent cellular attachment and function, it is anticipated that this technology will allow rapid investigation of many adhesion-prevention systems. As shown in Figure 20.4, very small amounts (specifically, nanograms per square centimeter) of barrier polymer, diblock copolymer poly(ethylene glycol)-b-poly(acrylic acid) adsorbing to surfaces can be detected and the kinetics of these interactions can be readily investigated. In addition, by simultaneously monitoring the energy loss, or dissipation, of the model system, the adsorption phenomena mechanism (e.g., rigid, tight binding vs. fully extended viscoelastic gels) can be studied. While very preliminary, this technique holds great potential for the rapid development and screening of self-assembled materials for use in the prevention of PSAs. The ability to quantitatively

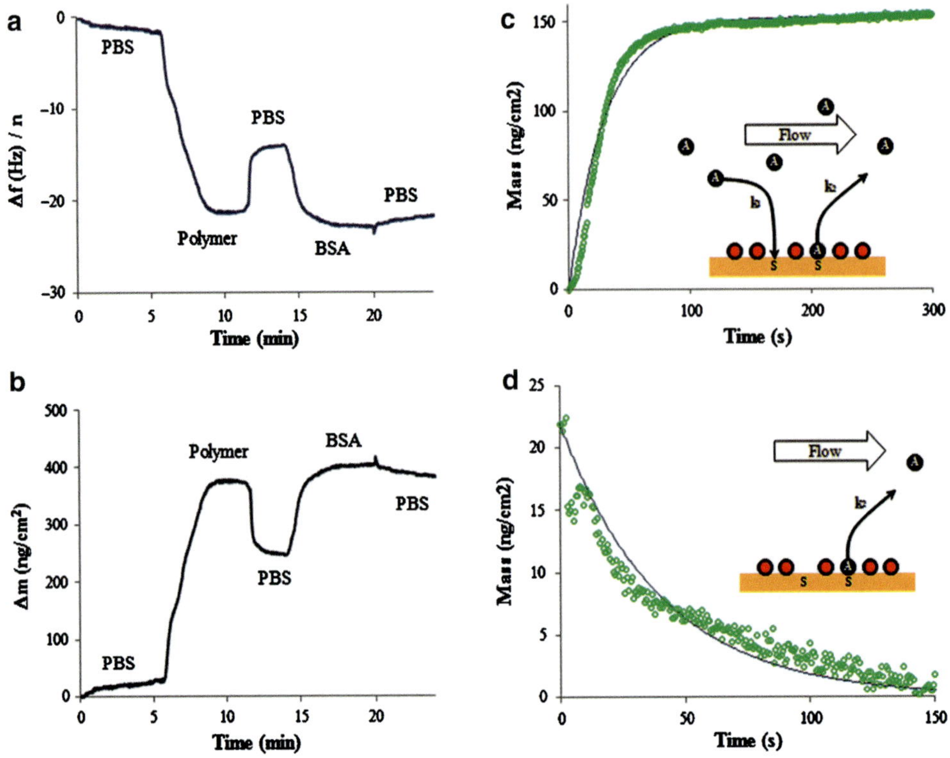

Figure 20.4. Representative quartz crystal microbalance with dissipation (QCM-D) curves for adsorption of adhesion-resistant polymer and subsequent protein adsorption. (a) Raw data of the time course of the frequency response (**Δf**). (**b**) Mass adsorption based on the Sauerbrey model, which assumes a linear fit between frequency response and adsorbed protein and polymer mass [111]. (c) Fit of the protein adsorption kinetics to a model substrate after preadsorption of an adhesion-resistant polymer. $[SA]_{max} = 152.1$ ng/cm^2 and $t_{½} = 20.9$ s. (**d**) Kinetic model fit of protein desorption from surface. $[SA]_0 = 21.9$ ng/cm^2 and $t_{½} = 27.8$ s.

compare interactions of proteins, cells, and tissues with materials and to identify the underlying mechanisms is essential for the logical design of the next generation of PSA barriers.

20.5. Conclusion

PSAs represent a significant clinical complication that affects a large patient population. To address this multifaceted problem, a range of methods was developed. Although use of physical material barriers provides the best results, no single method has proven to be effective in all instances, to date. This failure is most likely due to the complexity of PSA formation, the uniqueness of each surgical procedure and patient, and lack of understanding of the underlying mechanism of PSA etiology and successful prevention. PSA remains a serious clinical problem. New methodologies to improve material/device performance at a mechanistic level are a necessity for the success of PSA barriers in the future. In addition, novel materials must be designed and formulated to address all pertinent aspects of the PSA formation (e.g., physical, chemical, inflammatory, oxidative, and wound healing) in order to maximize the potential for clinical success. In this regard, biomaterial engineers could make crucial contributions toward eventual elimination of PSA formation.

References

1. Ellis, H., et al., Adhesion-related hospital readmissions after abdominal and pelvic surgery: a retrospective cohort study. Lancet. 1999. **353**(9163): p. 1476–80.
2. Risberg, B., Adhesions: preventive strategies. Eur J Surg Suppl. 1997. (577): p. 32–9.
3. Kutlay, J., et al., Comparative effectiveness of several agents for preventing postoperative adhesions. World J Surg. 2004. **28**(7): p. 662–5.
4. Matthews, B.D., et al., Assessment of adhesion formation to intra-abdominal polypropylene mesh and polytetrafluoroethylene mesh. J Surg Res. 2003. **114**(2): p. 126–32.
5. Nagelschmidt, M., T. Minor, and S. Saad, Polyethylene glycol 4000 attenuates adhesion formation in rats by suppression of peritoneal inflammation and collagen incorporation. Am J Surg. 1998. **176**(1): p. 76–80.
6. diZerega, G.S., and J.D. Campeau, Peritoneal repair and post-surgical adhesion formation. Hum Reprod Update. 2001. **7**(6): p. 547–55.
7. Boland, G.M., and R.J. Weigel, Formation and prevention of postoperative abdominal adhesions. J Surg Res. 2006. **132**(1): p. 3–12.
8. Weis, C., et al., Poly(vinyl alcohol) membranes for adhesion prevention. J Biomed Mater Res B Appl Biomater. 2004. **70**(2): p. 191–202.
9. diZerega, G.S., Biochemical events in peritoneal tissue repair. Eur J Surg Suppl. 1997. (577): p. 10-6.
10. Practice Committee of the American Society for Reproductive Medicine. Control and prevention of peritoneal adhesions in gynecologic surgery. Fertil Steril. 2006. **86**(5 Suppl): p. S1-5.
11. Al-Took, S., R. Platt, and T. Tulandi, Adhesion-related small-bowel obstruction after gynecologic operations. Am J Obstet Gynecol. 1999. **180**(2 Pt 1): p. 313–5.
12. Miller, G., et al., Etiology of small bowel obstruction. Am J Surg. 2000. **180**(1): p. 33–6.
13. Diamond, M.P., and M.L. Freeman, Clinical implications of postsurgical adhesions. Hum Reprod Update. 2001. **7**(6): p. 567–76.
14. Tulandi, T., and A. Al-Shahrani, Adhesion prevention in gynecologic surgery. Curr Opin Obstet Gynecol. 2005. **17**(4): p. 395–8.
15. Kresch, A.J., et al., Laparoscopy in 100 women with chronic pelvic pain. Obstet Gynecol. 1984. **64**(5): p. 672–4.
16. Swank, D.J., et al., Laparoscopic adhesiolysis in patients with chronic abdominal pain: a blinded randomised controlled multi-centre trial. Lancet. 2003. **361**(9365): p. 1247–51.
17. Dijkstra, F.R., et al., Recent clinical developments in pathophysiology, epidemiology, diagnosis and treatment of intra-abdominal adhesions. Scand J Gastroenterol Suppl. 2000. (232): p. 52-9.
18. Brill, A.I., et al., The incidence of adhesions after prior laparotomy: a laparoscopic appraisal. Obstet Gynecol. 1995. **85**(2): p. 269–72.
19. Cheong, Y.C., N. Bajekal, and T.C. Li, Peritoneal closure – to close or not to close. Hum Reprod. 2001. **16**(8): p. 1548–52.
20. Lyell, D.J., et al., Peritoneal closure at primary cesarean delivery and adhesions. Obstet Gynecol. 2005. **106**(2): p. 275–80.
21. Roset, E., M. Boulvain, and O. Irion, Nonclosure of the peritoneum during caesarean section: long-term follow-up of a randomised controlled trial. Eur J Obstet Gynecol Reprod Biol. 2003. **108**(1): p. 40–4.
22. Setzen, G., and E.F. Williams, 3rd, Tissue response to suture materials implanted subcutaneously in a rabbit model. Plast Reconstr Surg. 1997. **100**(7): p. 1788–95.
23. Merad, F., et al., Prophylactic abdominal drainage after elective colonic resection and suprapromontory anastomosis: a multicenter study controlled by randomization. French Associations for Surgical Research. Arch Surg. 1998. **133**(3): p. 309–14.
24. Memon, M.A., et al., The uses and abuses of drains in abdominal surgery. Hosp Med. 2002. **63**(5): p. 282–8.
25. Memon, M.A., M.I. Memon, and J.H. Donohue, Abdominal drains: a brief historical review. Ir Med J. 2001. **94**(6): p. 164–6.
26. Bertram, P., et al., Effects of intra-abdominal drainages on adhesion formation and prevention by phospholipids in a rat model. Drainages and adhesion formation. Eur Surg Res. 2003. **35**(2): p. 92–7.
27. Schein, M., To drain or not to drain? The role of drainage in the contaminated and infected abdomen: an international and personal perspective. World J Surg. 2008. **32**(2): p. 312–21.
28. Matsuzaki, S., et al., Effects of supplemental perioperative oxygen on post-operative abdominal wound adhesions in a mouse laparotomy model with controlled respiratory support. Hum Reprod. 2007. **22**(10): p. 2702–6.
29. Gutt, C.N., et al., Fewer adhesions induced by laparoscopic surgery? Surg Endosc. 2004. **18**(6): p. 898–906.

30. Milingos, S., et al., Adhesions: laparoscopic surgery versus laparotomy. Ann N Y Acad Sci. 2000. **900**: p. 272–85.
31. Pattaras, J.G., et al., Incidence of postoperative adhesion formation after transperitoneal genitourinary laparoscopic surgery. Urology. 2002. **59**(1): p. 37–41.
32. Muller, S.A., et al., Adhesion prevention comparing liquid and solid barriers in the rabbit uterine horn model. Eur J Obstet Gynecol Reprod Biol. 2005. **120**(2): p. 222–6.
33. Cooper, K., et al., Reduction of post-surgical adhesion formation with tranilast. J Surg Res. 2007. 141(2): p. 153–61.
34. Stramer, B.M., R. Mori, P Martin, and . The inflammation-fibrosis link? A Jekyll and Hyde role for blood cells during wound repair. J Invest Dermatol. 2007. **127**(5): p. 1009–17.
35. Knight, J.A., Review: free radicals, antioxidants, and the immune system. Ann Clin Lab Sci. 2000. 30(2): p. 145–58.
36. Peterhans, E., Oxidants and antioxidants in viral diseases: disease mechanisms and metabolic regulation. J Nutr. 1997. **127**(5 Suppl): p. 962S–5S.
37. Prakash Kumar, B., and K. Shivakumar, Alterations in collagen metabolism and increased fibroproliferation in the heart in cerium-treated rats: implications for the pathogenesis of endomyocardial fibrosis. Biol Trace Elem Res. 1998. **63**(1): p. 73–9.
38. Wilgus, T.A., et al., Hydrogen peroxide disrupts scarless fetal wound repair. Wound Repair Regen. 2005. **13**(5): p. 513–9.
39. ten Raa, S., et al., The role of neutrophils and oxygen free radicals in post-operative adhesions. J Surg Res. 2006. **136**(1): p. 45–52.
40. Portilla, F.d.l., et al., Prevention of peritoneal adhesions by intraperitoneal administration of vitamin E: an experimental study in rats. Diseases of the Colon & Rectum. 2004. **47**(12): p. 2157–61.
41. Demirbag, S., et al., Comparison of hyaluronate/carboxymethylcellulose membrane and melatonin for prevention of adhesion formation in a rat model. Hum Reprod. 2005. **20**(7): p. 2021–4.
42. Yuzbasioglu, M.F., et al., The effect of intraperitoneal catalase on prevention of peritoneal adhesion formation in rats. J Invest Surg. 2008. **21**(2): p. 65–9.
43. Johns, A., Evidence-based prevention of post-operative adhesions. Hum Reprod Update. 2001. **7**(6): p. 577–9.
44. Hellebrekers, B.W., et al., A role for the fibrinolytic system in postsurgical adhesion formation. Fertil Steril. 2005. **83**(1): p. 122–9.
45. Hellebrekers, B.W., et al., Short-term effect of surgical trauma on rat peritoneal fibrinolytic activity and its role in adhesion formation. Thromb Haemost. 2000. **84**(5): p. 876–81.
46. Hellebrekers, B.W., et al., Use of fibrinolytic agents in the prevention of postoperative adhesion formation. Fertil Steril. 2000. **74**(2): p. 203–12.
47. Whitting, H.W., and B.A. Young, The effect of varidase in carboxymethylcellulose jelly on peritoneal adhesion formation. Virchows Arch Pathol Anat Physiol Klin Med. 1966. **341**(2): p. 155–63.
48. D'Amico, G., Experimental research on the effect of varidase on peritoneal adhesions. Riv Patol Clin. 1954. **9**(1): p. 23–36.
49. Schutze, U., et al., Prophylaxis of peritoneal adhesions with streptokinase and streptodornase (Varidase). An experimental study in animals (author's transl). MMW Munch Med Wochenschr. 1977. **119**(4): p. 123–6.
50. Hill-West, J.L., R.C. Dunn, and J.A. Hubbell, Local release of fibrinolytic agents for adhesion prevention. J Surg Res. 1995. **59**(6): p. 759–63.
51. Jewett, T.C., Jr, et al., Effects of fibrinolytic enzymes on experimentally induced peritoneal adhesions. Surgery. 1965. **57**: p. 280–4.
52. Okamoto, Y., S. Takai, and M. Miyazaki, Oral administration of a novel chymase inhibitor, NK3201, prevents peritoneal adhesion formation in hamsters. Jpn J Pharmacol. 2002. **90**(1): p. 94–6.
53. David-Raoudi, M., et al., Differential effects of hyaluronan and its fragments on fibroblasts: relation to wound healing. Wound Repair Regen. 2008. **16**(2): p. 274–87.
54. Turley, E.A., Hyaluronan and cell locomotion. Cancer Metastasis Rev. 1992. **11**(1): p. 21–30.
55. Moore, A.R., et al., The chemotactic properties of cartilage glycosaminoglycans for polymorphonuclear neutrophils. Int J Tissue React. 1989. **11**(6): p. 301–7.
56. Gao, F., et al., Hyaluronan oligosaccharides are potential stimulators to angiogenesis via RHAMM mediated signal pathway in wound healing. Clin Invest Med. 2008. **31**(3): p. E106–16.
57. Sawada, T., et al., Adhesion preventive effect of hyaluronic acid after intraperitoneal surgery in mice. Hum Reprod. 1999. **14**(6): p. 1470–2.
58. Shushan, A., et al., Hyaluronic acid for preventing experimental postoperative intraperitoneal adhesions. J Reprod Med. 1994. **39**(5): p. 398–402.

59. Detchev, R., et al., Prevention of de novo adhesion by ferric hyaluronate gel after laparoscopic surgery in an animal model. JSLS. 2004. **8**(3): p. 263–8.
60. Johns, D.B., et al., Reduction of postsurgical adhesions with intergel adhesion prevention solution: a multicenter study of safety and efficacy after conservative gynecologic surgery. Fertil Steril. 2001. **76**(3): p. 595–604.
61. Wiseman, D.M., Possible Intergel Reaction Syndrome (pIRS). Ann Surg. 2006. **244**(4): p. 630–2.
62. Yeo, Y., et al., In situ cross-linkable hyaluronic acid hydrogels prevent post-operative abdominal adhesions in a rabbit model. Biomaterials. 2006. **27**(27): p. 4698–705.
63. Guida, M., et al., Effectiveness of auto-crosslinked hyaluronic acid gel in the prevention of intrauterine adhesions after hysteroscopic surgery: a prospective, randomized, controlled study. Hum Reprod. 2004. **19**(6): p. 1461–4.
64. Pellicano, M., et al., Effectiveness of autocrosslinked hyaluronic acid gel after laparoscopic myomectomy in infertile patients: a prospective, randomized, controlled study. Fertil Steril. 2003. **80**(2): p. 441–4.
65. Mazzone, A., et al., Pharmacological effect of hyaluronic acid (HA) on phagocytes: hypothesis for an HA-induced monocyte chemotactic factor for neutrophils. Clin Ther. 1986. **8**(5): p. 527–36.
66. Bulpitt, P., and D. Aeschlimann, New strategy for chemical modification of hyaluronic acid: preparation of functionalized derivatives and their use in the formation of novel biocompatible hydrogels. J Biomed Mater Res. 1999. **47**(2): p. 152–69.
67. Jia, X., et al., Prolongation of sciatic nerve blockade by in situ cross-linked hyaluronic acid. Biomaterials. 2004. **25**(19): p. 4797–804.
68. Ito, T., et al., The prevention of peritoneal adhesions by in situ cross-linking hydrogels of hyaluronic acid and cellulose derivatives. Biomaterials. 2007. **28**(6): p. 975–83.
69. Diamond, M.P., Reduction of adhesions after uterine myomectomy by Seprafilm membrane (HAL-F): a blinded, prospective, randomized, multicenter clinical study. Seprafilm Adhesion Study Group. Fertil Steril. 1996. **66**(6): p. 904–10.
70. Kelekci, S., et al., The efficacy of a hyaluronate/carboxymethylcellulose membrane in prevention of postoperative adhesion in a rat uterine horn model. Tohoku J Exp Med. 2004. **204**(3): p. 189–94.
71. FDA Panel Recommends Against Approval of Genzyme General's Sepracoat 1997 [cited 2008 September 16, 2008]; Available from: www.prnewswire.com.
72. Hills, B.A., B.D. Butler, and R.E. Barrow, Boundary lubrication imparted by pleural surfactants and their identification. J Appl Physiol. 1982. **53**(2): p. 463–9.
73. Muller, S.A., et al., Efficacy of adhesion prevention and impact on wound healing of intraperitoneal phospholipids. J Surg Res. 2001. **96**(1): p. 68–74.
74. Muller, S.A., et al., Influence of intraperitoneal phospholipid dosage on adhesion formation and wound healing at different intervals after surgery. Langenbecks Arch Surg. 2001. **386**(4): p. 278–84.
75. Treutner, K.H., et al., Prevention of postoperative adhesions by single intraperitoneal medication. J Surg Res. 1995. **59**(6): p. 764–71.
76. Baxter, ADEPT Instructions for Use: Deerfield, IL.
77. Verco, S.J., et al., Development of a novel glucose polymer solution (icodextrin) for adhesion prevention: preclinical studies. Hum Reprod. 2000. **15**(8): p. 1764–72.
78. Hosie, K., et al., Fluid dynamics in man of an intraperitoneal drug delivery solution: 4% icodextrin. Drug Delivery. 2001. **8**(1): p. 9–12.
79. Menzies, D., et al., Use of icodextrin 4% solution in the prevention of adhesion formation following general surgery: from the multicentre ARIEL Registry. Ann R Coll Surg Engl. 2006. **88**(4): p. 375–82.
80. van den Tol, P., et al., Icodextrin reduces postoperative adhesion formation in rats without affecting peritoneal metastasis. Surgery. 2005. **137**(3): p. 348–54.
81. Brown, C.B., et al., Adept (icodextrin 4% solution) reduces adhesions after laparoscopic surgery for adhesiolysis: a double-blind, randomized, controlled study. Fertil Steril. 2007. **88**(5): p. 1413–26.
82. Wallwiener, M., et al., Innovative barriers for peritoneal adhesion prevention: liquid or solid? A rat uterine horn model. Fertil Steril. 2006. **86**(Suppl 4): p. 1266–76.
83. Krsko, P., and M. Libera, Biointeractive hydrogels. Materials Today. 2005. **8**(12): p. 36–44.
84. Hildebrand, H.F., et al., Surface coatings for biological activation and functionalization of medical devices. Surface and Coatings Technology. 2006. **200**(22–23): p. 6318–24.
85. Lundorff, P., et al., Clinical evaluation of a viscoelastic gel for reduction of adhesions following gynaecological surgery by laparoscopy in Europe. Hum Reprod. 2005. **20**(2): p. 514–20.
86. Nehez, L., et al., Prevention of postoperative peritoneal adhesions: effects of lysozyme, polylysine and polyglutamate versus hyaluronic acid. Scand J Gastroenterol. 2005. **40**(9): p. 1118–23.
87. Nehez, L., et al., Differently charged polypeptides in the prevention of post-surgical peritoneal adhesions. Scand J Gastroenterol. 2007. **42**(4): p. 519–23.

88. Kapadia, M.R., D.A. Popowich, and M.R. Kibbe, Modified prosthetic vascular conduits. Circulation. 2008. **117**(14): p. 1873–82.
89. Adam, D.J., et al., Antiplatelet and anticoagulant therapy to prevent bypass graft thrombosis in patients with lower extremity arterial occlusive disease. Int Angiol. 2001. **20**(1): p. 90–8.
90. Kenny, D.A., et al., Experimental comparison of the thrombogenicity of fibrin and PTFE flow surfaces. Ann Surg. 1980. **191**(3): p. 355–61.
91. Patel, M., et al., Experimental evaluation of ten clinically used arterial prostheses. Ann Vasc Surg. 1992. **6**(3): p. 244–51.
92. Harris, E.S., R.F. Morgan, and G.T. Rodeheaver, Analysis of the kinetics of peritoneal adhesion formation in the rat and evaluation of potential antiadhesive agents. Surgery. 1995. **117**(6): p. 663–9.
93. Hellebrekers, B.W., et al., Effects of five different barrier materials on postsurgical adhesion formation in the rat. Hum Reprod. 2000. **15**(6): p. 1358–63.
94. Montz, F.J., B.J. Monk, and S.M. Lacy, The Gore-Tex Surgical Membrane: effectiveness as a barrier to inhibit postradical pelvic surgery adhesions in a porcine model. Gynecol Oncol. 1992. **45**(3): p. 290–3.
95. Haney, A.F., and E. Doty, A barrier composed of chemically cross-linked hyaluronic acid (Incert) reduces postoperative adhesion formation. Fertil Steril. 1998. **70**(1): p. 145–51.
96. Himeda, Y., et al., Application of biocompatible gel of hyaluronic acid in adhesion prevention. J Gynecol Surg. 2004. **20**(2): p. 39–46.
97. Stuart, M., Breaking the surgical adhesion barrier. Start-Up. 2005. (April): p. 16–22.
98. Ferland, R., D. Mulani, and P.K. Campbell, Evaluation of a sprayable polyethylene glycol adhesion barrier in a porcine efficacy model. Hum Reprod. 2001. **16**(12): p. 2718–23.
99. Dunn, R., et al., Evaluation of the SprayGel adhesion barrier in the rat cecum abrasion and rabbit uterine horn adhesion models. Fertil Steril. 2001. **75**(2): p. 411–6.
100. Rodgers, K., et al., Evaluation of polyethylene glycol/polylactic acid films in the prevention of adhesions in the rabbit adhesion formation and reformation sidewall models. Fertil Steril. 1998. **69**(3): p. 403–8.
101. Zeng, Q., et al., Efficacy and safety of Seprafilm for preventing postoperative abdominal adhesion: systematic review and meta-analysis. World J Surg. 2007. **31**(11): p. 2125–31; discussion 2132.
102. Shinohara, T., et al., A simple and novel technique for the placement of antiadhesive membrane in laparoscopic surgery. Surg Laparosc Endosc Percutan Tech. 2008. **18**(2): p. 188–91.
103. Mettler, L., et al., A randomized, prospective, controlled, multicenter clinical trial of a sprayable, site-specific adhesion barrier system in patients undergoing myomectomy. Fertil Steril. 2004. **82**(2): p. 398–404.
104. Jackson, J.K., et al., Paclitaxel-loaded crosslinked hyaluronic acid films for the prevention of postsurgical adhesions. Pharm Res. 2002. **19**(4): p. 411–7.
105. Yagmurlu, A., et al., Reduction of surgery-induced peritoneal adhesions by continuous release of streptokinase from a drug delivery system. Eur Surg Res. 2003. **35**(1): p. 46–9.
106. Yeo, Y., et al., Prevention of peritoneal adhesions with an in situ cross-linkable hyaluronan hydrogel delivering budesonide. J Control Release. 2007. **120**(3): p. 178–85.
107. Leach, R.E., et al., Reduction of postsurgical adhesion formation in the rabbit uterine horn model with use of hyaluronate/carboxymethylcellulose gel. Fertil Steril. 1998. **69**(3): p. 415–8.
108. Cheung, M., et al., Development of a Swine Model for the Evaluation of Novel Compounds in the Prevention of Pelvic Adhesions. In ISPE Great Lakes Chapter Meeting, 2008, Chicago, IL.
109. Medley, J.M., et al., In vitro, QCM-D evaluation of diblock copolymers for the rational design of self-forming postsurgical adhesion barriers. Biomacromolecules (under review).
110. Jaiswal, A., Introduction to Data Analysis. In Q-Sense Users Meeting, 2007, Stanford, CA.
111. Sauerbrey, G., Zeitschrift für Physik. 1959. **155**: p. 206.
112. Moreira, H., Jr, et al., Use of bioresorbable membrane (sodium hyaluronate + carboxymethylcellulose) after controlled bowel injuries in a rabbit model. Dis Colon Rectum. 2000. **43**(2): p. 182–7.v

Index

O

P

Printing and Binding: Stürtz GmbH, Würzburg